Walther Schmid
Arnold E. Stütz (eds.)

Timely Research Perspectives in Carbohydrate Chemistry

Springer-Verlag Wien GmbH

Prof. Dr. Walther Schmid
Institute of Organic Chemistry, University of Vienna,
Vienna, Austria

Prof. Dr. Arnold E. Stütz
Institute of Organic Chemistry, Technical University of Graz,
Graz, Austria

Cover illustration:
M. Walter and T. K. Lindhorst, „Reactivity ...", Fig. 2, page 126, this volume
Typesetting: Thomson Press (India) Ltd., New Delhi

Printed on acid-free and chlorine-free bleached paper

SPIN: 10861929

With numerous Figures and Tables

CIP data applied for

Special Edition of
Monatshefte für Chemie/Chemical Monthly Vol. 133, No. 4, 2002

ISBN 978-3-211-83777-1 ISBN 978-3-7091-6130-2 (eBook)
DOI 10.1007/978-3-7091-6130-2

Editorial

Accompanying the rapid increase of knowledge on the extreme importance and vital roles of carbohydrates in living systems, carbohydrate chemistry has made its way from a quite underrated area of natural products research right into the focus of modern organic and bioorganic chemistry within a quarter of a century. Just by flicking through the graphical abstracts of major journals, large numbers of papers on carbohydrate chemistry can be found almost anywhere in the contemporary literature.

The complexity and polyfunctionality of carbohydrates as a testing ground for modern organic methodology, carbohydrates as building blocks, carbohydrate related mimetics of natural products as enzyme probes, diagnostics, and therapeutics, carbohydrates in material sciences, all these and many more exciting topics demonstrate this lasting boom in glycosciences. Not many other areas are as rich in complexity, subtle reactivity differences, unusual behaviour of compounds and, as a result, so full of challenges, surprises, and rewards. Consequently, it was a great pleasure to accept the proposal by the *Managing Editor* of the Austrian journal *Chemical Monthly*, Prof. *Heinz Falk*, to compile a representative survey of timely research contributed by international leaders in the field as well as by outstanding young talents. This collection is intended to give evidence for the statements made above as well as to provide informative and entertaining reading for the many who are interested in our field, from advanced students to seasoned researchers.

The topic span from an interesting and wide range of contributions in the section synthetic methodology to important subunits of biologically significant lipopolysaccharides, sugar, and nucleoside mimetics and their biological activities as well as applications of saccharides as scaffolds for compound libraries and the application of carbohydrates as chiral auxiliaries.

This special issue of *Chemical Monthly* features four invited reviews: *B. La Ferla* has surveyed applications of lipases and esterases in selective biocatalytic carbohydrate protection and deprotection reactions. *F. Nicotra* and his group have exploited carbohydrates as scaffolds for library generation. The liberation of carbonyl functions from C-nitro groups in sugars, an important step in the chain extension of sugars, has been treated by *L. Petruš* and coworkers. Last but not least, *T. M. Wrodnigg* has provided a review on important pyrrolidine-related iminoalditols, their natural sources, synthetic approaches, and glycosidase inhibitory activities.

In the article section, *K. Dax* and coworkers have elaborated a system to categorize reactions occurring during attempted nucleophilic displacement reactions at C-2 of pyranosides with diethylaminosulfur trifluoride. *A. Fairbanks* and his group describe an interesting intramolecular aglycon delivery approach employing allyl protecting groups. *R. Madsen* and colleagues have taken advantage of a transition metal mediated fragmentation of 6-deoxyiodo glycopyranosides leading to key intermediates for RCM methodology and other approaches to natural products and analogues. *T. Lindhorst* and *M. Walter* observed differences in kinetics during a "reactivity tournament" of sugar derived isothiocyanates.

Carbohydrate photochemistry, exemplified by transformations of levoglucosan imides to nitrogen containing heterocycles, is reported by *J. Thiem* and coworkers. In the carbohydrate mimetics section, an account on nucleoside analogues and their biological activities was contributed by the *M. Jung* group. Thiosugars are the topics in both the contributions by *S. Witczak* and collaborators and *W. Schmid* and coworkers. *R.V. Stick* and *A. G. Watts* have chased and discussed common features in the mechanistic principles of retaining glycosyl hydrolases and retaining glycosyl transferases. Some new xyloside mimetics related to 1,5-dideoxy-1,5-iminoxylitol were probed as reversible inhibitors of β-xylosidases by *S. G. Withers* and collaborators.

Immunologically interesting sub-units of chlamydial lipopolysaccharides have been prepared by *P. Kosma* and coworkers. *H. Kunz* and colleagues took advantage of carbohydrates as chiral auxiliaries in their synthetic work on tetrahydroquinoline alkaloids. Application of modern NMR spectroscopy is demonstrated by *H. Kählig* and coauthors who employed diffusion difference NMR spectroscopy for the analysis of carbohydrate mixtures.

It is a special feeling to find so much appreciation for our 'trade' here in Austria; we, the guest editors, would like to thank *Heinz Falk* and *Hermann Kalchhauser* as well as the Springer publishing team for their kind support, patience, and help with this project. We hope that the readers will enjoy this collection of timely topics in carbohydrate chemistry.

Walther Schmid
Arnold E. Stütz
Issue Editors

Contents

Invited Reviews

Contributions

Abstracted/Indexed in:
Current Contents, SCI, ASCA, and Current Abstracts of Chemistry,
Chemical Abstracts, Polymer Contents, ADONIS

Invited Review

Lipases as Useful Tools for the Stereo- and Regioselective Protection and Deprotection of Carbohydrates

Barbara La Ferla*

University of Milano-Bicocca, Department of Biotechnology and Biosciences, I-20126 Milano, Italy

Summary. Stereo- and regioselective acylation and deacylation of carbohydrates can be achieved using lipases as biocatalysts. These enzymes are extremely versatile, quite stable in aqueous and organic solvents, easily available, and easy to handle. Recently, they have become the object of site directed mutagenesis which suggests that they soon could be 'built *ad hoc*' for specific necessities.

Keywords. Biocatalyst; Lipases; Protective groups; Carbohydrates.

Introduction

Among all synthetic transformations, the proper introduction and removal of protecting groups is one of the most widely adopted in synthetic organic chemistry. Chemo-, regio-, and stereoselection are daily beneath our eyes; all chemical reactions nature has realized have these amazing characteristics. It is therefore not surprising that many chemists make efforts to somehow try to lean towards nature's perfection, even if this target still represents a 'beautiful dream'.

Many classical chemical methods have been developed for the manipulation of protecting groups under mild conditions [1, 2]. Such methods are necessary, in particular, in the construction of complex, polyfunctional molecules, among which are oligosaccharides and their conjugates. Nevertheless, the synthesis of these and other complex polyfunctional molecules still presents severe problems which can be solved only with great difficulties and long multistep strategies. Attempts to exploit the same or similar tools as nature to perform comparable reactions have led to an extended use of biocatalysts which have substantially enriched the arsenal of available protecting group techniques, offering viable alternatives to classical methods [3, 4].

* E-mail: barbara.laferla@unimib.it

More specifically, the protection/deprotection of carbohydrates, necessary for the synthesis of more complex saccharidic structures or other glycoconjugates, represent challenging goals. In fact, the numerous hydroxyl groups with 'similar' reactivity that characterize these compounds need to be manipulated selectively during the synthesis. Although numerous chemical techniques are available to mask and liberate hydroxyl groups [1, 2, 5, 6], the development and exploitation of enzymatic methods has been progressing and becoming a common tool for workers in this field.

Lipases as Biocatalysts

Many biocatalysts are used in the carbohydrate field, among them hydrolitic enzymes such as lipases, which are particularly suitable for the regioselective acylation and deacylation of hydroxyl groups.

Lipases hydrolyze fat in the digestive tract but are extremely flexible for the acylation or deacylation of a wide range of unnatural substrates. Unlike most other enzymes, they can accommodate a wide range of substrates, are quite stable in non-aqueous organic solvents, and can be used to either hydrolyze or esterify depending on the solvent system used.

A recent review [7] extensively treats these enzymes with respect to their occurrence, preparation, analysis, structure, and mechanism. Moreover, it deals with the application of these enzymes in oleochemistry as well as in detergents, paper, and food industries.

Ester hydrolysis

A freely dissolved lipase in absence of an aqueous/lipid interface resides in its inactive state, and part of the enzyme molecule covers the active site. When the enzyme contacts the interface of a biphasic water–oil system, a short α-helix, the lid, is folded back. Thus, opening its active site the lipase is rearranged into its active state. As a consequence, lipase-catalyzed hydrolyses should be performed in a biphasic medium. It is sufficient to employ a substrate alone at elevated concentrations, such that it constitutes the second organic phase, or, alternatively, it may be dissolved in a water-immiscible organic solvent such as hexane, a dialkyl ether, or an aromatic liquid. Since most substrates are only sparingly soluble in water, the reaction mixture will be an emulsion or a suspension. Especially with solid substrates (suspension) the addition of cosolvents (5–20%), such as low aliphatic alcohols (methanol, ethanol, *tert*-butanol), water-soluble ethers (*THF*, dioxane), low aliphatic ketones (acetone), or *DMSO* and *DMF*, will be advantageous. Some cosolvents must be used in very low percentage as they may reduce the activity of the enzyme (Table 1).

Another important feature which must be taken into consideration during ester hydrolysis is the liberation of the carboxylic acid lowering the *pH* of the reaction mixture (which is neutral at the beginning of the reaction). It is essential to keep the *pH* value constant during the reaction because of lower *pH* could stop the reaction and may damage the enzyme. Thus, a phosphate buffer is usually adopted to keep the *pH* constant.

Table 1. Solvent conditions used in biocatalyzed ester hydrolysis

Solvent system		Cosolvent (5–20%)
	hexane	aliphatic alcohol (MeOH,
Water	dialkyl ether	EtOH, *t*BuOH)
buffter	aromatic	water-soluble ether (*THF*,
–	liquid	dioxane)
		aliphatic ketone (acetone)
		DMSO, *DMF*

Table 2. Solvents in transesterification

Organic solvents
tert-butanol
methyl-*tert*-butyl ether
hexane
THF
methylene chloride
dioxane
acetone
acetonitrile
vinyl acetate
benzene/pyridine
DMF
pyridine

Transesterification (acylation)

Several studies have indicated that many enzymes, among them lipases such as porcine pancreatic lipase (PPL), are more thermostable in organic solvents than in water. For example, PPL remains active for many hours (half-life longer than 12 h) when incubated at 100°C in an almost anhydrous mixture of heptanol and tributyrin (Fig. 1) [8]. Lower stability was observed in the presence of 0.8% (w/v) water, whereas inactivation occurred almost instantaneous in aqueous buffer or water.

The high thermal stability of enzymes in organic solvents, especially in hydrophobic ones and at low water contents, was attributed to increased conformational rigidity and to the absence of nearly all covalent reactions which cause irreversible thermo-inactivation in water [9]. It should be mentioned that, as demonstrated by FTIR spectroscopic studies on lysozyme and subtilisin, enzyme structure is much more similar to the native form in pure organic solvents such as acetonitrile, *THF*, or 1-propanol than in aqueous solvent mixture [10]. This behaviour was found to be kinetically controlled, that is, due to the inherent restriction on protein

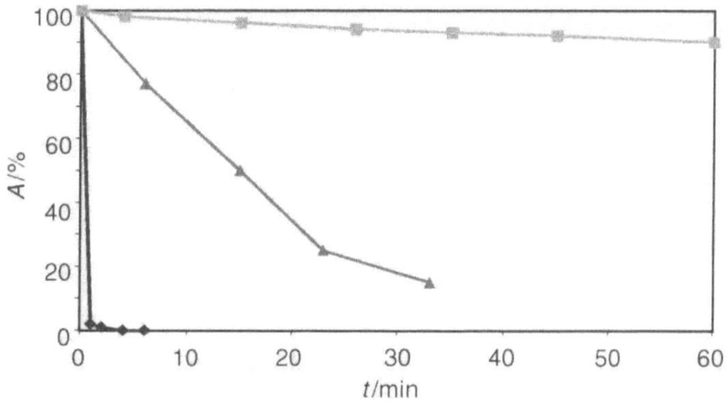

Fig. 1. Activity decay of dried powder of porcine pancreatic lipase; -◆-: H_2O or 0.1 *M* phosphate buffer; ▲: 2 *M* solution of heptanol + 0.8% H_2O; -■-: 2 *M* solution of heptanol + 0.015% H_2O

conformational mobility in anhydrous media in contrast to aqueous-organic solvent mixtures.

The solvent also affects the selectivity of enzyme. Indeed, a great number of papers have reported that the enantio-, prochiral, and regioselectivity of enzymes can be affected, sometimes very remarkably, by the nature of the organic solvent used as the reaction medium [11]. *Klibanov* and coworkers first coined the term 'medium engineering' referring to the possibility of influencing enzyme properties by changing the nature of the solvent in which the reaction is carried out. An example of different enzymatic reactivity in different solvents is that observed for lipase from *Candida antarctica* on the 2-azidodeoxy derivative of *β-D-Gal*(1 → 3)-D-*Glc*NAc [60]. Various hypotheses have been formulated to rationalize this phenomenon. For instance, the solvent, depending on its polarity, could modify the enzyme's conformation and, thus, influence the selectivity by altering the molecular recognition process between substrate and enzyme [12]. According to another theory, selectivity depends on the energetics of substrate solvation [13], whereas a third model envisages that solvent molecules could bind within the active site and, depending on their structure, interfere with the association or transformation of one enantiomer more than with that of the other [14, 15]. However, both the hypotheses based on the physico-chemical properties of the solvents and that based on solvent structure are unsatisfactory from the point of view of predictive value. At present, no link among the various hypotheses appears to exist, even though it is likely that the solvent influences enzymatic selectivity through more than a single mechanism. Table 2 reports a list of the organic solvents used in enzymatic acylation of carbohydrates.

Lipases as hydrolytic enzymes on acylated carbohydrates

First studies on the application of lipases in hydrolysis date back to 1969 when *Fink* and *Hay* [16] and *Sachder* and *Starkovsky* [17] used lipases from wheat germ (WGL) on penta-O-acetyl-D-glucose (**1**). The results were poor (only a mixture of partially deacetylated glucoses was obtained), but the authors made the interesting observation that C-1 and C-6 acyl groups were preferentially removed. The hydrolytic application of lipases was abandoned for about seventeen years. Then, in 1986 *Klibanov*, *Shaw* and, independently, *Wong* published their results [18–20].

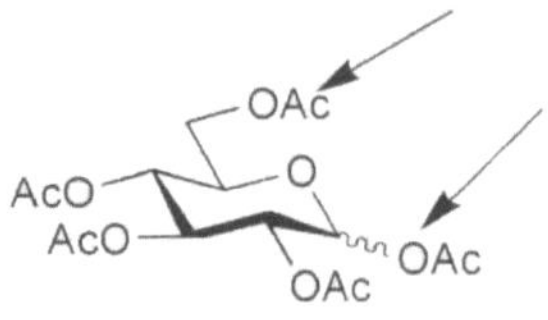

1

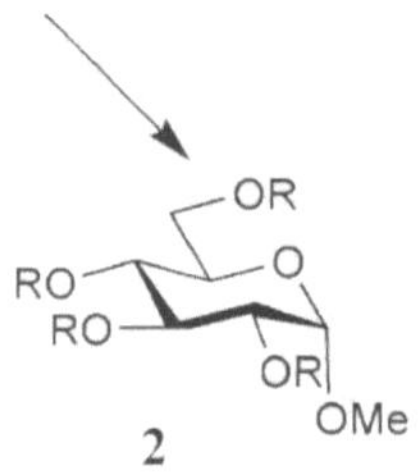

2

Lipase from *Candida cilindracea* (CCL) was the most adequate for methyl *α*- and *β*-2,3,4,6-tetra-O-acylpyranosides (**2**) [20], resulting in a 80–90% deacylation of O-6 (Fig. 2). Pentanoyl esters were preferred for technical reasons despite the fact the lipase best reacted with octanoyl derivatives. Moreover, the reaction was much cleaner for glucose with respect to other pyranosides tested and proceeded five times faster for α- than for β-glycosides. The hydrolysis on β-D-penta-O-acetylglucose [18, 19] was studied using *Aspergillus niger* lipase (ANL). In particular, it was analyzed how reaction conditions (time, temperature, and solvent system) influenced the product composition, *i.e.* the ratio of tetra-, tri-, and diacetylated glucose derivatives obtained. Particularly interesting was the observation that the same lipase was able to catalyze also the alcoholysis of penta-O-acetyl-β-D-glucose affording quantitatively the 2,3,4,6-tetra-O-acetylglucose.

In further studies carried out on different pyranosides [21] bearing an acetate at the anomeric position, porcine pancreas lipase (PPL) was found to hydrolyze the anomeric acetate selectively in 54–96% yield (**3–9**). On pyranosides lacking the acyl group at C-1, the primary hydroxy group was removed in good yield using lipase *Candida cylindracea* (CCL) (**10**).

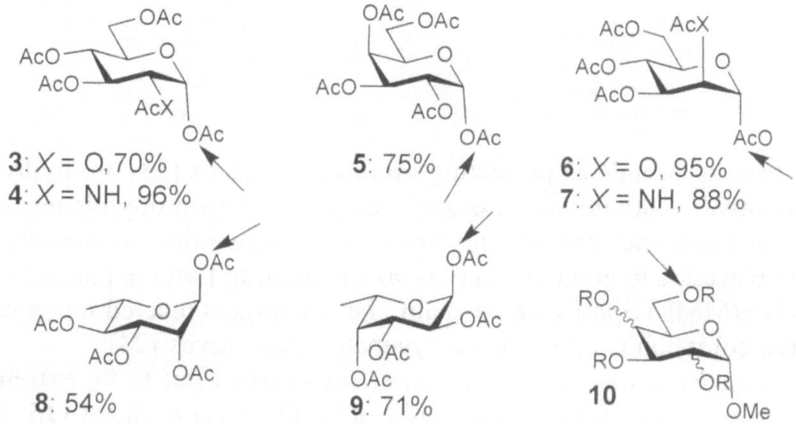

More recently, progress has been achieved in the regioselective deacetylation of penta-O-acetyl-α-D-glucose at positions other than C-1 and C-6 [22]. Testing different lipases at different *pH* values revealed a very interesting behaviour of CCL: this enzyme was not selective for the anomeric position and showed a dramatic change in regioselectivity at different *pH* (Scheme 1). Particularly interesting is the product selectively deprotected at O-4, a useful intermediate for the synthesis of many 1–4 linked oligosaccharides.

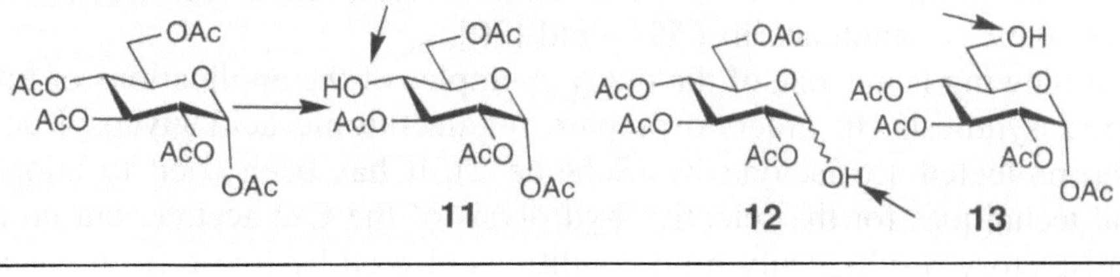

Lipase	*pH*	Yield/%	**11**	**12** α/β	**13**	Others
PFL	7	98	1	91	–	8
PFL	5	98	1	94	–	4
CCL	7	78	71	–	24	5
CCL	5	80	4	–	96	–

Scheme 1

The results so far observed are obviously of great interest, but none of them seem to be of fundamental importance for a synthetic chemist. Things become more attractive when differently protected substrates carrying various and liable functionalities showed to be well tolerated and region- and stereoselectively de-O-acylated by these enzymes. Among them are the 1,6-anhydro-2,3,4-tri-O-acyl sugars **16**, substrates bearing different functionalities such as acetals (*e.g.* **14**), glycitols such as **17** and **18**, and glycals (*e.g.* **15**).

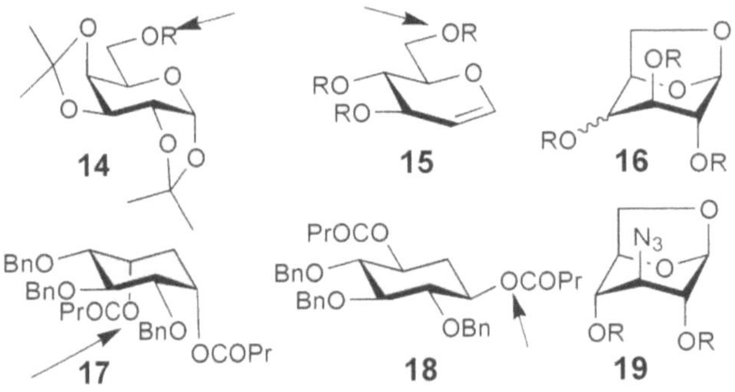

Substrates bearing acetals as protecting groups such as **14** [23] are hydrolyzed by CCL quantitatively. Studies on 2,6-di-O-acetyl-3,4-O-isopropylidenegalactosides with different anomeric substitutents showed the possibility to hydrolyze selectively the O-6 acetate in good yields (75–90%) adopting Porcine pancreas (PPL) or *Mucor miehei* (MML) lipases. Once again, the α-anomers reacted much faster than β-configured compounds and C-1 deoxygenated derivatives [24].

Lipase from *Pseudomonas fluorescens* (PFL) turned out to be extremely efficient and regioselective for the hydrolysis of tri-O-acetyl-D-glucal (**15**, Fig. 4) at C-6 (90% yield). Less selective was the reaction with tri-O-acetyl galactal, which afforded a complex mixture of di- and monoacetylated products [25, 26].

1,6-Anhydropyranoses, being useful starting materials in carbohydrate chemistry, have been the objects of studies by many groups [27–30]. Interesting results obtained on differently protected substrates **20** and adopting different lipases are summarized in Table 3.

Of great importance seems to be the stereochemistry at C-4. When the hydroxyl group is equatorially oriented, only CCL and PPL are effective in hydrolyzing the C-2 butanoate in high yields (90%) and, on prolonged reaction time, PPL can also hydrolyze the C-4 butanoate in 65% yield [30].

The following is just one of the many examples of the applications of lipases in a complex synthesis. In order to prepare D-galacturonic acid glycosyl acceptors from the protected 1,6-lactone **21** (Scheme 2), it has been tried to adopt many classical techniques for the selective hydrolysis of the C-2 acetate, but no attempt gave satisfactory results. Only mild conditions offered by lipase from wheat germ afforded the desired compound **22** in 60% yield [31].

Lipase from *Candida cilindracea* (CCL) was also adopted for the stereoselective hydrolysis of (*R*)-configured propionates of cyclohexitols **17** and **18** [32]. Lipases being able to hydrolyze esters on pyranosides are also efficient on protected five-membered rings and exhibit a similar behaviour. The primary acetyl groups of furanosides **23–26** are cleaved with good yield using CCL [21], whereas the

Table 3. Hydrolysis of 1,6-anhydropyranoses; WGL = wheat germ; *RJL* = *Rhizopus javanicus*; CVL = *Chromobacterium viscosum*; PSL = *Pseudomonas Sp*; MML = *Mucor mihei*

Substrate			Lipase	Products			Yield/%
R^1	R^2	R^3		R^1	R^2	R^3	
OAc	OAc	OAc	PPL	OAc	OAc	OH	63–69
OAc	OAc	OAc	WGL	OAc	OH	OAc	42–67
OAc	OAc	OAc	RJL	OH	OAc	OAc	47
OBu	OBu	OBu	CVL/PSL/MML	OBu	OBu	OH	91
OBu	OBu	OBu	CCL	OH	OBu	OH	47
OAc	N_3	OAc	CCL	OAc	N_3	OH	85–90

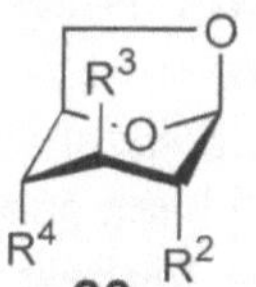

20

peracetylated furanoses **27** and **28** are deprotected at the anomeric position by lipase from *Aspergillus niger*. It is also interesting the observation that the regioselectivity of lipases is partially influenced by the presence of different acyl groups.

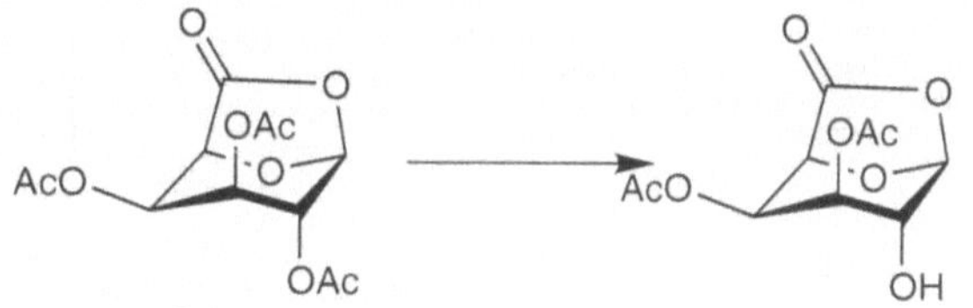

21 **22**

Scheme 2

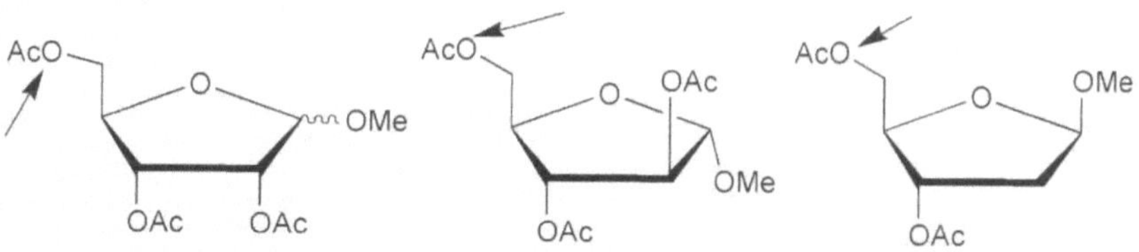

23: *α-D*-ribo, 85%
24: *β-D*-ribo, 96%

25: *α-D*-arabino, 98%
Candida cilindracea

26: *β-D*-2-deoxyribo, 63%

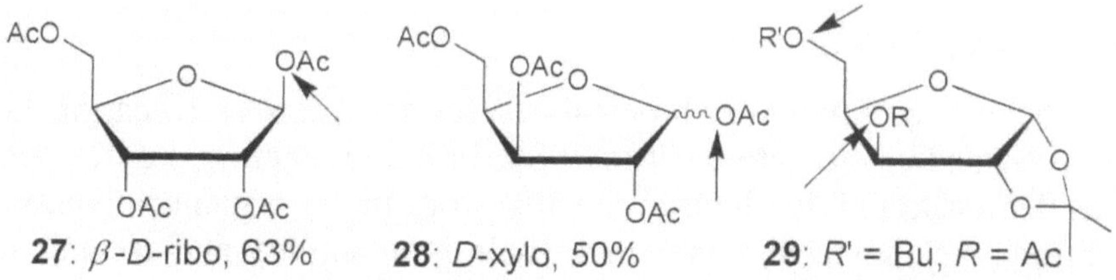

27: *β-D*-ribo, 63%
28: *D*-xylo, 50%
Aspergillus niger

29: *R'* = Bu, *R* = Ac
30: *R'* = *Piv*, *R* = Bu
CCL

For example, CCL selectively removes the primary butyryl ester in presence of the secondary acetate on compound **29** in a 90% yield; if the bulky pivaloate ester is present on the primary group (**30**), the lipase prefers to hydrolyze the secondary butyryl chain [23].

Disaccharides have also been the object of hydrolytic reactions with lipases. First of all, sucrose **31** has been used as a substrate. Most enzymes evidenced a preference for hydrolysis at the fructosidic moiety. Using lipase from wheat germ (WGL) [33], the octa-O-acetyl sucrose was deacetylated at positions O-1′,4′,6′ affording the pentaacetyl derivative in 40% yield. Lipase from CCL preferentially hydrolyzes the O-4′ ester [34, 35], whereas lipase from *Candida antartica* deacetylates the O-6′ position [36]. Quite surprisingly, lipase from *Aspergillus niger* acts on the glucose moiety, affording a 1:1-mixture of heptaacetyl sucroses with the free hydroxyl groups at C-4 and C-6.

Other common per-O-acetylated disaccharides (maltose (**32**), cellobiose (**33**), lactose (**34**), and melibiose (**35**)) have been submitted to this methodology [37–39]. Lipase from *Aspergillus niger* gave the highest reaction rate and afforded selectively the corresponding heptaacetates with a free hydroxyl group at C-1 after 30 min. Prolonged reaction time on cellobiose and lactose per-O-acetates gave the hexaacetates with free hydroxyl groups at C-1 and C-2, whereas maltose and melibiose gave a complex mixture of products [39].

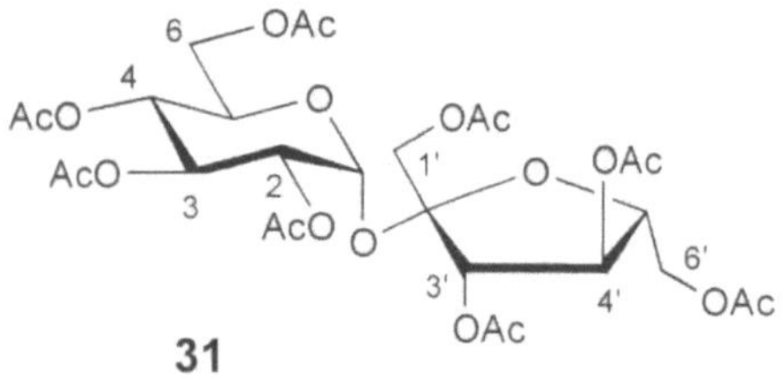

31

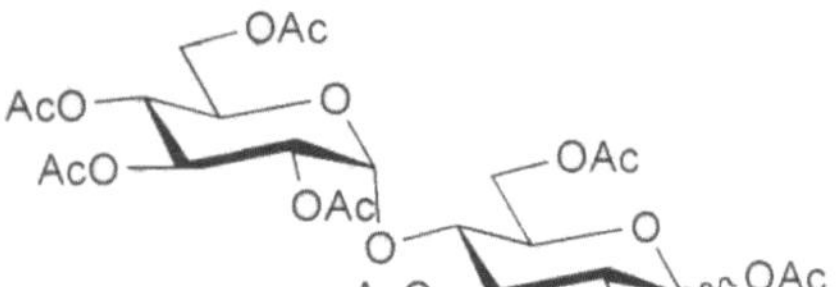

32

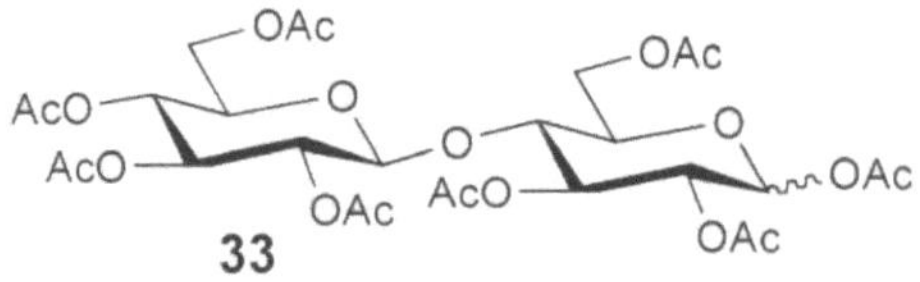

33

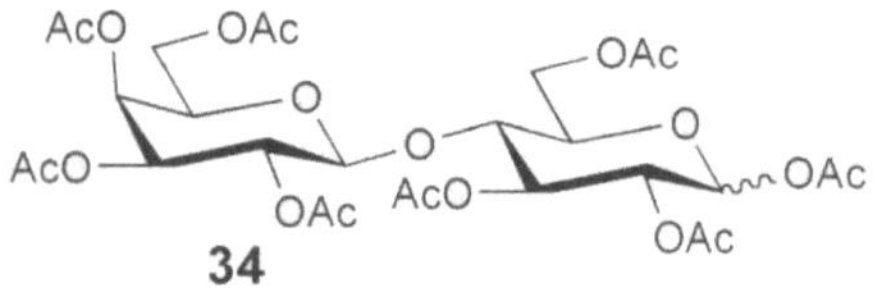

34

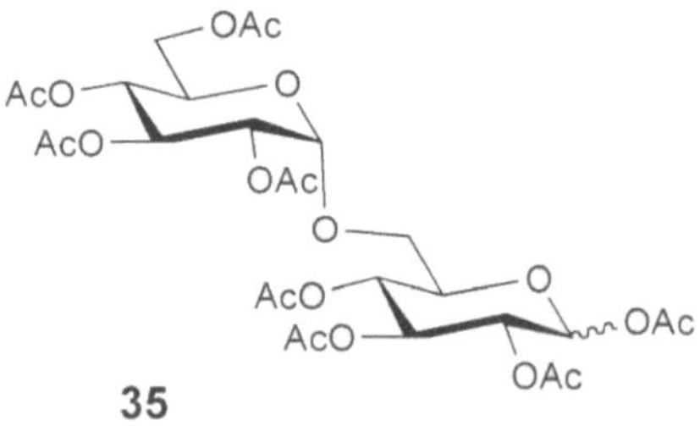

35

Lipases as acylating biocatalysts

Studies on the use of lipases as biocatalysts for the transesterification between sugars and activated esters date back to 1986. In his pioneering studies, *Klibanov* [40, 41] has demonstrated that lipases can also work in dry organic solvents. Under these conditions the enzymes reverse their hydrolytic activity and are able to catalyze acyl transfer reactions from activated esters to suitable acceptors. First studies employing pyridine as solvent and activated esters showed that lipase from Porcine

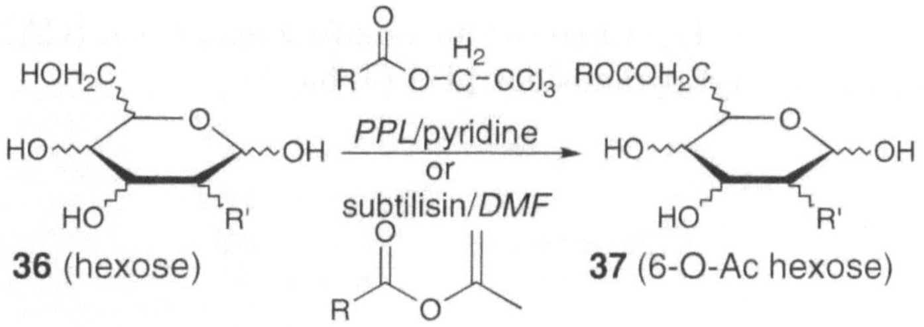

Hexose	Enzyme	*R*	Yield/%	Ref.
Glucose	*PPL*	CH_3	76	[73]
Glucose	*PPL*	C_3H_7	50	[73]
Galactose	*PPL*	CH_3	60	[73]
Mannose	*PPL*	CH_3	85	[73]
Sialic acid [a]	*PPL*	CH_3	64	[74]
Glucose	Subtilisin	C_3H_7	61–64	[76]
Mannose	Subtilisin	CH_3	40	[77]
Sialic acid	Subtilisin	CH_3	73	[77]

[a] 9-O-acetyl-N-neuraminic acid was formed

Scheme 3

pancreas (PPL) could regioselectively transfer acyl groups to the primary hydroxyl groups (Scheme 3). The protease subtilisin was tested as well.

Acyl donors that proved to be suitable for the process and which are commonly adopted are 2,2,2-trichloroethyl, 2,2,2-trifluoroethyl [42], enol, [43, 44] and oxime [45] esters. In particular, enol esters have the advantage that the liberated enol tautomerizes to a ketone or a aldehyde, thereby shifting the equilibrium. Glucoses already protected in the primary hydroxyl group have been selectively acylated at C-2 or C-3 according to different lipases employed (Table 4) [46].

Many research groups have concentrated their studies on anomerically protected pyranosides which are easier to handle and better soluble in organic solvents

Table 4. Acylation of C-6 protected glucoses; CVL = *Chromobacterium viscosum*; ANL = *Aspergillus niger*

R	Lipase	Acylating agent	C-2	C-3	Yield/%
Bu	CVL/ANL	trichloroethylBu	OH	OBu	80
Bu	PPL	"	OBu	OH	51
Trt	CVL	"	OH	OBu	88
TBDPS	CCL	"	OBu	OH	75

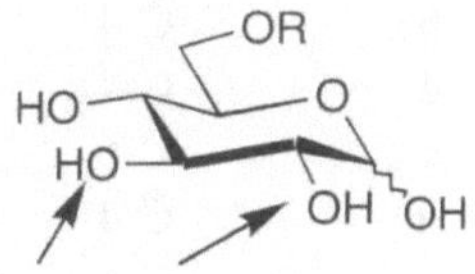

38

than the free sugars [46, 47]. Lipase from *Candida antarctica* (CAL) was used on methyl, phenyl, and octyl pyranosides [47] (Table 5).

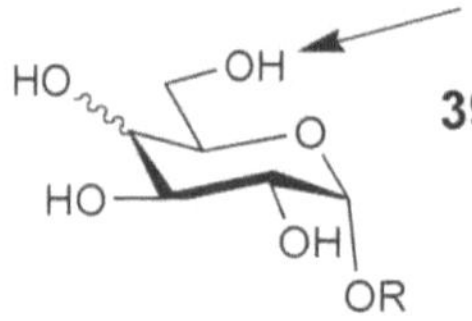

39: α-glycosides
vinylacetate
74-98%

This enzyme is able to acylate regioselectively not only primary but also secondary hydroxyl groups, and the study seems to outline that lipase activity is dramatically influenced by the stereochemistry of the anomeric position. As a matter of fact, α-glycosides (**39**) are acylated only at C-6, whereas β-glycosides (**40**) are partially acylated also at positions C-2 and/or C-3 as summarized in Table 5. D- and L-fucopyranosides, rhamnopyranosides [47, 48], and mannopyranosides under lipase acylation conditions all afforded selectively protected products (Table 6).

Lipase from *Pseudomonas fluorescens* (PFL) converts both D-6-deoxysugars (Table 6, entries 1,2) to the 2-monobutyrates with high regioselectivity, whereas

Table 5. Acylation of anomerically protected pyranosides

X	Galactosides		Glucosides	
	Yield/%	Acetylated	Yield/%	Acetylated
	24	C-6		
Me	44	C-6, C-3	97	C-6, C-3
	31	C-6, C-2		
Ph	92	C-6, C-3	77	C-6, C-3
	6	C-6, C-2	23	C-6, C-2
Octyl	–	–	99	C-6

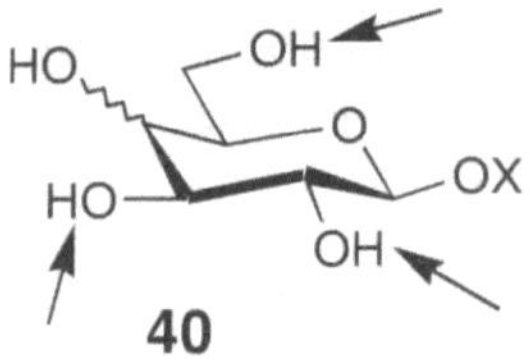

40

Table 6. Acylation of fuco-, rhamno-, and mannopyranosides; a: *THF*:*Py* = 4:1; b: *THF*; c: acetone: Py = 10:1; d: acetone; *TFEB* = trifluoroethylbutyrate

Entry	Substrate	Lipase	Solvent	Product	Yield/%	Acylating agent
1	Me-*D*-fuco	PFL	a	C-2	88	*TFEB*
2	Me-*D*-rhamno	PFL	b	C-2	42	*TFEB*
3	Me-*L*-fuco	PFL	a	C-4	45	*TFEB*
4	Me-*L*-rhamno	PFL	b	C-4	63	*TFEB*
5	*Bn-D*-rhamno	CAB	d	C-4	20	Vinyl acetate
6	*Bn-D*-manno	CAB	c	C-6	97	Vinyl acetate
7	*Bn-L*-rhamno	CAB	d	C-4	90	Vinyl acetate
8	*Bn-L*-manno	CAB	c	C-6 (C-4,6)	64 (29)	Vinyl acetate

the *L*-enantiomers (Table 6, entries 3,4) are esterified at C-4. The same result is obtained using lipase from *Candida antarctica* (CAB) on benzyl *L*-rhamnopyranosides (Table 6, entry 7), but unlike with PFL, also the *D*-enantiomer is acylated at C-4 (Table 6, entry 5). As expected, mannosides are esterified at the primary position, the *L*-enantiomer affording also 29% of diacetylated compound (Table 6, entry 8).

As already observed for the hydrolitic methodology, lipases are able to accept also sugars with functional groups such as acetals and azides, glycals, disaccharides, glycoconjugates, and other complex unnatural moieties as substrates.

Among the monosaccharides bearing acetals as protecting groups, the 4,6-O-benzylidene pyranosides are the most studied so far [47, 49]. Different lipases have been tested on protected glucose and galactopyranosides with different anomeric substituents (Table 7).

Table 7. Lipase selectivity on 4,6-O-benzylidene pyranosides

Entry	Lipase	*X*	*Y*	C-2	C-3
1	PFL	SEt	H	–	100
2	PPL	SEt	H	–	100
3	CCL	SEt	H	64	36
4	PFL	SPh	H	14	86
5	PPL	SPh	H	18	82
6	PFL	SePh	H	47	53
7	PPL	SePh	H	35	65
8	CAB	OMe	H	21	79
9	CAB	H	OMe	58	–

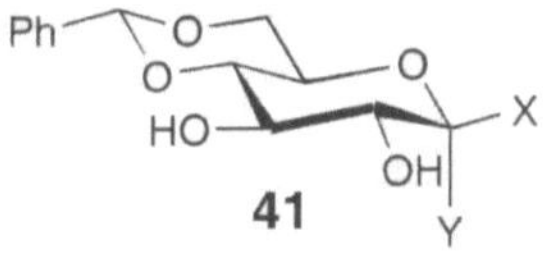

41

The thioethyl β-*D*-glucoside **41** (Table 7, entries 1,2) was well accepted by PPL; PFL afforded mainly the C-3 acetylated compound, whereas CCL showed the opposite regioselectivity (Table 7, entry 3). If the anomeric substituent is more hindered like SPh and SePh, the regioselectivity is sensibly decreased (Table 7, entries 4–7). CAB reveals a preference for the acetylation of C-3 in the methyl β-*D*-glucopyranoside (Table 7, entry 8), whereas the regiochemistry is completely reversed for the α-anomer (Table 7, entry 9). *Pseudomonas Sp* lipase (PSL) is able to selectively acylate the primary hydroxyl group of 3,4-O-isopropylidene-*D*-galactopyranosides **43** [24]; the 1-deoxy derivative showed the highest acylating rate along with the α-glycosides (Table 8).

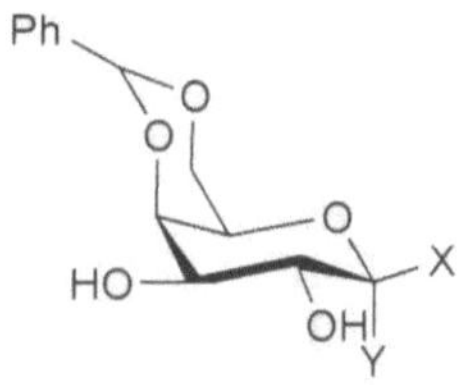

42: very poor substrate preference for C-2

Glycals are very useful precursors in the synthesis of complex oligosaccharides as well as for the formation of C-2 functionalized compounds. *Holla* [25], as cited

Table 8. Acylation of 3,4-O-isopropylidene galactopyranosides; solvent: vinyl acetate: *THF* = 7:3

Substrate	t/h	Lipase	Yield/%
1a; X = OMe, Y = H	48	PSL	90
1b; X = H, Y = OMe	24	PSL	93
1c; X = OBn, Y = H	48	PSL	87
1d; X = H, Y = H	24	PSL	94

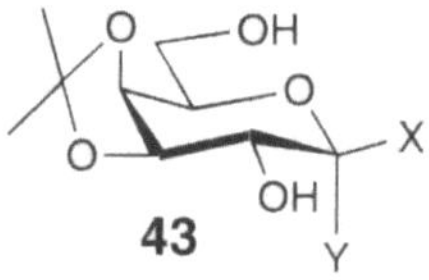

43

in the hydrolysis paragraph, dedicated part of his work to the study of enzymatic protection and deprotection of these compounds. *Candida* lipases OF and S-VII were found to be suitable for the selective acetylation of the primary hydroxyl group, whereas PPL afforded the 3,6-diacetylated compounds. For the complete differentiation of the three hydroxyl groups, also benzoylation and chloroacetylation were investigated (Scheme 4). Combining the use of the cited enzymes and lipase AY-20 with different acylating agents it was possible to obtain the completely differentiated glycals (Table 9).

Furanosides and furanoses, although more flexible in solution than pyranoses, can also be regioselectively protected by lipases. As expected, the primary hydroxyl

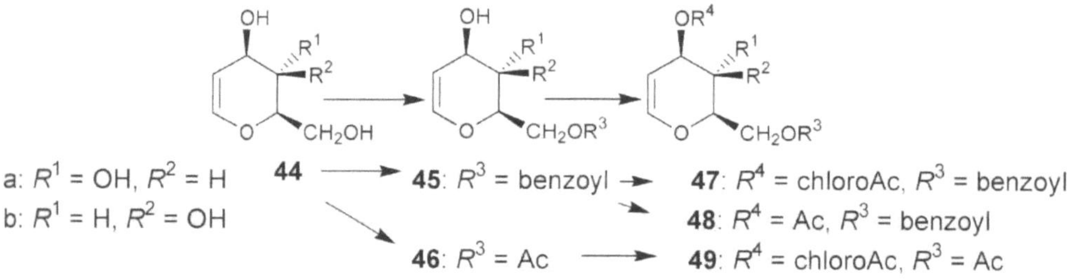

Scheme 4

Table 9. Conditions for the generation of completely differentiated glycals; *VA* = vinyl acetate; *VB* = vinyl benzoate; *VCA* = vinyl chloroacetate

Entry	Compound	Lipase	Acylating agent, conditions	Product	Yield/%
1	**44a**	OF	*VA*, 20% EtOAC	**46a**	90
2	**44a**	AY-20	*VB*, 30–40% *THF*	**45a**	70
3	**44b**	S-VII	*VA*, 1–4% H_2O	**46b**	93
4	**44b**	AY-20	*VB*, 5–10% H_2O	**45b**	67
5	**46a**	PPL	*VCA*, 20–30% *DME*	**49a**	83
6	**46b**	PPL	*VCA*, 20–30% *DME*	**49b**	80
7	**45a**	PPL	*VCA*, 20–30% *DME*	**47a**	82
8	**45b**	PPL	*VCA*, 20–30% *DME*	**47b**	80
9	**45a**	PPL	*VA*	**48a**	92
10	**45b**	PPL	*VA*, 20–30% *DME*	**48b**	80

group is always the first to be acylated. D-Ribo-, D-xylo-, and D-arabinofuranosides were acylated using PPL [21] affording **50**, **51**, and **52** in good yields, whereas 2-deoxyribofuranoside **53** afforded a mixture of regioisomers. The problem was overcome using the protease subtilisin instead of a lipase [50, 51]. The free sugars D-2-deoxyribose and D-ribose were propionylated at C-5-OH using CAL, propionic anhydride in *THF* affording **54** and **55**, respectively [52].

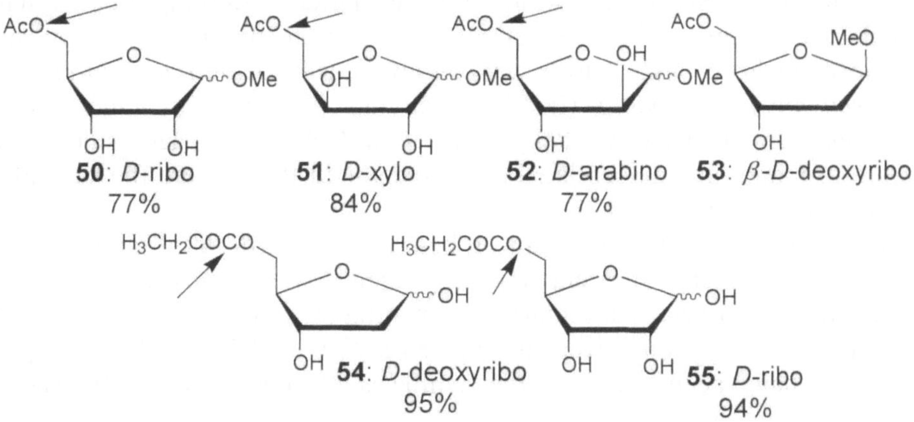

Lipases acylate the free primary hydroxyl groups of furanosides partially protected with acetals [53]. If the primary hydroxyl group is already protected, some lipases are able to discriminate between the secondary hydroxyls [54]. Of particular interest is the possibility of using lipases on more complex oligosaccharides and glycoconjugates both in hydrolysis (as previously reported) and in esterification. It must be pointed out that most, if not all, results obtained on lipase selective protection/deprotection of complex and unnatural compounds are empirical, obtained often after a careful screening of many lipases under different exper-imental conditions. This work is generally tedious and time consuming, but once the right conditions are found these reactions are mild and regioselective, and it is difficult (sometimes impossible) to achieve the same result by means of chemical methods.

Very few disaccharides have been subjected to a systematic study applying lipases as acylating catalysts; in his pioneering investigation [55], *Klibanov* has shown that not a lipase but a protease (subtilisin) was able to catalyze the acylation of disaccharides such as sucrose **56**, lactose **57**, cellobiose **58**, and maltose **59** dissolved in *DMF*.

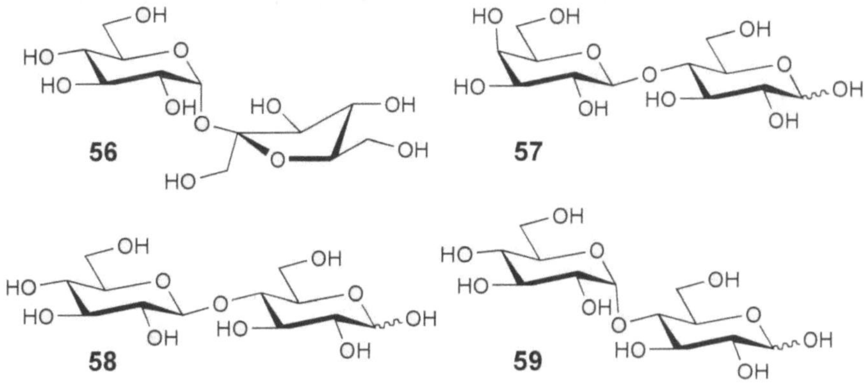

The acylation is regioselective; the non reducing sucrose is esterified at the 1′-OH, whereas on the other disaccharides the acyl group is always transferred to the primary hydroxyl group of the nonreducing end. This means that the enzyme is able to discriminate between the two primary positions, which is not easily achieved with chemical methods. Some examples of acylation of simple derivatives of these sugars have been reported in the literature [36, 56, 57]. A systematic study on the acylation of disaccharides adopting lipases has been carried out recently [58]. Selective esterification of dodecyl glycosides of cellobiose, maltose, and lactose was achieved using *Candida antarctica* (CAL) and *Pseudomonas cepacia* (PCL) employing different acylating agents. All substrates were acylated at the primary C-6′-OH; only β-lactoside afforded a mixture of mono- and diacetylated derivatives with PCL where the second acetate is transferred on the 2′-OH.

A further study was carried out on benzyl β-lactoside **60** (Table 10) using the same two lipases and different acylating agents, affording different mono- and diacylated derivatives; particularly interesting is the product in which two different protecting groups were introduced at C-6′ and C-2′ (**61**). Such protected derivatives have been used for the synthesis of human milk oligosaccharides [59].

Table 10. Regioselectivity of benzyl β-lactoside

Acylating agent	Product	CAL	PCL
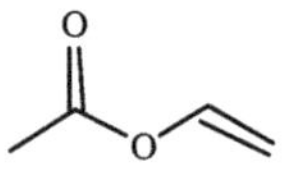	6′ OAc	75%	67%
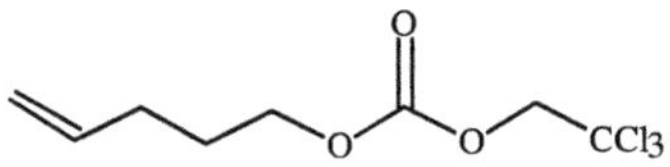	6′ O*Pent*	73%	64%
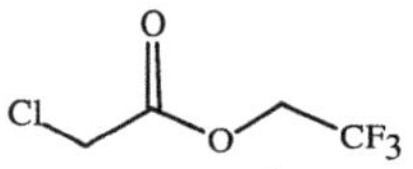	6′ OAcCl	81%	30%
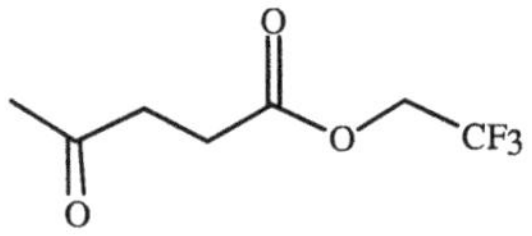	6′ O*Lev* 2′,6′ di-O*Lev*	73% 19%	66% 23%

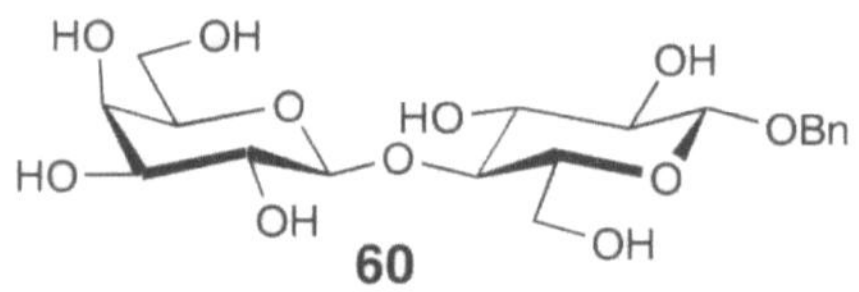

60

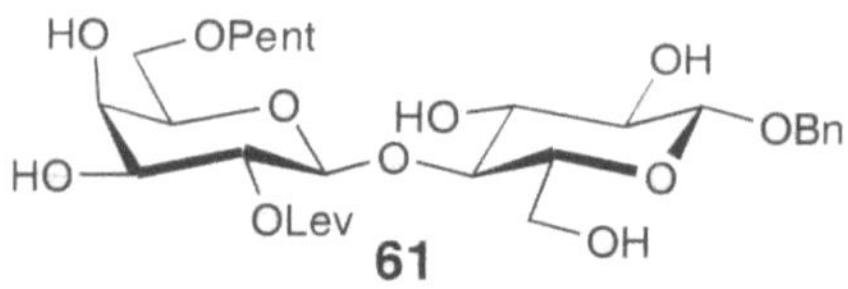

61

Table 11. Enzymatic selective acylation of disaccharides **52** and **53**; for abbreviations, see Table 9

Entry	Compound	Conditions	/°C, *t*/h	Products	Acylation position	Yield/%
1	**62**	*THF, VA*	40°, 24 h	**64**	6′-OAc	90
				64	6′-OAc	8
2	**62**	CH_3CN, *VA*	40°, 24 h	**65**	6,6′-OAc	70
				66	6,2′,6′-OAc	16
3	**63**	*TMF, VA*	45°, 40 h	**70**	6′-OAc	90
4	**63**	CH_3CN, *VA*	45°, 40 h	**70**	6′-OAc	90
5	**64**	CH_3CN, *VCA*	28°C, 4 h	**67**	6′-OAc, 6-OChloroAc	90
6	**64**	CH_3CN, *VCA*	35°C, 1 h	**67**	6′-OAc, 6-OChloroAc	67
				68	6′-OAc, 6,2′-OChloroAc	29
7	**65**	CH_3CN, *VCA*	40°C, 12 h	**69**	6′,6-OAc, 2′-OChloroAc	91

CAB was also employed for a systematic study of the regioselective protection of the 2-azidodeoxy derivatives **62** and **63**, synthetic equivalents of *β*-D-*Gal*(1-3)-D-*Glc*NAc and *β*-D-*Gal*(1-4)-D-*Glc*NAc [60]. The latter compounds are among the most important components of glycoproteins and glycolipids; therefore, it is extremely important to protect them selectively for the synthesis of their complex derivatives. The 2-azidodeoxy compounds were submitted to selective acylation using CAB under different conditions (Table 11).

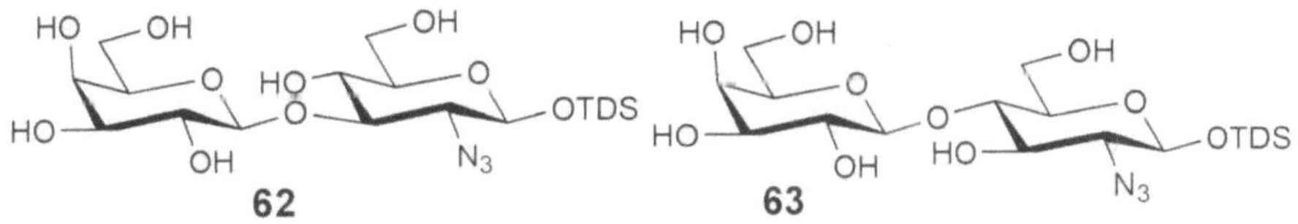

The regioselectivity depended on the solvent as well as on the nature of substrate and acylating agent. The azido derivative **62** in CH_3CN afforded as main product the 6,6′-diacetylated compound (Table 11, entry 2); as lipases usually acylate only the non-reducing moiety of a disaccharide, this behaviour appeared quite surprising. The possibility of modulating the regioselectivity in different solvents was then exploited for the preparation of compounds with two diverse acyl groups on the two primary hydroxyl groups (Table 11, entries 5,6), a result that can be achieved chemically only in many steps.

Finally, some selected examples of the use of regioselective lipase acylations of complex glycoconjugated compounds are reported to stress once more the versatility of these enzymes.

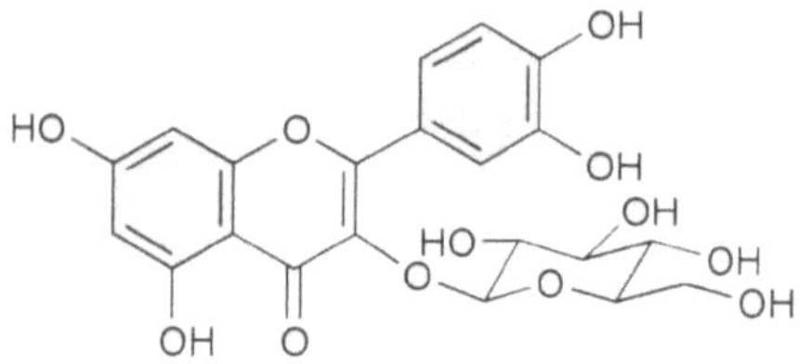

Isoquercitrin (**71** [47])
CAB 79% 3'',6''-di-*O*-Ac

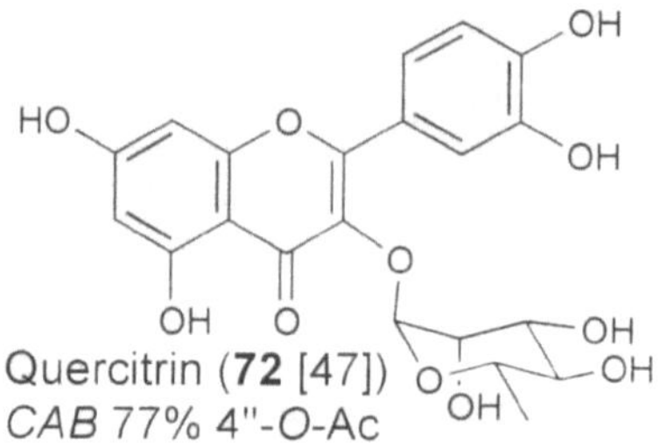

Quercitrin (**72** [47])
CAB 77% 4''-*O*-Ac

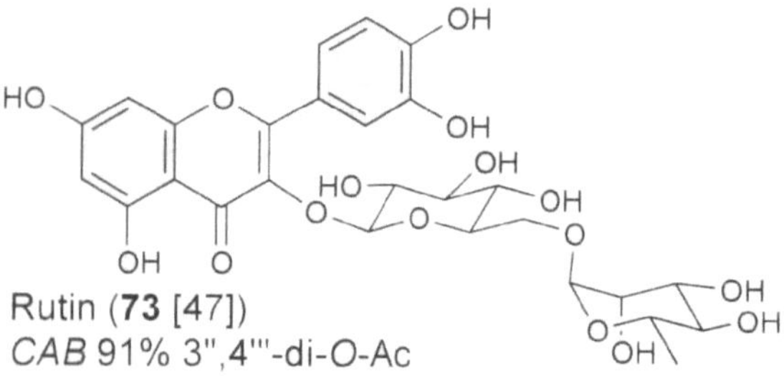

Rutin (**73** [47])
CAB 91% 3'',4'''-di-*O*-Ac

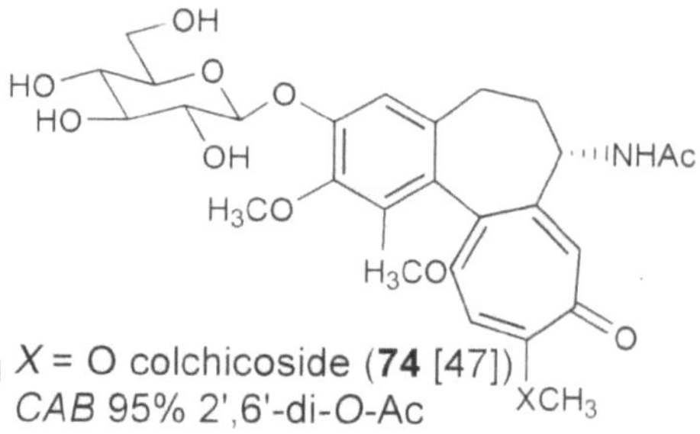

X = O colchicoside (**74** [47])
CAB 95% 2',6'-di-*O*-Ac
X = S thiocolchicoside (**65**)
CAB 93% 2',6'-di-*O*-Ac

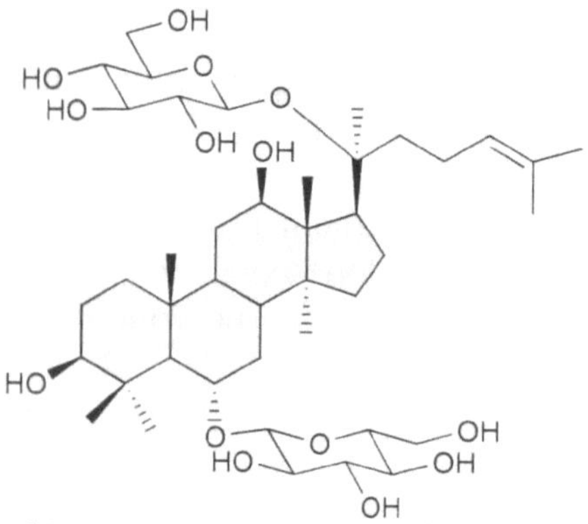

Ginsenoside Rg1 (**76** [61])
CAB 6'-*O*-Ac/6,6'-di-*O*-Ac 22/1

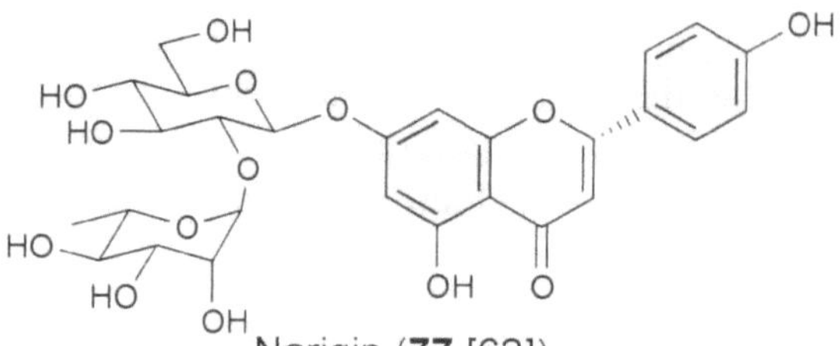

Narigin (**77** [62])
CAB

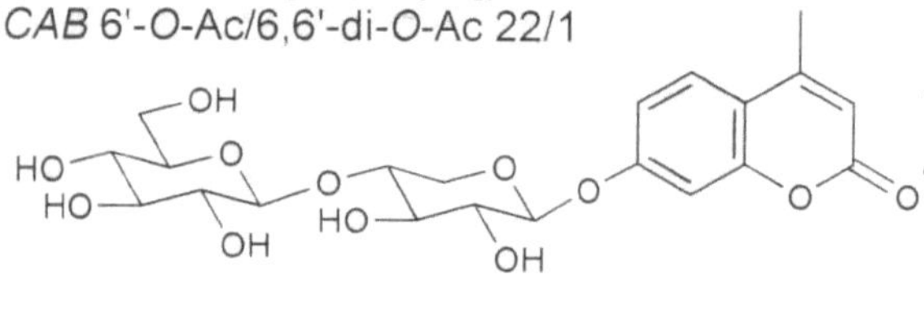

78 [63]
Amano *PS* 89% 6''-*O*-Ac

Lipases have also been used for the resolution of racemic *β*-D,L-amicetosides **79** (Scheme 5); the pure free D-enantiomer, amicetose (a deoxy sugar), is a constituent of the antibiotic substances amicetin and axemomycin. Among the many lipases tested, PPL, *Candida rugosa*, and *Pseudomonas PS*, the latter gave the best result (Scheme 5) [64].

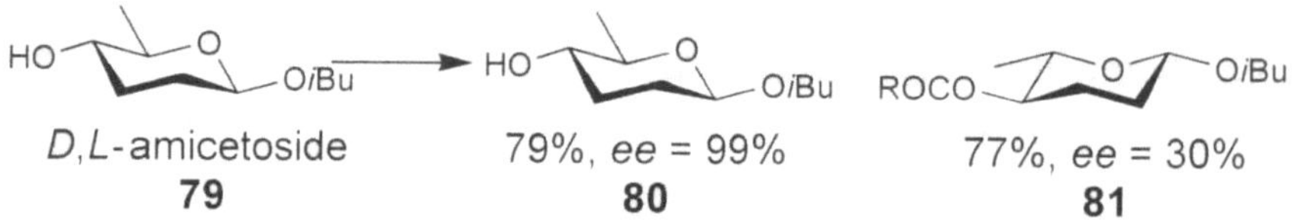

Scheme 5

Conclusions

Quite a lot of work has been done in the study and application of lipases in carbohydrate chemistry; lipases have already found important applications, especially in the industrial synthesis of fatty acid esters of carbohydrates, compounds that have important applications in detergents, food, cosmetics, and pharmaceuticals because of their properties as non-ionic surfactants. In organic synthesis these enzymes are the most extensively used as biocatalysts; moreover, the increasing knowledge of their structure and function has already led to site directed mutagenesis of *Rhizopus delemar* lipase to shift the preference of the mutants for the hydrolysis of medium-chain triglycerides [65]. This suggests that very soon these already versatile chemical tools could be designed and built as required for specific substrates and acylating agents.

References

[1] Green TW, Wuts PJM (1991) Protective Groups in Organic Synthesis, 2nd edn. Wiley, New York
[2] Kunz H, Waldmann H (1991) In: Trost BM, Flemming I, Winterfeldt E (eds) Comprehensive Organic Synthesis, vol 6. Pergamon Press, Oxford, p 631
[3] Waldmann H (1991) Kontakte (Darmstadt), 2, p 33
[4] Waldmann H (1993) J Prakt Chem **335**: 109
[5] Haines H (1976) Adv Carbohydr Chem Biochem **33**: 11; ibid (1981) **39**: 13
[6] Shaw JF, Klibanov AM (1987) Biotechnol Bioeng **29**: 648
[7] Schmid RD, Verger R (1998) Angew Chem Int Ed Engl **37**: 1608
[8] Klibanov AM (1984) Science **224**: 1249
[9] Volkin DB, Staubli A, Langer R, Klibanov AM (1991) Biotechnol Bioeng **37**: 843
[10] Griebenow K, Klibanov AM (1996) J Am Chem Soc **118**: 11695
[11] Wescott R, Klibanov AM (1994) Biochim Biophys Acta **1206**: 1
[12] Fitzpatrick PA, Klibanov AM (1991) J Am Chem Soc **113**: 3166
[13] Terradas F, Teston-Henry M, Fitzpatrick PA, Klibanov AM (1993) J Am Chem Soc **115**: 390
[14] Hirose Y, Kariya K, Sasaki J, Kurono Y, Ebike H, Achiwa K (1992) Tetrahedron Lett **33**: 7157
[15] Nakamura K, Takebe J, Kitayama T, Ohno A (1991) Tetrahedron Lett **32**: 4941; Secundo F, Riva S, Carrera G (1992) Tetrahedron Asymm **3**: 267
[16] Fink L, Hay GW (1969) Can J Biochem **47**: 353
[17] Sachder HS, Starkovsky NA (1969) Tetrahedron Lett **9**: 733
[18] Shaw JF, Klibanov AM (1987) Biotechnol Bioeng **29**: 648
[19] Shaw FJ, Liaw ET (1986) Biocatalysis in Organic Med. In: Laane C, Tramper J, Lilly MD (eds) Proceedings of the International Symposium, Wageningen, The Netherlands, 7–10 December
[20] Sweers HM, Wong CH (1986) J Am Chem Soc **108**: 6421
[21] Hennen WJ, Sweers HM, Wang YF, Wong CH (1988) J Org Chem **53**: 4939
[22] Bastida, Fernandez-Lafuente R, Fernandez-Lorente G, Guisan JM, Pagani G, Terreni M (1999) Biorg Med Chem Lett **9**: 633
[23] Kloosterman M, Moosmuller EW, Schoemaker HE, Meijer EM (1987) Tetrahedron Lett **28**: 2989
[24] Barili PL, Catelani G, D'Andrea F, Mastrorilli E (1997) J Carbohydr Chem **16**: 1001
[25] Holla EW (1989) Angew Chem Int Ed Engl **28**: 220
[26] Matsui T, Kita Y, Matsushita Y, Nakayama M (1992) Chem Exp **7**: 45
[27] Koosterman M, De Nijs MP, Weijnen JG, Schoemaker HE, Meijer EM (1989) J Carbohydr Chem **8**: 333
[28] Zemek J, Kucar S, Aderle D (1987) Collect Czech Chem Commun **52**: 2347

[29] Csuk R, Glanzer BJZ (1988) Z Naturforsch **43b**: 1355
[30] Ballesteros A, Bernabe M, Cruzado C, Martin-Lomas M, Otero C (1989) Tetrahedron **45**: 7077
[31] Vogel P, Kramer S, Ott AJ (1997) Liebigs Ann Chem 1425
[32] Honig H, Senfer-Wasserthal P, Stutz AE, Zenz E (1989) Tetrahedron Lett **30**: 811
[33] Kloosterman M, Schoemaker HE, Mejer EM, de Vries NK (1989) Xth Enzyme Engineering Conference, Koshikojima, Japan, September 1989. New York Academy of Science
[34] Kloosterman M, Weijnen JGJ, de Vries NK, Mentech J, Caron L, Descotes G, Schoemaker HE, Mejer EM (1989) Carbohydr Chem **8**: 693
[35] Change K-Y, Wu S-H, Wang KT (1991) Carbohydr Res **222**: 121
[36] Palmer DC, Terradas F (1994) Tetrahedron Lett **35**: 1673
[37] Khan R, Gropen L, Konowicz PA, Matulovà M, Paoletti S (1993) Tetrahedron Lett **34**: 7767
[38] Ong G-T, Change K-Y, Wu S-H, Wang K-T (1994) Carbohydr Res **265**: 311
[39] Gardossi L, Khan R, Konowicz PA, Gropen L, Paulsen BS (1999) J Mol Cat B **6**: 89
[40] Klibanov M (1986) Chemtech 354
[41] Klibanov M (1985) Proc Natl Acad Sci USA **82**: 3192
[42] Therisod M, Klibanov AM (1986) J Am Chem Soc **108**: 5638
[43] Castain-Degueil M, de Jeso B, Drouillard S, Maillard B (1987) Tetrahedron Lett **28**: 953
[44] Wang YF, Lalonde JJ, Momongan M, Bergbreiter DE, Wong CH (1988) J Am Chem Soc **110**: 7200
[45] Ghogare G, Kumar S (1989) J Chem Soc Chem Commun 1533
[46] Therisod M, Klibanov AM (1987) J Am Chem Soc **109**: 3977
[47] Danieli B, Luisetti M, Sampognaro G, Carrea G, Riva S (1997) J Mol Cat B **3**: 193
[48] Ciuffreda P, Colombo D, Ronchetti T, Toma L (1990) J Org Chem **55**: 4187
[49] Gridley JJ, Hacking AJ, Osborn HMI, Spackman DG (1997) Synlett 1397
[50] Wong C-H, Chen S-T, Hennen W-J, Bibba JA, Wang Y-F, Liu JL-C, Pantoliano MW, Whitlow M, Bryan PN (1990) J Am Chem Soc **112**: 945
[51] Zhong Z, Liu JL-C, Dinterman LM, Finkelman MAJ, Mueller WT, Rollence ML, Whitlow M, Wong C-H (1991) J Am Chem Soc **113**: 683
[52] Prasad AK, Sørensen MD, Parmar VS, Wengel J (1995) Tetrahedron Lett **36**: 6163
[53] Kloosterman M, Schoemaker HE, Kloosterman-Castro EN, Meijer EM (1987) In: Lichtenthaler FW, Nett KK (eds) Book of Abstracts of the XXth European Carbohydrate Symposium. Darmstadt, D-17
[54] Nicotra F, Riva S, Secundo F, Zuchelli L (1989) Tetrahedron Lett **30**: 1703
[55] Riva S, Chopineau J, Kieboom APJ, Klibanov AM (1988) J Am Chem Soc **110**: 584
[56] Khan R, Gropen L, Konowicz PA, Matulovà M, Paoletti S (1993) Tetrahedron Lett **34**: 7767
[57] Cai S, Hakomori S, Toyokumi T (1992) J Org Chem **57**: 3431
[58] Lay L, Panza L, Riva S, Khitri M, Tirendi S (1996) Carbohydr Res **291**: 197
[59] La Ferla B, Lay L, Poletti L, Russo G, Panza L (2000) J Carbohydr Chem **19**: 331
[60] La Ferla B, Lay L, Panza L, Russo G (2000) Tetrahedron Asymm **11**: 3647
[61] Danieli B, Riva S (1994) Pure Appl Chem **66**: 2215
[62] Riva S, Danieli B, Luisetti M (1996) J Nat Prod **59**: 618
[63] Yasukochi T, Fukase K, Suda Y, Takagaki K, Endo M, Kusumoto S (1997) Bull Chem Soc Jpn **70**: 2719
[64] Noecker LA, Martino JA, Foley PJ, Rush DM, Giuliano RM, Villani Jr FJ (1998) Tetrahedron Asymm **9**: 203
[65] Joerger RD, Haas MJ (1994) Lipids **29**: 377

Received September 11, 2001. Accepted November 19, 2001

Invited Review

Carbohydrate-Based Scaffolds for the Generation of Sortiments of Bioactive Compounds

Francesco Peri, Laura Cipolla, Eleonora Forni, and **Francesco Nicotra***

Department of Biotechnology and Biosciences, University of Milano-Bicocca, I-20126 Milano, Italy

Summary. The polyfunctionality and conformational rigidity of carbohydrates make this class of compounds ideal scaffolds for the production of sortiments[1] of bioactive compounds. Examples of carbohydrate-derived peptidomimetics of biological interest, such as somatostatin agonists and integrin antagonists, are presented. In order to have access to solid phase supported sortiments of compounds, orthogonally protected or unprotected carbohydrates were linked to polymers and reacted in the solid phase employing different regioselective strategies. Original bicyclic and tricyclic glycidic scaffolds were easily obtained starting from natural sugars such as *D*-arabinose and *D*-fructose. Manipulation of these conformationally blocked compounds afforded different carbohydrate-based derivatives, among which azidoacids are useful precursors of β-turn peptidomimetics.

Keywords. Carbohydrates; Combinatorial chemistry; Scaffolds; Peptidomimetics.

Introduction

The relevant role played by carbohydrates in biological recognition phenomena lies in the unique feature of this class of natural compounds which combine high combinatorial potential and conformational rigidity. Monosaccharides, if compared to amino acids, contain more than one stereocentre (in general 5 for hexoses), are present in the pyranosidic or furanosidic form and, interestingly, the cyclic structure induces a unique conformational rigidity fundamental for biological interactions of oligosaccharides. It is now clear that the complex structure of oligosaccharides is only partially involved in the weak binding with their receptors; the remaining part appears to act as a scaffold that orients the binding determinants in the appropriate conformation and provides a connection to the aglycons. An interesting example is

---

* Corresponding author. E-mail: francesco.nicotra@unimib.it

[1] Following the argumentation of *Roald Hoffmann* (Hoffmann R (2001) Angew Chem **113**: 3439), the terminus 'sortiment' is to be preferred over 'library' in this context

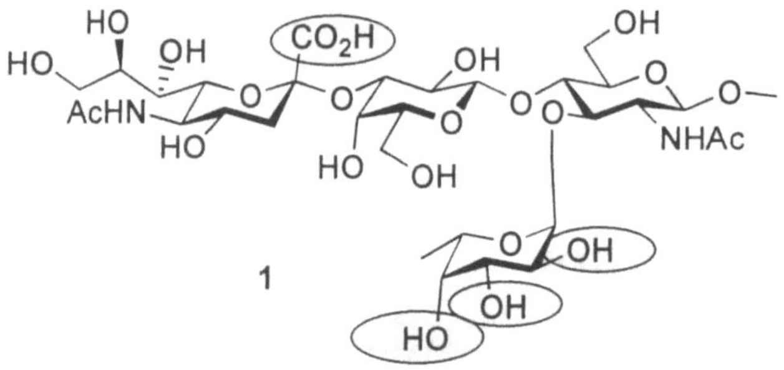

1

Fig. 1. The tetrasaccharide sialyl *Lewis*^{x}

represented by the tetrasaccharide sialyl *Lewis*^{x} (*sLe*^{x}, Fig. 1) that interacts with selectins, the cell adhesion molecules expressed at the surface of vascular endothelial cells after injury. This recognition triggers the anti-inflammatory response and the metastasis formation, *sLe*^{x} being a determinant of neutrophils and metastatic tumor cells. Extensive studies have shown that the structural requirements for the biological activity of this tetrasaccharide are mainly the presence of the carboxylic group of neuraminic acid and the hydroxyl groups of *L*-fucose (as highlighted in Fig. 1). This observation has prompted the synthesis of a variety of analogues in which the glycidic parts of the molecule not involved in the recognition are replaced by spacers with lower molecular, and hence synthetic, complexity and with improved metabolic stability. The studies have pointed out that the conformational rigidity induced by the glycidic rings is fundamental for the biological activity. The substitution of glycidic units with acyclic moieties causes a drop in the biological activity due to the gain of conformational freedom; on the contrary, the insertion of rigid scaffolds with limited conformational mobility restores an adequate biological activity. For instance, the 3,4-disubstitututed N-acetylglucosamine moiety of *sLe*^{x} can be substituted by 1,2-*trans*-cyclohexanediol (Fig. 2) in the synthesis of a highly effective mimic [1]. Furthermore, the orientation of the carboxylic group of neuraminic acid is quite important for an effective adhesion to E-selectins. Compound **2** (Fig. 2) lacks the pyranosidic ring of neuraminic acid and has a lower biological activity than compound **3** in which the methylcyclohexyl substituent decreases the mobility of the carboxylic group [2].

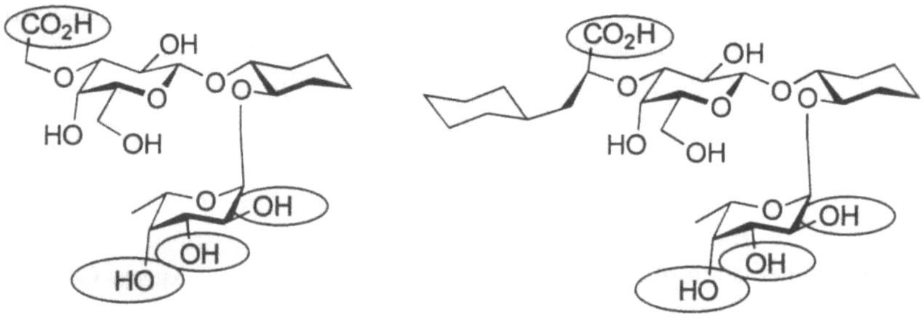

2

IC_{50}=500 μM

3

IC_{50}=80 μM

Fig. 2. Simplified mimic of sialyl *Lewis*^{x}

From these observation it is clear that the structural diversity and conformational rigidity of carbohydrates render these compounds interesting scaffolds capable to link to different pharmacophoric groups and orient them in a precise spatial arrangement.

Glycidic Scaffolds

An interesting example of the use of monosaccharides as scaffolds for the synthesis of bioactive compounds is represented by the work of *Hirschmann et al.* [3] describing the synthesis of bioactive somatostatin mimics from the natural sugar *D*-glucose. The peptide hormone somatostatin (**4**, Fig. 3) is a cyclic tetradecapeptide which inhibits the release of several hormones including the growth hormone (GH). Since its biological activity is due to the interaction of the β-turn type I fragment *Phe-Trp-Lys-Thr* (highlighted in Fig. 3) with the receptor, the aminoacidic side chains responsible for the interaction have been attached to glycidic scaffolds in a proper spatial arrangement. Figure 4 describes a prototype of a carbohydrate-derived somatostatin mimic (**5**) prepared from commercially available 2,3,4,6-tetra-O-acetyl-α-*D*-glucopyranosyl bromide. Employing different glycidic starting materials, other analogues lacking the benzyloxy group alternatively in position 2, 3, or 4 of the sugar have also been synthesized [3]. More recently, this approach has been extended to a model study in solution for the preparation of solid-phase sortiments of compounds based on the α-*D*-glucose scaffold [4].

Another example of the use of carbohydrates to mimic non-carbohydrate bioactive compounds has been reported by *Nicolau et al.* who synthesized mimics such as compound **6** (Fig. 5) of the *Arg-Gly-Asp* (RGD) sequence [5]. The RGD sequence is responsible for the adhesion of extracellular matrix glycoproteins, such as fibronectin, fibrinogen, or vitronectin, to endothelial cells (Fig. 6), a phenomenon that promotes, *inter alia*, angiogenesis in the case of $\alpha_V\beta_3$-integrins and platelet

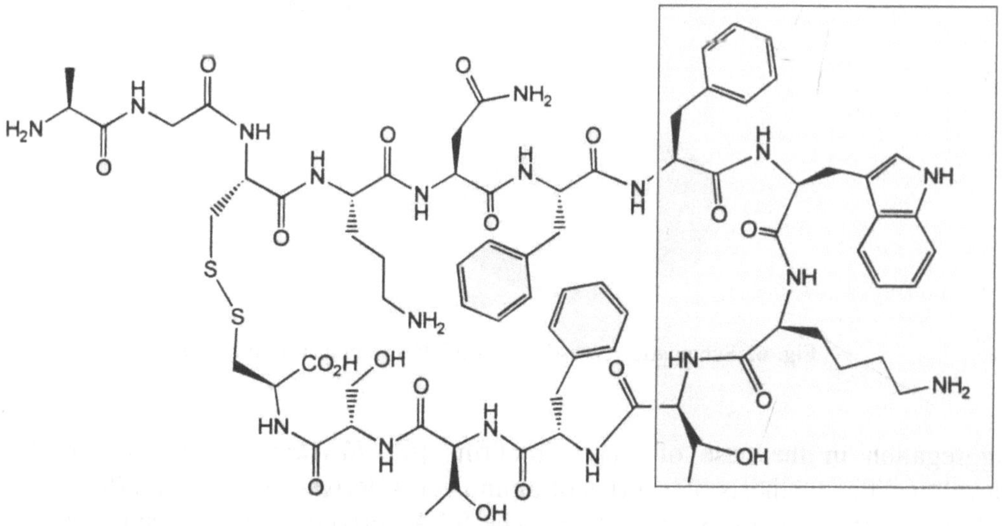

4

Fig. 3. The hormone somatostatin

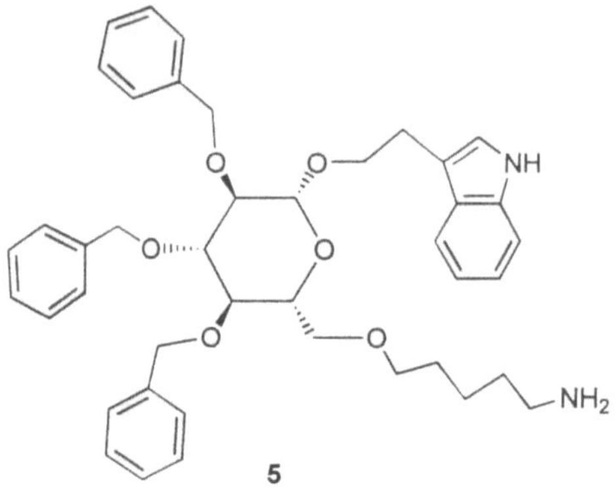

Fig. 4. A carbohydrate derived somatostatin mimic

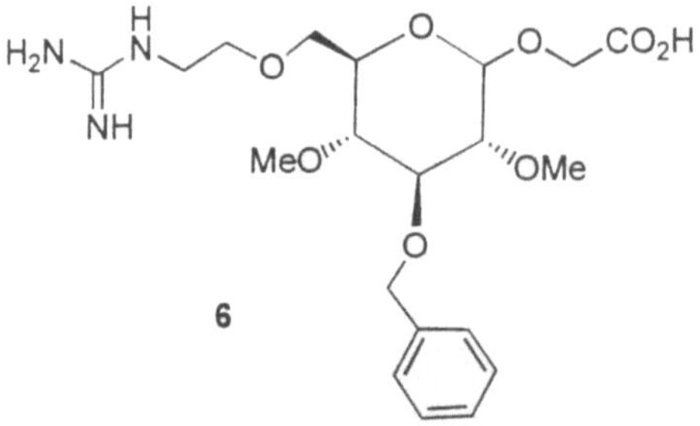

Fig. 5. Carbohydrate derived mimic of the RGD sequence

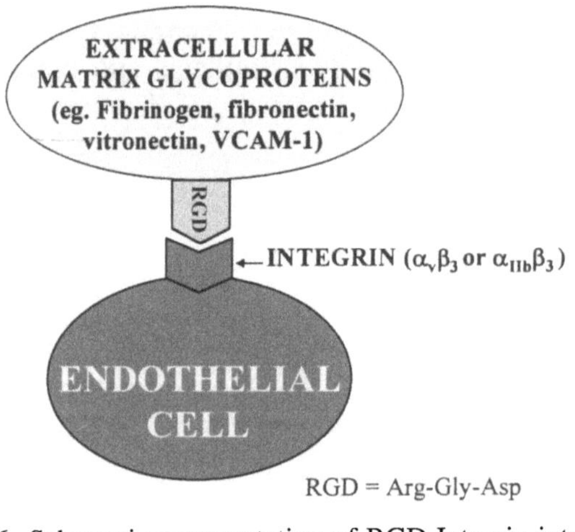

Fig. 6. Schematic representation of RGD-Integrin interaction

aggregation in the case of $\alpha_{IIb}\beta_3$-integrins [6]. *Kessler* and co-workers have developed the synthesis of a series of amino acids derived from sugars (SAA) with different turn-inducing properties [7]. The SAA derived from glucosyluronic acid methylamine (*Gum*) was incorporated in cyclic RGD peptides that proved active as selective $\alpha_V\beta_3$-integrin antagonists [8]. In all these examples, the potentials of the glycidic scaffold have only partially been exploited, since only a couple of hydroxyl

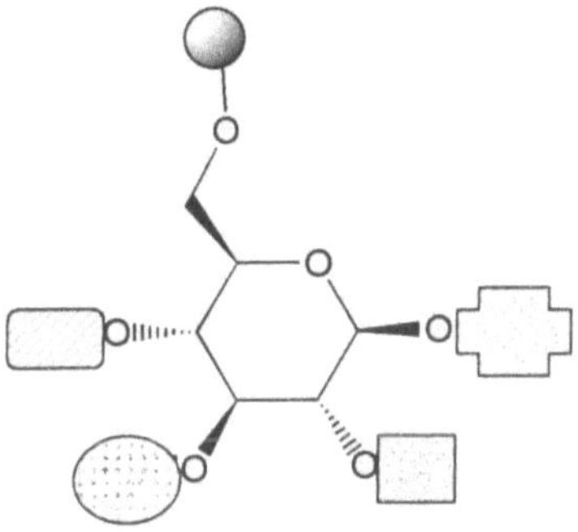

Fig. 7. Different pharmacophores linked to a solid phase supported carbohydrate

groups out of five were involved in the derivatization. Examples in which the whole potential diversity of the sugar is exploited in the preparation of glycidic scaffolds have been reported by *Kunz* [9], *Hirschmann* [4], and *Hanessian* [10] who were able to differentiate each hydroxyl group of the sugar. The problem consists mainly in the orthogonal protection of numerous hydroxyl groups; once this problem will be completely solved, orthogonally protected carbohydrates could be used in combinatorial chemistry, and compound sortiments in which the diversity consists in the different arrangement of pharmacophores linked to the hydroxyl groups of carbohydrates will be efficiently produced (Fig. 7). It is interesting to note that in addition to hydroxyl groups, natural sugars contain other functionalities such as the amino or the carboxylic group which further enhance the potential diversity.

The development of orthogonalities among the numerous hydroxyl protecting groups of sugars is a particularly tedious task. Furthermore, if the combinatorial approach is developed in the solid phase, the linkage of the sugar to the solid support must be retained throughout the synthesis and cleaved selectively at the end of the process. As each derivatization step requires two reactions (deprotection and derivatization), a hexose linked to the resin *via* a hydroxyl group has still four positions to be manipulated, which means eight reactions. In oligosaccharide synthesis the protective groups are classified as temporary and permanent protections, the latter being removed at the end of the synthesis without affecting the glycosidic linkages. A permanent protection should link the glycidic scaffold to the resin and ensure complete orthogonality to the commonly used temporary protecting groups and to the derivatization conditions.

A variety of groups including silanes [11], thioethers [12], benzylidene acetals [13], succinamides [14], photolabile esters [15], *p*-acylaminobenzyl esters [16], alkenes [17], *tris*-(alkoxy)-benzyl amines (*BAL*) [18], and phenoxyacetates [19] have been employed to anchor monosaccharides to polymers in the first step of the solid-phase oligosaccharide synthesis. Most of these linkers interfere with some common deprotection or derivatization conditions, thus limiting the versatility and flexibility in synthetic planning.

Examples of solid phase supported orthogonally protected glucoderivatives are reported in Schemes 1–3. Starting from *D*-glucose (**7**) and exploiting different protections as shown in the Schemes, phenylthio 3-O-allyl-4-O-*p*-methoxybenzyl-*β*-*D*-glucopyranoside (**19**) and phenylthio 2-O-allyl-4-O-*p*-methoxybenzyl-*β*-*D*-glucopyranoside (**20**) linked to a polystyrene-divinylbenzene-NH_2 resin through a

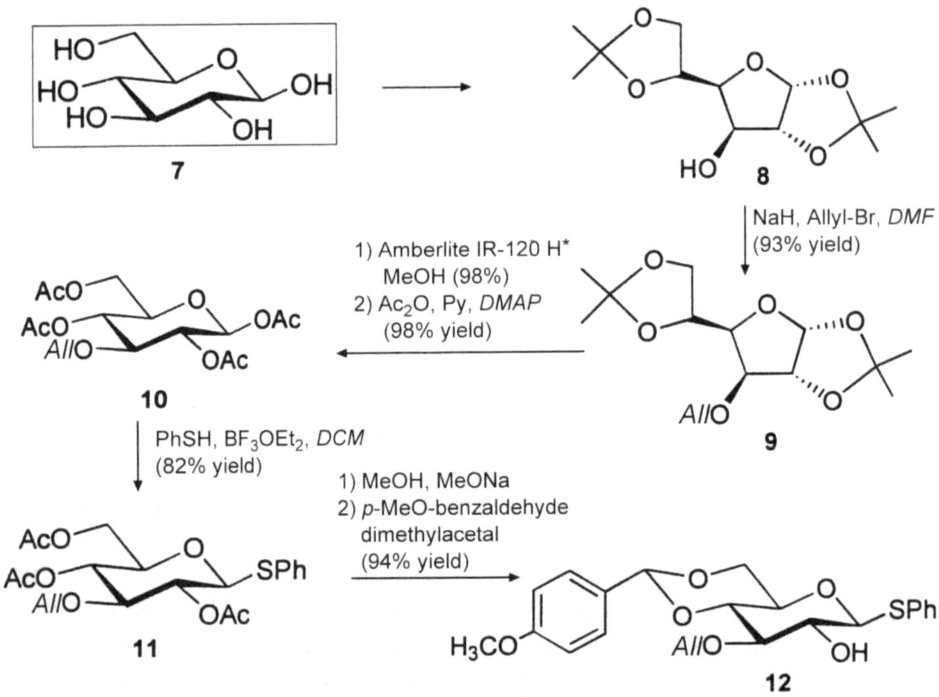

Scheme 1

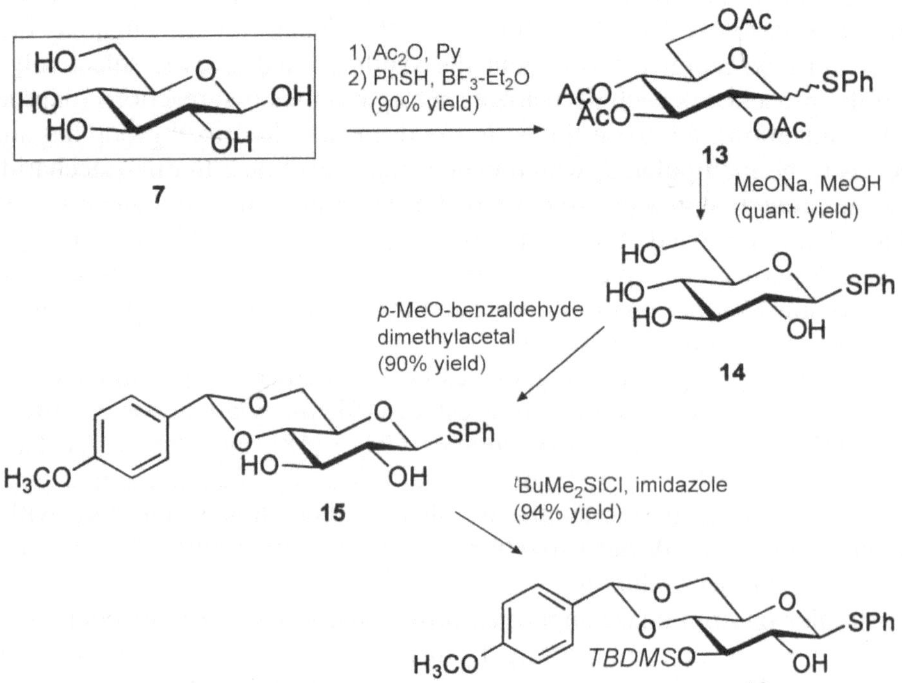

Scheme 2

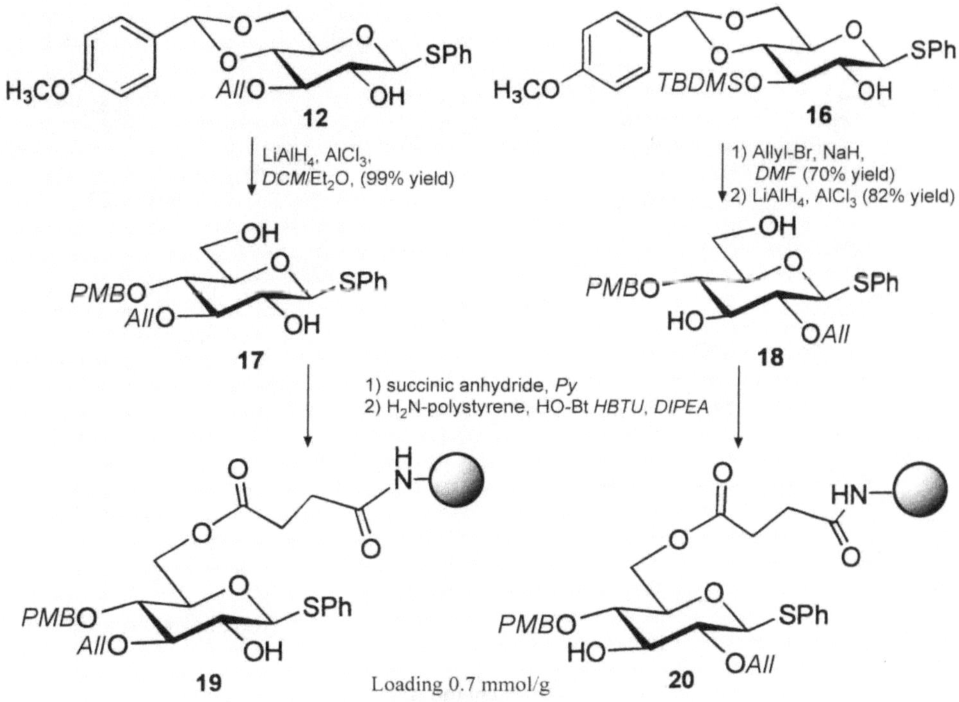

Scheme 3

succinamide linker have been synthesized [20]. In these compounds, a hydroxyl group (in position 2 or 3) is deprotected and a second one (3 or 2) is protected as allyl ether which was selectively hydrolyzed on solid phase by transition metal catalyzed isomerization of the double bond followed by acid treatment. The third OH (at position 4) is protected as *p*-methoxybenzyl ether that was cleaved selectively on the solid phase either with oxidants such as *DDQ* or under acidic conditions (50% *TFA* in CH_2Cl_2). Finally, the anomeric position was selectively functionalized on the solid phase by reaction of the thiophenyl group activated as sulfoxide with an alcohol in the presence of triflic anhydride and a base. After regioselective decoration of the four independent positions, the succinamide linker, which turned out to be stable under the deprotection and functionalization conditions, was hydrolyzed with bases.

Schemes 1–3 clearly indicate the complexity underlying the preparation of orthogonally protected substrates to be linked to a solid support. It is clear that the possibility of a regioselective derivatization of the free hydroxyl groups of unprotected sugars bound to solid supports would be much more interesting and effective. In order to explore this opportunity, we investigated the behaviour of the monosaccharides methyl α-*D*-glucopyranoside (**21**), methyl α-*D*-mannopyranoside (**22**), and methyl β-*D*-galactopyranoside (**23**) linked to a trityl resin by the primary hydroxyl group. When treated with Bu_2SnO and then with benzoyl chloride, the benzoylation of the stannylidene intermediates (Scheme 4) occurred in positions 2, 3, and again 3, respectively, with a regioselectivity above 98% [21]. These results are promising in the perspective of decorating polymer-bound monosaccharides regioselectively with a series of functional groups (including other monosaccharides in the stannylene-assisted regioselective glycosylation). In this context, new

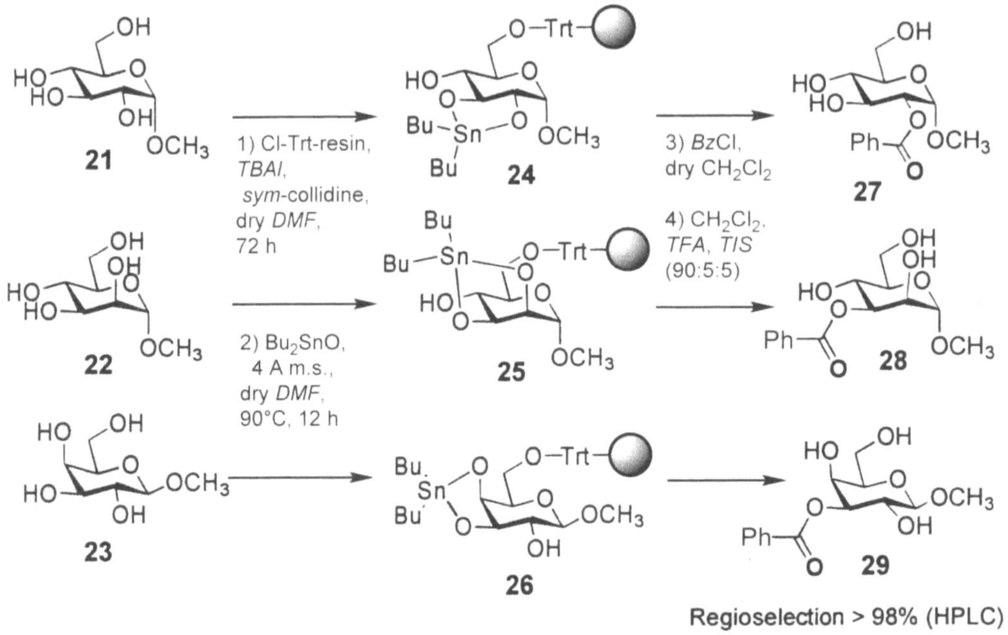

Scheme 4

opportunities can arise from lipase-catalyzed acylation of sugars on solid phase, once the problems related to the permeability of the resin to the enzymes and to solid-phase kinetics are solved.

Conformationally Constrained Glycidic Scaffolds

In the development of glycidic scaffolds for the production of sortiments of compounds of pharmaceutical interest, two main aspects must be taken into account. First of all, the originality of the scaffold is important in order to allow patenting, and natural sugars are of course not ideal substrates from this point of view. Secondly, the conformational rigidity should be ideally enhanced; this strategy affords much more active compounds once the correct conformation has been achieved. Having both of these aspects in mind, we studied the possibility to synthesize original polycyclic scaffolds starting from perbenzylated sugars by an easy debenzylation-iodocyclization procedure [22]. Starting from polybenzylated sugars with an anomeric appendage with a C=C double bond, a cyclic iodoether was obtained by treatment with iodine. The mechanism of this iodocyclization consists in the attack of the intermediate iodonium ion by the benzyloxy group in γ-position with respect to the double bond with loss of the benzyl group (Scheme 5). Two examples of application of this reaction are reported in Schemes 6 [23] and 7 [24]. The allylic substituent has been introduced at the anomeric centre of polybenzylated acetyl arabinofuranoside **30** and methyl fructofuranoside **36** affording the corresponding allyl C-glycosides **31** and **37**. Treatment of **31** and **37** with iodine results in the iodocyclization. In the case of the arabinoderivative **31**, in which a mixture of α- and β-C-glycosides is obtained, it is possible to effect a kinetic resolution due to the easier formation of the *cis*-fused bicyclic structure **32**. Besides the formation of a second cycle, the iodocyclization

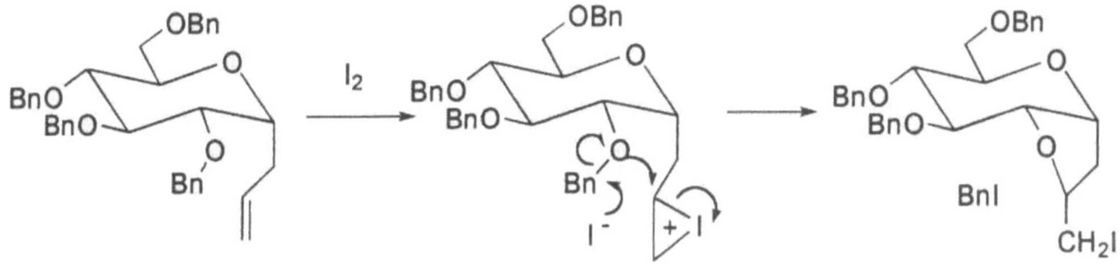

Scheme 5

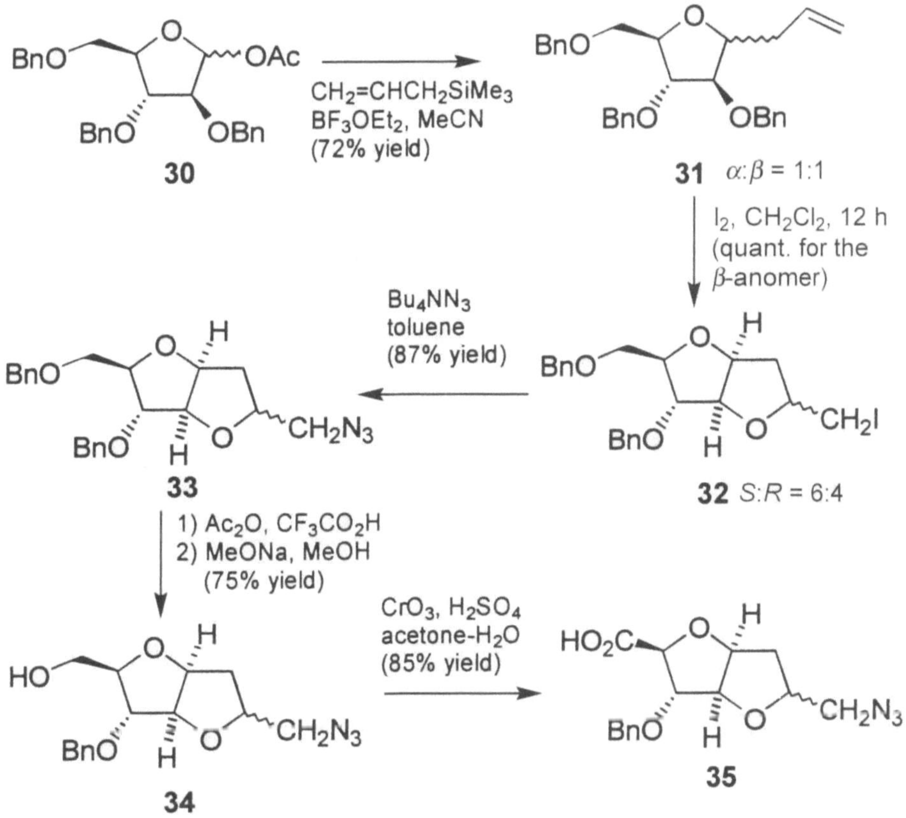

Scheme 6

reaction gives rise to a new useful functional group, the alkyl iodide, which can be exploited for further functionalizations. In the examples reported in Schemes 6 and 7, treatment of the iodides **32** and **38** with tetrabutyl ammonium azide afforded the azides **33** or **39**, precursors of the corresponding amines. A carboxylic function was subsequently introduced in these glycidic scaffolds in order to obtain sugar-aminoacids. The primary hydroxyl group of **33** and **39** was selectively deprotected by acetolysis of the benzyl ether followed by basic hydrolysis of the acetate. The primary hydroxyl group of the obtained compounds was oxidized using the *Jones* reagent affording the azidoacids **35** and **40**.

The conformational analysis of azidoacidic scaffold (*S*)-**35** and (*R*)-**35** was performed by NMR spectroscopy and molecular modelling. Both compounds

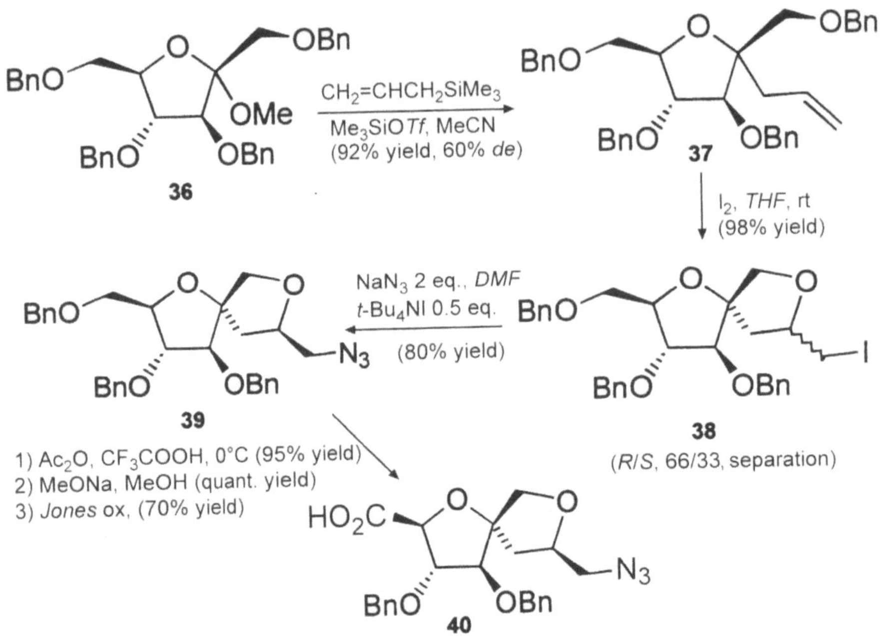

Scheme 7

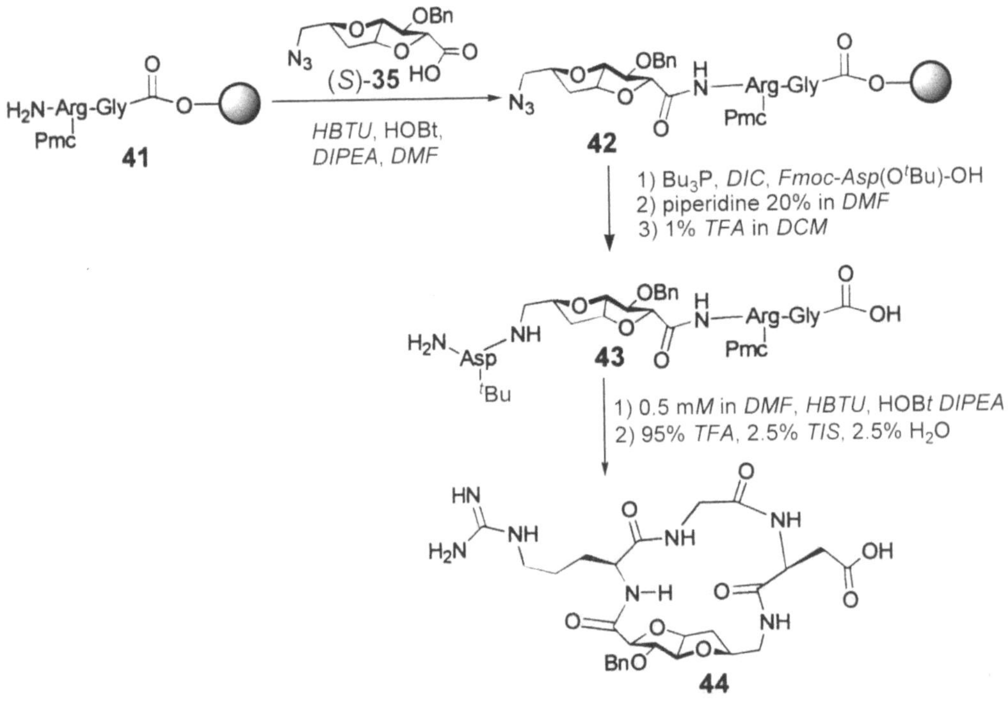

Scheme 8

turned out to be good isosters of the dipeptide in position $i+1$ and $i+2$ of a β-turn. Compound (*S*)-**35** was incorporated as amino acid in a cyclic RGD peptide according to the synthetic procedure shown in Scheme 8 [23]. Compound **44** was tested as inhibitor of the adhesion of fetal bovine aortic endothelial GM7373 cells to their ligands: fibroblast growth factor 2 (FGF-2), vitronectin (VN), and fibronectin (FN). A selective inhibition of the adhesion of FGF-2 and VN was observed, indicating a role of **44** as $\alpha_V\beta_3$ integrin antagonist [25].

Alternatively, the cyclic iodoethers obtained by iodocyclization-debenzylation undergo reductive elimination if treated with Zn and acids, restoring the original double bond (Scheme 9). The final result in this case is the selective debenzylation on the hydroxyl group in γ-position to the double bond. The free hydroxyl group can therefore be manipulated in order to perform further transformations. In the case reported in Scheme 9, the free hydroxyl group of compound **45** has

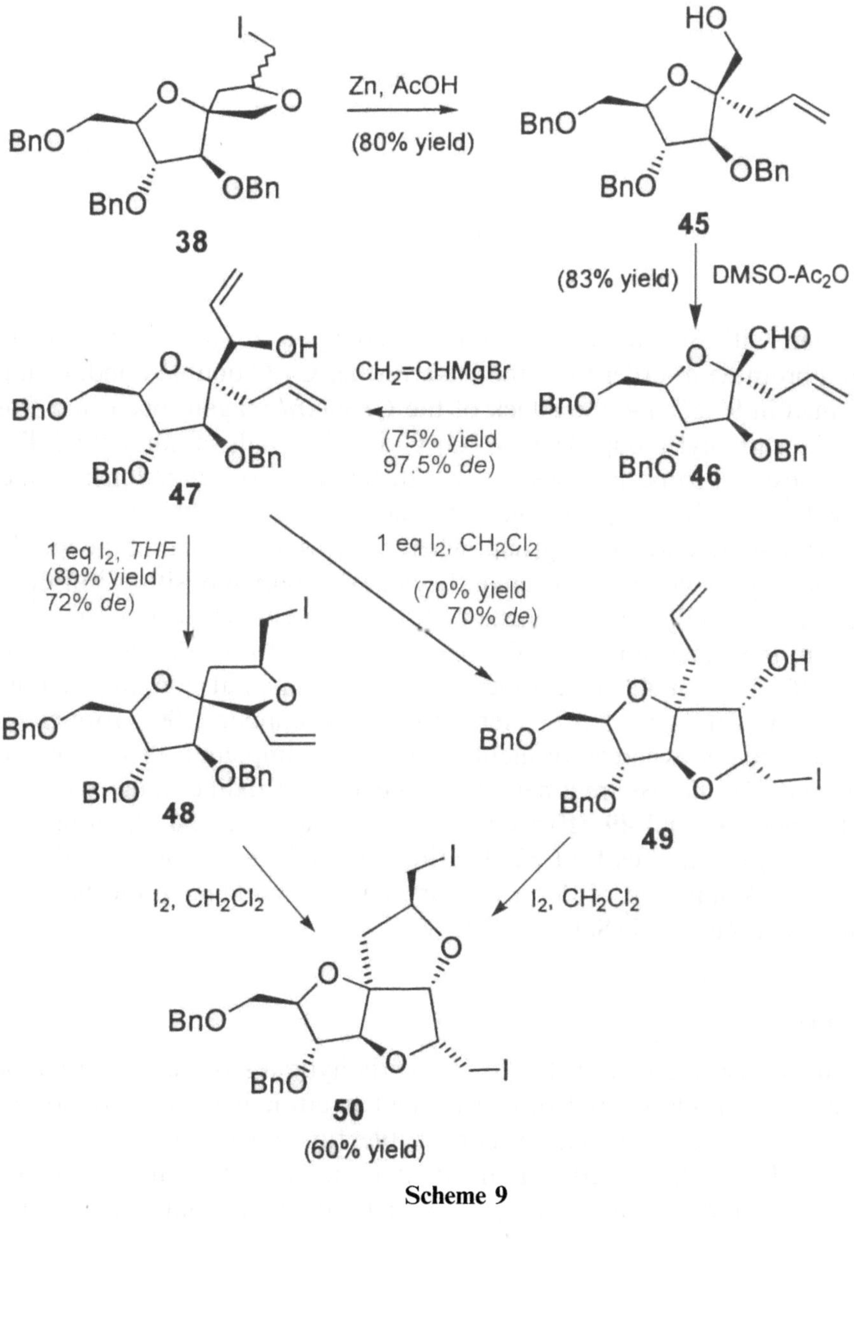

Scheme 9

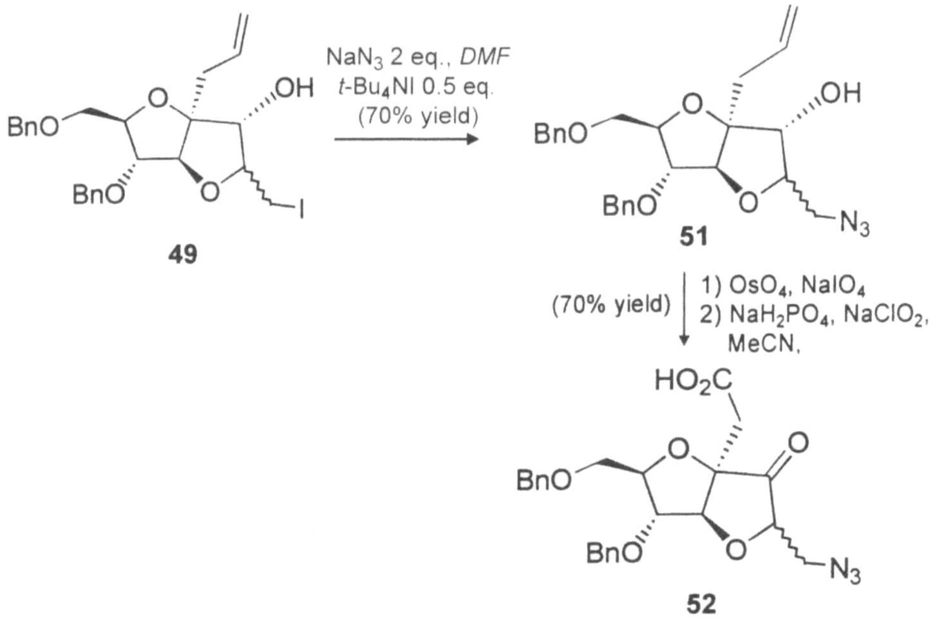

Scheme 10

been oxidized with *Swern* reagent, and the aldehyde **46** was treated with vinyl magnesium bromide in order to introduce a further C=C double bond. Compound **47** is obtained in 97.5% *de*, the attack of the *Grignard* reagent occurring from the *re*-face of the carbonyl group. The two C=C double bonds of compound **47** are in the correct position to be involved in two different iodocyclizations; in fact, the double bond of the allylic appendage at the anomeric centre can generate a cyclic iodoether with the free hydroxyl group, whereas the double bond of the vinyl group can originate a cycle with the oxygen of the benzyl group in position 3 of the parent fructoside. Interestingly, the stereochemistry of the vinylation is correct for two cyclization in the same molecule. We have been able to obtain alternatively each bicyclic scaffold **48** or **49** depending on the experimental conditions. For short reaction times in *THF* with 1 equivalent of iodine, compound **48** is formed chemo- and stereoselectively, whereas in dichloromethane compound **49** was obtained in 70% yield and 71% *de*. By treatment with an excess of iodine in dichloromethane, as expected both **48** and **49** afforded the tricyclic scaffold **50**. Alternatively, the polyfunctional glycidic scaffold **52**, bearing five different functionalities (a carboxylic acid, a ketone, an azide, a primary and a secondary protected hydroxyl group), has been obtained (Scheme 10).

Conclusions

The present review focuses on the use of carbohydrates or carbohydrate-derived compounds as scaffolds. Carbohydrates present an attractive option for non-peptide scaffolding as they contain readily convertible functional groups on a pyran or furan ring with a well-defined spatial orientation. Playing with carbohydrates, it is possible to produce an unexpected variety of new conformationally rigid polyfunctional molecules. Searching for new pharmacologically active compounds, sortiments of molecules have been obtained in solid phase from regioselective decoration of carbohydrates with pharmacophores. The bioactive compounds containing a carbohydrate or carbohydrate-derived core have generally a reduced conformational mobility. Consequently, they are also excellent candidates for structure-activity investigations.

References

[1] Kolb HC, Ernst B (1997) Chem Eur J **3**: 1571

[2] a) Simanek EE, McGarvey GJ, Jablonowski JA, Wong C-H (1998) Chem Rev **98**: 833; b) Sibata K, Hiruma K, Kante O, Wong C-H (2000) J Org Chem **65**: 2393; c) Jahnke W, Kolb HC, Blommers MJJ, Magnani JL, Ernst B (1997) Angew Chem Int Ed **36**: 2603; d) Thoma G, Kinzy W, Bruns C, Patton JT, Magnani JL, Bantli R (1999) **42**: 4909

[3] a) Hirschmann R, Nicolau KC, Pietranico S, Salvino J, Laehy EM, Splengeler PA, Furst G, Smith III AB, Strader CD, Cascieri MA, Candelore MR, Donaldson C, Vale W, Maechler L (1992) J Am Chem Soc **114**: 9217; b) Hirschmann R, Nicolau KC, Pietronico S, Leay EM, Salvino J, Arison B, Cichy MA, Spoors PG, Shakespeare WC, Sprengeler PA, Hamley P, Smith III AB, Reisine T, Raynor K, Maechler L, Donaldson C, Vale W, Freidinger RM, Cascieri MR, Strader CD (1993) J Am Chem Soc **115**: 12550; c) Hirschmann R, Wenquing Y, Cascieri M, Strader CD, Maechler L, Cichy-Knight MA, Hynes J Jr, van Rijn RD, Sprengeler P, Smith III A (1996) J Med Chem **39**: 2441; d) Hirschmann R, Hynes J Jr, Cichy-Knight MA, van Rijn RD, Sprengeler PA, Spoors PG, Shakespeare WC, Pietronico-Cole S, Barbosa J, Liu J, Yao W, Rohrer S, Smith III AB (1998) J Med Chem **41**: 1382

[4] Hirschmann R, Ducry L, Smith AB III (2000) J Org Chem **65**: 8307

[5] Nicolau KC, Trujillo JI, Chibale K (1997) Tetrahedron **53**: 8751

[6] a) Aumalley M, Gurrah M, Müller G, Calvete J, Timpl R, Kessler H (1991) FEBS Lett **291**: 50; b) Müller G, Gurrath M, Kessler H, Timpl R (1992) Angew Chem **31**: 326; c) Haubner R, Gratias R, Diefenbach B, Goodman SL, Jonczyk A, Kessler H (1996) J Am Chem Soc **118**: 7461; d) Giannis A, Rübsam F (1997) Angew Chem **109**: 606; e) Bach ACII, Espina J, Jackson SA, Stouten PFW, Duke JL, Mousa SA, De Grado WF (1996) J Am Chem Soc **118**: 293

[7] a) Von Roedern EG, Kessler H (1994) Angew Chem Int Ed **33**: 687; b) Von Roedern EG, Lohof G, Hessler G, Hoffmann M, Kessler H (1996) J Am Chem Soc **118**: 10165

[8] a) Lohof E, Planker E, Mang C, Burkhart F, Dechantsreiter MA, Haubner R, Wester HJ, Schwaiger M, Holzemann G, Goodman SL, Kessler H (2000) Angew Chem Int Ed **39**: 2761; b) Haubner R, Wester HJ, Burkhart F, Senekowitsch-Schmidtke R, Weber W, Goodman SL, Kessler H, Schwaiger M (2001) J Nuc Med **42**: 326; c) Gibson C, Sulyok GAG, Hahn D, Goodman SL, Holzemann G, Kessler H (2001) Angew Chem Int Ed **40**: 165

[9] Kallus C, Opatz T, Wunberg T, Schmidt W, Henke S, Kunz H (1999) Tetrahedron Lett **40**: 7783

[10] Hanessian S, Saavedra OM, Xie F, Amboldi N, Battistini C (2000) Bioorg Med Chem Lett **10**: 439

[11] a) Danishefsky SJ, Mc Clure KF, Randolph JT, Ruggieri RB (1993) Science **260**: 1307; b) Doi T, Sugiki M, Yamada H, Takahashi T, Porco JA Jr (1999) Tetrahedron Lett **40**: 2141

[12] Zhu T, Boons G-J (1998) Angew Chem Int Ed **37**: 1898

[13] a) Chiu SHL, Anderson L (1976) Carbohydr Res **50**: 227; b) Rademann J, Schmidt RR (1996) Tetrahedron Lett **37**: 3989

[14] Lampe TFJ, Weitz-Schmidt G, Wong CH (1998) Angew Chem Int Ed **37**: 1707

[15] a) Rodebaugh R, Joshi S, Fraser-Reid B, Geysen HM (1997) J Org Chem **62**: 5660; b) Zehavi T, Patchornik A (1973) J Am Chem Soc **95**: 5673

[16] Fukase K, Nakai Y, Egusa K, Porco JA (1999) Synlett **7**: 1074

[17] a) Knerr L, Schmidt RR (1999) Synlett **11**: 1802; b) Melean LG, Haase W-C, Seeberger PH (2000) Tetrahedron Lett **41**: 4329; c) Andrade RB, Plante J, Melean LG (1999) Org Lett **11**: 1811
[18] Tolborg JF, Jensen KJ (2000) Chem Commun **2**: 147
[19] Wu X, Grathwohl M, Schmidt RR (2001) Org Lett **3**: 747
[20] Peri F, Nicotra F (unpublished results)
[21] Peri F, Cipolla L, Nicotra F (2000) Tetrahedron Lett **41**: 8587
[22] a) Nicotra F, Panza L, Russo G (1987) J Org Chem **52**: 5627; b) Cipolla L, Lay L, Nicotra F (1997) J Org Chem **62**: 6678
[23] Peri F, Cipolla L, La Ferla B, Nicotra F (2000) J Chem Soc Chem Commun 2303
[24] Forni E, Cipolla L, Caneva E, La Ferla B, Peri F, Nicotra F (2002) Tetrahedron Lett (in press)
[25] Peri F, Bassetti R, Caneva E, De Gioia L, La Ferla B, Presta M, Tanghetti E, Nicotra F (2002) J Chem Soc Perkin Trans 1 (in press)

Received September 10, 2001. Accepted November 2, 2001

Invited Review

Conversions of Nitroalkyl to Carbonyl Groups in Carbohydrates

Ladislav Petruš*, **Mária Petrušová**, **Duy-Phong Pham-Huu**, **Erika Lattová**, **Božena Pribulová**, and **Jozef Turjan**

Institute of Chemistry, Slovak Academy of Sciences, SK-84238 Bratislava, Slovakia

Summary. Since the development of the *Sowden* methodology in the middle of the twentieth century, several other efficient and complementary methods for the transformation of sugar nitromethyl groups to aldehyde functionalities in their free, hemiacetal, or otherwise derivatized forms have been developed. Hydrogen peroxide oxidation as well as ozonolysis of 1-deoxyalditol-1-nitronates in aqueous solution provide free aldoses, thus presenting alternatives to the well-known *Nef* reaction. When applied to 2,5- or 2,6-anhydro-1-deoxy-1-nitroalditols, also known also as glycosylnitromethanes, the *Nef* reaction fails, and the hydrogen peroxide oxidation overoxidizes the expected products to 2,5- or 2,6-anhydroaldonic acids. On the other hand, the ozonolysis of such compounds under *pH*-controlled conditions results in up to 85% of the interesting glycosylformaldehydes. The *Nef* reaction carried out in anhydrous low alcohols, however, has revealed a new conversion of glycosylnitromethanes to glycosylmethanal dialkyl acetals, even more interesting C-glycoside synthons. A similar acid-catalyzed methanolysis of 1-deoxyalditol-1-nitronates leads to methyl furanosides. Finally, a treatment of per-O-substituted glycosylnitromethanes with tributyltin hydride in boiling benzene causes their radical reduction under a nearly quantitative formation of glycosylmethanal oximes.

Keywords. Nitroalditols; C-Glycosyl compounds; *Nef* reaction; Ozonolysis; Sugar oximes.

Introduction

The nitro group is a rather scarcely occurring functionality in natural compounds, and few such compounds are known among carbohydrates. The functional group has become more frequently employed in synthetic carbohydrate chemistry since the introduction of the nitromethane methodology for the carbon skeleton elongation of aldoses known as the *Sowden* method. By convenient modifications of the *Henry* and *Nef* reactions [1, 2], both known since about 1890, *J. C. Sowden* and *H. O. L. Fischer* have developed an efficient synthesis of aldoses *via* intermediate 1-deoxy-1-nitroalditols [3]. The addition of the methanenitronate nucleophile **2** to an aldose **1** results in the formation of a mixture of an epimeric pair of products in

* Corresponding author. E-mail: chemlpet@savba.sk

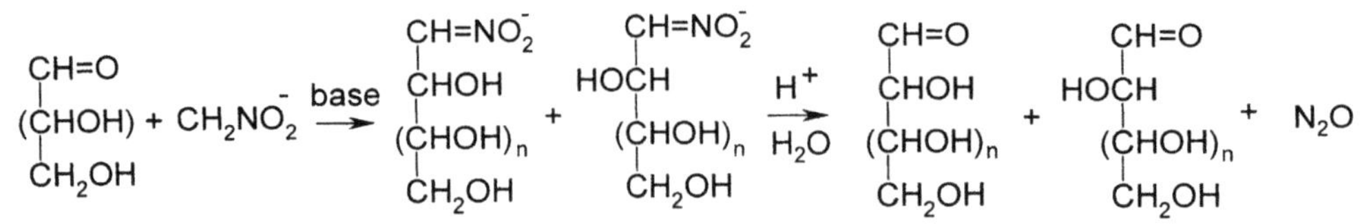

1 **2** **3a** **3b** **4a** **4b**

Scheme 1

their nitronate forms (**3a**,**b**). In a subsequent step these are subjected to a strong acid-catalyzed hydrolysis under the formation of two new aldoses by one carbon longer (**4a**,**b**) and nitrous oxide (Scheme 1).

Although the acid is used as a catalyst, an excess is necessary to quantitatively accomplish the hydrolysis of the alditolnitronates. The method has been shown to be universal for the synthesis of aldoses and, if coapplying formaldehyde in the addition step, *i.e.*, in the so-called nitroethanol synthesis, also for some 2-ketoses. A comprehensive review on the *Sowden* method has been published where details can be found [4]. Later, the methodology has been utilized also for syntheses of branched-chain aldoses [5] or keto derivatives of sugars [6]. However, 2,6-anhydro-1-deoxy-1-nitroalditols (and their less common 2,5-anhydro isomers) in their nitronate forms are resistant to the *Nef* reaction, and only the starting nitro compounds are recovered from the reaction mixtures [7].

The Hydrogen Peroxide Oxidation

In 1974, *V. Bílik* has introduced an alternative transformation of 1-deoxyalditol-1-nitronates to aldoses [8][1], showing that the treatment of an aqueous solution of a mixture of epimeric sodium 1-deoxyalditol-1-nitronates **3a**,**b** with hydrogen peroxide in the presence of molybdate, tungstate, or vanadate ions results in the formation of the corresponding aldoses **4a**,**b** and sodium nitrite (Scheme 2). The molybdate ions have been found to be the best catalyst for this transformation.

The transformation is believed to proceed *via* intermediates **5a**,**b** which spontaneously eliminate the nitrite ions giving rise to aldoses **4a**,**b**. Being able to form their hemiacetals, the aldoses are resistant to a further oxidation to aldonic acids. However, this is not a case when 2,5- or 2,6-anhydro-1-deoxy-1-nitroalditols (glycosylnitromethanes) in their nitronate forms are subjected to the oxidation. The respective products, 2,5- or 2,6-anhydroaldoses, usually cannot form stable hemi-acetals and undergo a further oxidation with hydrogen peroxide, even without the presence of transient metal anions, to the corresponding anhydroaldonic acids [11].

---

[1] The reaction has actually been developed in analogy to another carbohydrate reaction developed by *V. Bílik*, the stereoselective hydroxylation of glycals with peroxomolybdate or peroxotungstate ions providing *cis*-C-2-OH–C-3-OH aldoses [9]. However, the most famous carbohydrate reaction introduced by the same author in 1972 is a stereospecific, dimolybdate-catalyzed C-1-C-2-carbon skeleton rearrangement of aldoses which with isotopically uniform aldoses leads to their mutual C-2-epimerization and which has been coined the *Bílik* reaction [10].

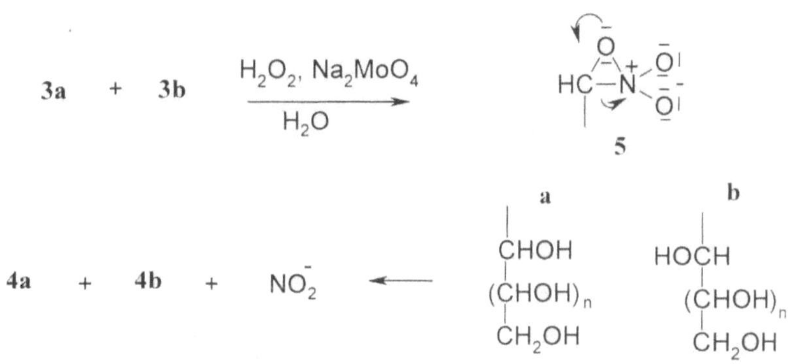

Scheme 2

The hydrogen peroxide oxidation, preferentially called the oxidative decomposition of 1-deoxy-1-nitroalditols in the literature, has been applied for the preparation of monosaccharide hexoses [8, 12], heptoses [13–15], octoses [16, 17], and disaccharide 4-O-hexosylhexoses [18]. Also, some ω-deoxyhexoses and heptoses have been prepared by this route [19].

Ozonolysis

In the early 1990s the third simple and universal method for the transformation of 1-deoxy-1-nitroalditols to aldoses with the same number of carbon atoms has been introduced [20]. The method utilizes the treatment of an aqueous solution of the intermediate sodium 1-deoxyalditol-1-nitronates with ozone at room temperature. Scheme 3 represents such an ozonolysis on the example of 1-deoxy-*D*-glycero-*D*-ido-heptitol-1-nitronate (**6**), leading to *D*-glycero-*D*-ido-heptose (**7**) and sodium nitrate. The yields are comparable with those of the hydrogen peroxide oxidation as well as those of the acid-catalyzed hydrolysis of the same substrates; the isolated yield of heptose **7** in the conversion shown in Scheme 3 is 85% [20]. This summary follows also from a comparative study of the three nitromethane methods applied to the *D*-mannose to *D*-glycero-*D*-galacto- and -*D*-talo-heptose transformation [21].

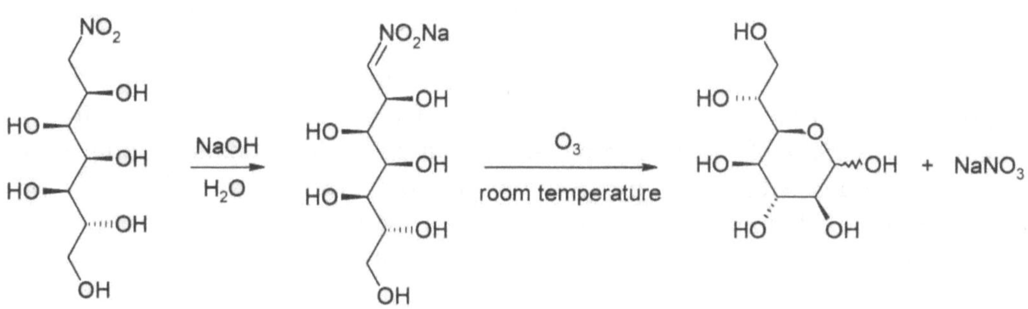

6 **7**

Scheme 3

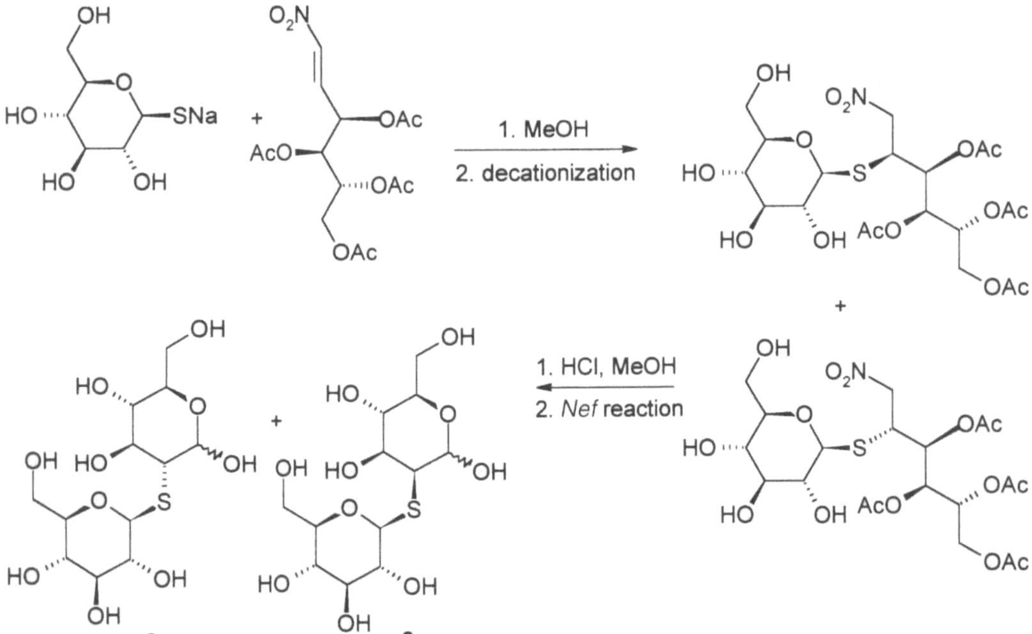

8 9

Scheme 4

Similarly as in the hydrogen peroxide oxidation, the hemiacetal functionality sufficiently protects the aldose from subsequent oxidation to the corresponding aldonic acid. Thus, no formation of acid derivatives of saccharides in ozonolysis of sodium 1-deoxyalditol-1-nitronates with ozone generated from pure oxygen has been observed[2].

Ozonolyzing alkaline alditolnitronates, monosaccharide hexoses and heptoses [20] as well as two disaccharidic 4-O-hexosyl-2-deoxy-2-acetamidohexoses have been prepared [22]. However, if electron-rich heteroatoms are present in the substrates a competitive oxidation occurs which is a significant limitation of the method. Thus, ozonolysis of the nitronate forms of the 1-thio-*D*-glucose-to-1,2-dideoxy-1-nitro-*D*-arabino-hex-1-enitol adducts failed, and neither 2-thiosophorose (**8**) nor its epimer **9** could be prepared (Scheme 4). A classical *Nef* reaction solved the problem, and high yields of the crystalline thiodisaccharides were obtained [23].

Even with 2-acetamido-1,2-dideoxyaditol-1-nitronates treated with ozone a partial degradation occurs, decreasing the yields of the corresponding 2-acetamidosugars by 10–15% compared with an analog conversion leading to unsubstituted aldoses. However, things do not improve using either *Nef* hydrolysis or hydrogen peroxide oxidation. The conditions of the former reaction bring about a significant acid-catalyzed N-deacetylation, whereas under those of the latter a

[2] When insufficiently pure oxygen is used for ozone generation, significant amounts of acidic by-products are formed, apparently from oxidation by cogenerated oxides of nitrogen.

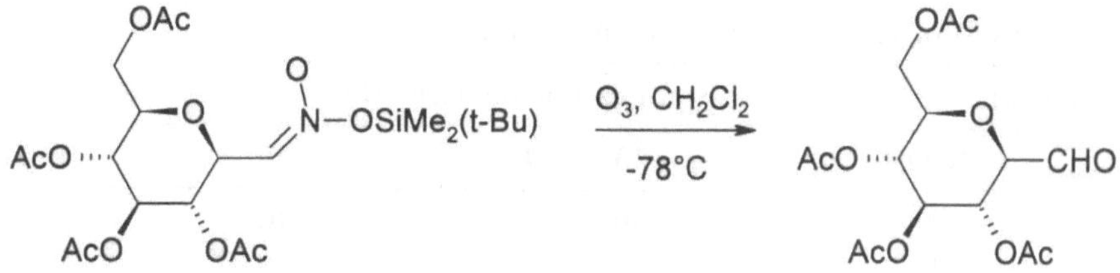

10 **11**

Scheme 5

competitive degradation of the 2-acetamidosugars to the corresponding one-carbon shorter aldose occurs [24].

Anyway, the ozonolysis of the 1-deoxy-1-nitroalditol-1-nitronates to aldoses is a side product of another endeavor. The challenge was a good access to glycopyranosylnitromethanes [25], despite of their known resistance to the *Nef* reaction. Thus, *O. Martin et al.* for the first time have introduced the ozonolysis of nitromethane derivatives of sugars already in 1989 [26], and the per-O-acetylated silylnitronate **10** was used for the conversion to the corresponding aldehyde **11** (Scheme 5).

Although the authors have claimed a practically quantitative yield, the method has still limitations due to a low stability of the silylnitronates (*e.g.*, trimethylsilylnitronates are too unstable to be prepared). A little later, a significant simplification of the method, the ozonolysis of the sodium nitronate forms of β-glycopyranosylnitromethanes at room temperature, has been introduced [27]. Thus, the ozonolysis of the sodium β-*D*-galactopyranosylmethanenitronate gives rise to the expected aldehyde **13** (Scheme 6). The compound, unable to be stabilized *via* a hemiacetal due to stereochemical reasons, is present in reaction mixture in an equilibrium with its aldehydrol form (**14**) in ratio of 1:4 (hydrate predominates) and is consecutively oxidized to the corresponding 2,6-anhydroaldonic acid **15**.

Whereas at the beginning of the reaction the *pH* is strongly alkaline at the end it is close to 7, and it is easy to control the ozonolysis and thus to prevent undesirable further oxidation. As part of the starting alkalinity necessary for generation of the nitronate form is consumed by acid **15**, the optimum conditions for the conversion

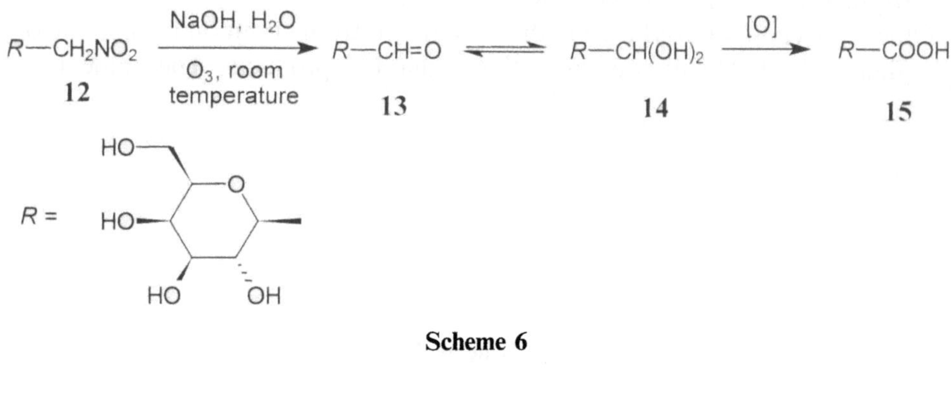

12 **13** **14** **15**

Scheme 6

Table 1. Ozonolysis of **12** in its sodium nitronate form

Reactants (mmol)[a]		Reaction products (%)[b]		
12	NaOH	**13** (+**14**)	**15**	**12**
1.0	1.0	80	10	10
1.0	1.1	85	10	5
1.0	1.2	75	25	0
1.0	1.3	65	35	0
1.0	1.4	55	45	0
1.0	1.5	40	60	0

[a] Total volume of reaction mixture: 4 cm^3; [b] values estimated from the ^{13}C NMR spectra

of **12** to **13** employ a 10% molar excess of sodium hydroxide based on **12** (Table 1). In addition to about 85% of aldehyde, including its hydrate, the final reaction mixture contains about 10% of aldonic acid **15** and about 5% of glycopyranosylnitromethane **12**. For a complete disappearance of the starting compound, a 20% excess of sodium hydroxide has to be used. In spite of a decreased yield of the aldehyde derivative (to about 75%) it is rather a practical step to avoid a contamination of the product with the starting glycopyranosylnitromethane as the aldonic acid is more easily separable later on. Thus, such raw aldehydes can be used for the preparation of homologous 2-(glycopyranosyl)-nitroethanes [27b].

A very efficient reaction, the ozonolysis of the secondary nitronate group, has been performed with an epimeric mixture of the 1:1-adducts of *β-D*-galactopyranosylnitromethane to formaldehyde [28]. The product obtained is 3,7-anhydro-*D*-glycero-*L*-manno-oct-2-ulose, a potential precursor of the 3-deoxy-*D*-galacto-oct-2-ulosonic acid *via* an intramolecular elimination which remains to be investigated.

The Radical Reduction

The primary nitro group originally considered to be inert with respect to the *TBTH/AIBN* radical denitration conditions [29, 30] has later been reported to undergo such a reaction with *TBTH/ABCN* [31]. Attempts to accomplish the denitration of per-O-acetylated *β-D*-glucopyranosylnitromethane (**16**, Scheme 7) with *TBTH* in boiling benzene in the presence of *ABCN* did not lead to the expected replacement of the nitro group by hydrogen but to its reduction to *β-D*-glucopyranosylmethanal oxime (**17**) in 90% yield [32].

Other *β*-glycopyranosylnitromethanes derived from common hexoses also gave high yields (85–97%) of the respective reduction products. The same products

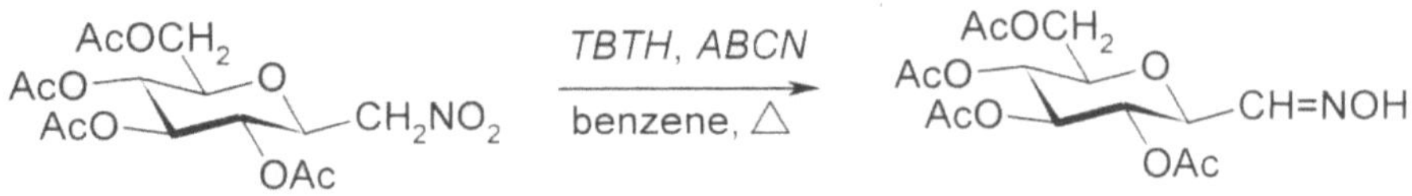

16 17

Scheme 7

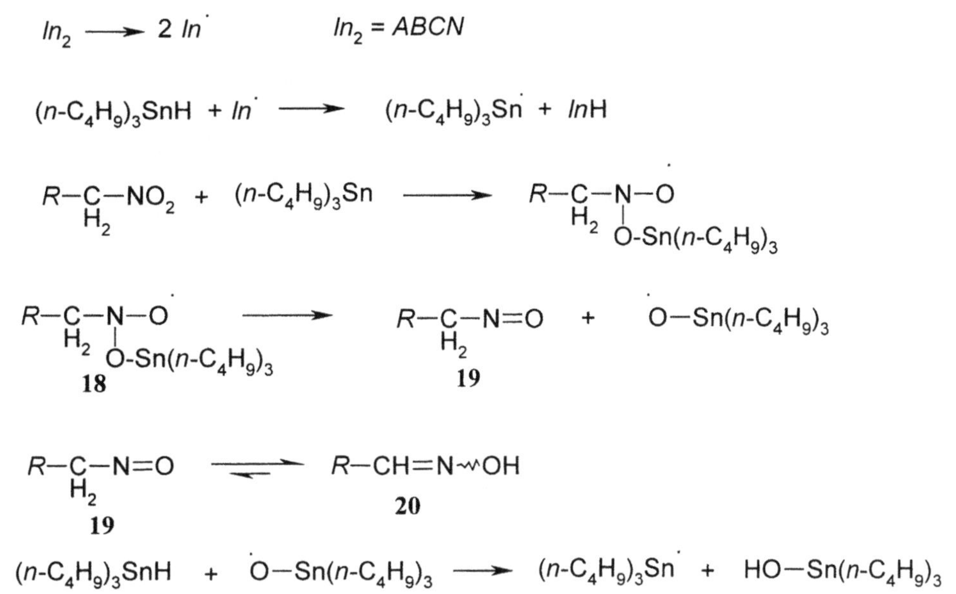

Scheme 8

were formed when *AIBN* instead of *ABCN* was used. High yields of the analog reduction products were obtained from per-O-acetylated β-glycopyranosylnitromethanes derived of *L*-rhamnose and N-acetyl-*D*-glucosamine as well as from α-*D*-ribofuranosylnitromethane peracetate or 2,3:4,6-di-O-isopropylidene-β-*D*-mannopyranosylnitromethane [33].

The resistance of the primary nitro group to the denitration reaction is apparently a result of a stronger carbon-nitrogen bond than in tertiary or activated secondary nitro compounds. As a consequence, the SnO–N linkage in the key intermediate **18** (Scheme 8) is preferentially split, giving rise to the nitrosomethyl compound **19** which is a less populated tautomeric form of the final product **20**. The glycosylmethanal oximes thus obtained are good starting materials for the synthesis of some glycopyranosyl cyanides with complementary anomeric configuration to those obtainable from glycosyl halides [34].

Acid-Catalyzed Methanolysis

A surprising transformation of glycosylnitromethanes to very conveniently protected corresponding aldehyde derivatives by a modified *Nef* reaction was even more serendipitous than in the foregoing case. The idea was to prepare anomeric *L*-arabinofuranosylnitromethanes **23a,b** by a preferred five-member ring closure of 1,2-dideoxy-1-nitro-*L*-arabino-hex-1-enitol (**22**) under kinetic conditions, *i.e.* by acid-catalyzed methanolysis of its per-O-acetate **21** (Scheme 9). The proposed reaction, however, did not stop at the stage expected but afforded a mixture of the corresponding *L*-arabinofuranosylmethanal dimethyl acetals **25** and **26** and a mixed internal methyl acetal **27** [35].

The compounds thus obtained are formally the products of the methanolysis of the anomeric *L*-arabinofuranosylmethylnitronic acids **24a,b**, a reaction analogous

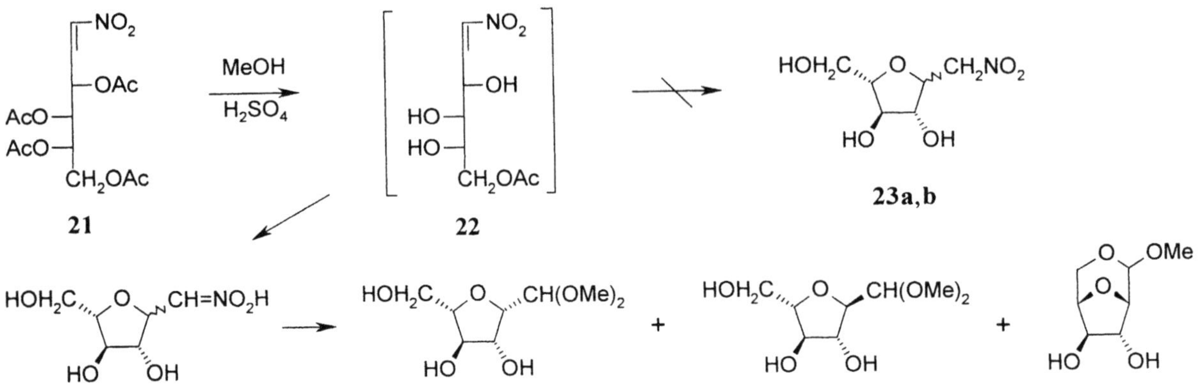

21 22 23a,b

24a,b 25 26 27

Scheme 9

to their hypothetic (not proceeding) hydrolysis, generally known as the *Nef* reaction. This means that the ring closure of 1,2-dideoxy-1-nitro-L-arabino-hex-1-enitol does not proceed as 1,2-addition but as 1,4-addition, giving rise to nitronic acids ready for the next step of the reaction. The switch from water to anhydrous methanol enabled to successfully transform common glycosylnitromethanes as β-*D*-glucopyranosylnitromethane, β-*D*-galactopyranosylnitromethane, or β-*D*-mannopyranosylnitromethane by the modified *Nef* reaction to the corresponding β-*D*-glycopyranosylmethanal dimethyl acetals in 80–94% yields [36]. Similar conversions were observed also in anhydrous ethanol, and the corresponding diethyl acetals were prepared.

A return to the acyclic 1-deoxy-1-nitroalditols and the application of *Nef* reaction under the new conditions has proposed a new method of preparation of low alkyl glycofuranosides, sometimes with very surprising results. For example, the only product from such a methanolysis of sodium 1-deoxy-*D*-mannitol-1-nitronate at −30°C isolated in 30% yield was methyl-β-*D*-mannofuranoside [36].

In summary, the extension of the *Nef* reaction to acidified lower alcohols turns the very easily available glycosylnitromethanes, since recently also as

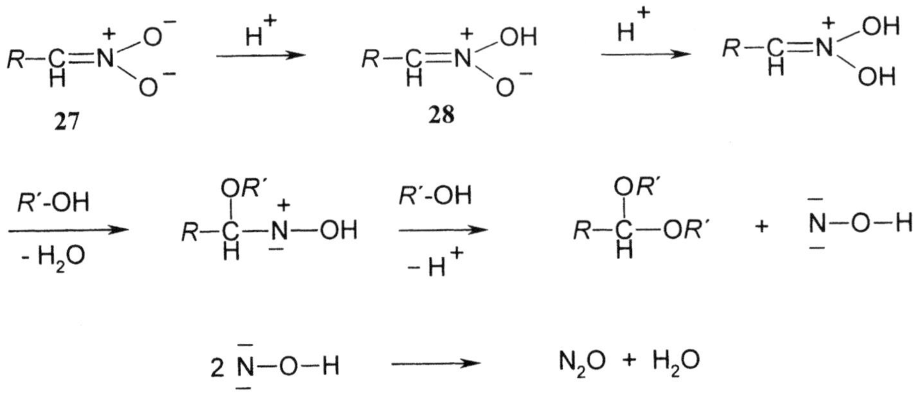

27 28

Scheme 10

per-O-protected acetals [37], into a role of quite attractive C-glycosyl synthons. Moreover, the structures of the reaction products also provide an unambiguous evidence of the mechanism of the *Nef* reaction that can be generally described as a solvolysis of the (substituted) methanenitronate anion **27** catalyzed by an excess of a strong acid (Scheme 10). This is necessary to further protonize the methanenitronic acid **28** to increase the electron deficit at its carbon atom to facilitate the addition of the neutral nucleophiles. Finally, being stronger nucleophiles than the water molecules, the aliphatic alcohols are expected to be more reactive under such an arrangement.

The fine differences in reactivity of acyclic and cyclic 1-deoxyalditol-1-nitronates under the conditions of the classic *Nef* reaction apparently go to the account of an electronic effect similar to the anomeric one. Whereas acyclic species can avoid such an unfavourable interaction by a simple rotation of a bond, the effect in cyclic glycosylnitromethanes can represent a serious obstacle of their reactivity and remains to be further studied and quantified.

Perspectives

The developments introduced for the transformation of the nitroalkyl to carbonyl functionality imply some other possible transformations and their synthetic applications. In addition to the proton and the electrophilic oxygen species exploited until now for the decomposition of the electron-abundant nitronate group, several other possibilities still await their chances. Very interesting synthetic applications can be expected *e.g.* from applications of carbenes or nitrenes. Especially, a rapidly developing field of syntheses of glycomimetics based on the C-glycosyl compounds can efficiently benefit from these new developments. Finally, the better availability of the nitroalkyl to carbonyl group transformation methods should also lead to a broader exploitation of the acyclic species in oligosaccharide synthesis.

Acknowledgements

The preparation of this paper was supported by grant No. 2/7144 of the Slovak Academy of Sciences and by the sponsorship of CMS Chemicals.

References

[1] Henry L (1895) C R Acad Sci **120**: 1265
[2] Nef JU (1894) Liebigs Ann Chem **280**: 263
[3] Sowden JC, Fischer HOL (1944) J Am Chem Soc **66**: 1312; (1945) J Am Chem Soc **67**: 1713; (1946) J Am Chem Soc **68**: 1511; (1947) J Am Chem Soc **69**: 1048, 1963
[4] Baer H (1969) Adv Carbohydr Chem Biochem **24**: 67
[5] a) Petruš L, Hricovíniová Z, Petrušová M, Matulová M (1996) Chem Pap **50**: 373; b) Hricovíniová Z, Hricovíni M, Petrušová M, Matulová M, Petruš L (1998) Chem Pap **52**: 238
[6] Wade PA, Giuliano RM (1990) In: Feuer H, Nielsen AT (eds) Nitro Compounds: Recent Advances in Synthesis and Chemistry. VCH, New York, p 137
[7] Köll P, Dromowicz M (1991) Abstr 6th Eur Carbohydr Symp. Edinburgh, B56
[8] Bílik V (1974) Collect Czech Chem Commun **39**: 1621
[9] a) Bílik V, Kučár Š (1970) Carbohydr Res **13**: 311; b) Bílik V (1972) Chem Zvesti **26**: 76

[10] Petruš L, Petrušová M, Hricovíniová Z (2001) Topics in Current Chemistry, vol 215. Springer, Berlin Heidelberg, p 15
[11] Dromowicz M, Köll P (1998) Carbohydr Res **311**: 103
[12] Bílik V (1975) Chem Zvesti **29**: 114
[13] Bílik V, Anderle D, Alföldi J (1974) Chem Zvesti **28**: 668
[14] Bílik V, Petruš L (1976) Chem Zvesti **30**: 395
[15] Bílik V, Petruš L, Zemek J (1976) Chem Zvesti **30**: 693
[16] Bílik V, Petruš L, Alföldi J (1976) Chem Zvesti **30**: 698
[17] Bílik V, Alföldi J, Matulová M (1986) Chem Zvesti **40**: 763
[18] Bílik V, Jurčová E, Sutoris V (1978) Chem Zvesti **32**: 252
[19] Bílik V, Petruš L, Stankovič L', Linek K (1978) Chem Zvesti **32**: 372
[20] Lattová E, Petrušová M, Petruš L (1991) Chem Pap **45**: 823
[21] Pham-Huu D-P, Petrušová M, Petruš L (1998) Chem Pap **52**: 235
[22] Lattová E, Petruš L (1992) Carbohydr Res **235**: 289
[23] a) Petrušová M, Lattová E, Matulová M, Petruš L (1992) Chem Pap **46**: 351; b) Petrušová M, Lattová E, Matulová M, Petruš L, BeMiller JN (1996) Carbohydr Res **283**: 73
[24] Turjan J, Petrušová M, Petruš L (unpublished results)
[25] a) Petruš L, Bystrický S, Sticzay T, Bílik V (1982) Chem Zvesti **36**: 103; b) Förtsch A, Kogelberg H, Köll P (1987) Carbohydr Res **164**: 391; c) Petrušová M, Fedoroňko M, Petruš L (1990) Chem Pap **44**: 267
[26] Martin OR, Khamis FE, Rao SP (1989) Tetrahedron Lett **30**: 6143
[27] a) Petrušová M, Stručková I, Matulová M, Petruš L (1993) Chem Pap **47**: 195; b) Petrušová M, BeMiller JN, Krihová A, Petruš L (1996) Carbohydr Res **295**: 57; c) Petrušová M, BeMiller JN, Petruš L (1996) Tetrahedron Lett **37**: 2341
[28] Petrušová M, Lattová E, Matulová M, Petruš L (1992) Chem Pap **46**: 120
[29] Ono N, Miyake H, Tamura R, Kaji A (1981) Tetrahedron Lett 1705
[30] Ono N, Kaji A (1986) Synthesis 693
[31] Witczak Z, Li Y (1995) Tetrahedron Lett **36**: 2595
[32] Pham-Huu D-P, Petrušová M, BeMiller JN, Petruš L (1998) Chem Pap **52**: 186
[33] Pham-Huu D-P, Petrušová M, BeMiller JN, Petruš L (1998) Synlett **12**: 1319; (2000) J Carbohydr Chem **19**: 93
[34] a) Lopez MTG, Heras FGD, Felix AS (1987) J Carbohydr Chem **6**: 273; b) Kini GD, Petrie CR, Hennen WJ, Dalley NK, Wilson BE, Robins EK (1987) **159**: 81; c) Hoffman MG, Schmidt RR (1985) Liebigs Ann Chem 2403
[35] Petruš L, Petrušová M, Matulová M (1996) Abstr 18th Int Carbohydr Symp. Milano, p 377
[36] Pribulova B, Petrušová M, Lattová E, BeMiller JN, Petruš L (2000) Abstr 20th Int Carbohydr Symp. Hamburg, p 103
[37] Pham-Huu D-P, Petrušová M, BeMiller JN, Köll P, Kopf J, Petruš L (1998) Carbohydr Res **306**: 45

Received November 12, 2001. Accepted November 20, 2001

Invited Review

From Lianas to Glycobiology Tools: Twenty-Five Years of 2,5-Dideoxy-2,5-imino-D-mannitol

Tanja M. Wrodnigg*

Glycogroup, Institut für Organische Chemie, Technische Universität Graz, A-8010 Graz, Austria

Summary. 2,5-Dideoxy-2,5-imino-D-mannitol (*DMDP*) has been isolated from several natural sources. Many synthetic approaches are available, and many derivatives have been synthesized and their biological activities have been investigated. An overview on isolation, syntheses, and biological data of *DMDP* as well as some closely related compounds will be given in this review.

Keywords. 2,5-Dideoxy-2,5-imino-D-mannitol; Glycosidase inhibitors; Iminosugars; Iminoalditols; Glycosidase inhibitory activities.

Introduction

In 1976, *Welter* and coworkers isolated a new alkaloid from the leaves of *Derris eliptica* (Leguminosae) (Fig. 1) with the molecular formula of $C_6H_{13}NO$. Giving a positive nihydrine reaction, IR and NMR data as well as the obtained optical rotation led to the conclusion that the structure must be 2,5-dihydroxymethyl-3,4-dihydroxypyrrolidine (*DMDP*) having either a (2*R*,3*R*,4*R*,5*R*) or an all-*S* configuration [1]. Soon the crystal structure of this compound was determinated, and the all-*R* configuration was confirmed [2].

Ten years before the isolation of this alkaloid and its examination as a glycosidase inhibitor, important synthetic and structural studies on sugar analogues with nitrogen instead of oxygen in the ring had been conducted by *Paulsen* and coworkers [3]. Studies of glycosidases date back as early as the eighteen-forties when *Liebig* and *Wöhler* made their first contributions to this emerging area [4]. Recently, these important enzymes have been investigated comprehensively by

* Present Address: Department of Chemistry, University of British Columbia, Vancouver, B. C., V6T 1Z1 Canada. E-mail: wrodnigg@chem.ubc.ca

Fig. 1. *Derris elliptica*

Legler [5], *Sinnot* [6], and the *Withers* group [7] and are believed to be well understood today.

Glycosidase inhibitors [8] are known as compounds which inhibit these carbohydrate processing enzymes. This class of compounds can mainly be divided into high and low molecular mass structures. The most important subgroup in the low molecular class is formed by polyhydroxylated derivatives of piperidines, pyrrolidines, indolizidines, pyrrolizidines, and nortropane alkaloids. Such compounds can be regarded as mimetics of the corresponding sugar with nitrogen instead of oxygen in the ring. Some of the most investigated compounds in this class are 2,5-dideoxy-2,5-imino-D-mannitol (**1**), 1-deoxynojirimycin (*DNJ*, **2**), castanospermine (**3**), and autraline (**4**), all being excellent reversible inhibitors of D-glucosidases (Fig. 2).

1-Deoxynojirimycin (**2**), which had been synthesized by *Paulsen* and his group in 1966 [9], was later found in the fermentation broths of a *Bacillus* species [10] as well as in that of *Streptomyces lavandulae* [11]. Castanospermine (**3**), which can be regarded as a bicyclic version of 1-deoxynojirimycin (**2**) also exhibiting a D-gluco

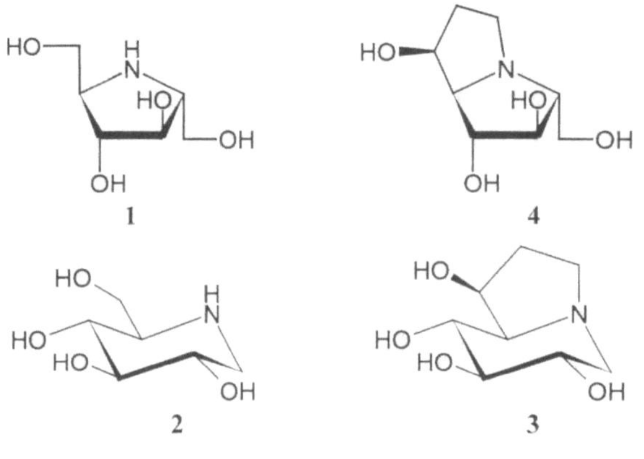

Fig. 2. *DMDP* (**1**), *DNJ* (**2**), castatonspermine (**3**), australine (**4**)

Fig. 3. Superposition (middle) of *DNM* (left) and *DMDP* (right)

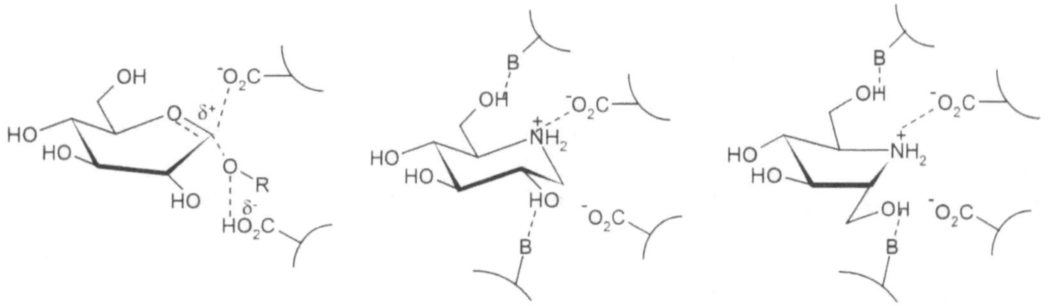

Fig. 4. Comparison of *DNJ* and *DMDP* with the carboxonium ion transition state

configuration pattern, was isolated from the Australian legume *Castanospermum australae* [12]. Australine (**4**) was also found in the fruit of this tree [13] and is related to *DMDP* in the same way as is **2** to **3**. Whereas in the case of **2** and **3** the D-gluco configuration is quite obvious, **1** and **4** are apparently rather α-D-fructofuranose analogues. Nevertheless, **1** is in some cases even more active against D-glucosidases than **2** or **3**. From the superposition of **2** and **1** (Fig. 3) it can be seen that all hydroxyl groups as well as the ring nitrogen match very well.

1 features a C_2 axis of symmetry in contrast to **4** in which one of the hydroxymethyl groups is locked in the bicyclic system which abolishes the symmetry and changes inhibitory properties relative to **1**. When compared to 1-deoxynoirimycin and castanospermine, the more flat five-membered ring in **1** is closer mimicking the half-chair conformation found in the oxocarbonium ion transition state of (enzymatic) glycoside hydrolysis [14] (Fig. 4).

The five-carbon-containing natural product 1,4-dideoxy-1,4-imino-D-arabinitol **5** (Fig. 5) is structurally closely related to *DMDP* and was isolated from two types of leguminose plants in 1985 [15]. Several synthetic approaches have been devised for this compound [16] which is an inhibitor of α-glucosidases, mammalian glycoprotein trimming glucosidases such as ER α-glucosidase II, plant α-mannosidase and *Golgi* located α-mannosidases I and II, and intestinal isomaltase and trehalase [22] (see Table 1).

Another closely related natural product is 2-hydroxymethyl-3,4-dihydroxy-6-methylpyrrolidine (6-deoxy-*DMDP*, **6**; Fig. 5). This compound was isolated from

Table 1. Biological activities of **1**, **2**, **3**, **4**, and **6** (D and L are indicated at the values); K_i and IC_{50} values in μM; NI: no inhibition, ng: not given, nd: not determined

Enzyme	pH	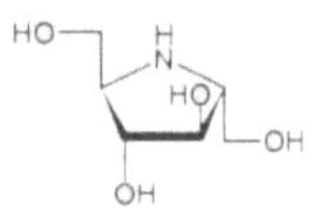 1	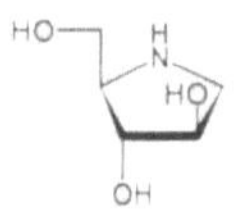 5	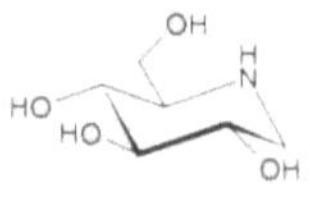 2	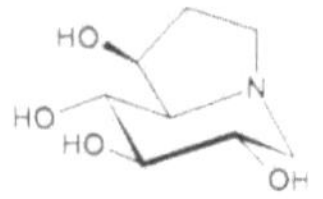 3	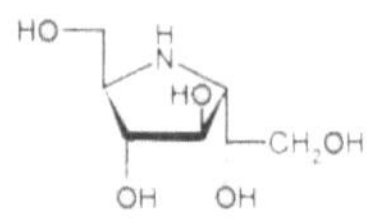 7	Ref.
α-Glucosidase	6.5	$K_i=0.73$	ng	$K_i=25$	$K_i \geqslant 1500$	ng	[37]
Yeast		$K_i=7$	ng	ng	ng	$K_i=54$	[24b]
	6.8	$IC_{50}=3.3$	$IC_{50}=0.18$	$IC_{50}=190$	ng	ng	[16a], [20]
Brewer's yeast	opt	$IC_{50}=0.0599$	ng	$IC_{50}=9.57$	ng	ng	[27]
Baker's yeast		$IC_{50}=15$	$IC_{50}=7.0$	ng	ng	$IC_{50}=85$ (DL)	[24a]
		$IC_{50}=3.6$	ng	ng	ng	$IC_{50}=270$ (DL)	[18]
α-Glucosidase	opt	$IC_{50}=300$	ng		ng	$IC_{50}=130$ (DL)	[18], [25]
Rice		NI	ng	$IC_{50}=1.44$	ng	ng	[27]
		$IC_{50}=200$	ng		ng	$IC_{50}=130$ (DL)	[24b], [24c]
α-Glucosidase (*saccharomyces sp*)	7.0	$K_i=330$	ng	ng	ng	$K_i=28$ (D)	[48]
α-Glucosidase (rabbit gut)	opt	NI	$IC_{50}=85$	ng	ng	NI	[24a]
α-Glucosidase	opt						
Rat intestinal maltase	6.8	$IC_{50}=290$	$IC_{50}=55$	$IC_{50}=0.36$	ng	$IC_{50}=400$ (DL)	[18], [22], [24b], [24c]
isomaltase	6.8	$IC_{50}=91$	$IC_{50}=5.8$	$IC_{50}=0.3$	ng	NI	[24b], [22]
sucrase		NI	ng	ng	ng	$IC_{50}=300$ (DL)	[24b]
α-Glucosidase	opt						
Rat liver lysosome	6.8	$IC_{50}=92$	$IC_{50}=100$	$IC_{50}=0.4$	ng	NI	[24b], [22]
ER glucosidase II		NI	$IC_{50}=20$	$IC_{50}=4.6$	ng	ng	[22]
α-Glucosidase							[27]
Porcine small intestine	6.8						
Starch		$IC_{50}=307$	ng	$IC_{50}=0.479$	ng	ng	
Maltose		$IC_{50}=9.57$	ng	$IC_{50}=0.00957$	ng	ng	
Sucrose		$IC_{50}=20.1$	ng	$IC_{50}=0.144$	ng	ng	
α-Glucosidase							
Bovine liver lysosome	4.0	$IC_{50}=170$	$IC_{50}=89$	$IC_{50}=0.4$	ng	ng	[22]
β-Glucosidase (*Asp. wentii*)	5.0	$K_i=57$	ng	$K_i=2.0$	$K_i=0.9$	ng	[37]
β-Glucosidase (bovine kidney)	5.0	$K_i=44$	ng	$K_i=250$	$K_i=25$	ng	[37]
β-Glucosidase (bitter almonds)	6.8	$IC_{50}=7.8$	$IC_{50}=200$	$IC_{50}=81$	ng	ng	[16a], [20]
β-Glucosidase	opt	$IC_{50}=11$	ng	ng	ng	$IC_{50}=3.8$ (DL)	[24b]

(*continued*)

Table 1 (*continued*)

Enzyme	pH	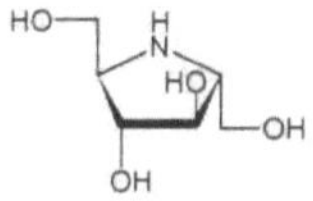 1	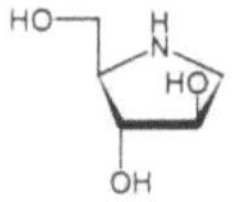 5	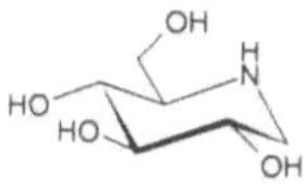 2	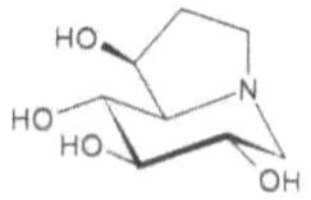 3	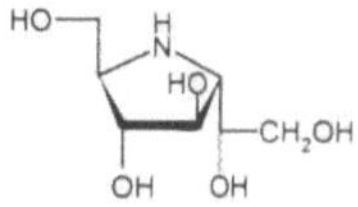 7	Ref.
(*Caldocellum saccharolyticum*)						$IC_{50}=3.8$ (DL)	[24c]
β-Glucosidase (almonds)	5.0	$K_i=1.7$	ng	$K_i=300$	$K_i=1.5$	ng	[37]
		$IC_{50}=13$	ng	ng	ng	$IC_{50}=23$	[18]
		$IC_{50}=2.40$	ng	$IC_{50}=153$	ng	ng	[27]
		$K_i=10$	ng	ng	ng	$K_i=1.5$	[24b]
	5.0	ng	ng	ng	ng	$K_i=24.5$ (D)	[47]
	opt	$K_i=10$	$K_i=280$	ng	ng	$K_i=1.5$ (DL)	[24a]
	opt	$IC_{50}=9$	$IC_{50}=965$	ng	ng	$IC_{50}=4$ (DL)	[24a]
	5.0	$K_i=50$	ng	ng	ng	$K_i=2.6$ (D)	[48]
	4.8	$K_i=5.2$	ng	ng	ng	$IC_{50}=23$ (DL)	[24c]
		$IC_{50}=7.3$	ng	ng	ng	ng	[68]
β-Glucosidase (rat intestine cellobiase)	4.0	$IC_{50}=34$	NI	$IC_{50}=520$	ng	ng	[22]
β-Galactosidase	opt	$IC_{50}=2.5$	ng	ng	ng	$IC_{50}=4.4$ (DL)	[24b], [24c]
(bovine liver)	opt	$IC_{50}=2.2$	ng	ng	ng	$IC_{50}=4.4$ (DL)	[18], [25]
cytosolic	6.8	$IC_{50}=3.3$	$IC_{50}=1000$	NI	ng	$IC_{50}=4.4$ (DL)	[22]
β-Galactosidase (rat intestinal lactase)	opt	$IC_{50}=3.6$	$IC_{50}=260$	$IC_{50}=26$	ng	$IC_{50}=4.0$ (DL)	[24b], [22], [24c]
α-Mannosidase (jack beans)	opt	NI	$IC_{50}=750$			$IC_{50}=695$ (DL)	[24a]
		NI	$IC_{50}=100$				[16a]
β-Mannosidase (snail)	opt	NI	ng	ng	ng	$IC_{50}=400$ (DL)	[24c]
α-Mannosidase					ng	ng	[22]
rat liver					ng	ng	
Golgi I	5.5	NI	$IC_{50}=53$	NI	ng	ng	
Golgi II	5.5	NI	$IC_{50}=46$	NI	ng	ng	
lysosomal	4.5	NI	$IC_{50}=110$	$IC_{50}=1000$	ng	ng	
soluble	6.5	$IC_{50}=260$	$IC_{50}=$ NI	NI	ng	ng	
rat epididymis		NI	$IC_{50}=84$	$IC_{50}=320$	ng	ng	
α,α-Trehalase					ng		
rat instestine	opt	$IC_{50}=360$	$IC_{50}=25$	$IC_{50}=42$	ng	$IC_{50}=2.0$ (DL)	[22], [24b], [24b]
porcine kidney		$IC_{50}=200$	$IC_{50}=4.8$	$IC_{50}=41$	ng	$IC_{50}=5.0$ (DL)	[22], [24c], [18]
		$IC_{50}=500$	ng	ng	ng	NI	[25], [24b]
		$K_i=150$	ng	ng	ng	ng	[25]

(*continued*)

Table 1 (*continued*)

Enzyme	pH	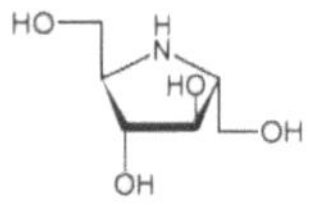 1	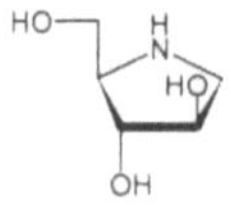 5	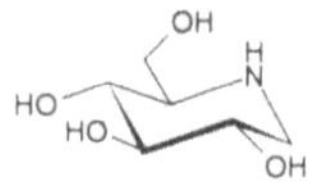 2	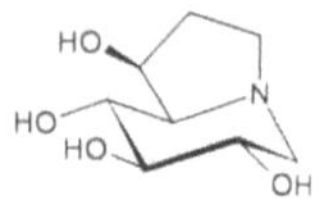 3	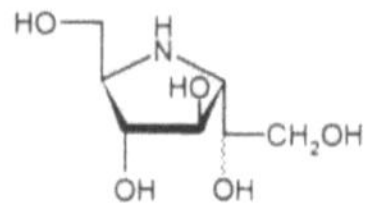 7	Ref.
Trehalase (*Corynebacterium sp.*)	ng	$IC_{50}=0.35$	ng	ng	ng	ng	[30]
Trehalase (*Plutella xylostella*)	ng	$IC_{50}=10$	ng	ng	ng	ng	[30]
Sucrase (baker's yeast)	ng	$IC_{50}=0.957$	ng	NI	ng	ng	[27]
β-Fructofuranosidase	4.6	$IC_{50}=52.5$	ng	NI	nd	ng	[20]
Yeast	5.0	$K_i=6.8$	ng	$K_i \geqslant 5000$	nd	ng	[37]
	6.0	$K_i=3.5$	ng	nd	nd	ng	[37]
	7.0	$K_i=1.1$	ng	nd	nd	ng	[37]
Invertase (*Pleum pratense*)	5.5	$K_i=78$	$K_i=1070$	ng	ng	$K_i=77$ (DL)	[24a]
β-Xylosidase (*Asp. niger*)	5	$IC_{50}=250$	NI	$IC_{50}=400$	ng	ng	[16a], [20]
α-L-Fucosidase (bovine epidermis)	opt	$IC_{50}=110$	ng	ng	ng	ng	[24c]
Amyloglucosidase (*Rhizophus*)	ng	$IC_{50}=38.3$	ng	$IC_{50}=613$	ng	ng	[27]
Amyloglucosidase (*A. niger*)	opt	$IC_{50}=19$	ng	ng	ng	$IC_{50}=180$	[18], [24c]
Thioglucosidase (mustard myrosinase):							[62]
singrin acetate	5	$IC_{50}=9.3$	NI	NI	$IC_{50}=10$	ng	
phosphate	7	$IC_{50}=33$	NI	NI	$IC_{50}=180$	ng	
progoitin acetate	5	$IC_{50}=35$	NI	NI	$IC_{50}=36$	ng	
citrate	5	$IC_{50}=150$	NI	NI	$IC_{50}=76$	ng	
phosphate	7	$IC_{50}=51$	NI	NI	$IC_{50}=340$	ng	
Thioglucosidase (*Brevicoryne brassicae*)							[62]
singrin acetate	5	$IC_{50}=240$	NI	NI	$IC_{50}=90$	ng	
phosphate	7	$IC_{50}=100$	$IC_{50}=280$	$IC_{50}=970$	$IC_{50}=43$	ng	
progoitin acetate	5	$IC_{50}=140$	NI	NI	$IC_{50}=58$	ng	
citrate	5	$IC_{50}=400$	$IC_{50}=570$	$IC_{50}=230$	$IC_{50}=6.0$	ng	
phosphate	7	$IC_{50}=87$	$IC_{50}=340$	$IC_{50}=850$	$IC_{50}=34$	ng	
Glucosidase I (mung bean seedlings)	6.5	$IC_{50}=40$	ng	$IC_{50}=5$–7	$IC_{50}=2$–3		[68]

(*continued*)

Table 1 (*continued*)

Enzyme	pH	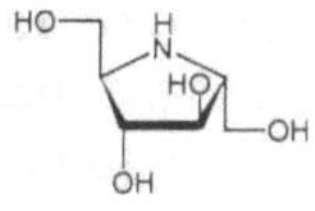 1	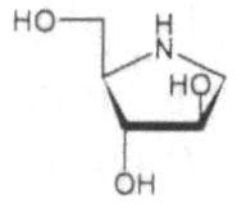 5	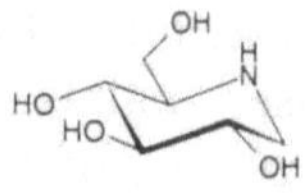 2	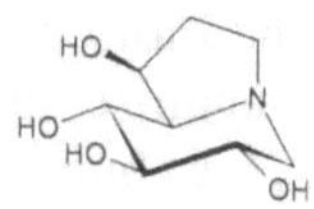 3	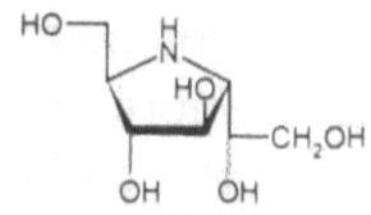 7	Ref.
Activity of mouse small intestinal mucosal homogenates:							[61]
p-nitrophenyl-α-D-glucoside	6	$IC_{50} = 300$	$IC_{50} = 47$	$IC_{50} = 0.83$	$IC_{50} = 2.8$	ng	
maltose	6.4	$IC_{50} = 200$	$IC_{50} = 35$	$IC_{50} = 0.37$	$IC_{50} = 0.83$	ng	
trehalose	6	$IC_{50} = 320$	$IC_{50} = 22$	$IC_{50} = 67$	$IC_{50} = 9.8$	ng	
sucrose	6	$IC_{50} = 42$	$IC_{50} = 23$	$IC_{50} = 0.06$	$IC_{50} = 0.042$	ng	
isomaltose	6.4	$IC_{50} = 23$	$IC_{50} = 4.0$	$IC_{50} = 0.12$	$IC_{50} = 3.1$	ng	
turanose	6	$IC_{50} = 71$	$IC_{50} = 28$	$IC_{50} = 0.14$	$IC_{50} = 0.23$	ng	
palatinose	6	$IC_{50} = 56$	$IC_{50} = 13$	$IC_{50} = 0.25$	$IC_{50} = 43$	ng	
p-nitrophenyl-β-D-glucoside	6	$IC_{50} = 10$	NI	NI	$IC_{50} = 17$	ng	
cellobiose	6	$IC_{50} = 26$	NI	NI	$IC_{50} = 8.5$	ng	
gentiobiose	6	$IC_{50} = 2.2$	$IC_{50} = 100$	$IC_{50} = 44$	$IC_{50} = 0.73$	ng	
p-Nitrophenyl-β-D-galactoside	6	$IC_{50} = 2.0$	$IC_{50} = 210$	$IC_{50} = 220$	$IC_{50} = 5.5$	ng	
Lactose	6	$IC_{50} = 2.1$	NI	$IC_{50} = 57$	$IC_{50} = 0.58$	ng	

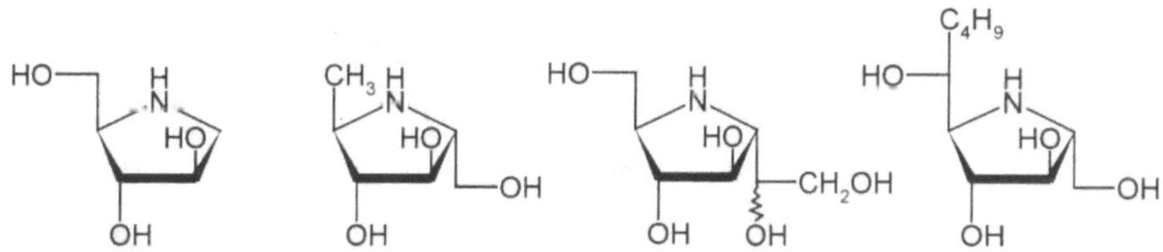

5 6 7 8

Fig. 5. Structures of *DMDP*-related natural compounds

seeds of the African legume *Angylocalyx pynaerii*. Its configuration was proved by stereospecific synthesis, and it could be shown to be a selective inhibitor of β-mannosidase from *Aspergillus niger* with an IC_{50} of 380 μM without any effect on α-mannosidase from jack beans and several other glycosidases [17].

A homologue of *DMDP* containing seven carbons, 2,5-dideoxy-2,5-imino-D-glycero-D-manno-heptitol (**7**, Fig. 5), synthesized by *Wong* and coworkers [47] prior to its discovery in nature, was isolated from *Hyacinthoides nonscripta* together with **1** and **5**. The absolute configuration at C-6 in compound **7** could not be determinated at this time, so all biological studies were referred to 2,5-dideoxy-2,5-imino-*DL*-glycero-*D*-manno-heptitol (**7**) [24a]. This inhibitor could be shown to be more potent against almond β-glucosidase (see Table 1) and to be more active against sucrase, maltase, and lactase than the parent compound. Recently, the 6-C-butyl derivative of *DMDP* (**8**, Fig. 5) was found in *Adenophora triphylla* var. *japonica* (Campanilacae) and showed interesting inhibitory activity against β-glucosidase and amyloglucosidase [18] (Table 2).

Broussonetines (Fig. 6) were discovered as a family of compounds having *DMDP*-type structures and could be shown to be β-galactosidase and β-mannosidase inhibitors. These compounds contain the same five-membered ring structure as *DMDP* with diversely modified lipophilic side chains at position C-1 such as in, for example, **9** and **10** [19].

Table 2. IC_{50} values of different derivatives (μM)

Enzyme	*pH*	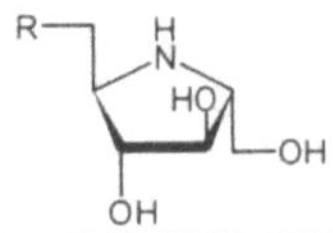 R=(CH2)2CHOH(CH2)2CHOHCH3	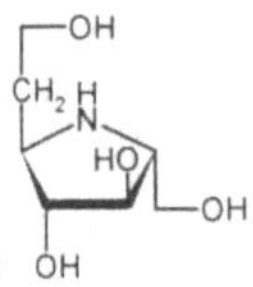	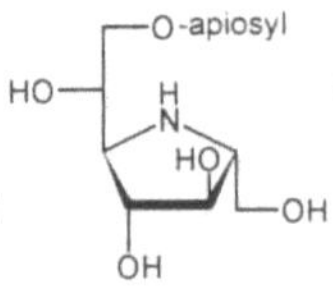	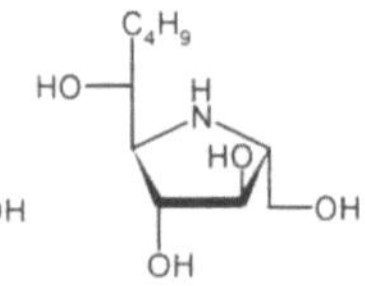	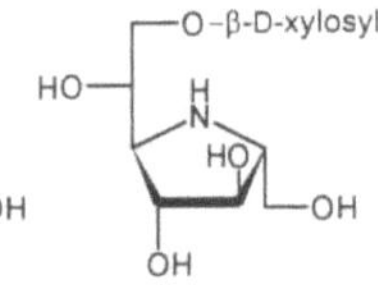	Ref.
α-Glucosidase (rice)	opt	NI	$IC_{50}=2.2$	NI	NI	NI	[24b]
α-Glucosidase rat intestinal maltase	opt	NI	$IC_{50}=2.5$	NI	NI	NI	[24b]
sucrase		NI	$IC_{50}=11$	NI	NI	NI	[24b]
α-Glucosidase rat liver lysosomal	opt	ng	$IC_{50}=7.2$	ng	ng	ng	[24b]
β-Glucosidase (almond)	opt	NI	ng	NI	$IC_{50}=68$	$IC_{50}=4.6$	[18], [24c]
β-Glucosidase *C. saccharolyticum*	opt	NI	$IC_{50}=380$	NI	ng	$IC_{50}=0.34$	[24b], [24c]
β-Galactosidase (bovine liver)	opt	NI	$IC_{50}=60$	$IC_{50}=40$	$IC_{50}=390$	$IC_{50}=24$	[24c], [24b] [18]
β-Galactosidase (rat intestinal lactase)	opt	NI	$IC_{50}=320$	$IC_{50}=1.6$	ng	$IC_{50}=0.18$	[24b], [24c]
β-Mannosiase (snail)	opt	NI	ng	NI	ng	$IC_{50}=140$	[24c]
α,α-Trehalase (rat instestine)	opt	NI	$IC_{50}=380$	NI	ng	NI	[24b]
α-*L*-Fucosidase (bovine epidermis)	opt	$IC_{50}=110$	ng	NI	ng	NI	[24c]
Amyloglucosidase (*A. niger*)	opt	$IC_{50}=40$	ng	$IC_{50}=60$	$IC_{50}=40$	$IC_{50}=100$	[24c] [18]

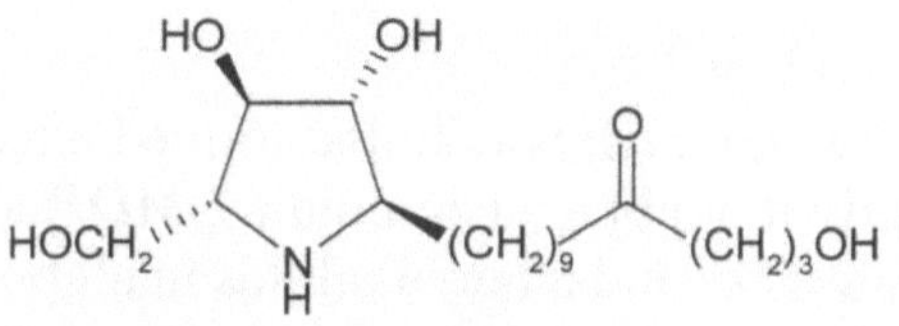
9

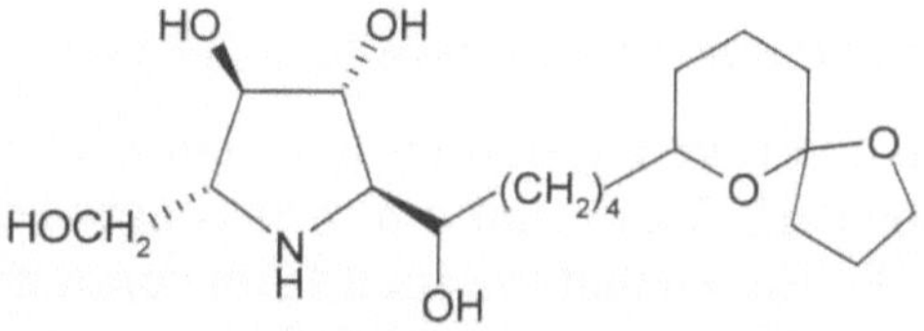
10

Fig. 6. Structures of two broussonetines

Natural Sources and Isolation

After the first isolation and recognition of the biological activities of *DMDP* and related compounds and the understanding of the potential biological importance of these structures, further screening of all kinds of sources became a research topic in the nineteen-eighties. The compound could be found in plants, animals, and microorganisms, demonstrating to be a fairly common metabolite.

Plant sources

Because of its first discovery in a leguminose plant, many others were screened for the presence of *DMDP*. The seeds of *Lonchocarpus sericeus* (Leguminosae) [20] and the leaves of the large liana *Omphalea diandra L.* [21] as well as the leaves of *Derris malaccensis* [22] were found to contain *DMDP* and other alkaloids such as deoxynojirimycin and derivatives. More recently, *DMDP* was found, besides other iminoalditols such as six-membered homologues with D-gluco, D-manno, or D-galacto configurations, in extracts of the plant *Aglaonema treubii* Engl. (Araceae) [23].

Since livestock have been poisoned by grazing the leaves of bluebells, *Hyacinthoides nonscripta* was screened for alkaloids, and *DMDP* as well as **5** and the homo-derivative **7** could be isolated from the British plant. Other species of Hyacinthaceae such as *Hyacinthus orientalis* and the bulbs of *Scilla campanulata*, were also investigated and found to be rich in alkaloids such as *DMDP* and related compounds, for example the 6-deoxy-homo-*DMDP* which had not been found in nature before [24]. Furthermore, *DMDP* and many other iminosugar derivatives were found in the plant *Lobelia sessilifolia* and the roots of *Adenophora spp.* (Campanulaceae) [25].

Kite and coworkers have screened many species of *Nephthytis* Schott, *Anchomanes* Schott, *Pseudohydrosme* Engl., *Aglaonema* Schott, and *Aglaodorum* Engl. where they found reasonable amounts of *DMDP* together with six-membered ring iminoalditols [26]. In their search for α-glucosidase inhibitors, *Kim* and his group became interested in *Commelina communis* L., the dried aerial parts of which have been used in traditional medicine as an antipyretic for noninfectious fever and to treat ascites, edema, and hordeolum. This herb was also very popular in Korea for the treatment of diabetes. Consequently, the extract was screened for glycosidase inhibitory activity, and *DMDP* together with four piperidine-related alkaloids were found [27].

Animal and microbial sources

The larvae of the day-flying moth *Urania fulgens*, which was found to feed exclusively on the liana *O. diandra*, was consequently thought to accumulate *DMDP*, and it could be shown that the adult form contains *DMDP* and other alkaloids in all body parts including their wings [28]. This was the first example of the occurrence of such a compound in animals. Several other moths were found to contain this alkaloid, such as adults of *Alcides metaurus*, *Lyssa macleayi*, and *Urapteroides astheniata* [29]. *DMDP* was also isolated from the fermentation broth of *Steptomyces sp.* KSC-5791 when it was screened for trehalase inhibitor activity, showing the general presence of *DMDP* not only in plants and animals but also in microorganisms [30].

Syntheses

Because of the biological importance of *DMDP*, great interest in synthetic approaches arose. The first synthesis was reported by *Fleet* who employed D-glucose as the starting material in 1985 [31] (Scheme 1). Protection with acetone followed by O-3 benzylation gave the protected sugar **11**. The 5,6-O-isopropylidene group was exchanged by a carbonate residue to yield **12**. Acidic cleavage of the 1,2-O-isopropylidene acetal in methanol furnished the α,β-glucosides. The α-isomer with O-2 available for triflate formation and subsequent S_N2 substitution at C-2 with

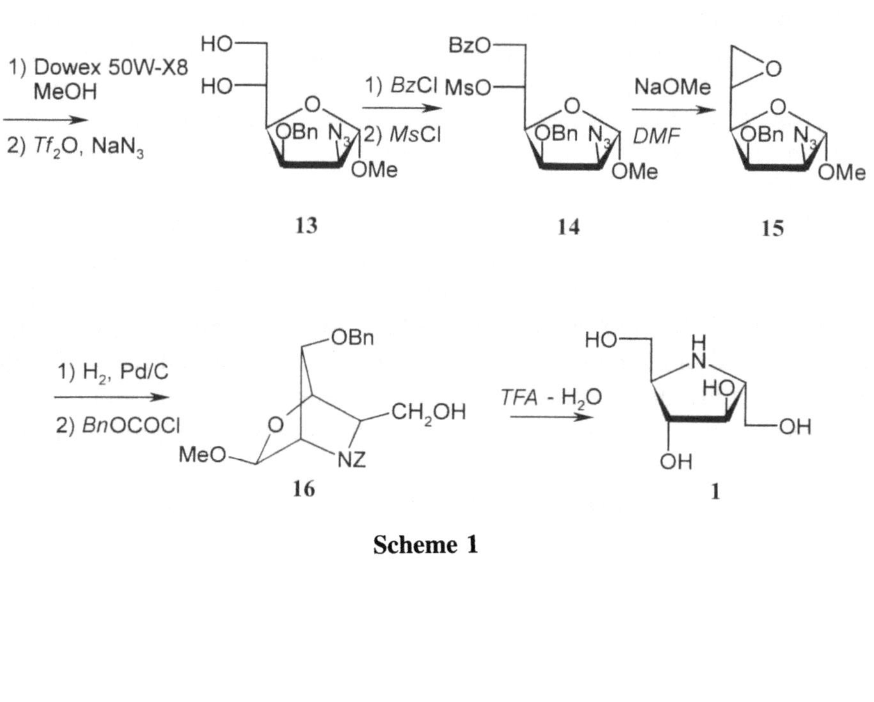

Scheme 1

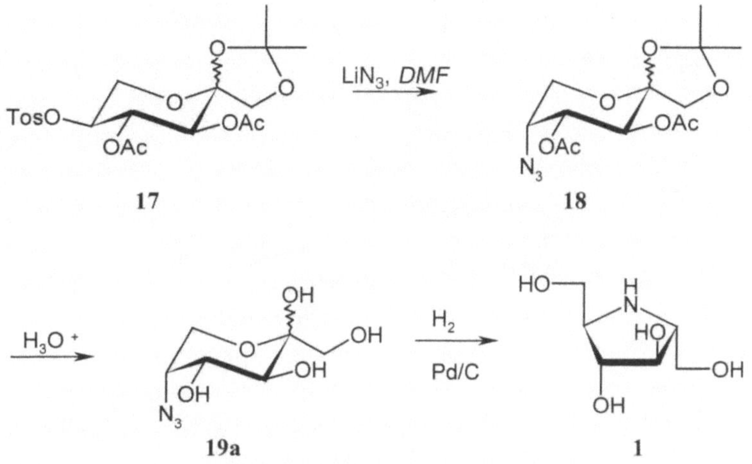

Scheme 2

sodium azide gave key intermediate **13**. The azidodeoxymannofuranoside can undergo subsequent cyclization *via* C-5 during an intramolecular nucleophilic attack with overall retention of configuration at this center by a nitrogen nucleophile generated from the azide yielding the bicyclic amine **16**. This approach was carried out with the epoxide **15** which subsequently suffered attack of the nitrogen to the desired bicyclic compound **16**. Subsequent deprotection converted **16** into *DMDP*. **13** can also undergo cyclization *via* C-6 leading to 1-deoxymannojirimycin.

At about the same time, *Card* and *Hitz* [32] who were interested in sucrose transport and inhibitors of invertase (β-*D*-fructofuranosidase) reported a different approach (Scheme 2). Starting from tosylate **17**, readily available from *L*-sorbose in three steps, they introduced the azide by substitution of the tosylate to obtain **18**. Deprotection under acidic conditions gave 5-azido-5-deoxy-*D*-fructose **19a** which underwent cyclization during hydrogenation to form *DMDP* in an overall yield of 68% for these three steps.

A double reductive amination of suitable dicarbonyl sugars, such as 5-keto-*D*-fructose, was introduced by *Reitz* and *Baxter*. With benzylamine as the amino compound the reaction gave *DMDP* only as a side product (6%), the corresponding *D*-gluco-epimer being the main product [33]. This unnatural isomer was also synthesized by an aldolase catalyzed carbon–carbon bond formation by *Wong* and coworkers [34].

The first chemoenzymatic synthesis had been conducted by *Whitesides* and coworkers in 1991 [35] who used RAMA (rabbit muscle aldolase, EC 4.1.2.13) to synthesize polyhydroxylated amines from C_3 fragments (Scheme 3). The synthesis of the key intermediate **23**, an azidoaldehyde, started with the methylenation of cinnamaldehyde **20** to epoxide **21**, followed by nucleophilic opening with sodium azide to **22**. Subsequent ozonolysis provided **23** for the RAMA-catalyzed reaction with dihydroxy acetone phosphate (*DHAP*) from which a 1:1-mixture of C-5 epimers **19a** and **19b** was obtained. Separation of the isopropylidene protected ketoses by silica gel chromatography and subsequent hydrogenation of the 5-azidodeoxy-*D*-fructose **19a** cleanly afforded the C_2-symmetric pyrrolidine **1** in approximately 10% overall yield.

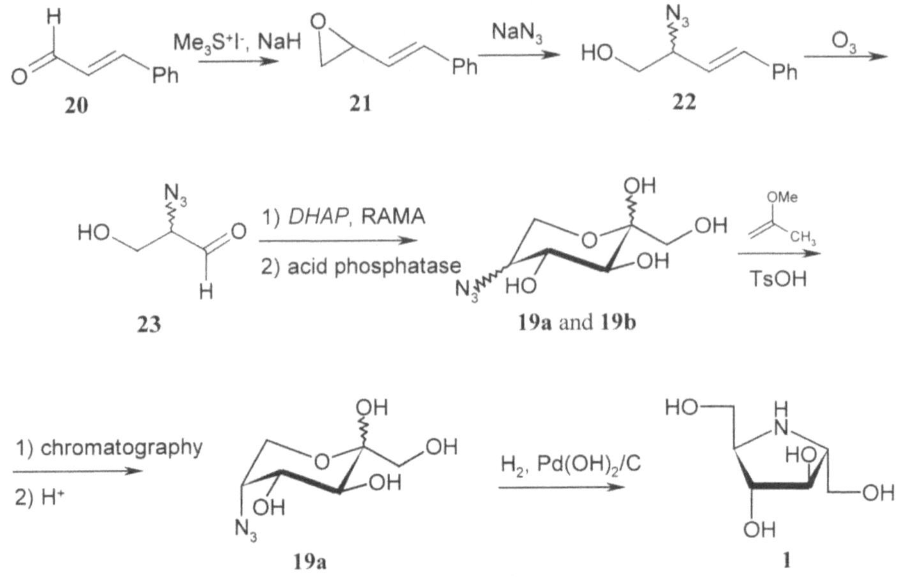

Scheme 3

Duréault and coworkers [36] employed *D*-mannitol (**24**) as the starting material (Scheme 4). Regioselective opening of the easily available *L*-ido-bisepoxide **25** with benzyl alcohol gave 1,6-di-O-benzyl-*L*-iditol derivative **26**. Ditosylation of the secondary alcohols to furnish **27** followed by cleavage of the 3,4-O-isopropylidene group to **28** and subsequent dibenzylation employing benzyltrichloroacetamidate gave the key indermediate **29**. The pyrrolidine ring **30** was obtained by heating of **29** in benzylamine for 12 h at 140°C. Hydrogenolysis of the benzyl groups gave **1** in an overall yield of 20%.

Another chemoenzymatic approach has been reported by *Stütz* and coworkers (Scheme 5) who exploited glucose isomerase (E.C. 5.3.1.5) in the key step [37]. 5-Azido-glucose **32**, easily available in six steps from *D*-glucurono-6,3-lactone **31**, was isomerized quantitatively to the 5-azidodeoxyfructose **19a** employing this enzyme. Conventional hydrogenation with concomitant intramolecular reductive amination furnished **1**. The N-(5-carboxypentyl) derivative **34** was prepared by reductive alkylation with adipic methyl ester hemialdehyde and subsequent saponification of the ester. Coupling to aminohexylsepharose with water-soluble carbodiimide and N-hydroxysuccinimide for activation led to an immobilized inhibitor which could be used as affinity ligand for the purification of glycosidases as previously demonstrated with the corresponding 1-deoxynojirimycin derivative attached to sepharose. The 2-^{2}H-derivative of *DMDP* **33** was prepared employing deuterium gas in methanol-d_1 as the solvent [38].

The N-methyl derivative **139** was made available by reductive amination of **1** with formaldehyde [39], and the iodide of the N,N-dimethyl derivative **140** by N-methylation with iodomethane [40] (Fig. 7).

Liu and coworkers reported a synthesis starting from arabinofuranose derivative **35** [41] (Scheme 6). *Wittig* methenylation to **36** and subsequent *Swern* oxidation gave an unstable unsaturated ketone which was immediately converted to oxime **37**. Reduction and protection of the nitrogen with benzyloxycarbonyl chloride (*Cbz*)

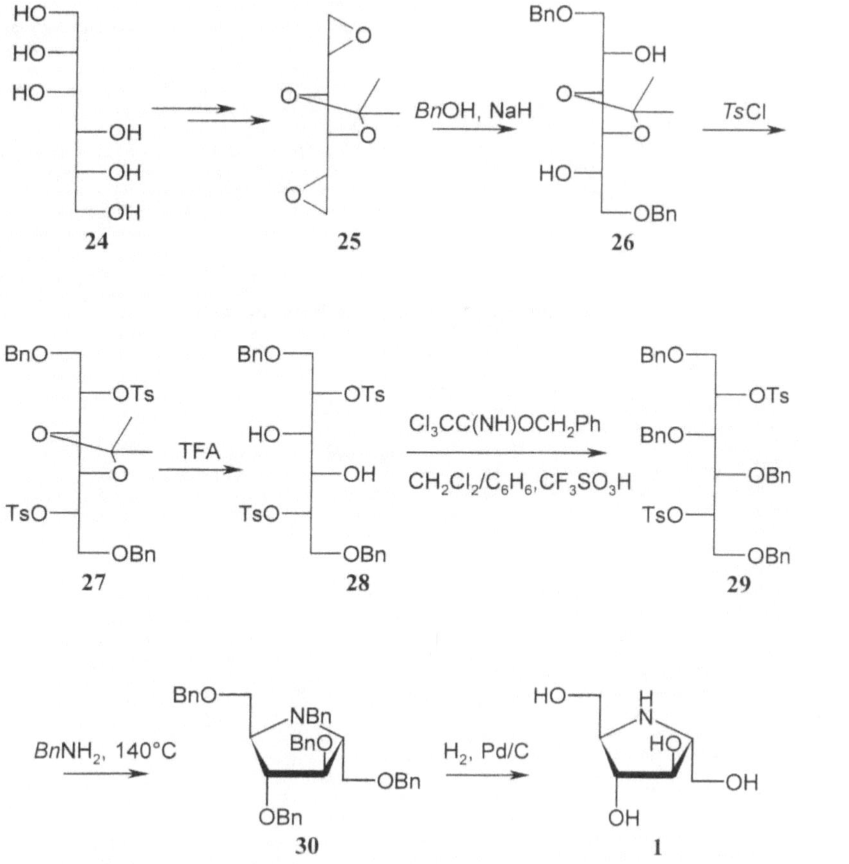

Scheme 4

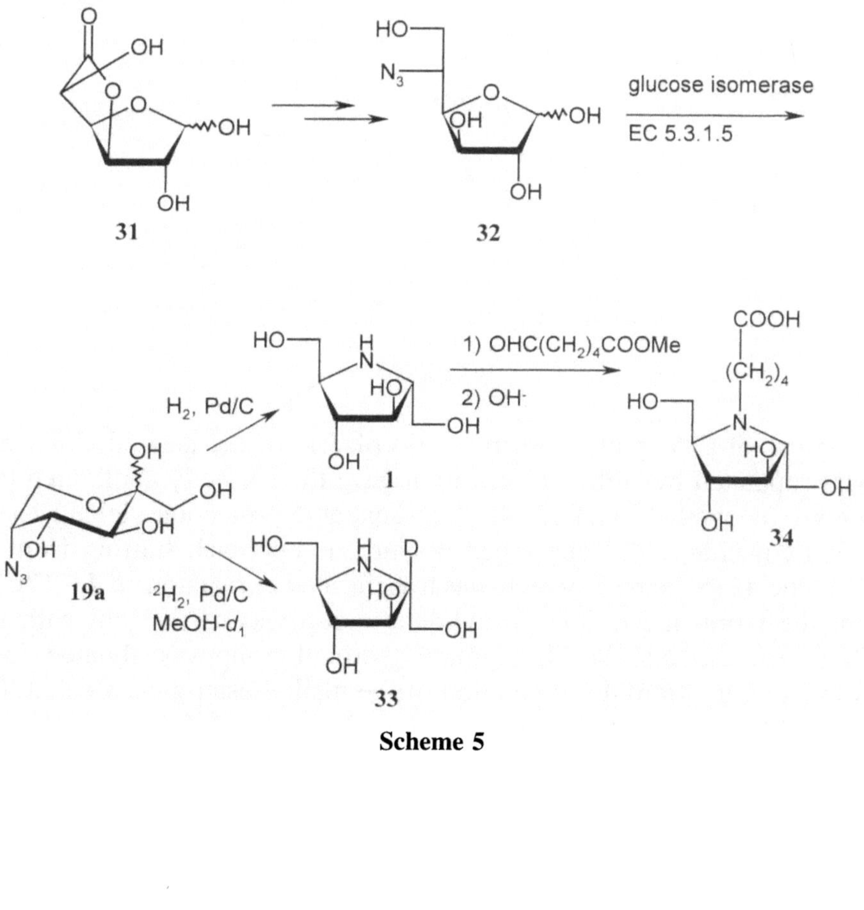

Scheme 5

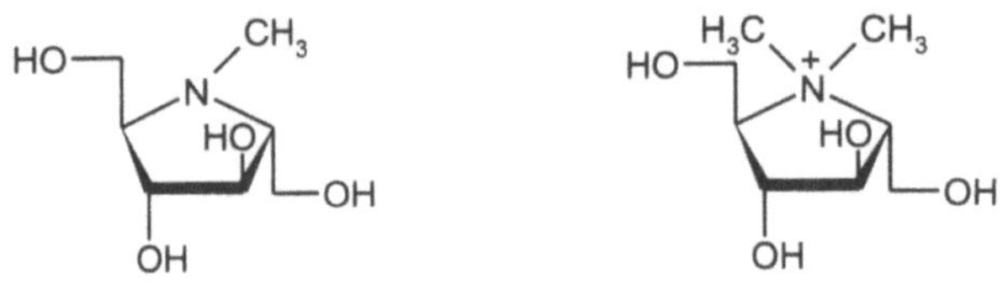

Fig. 7. Structures of compounds **139** and **140**

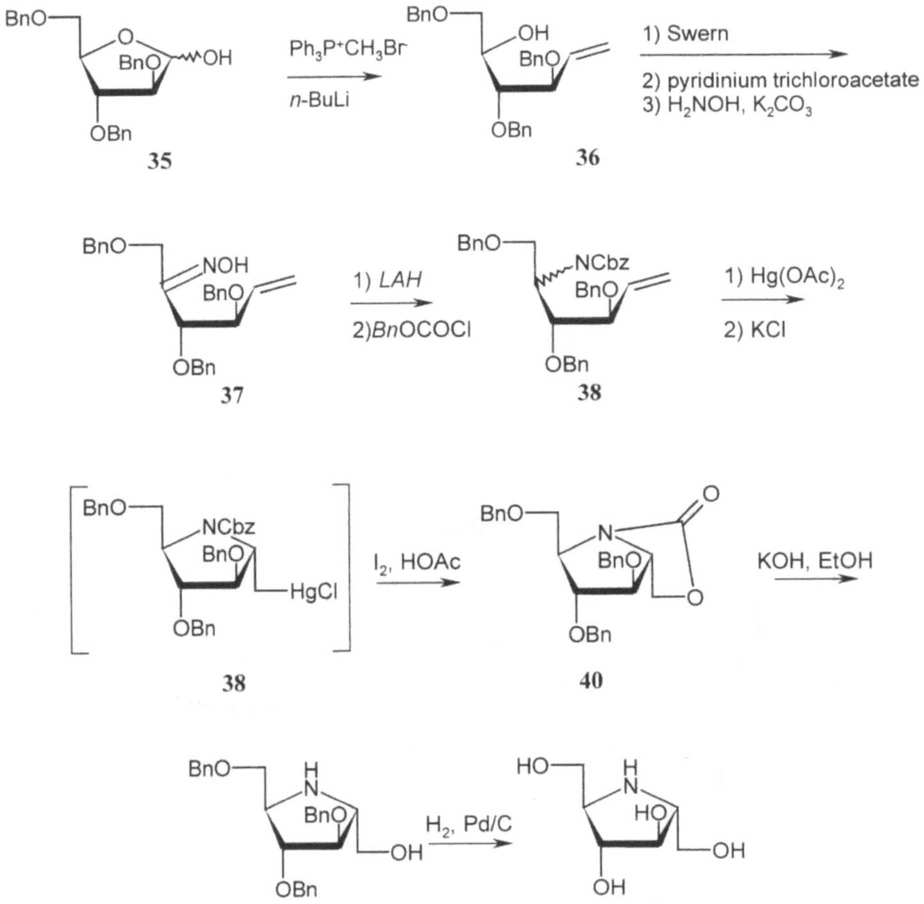

Scheme 6

gave a mixture of C-5 epimers **38** in a ratio of 7:1 for the desired *Cram* product which was separated by HPLC. Cyclization gave the correctly configured product **39** which was converted to **41** *via* **40**. Subsequent deprotection yielded **1**.

Lee and coworkers [42] developed a synthetic approach starting from D-glucono-δ-lactone **42** (Scheme 7) which was transformed to mannonate **43**. The deprotection of the terminal 5,6-O-isopropylidene group was carried out with Dowex 50W-X8 H^+ to yield 96% **44**. The primary hydroxyl group was silylated, followed by O-mesylation to afford **45**. Reduction of the methyl ester gave alcohol **46**. The remaining acid-labile protecting groups were cleaved with Dowex 50W-X8 H^+,

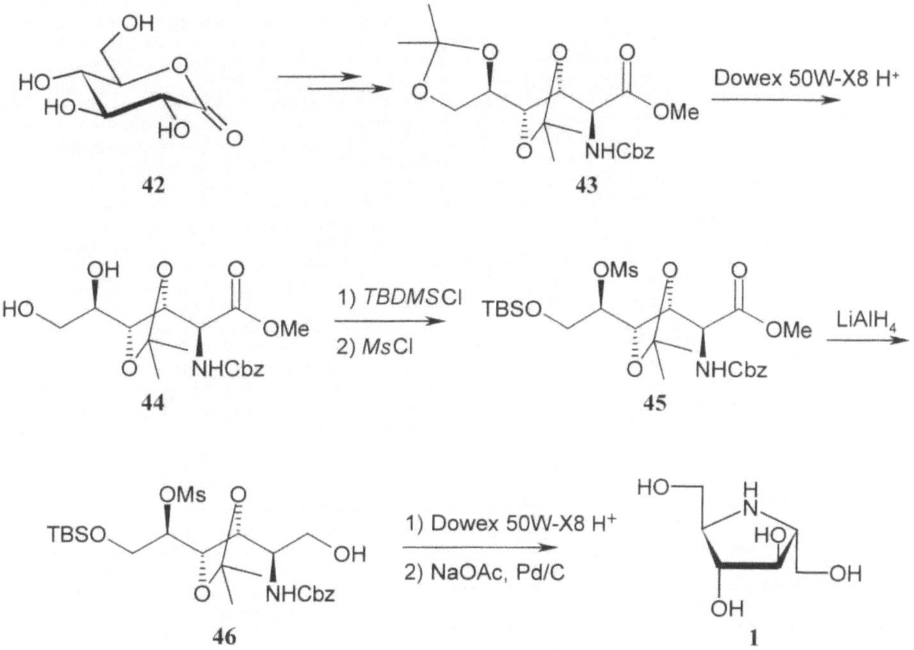

Scheme 7

and removal of the carbamate gave **1** in 67% yield from the di-O-isopropyl derivative **43**.

Ikota [43] used a dihydroxylation protocol for the functionalization of **47** (Scheme 8). The dihydroxylation was carried out with potassium osmate employing hydroquinidine-9-phenanthryl ether as a chiral ligand in the presence of $K_3Fe(CN)_6$ and K_2CO_3 in *tert*-BuOH-H_2O to give precursor **48** as the main product which could be separated from its isomer after O-methoxymethylation of the hydroxyl groups. Reduction to alcohol **50** followed by *Swern* oxidation and subsequent reaction of the resulting aldehyde with vinylmagnesium bromide gave a mixture of isomers in a ratio of 9:1 in favour of the desired intermediate **51**. This was, after separation by column chromatography, mesylated and cyclized with potassium *tert*-butoxide to furnish intermediate **52**. Ozonolysis and reduction to **53** followed by acidic removal of the protecting groups gave **1**.

Merrer and coworkers (Scheme 9) found that isomerization of partially protected polyhydroxypiperidines such as 1-deoxynojirimycin derivative **54** can be achieved by activating C-2 to give **55** which can undergo various transformations [44]. Activation of derivative **54** was carried out either by mesylation of the free hydroxyl groups or under *Mitsunobu* conditions (Ph_3P-*DEAD*-$PhCO_2H$, *THF*) at 0°C. In both cases a mixture of starting material and ring-contracted compound was obtained in favour of the pyrrolidine **56** in ratios of 31%:51% and 29%:64%, respectively. This has been explained by the ease of neighbouring nitrogen participation, since the leaving group is in equatorial position, and subsequent ring opening of the aziridinium ring occurs at the less substituted side.

Efforts were also made to trap the cyclic imine resulting from the reduction and ring closure of the corresponding azidodeoxyfructose derivative **19a** (Scheme 10),

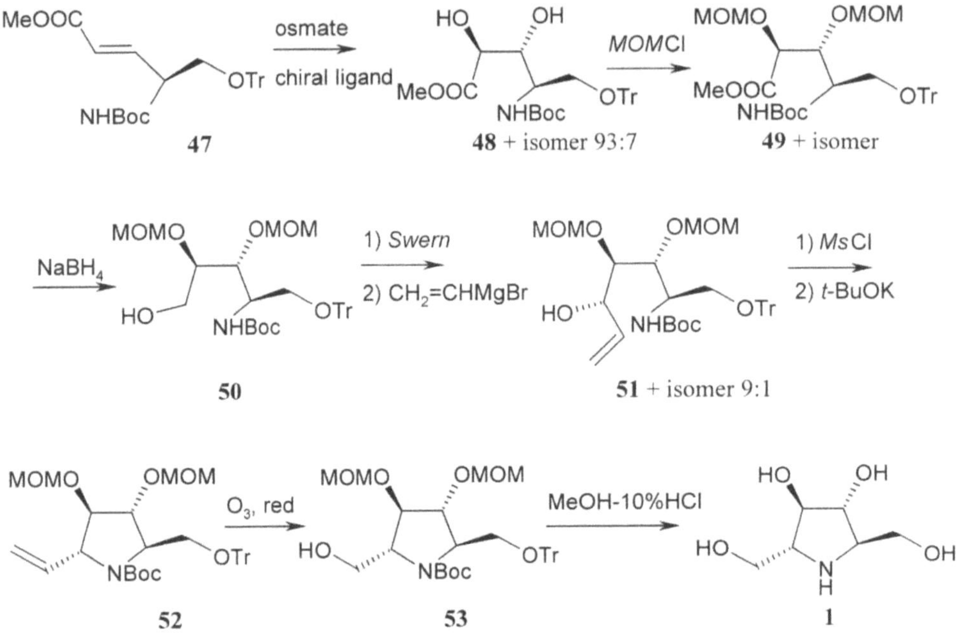

Scheme 8

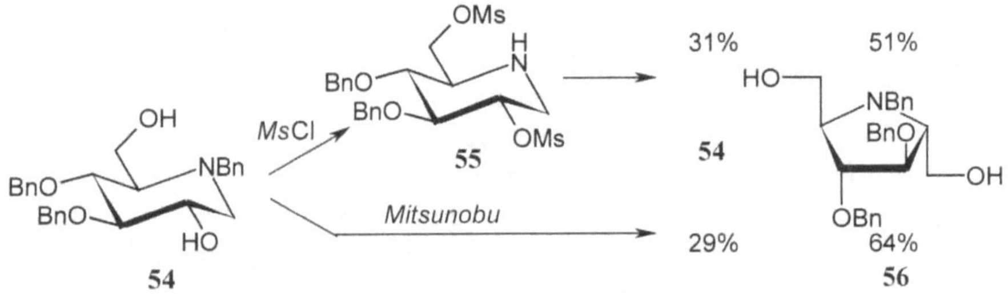

Scheme 9

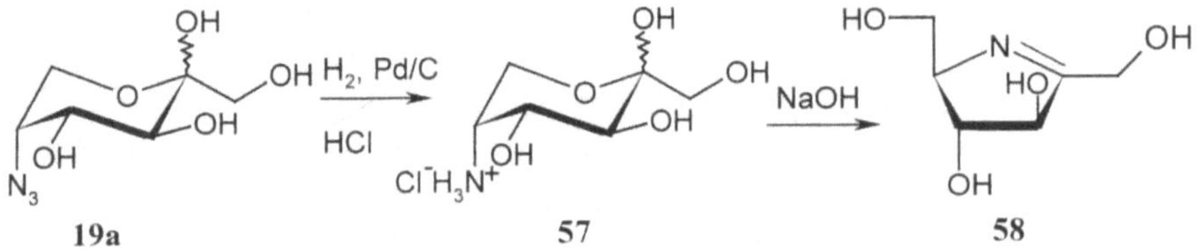

Scheme 10

whose shape and charge were assumed to mimic the transition state stabilized by glycoprocessing enzymes [45]. Thus, the aminosugar intermediate was trapped as the corresponding hydrochloride **57**. When the salt was transferred into basic solution, the compound immediately formed an equilibrium in which the cyclic imine **58** was the major species present.

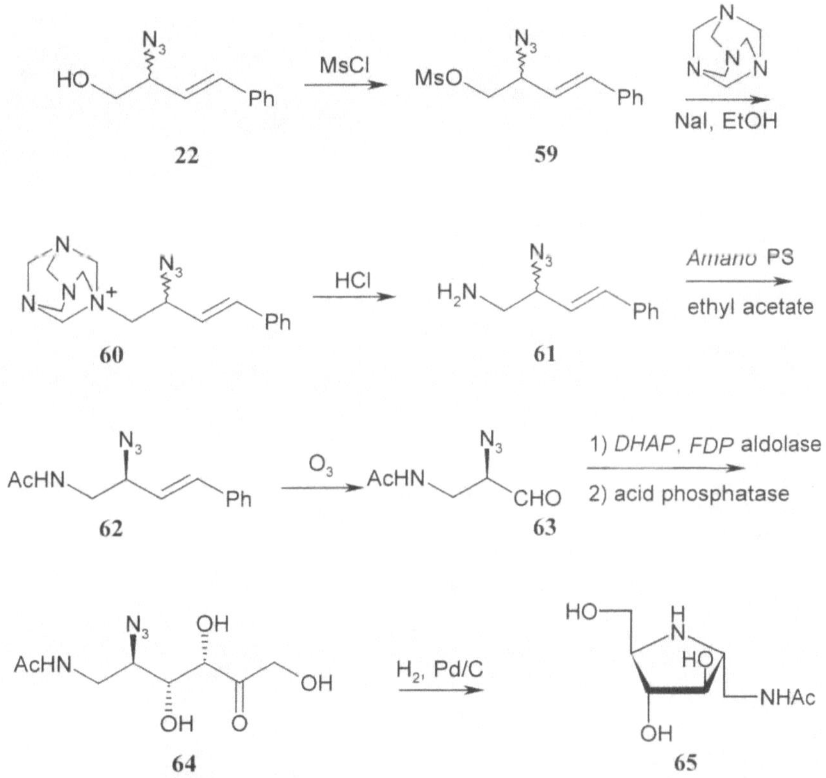

Scheme 11

Following the work of *Whitesides* and his group, the 1-acetamidodeoxy derivative of *DMDP* **65** was synthesized by *Wong* and coworkers employing a chemoenzymatic approach [46] (Scheme 11). The synthesis of the azidoaldehyde **63**, the key intermediate for the enzymatic step, started from azidoalcohol **22** whose mesylation gave **59** which by treatment with hexamethylenetetramine afforded ammonium salt **60**. The primary amine **61** was obtained by hydrolysis of **60** in concentrated hydrocloric acid. Enzymatic resolution exploiting *Pseudomonas* lipase in ethyl acetate gave the desired stereomer **62** in a yield of 34% based on a mixture of the primary amines **61**. Ozonolysis in methanol followed by reductive work-up employing dimethyl sulfide produced the azidoaldehyde derivative **63** required for the *FDP*-aldolase catalyzed condensation with *DHAP*. The phosphates were removed with acid phosphatase, and subsequent hydrogenation led to the desired 1-acetamido derivative **65**.

The *DMDP* homologue **7** was also synthesized by *Wong et al.* [47] who started from 2,3,5-tri-O-benzylarabinose **35** (Scheme 12). *Wittig* reaction gave the *E*-configured **66** in good yield. Reduction of the ester to the corresponding alcohol and subsequent protection of the latter as silylether furnished **67** bearing a free hydroxy group at position C-6. This was activated with chloromethylsulfonyl chloride, and a S_N2 reaction gave **68** in 60% yield. *Zemplen* deprotection gave **69** and was followed by chloromethylsulfonation and subsequent removal of the silyl protecting group to obtain **70**. Sharpless epoxidation employing *D*-(–)-diethyl tartrate led to the epoxide **71** in a highly stereoselective reaction. Replacement of the

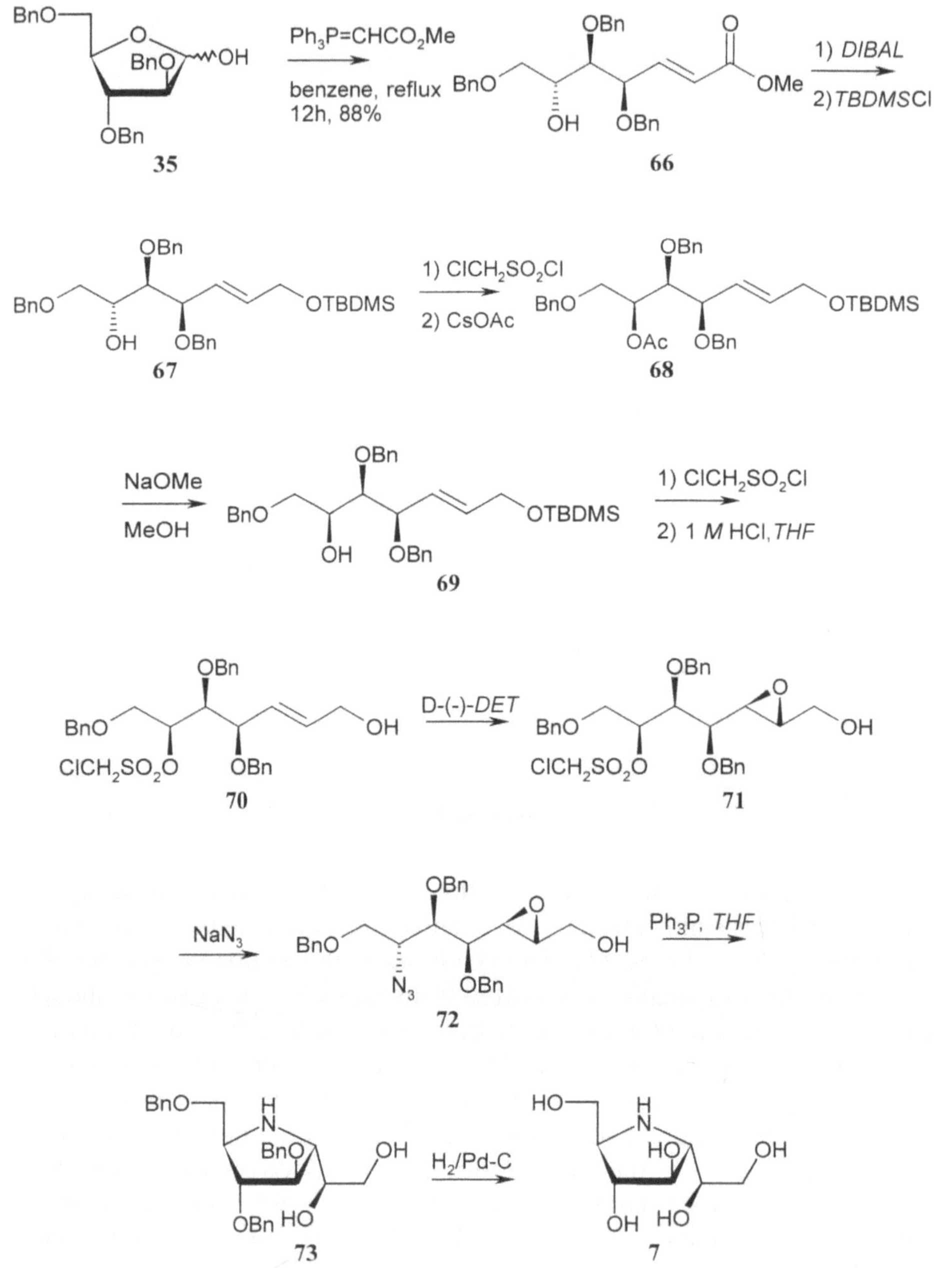

Scheme 12

chloromethylsulfonate with azide allowed the desired introduction of the azido group at this carbon under formal overall retention of configuration at this center to yield **72**. Reduction of the azide to the amine, ring closure to **73**, and removal of the protecting groups furnished **7**.

The same group has reported the synthesis of several N-alkylated derivatives [48] of **7** as well as of the 1-acetamidodeoxy derivative **65** (Scheme 13). Starting from the precursor **73**, straightforward derivatization, N-*Boc* protection to **74**, and introduction of the acetamido group at C-1 led to derivative **77** which was either converted to the known compound **65** or, by deprotection of the nitrogen with *TFA*,

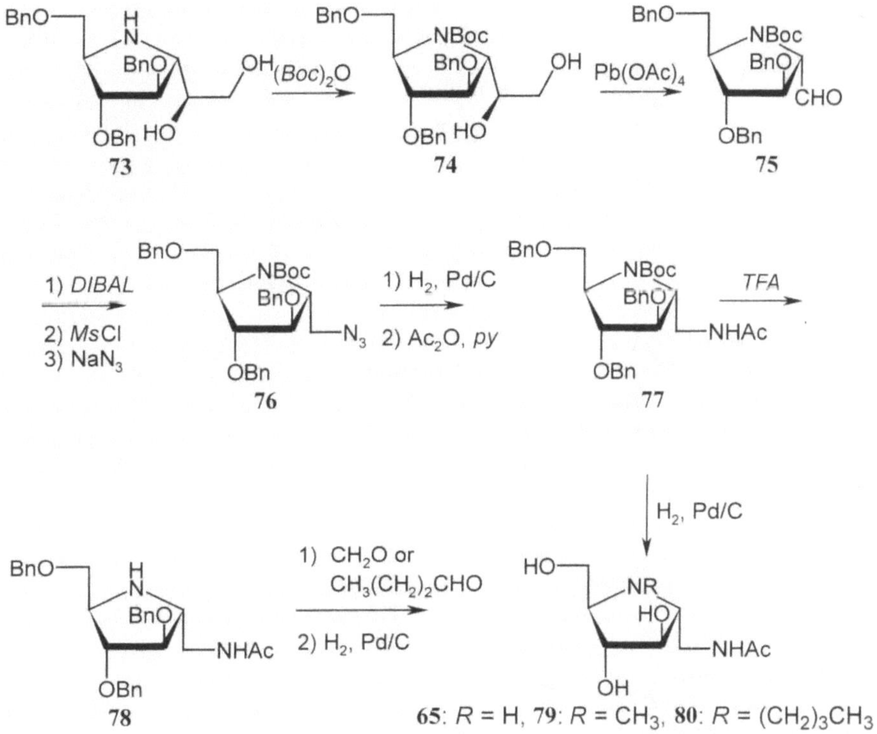

Scheme 13

to **78**. N-Alkylation employing two different aldehydes followed by deprotection of the hydroxyl groups led to N-alkylated derivatives **79** and **80**. These three products exhibited K_i values with β-N-acetylglucosaminidase in the low micromolar range.

The acetamidodeoxy homologues were synthesized in a similar manner (Scheme 14). Tosylation of the primary hydroxyl group of **74**, followed by substitution employing sodium azide and protection with benzyl bromide, furnished **81**. Reduction of the azide and acetylation of the resulting amine gave acetamide **82** which underwent hydrogenolysis yielding derivative **85**. On the other hand, deprotection of the nitrogen, reductive alkylation employing aldehydes, and hydrogenolysis gave the two N-alkylated compounds **83** and **84**.

A dimeric form of *DMDP* featuring a nitrogen bridge between the to rings was prepared by coupling aldehyde **75** under reductive amination conditions employing ammonium acetate and sodium cyanoborhydride to yield **86**.(Scheme 15)

The synthesis of (difluoromethyl)-phosphonate iminosugars was investigated by *Guillerm* and coworkers [49] when they became interested in using such compounds as glycosyl transferase inhibitors (Scheme 16). 2,3,5-Tri-O-benzyl-*L*-xylose **87** was reacted with benzylamine to yield furanosylamine **88** which was exposed to (diethylphosphinoyl)-difluoromethyllithium furnishing a separable mixture of the two possible diastereomers **89a** and **89b** with moderate diastereoselectivity. For the final cyclization step, the hydroxyl group was activated as the methanesulfonate which underwent intramolecular nucleophilic displacement by the amino group leading to **90**.

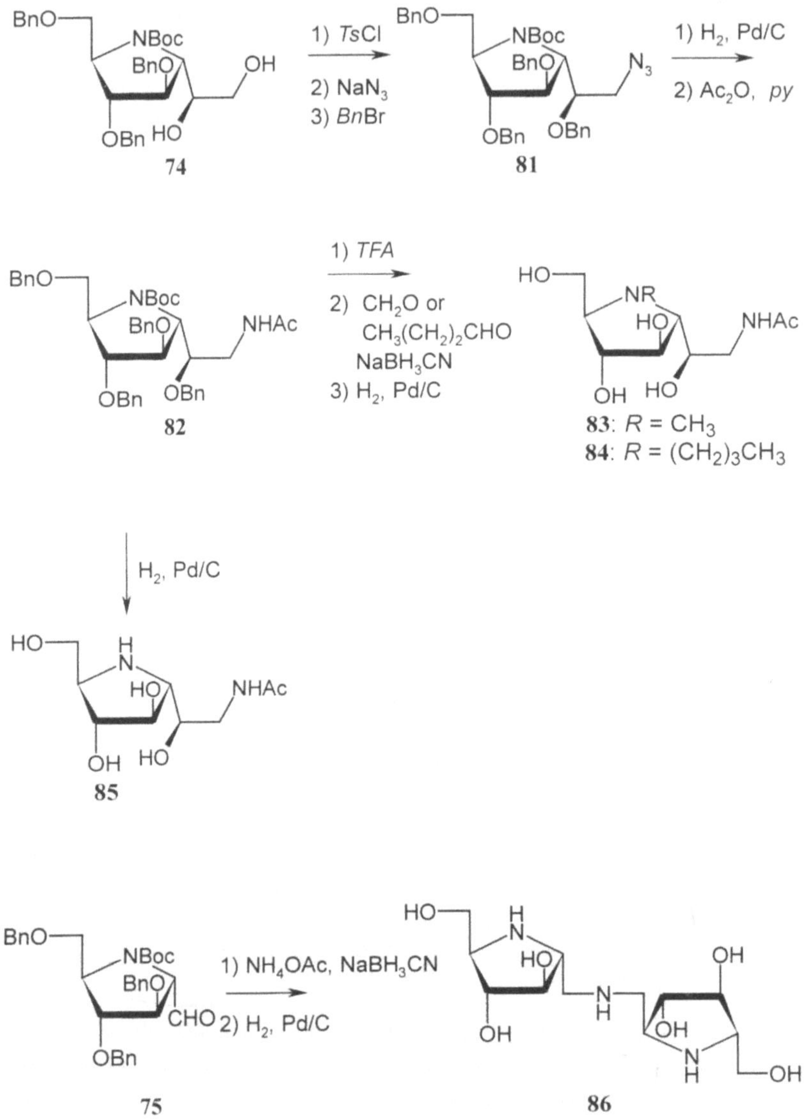

Scheme 15

Eustache and his group [50] were also interested in such 1-phosphonate derivatives in the context of glycosyltransferases. Starting from 2,3,5-tri-O-benzyl-D-arabinofuranose (**35**, Scheme 17) which underwent *Wittig* olefination and subsequent *Mitsunobu* reaction employing *p*-nitrobenzoic acid followed by phthalimide protection of the nitrogen, **91** was obtained. The nitrogen protecting group was then exchanged to N-*Cbz* protected **92**. Ring closure was achieved by treatment with *NBI*, and reaction of iododeoxy compound **93** with triethylphosphite followed by deprotection yielded the desired 1-phosphonate **94**.

A C-1 aryl derivative was synthesized recently by *Correia* and coworkers [51] during their synthetic studies of codonopsine and codonopsinine, two pyrrolidine alkaloids (Scheme 18). The key step was the *Heck* reaction on endocyclic ene-carbamates **95** and **96**, which was carried out in the presence of a diazonium salt

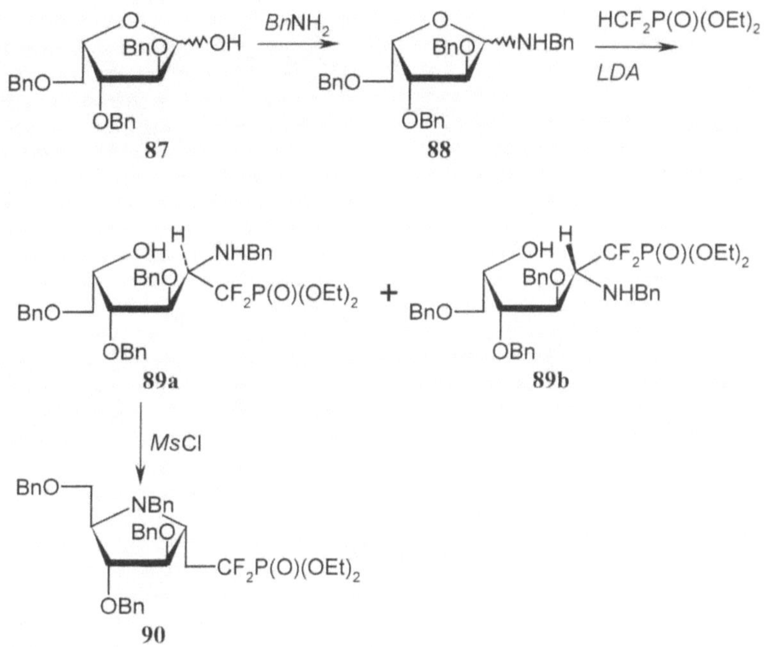

Scheme 16

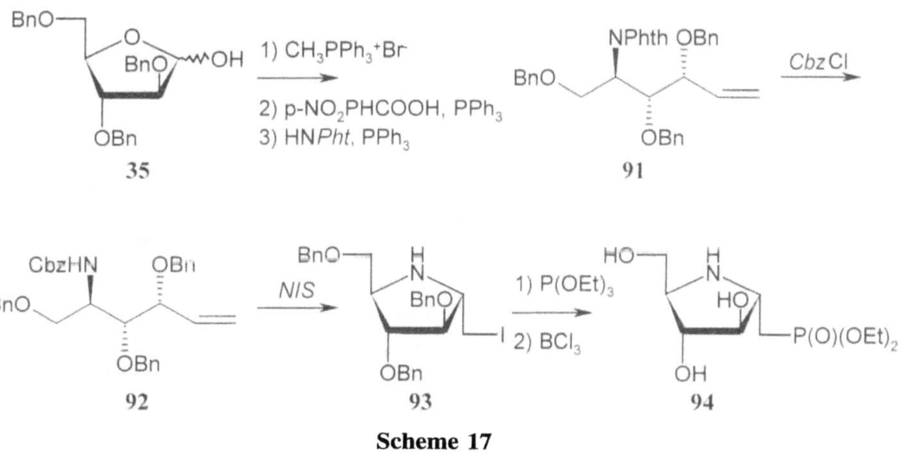

Scheme 17

and 2,6-di-*t*-butylpyridine to yield the C-1-arylated compounds **97** and **98** in a highly regio- and stereoselective manner. The silyl and trityl groups in **97** and **98**, respectively, were removed to yield intermediate **99** which was converted to a single epoxide **100** employing *m-CPBA*. Subsequent acidic hydrolysis led to **101**.

Fleet and coworkers synthesized the achiral aliphatic version of *DMDP* starting from dimeric dihydroxyacetone **102** by reductive amination employing ammonium chloride. **103** showed no biological activities with the common glycosidases [52] (Scheme 19).

The first O-glycosylated derivative of *DMDP* was synthesized by *Stütz* and coworkers [53] (Scheme 20) by reacting intermediate 5-azidodeoxyfructose **19a** with

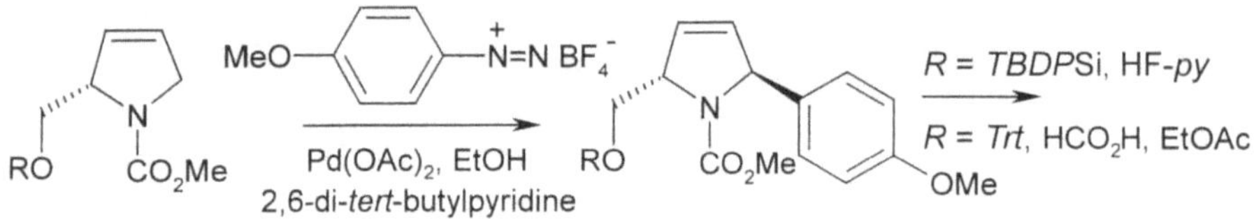

95: *R* = *TBDPS*
96: *R* = trityl

97: *R* = *TBDPS*
98: *R* = trityl

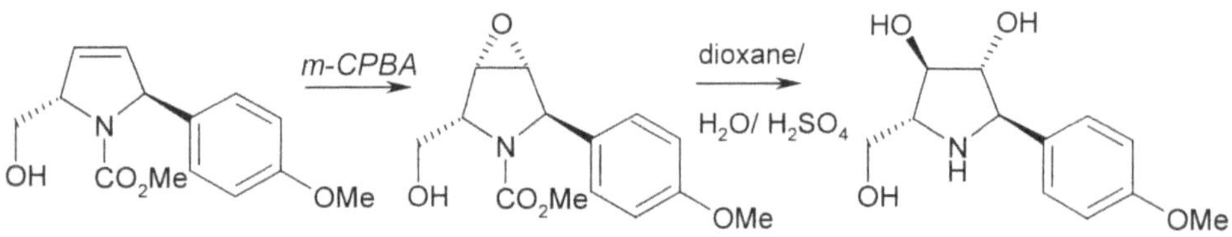

99 **100** **101**

Scheme 18

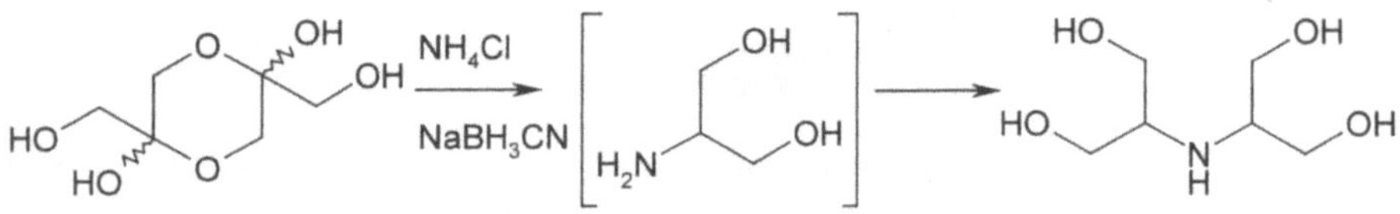

102 **103**

Scheme 19

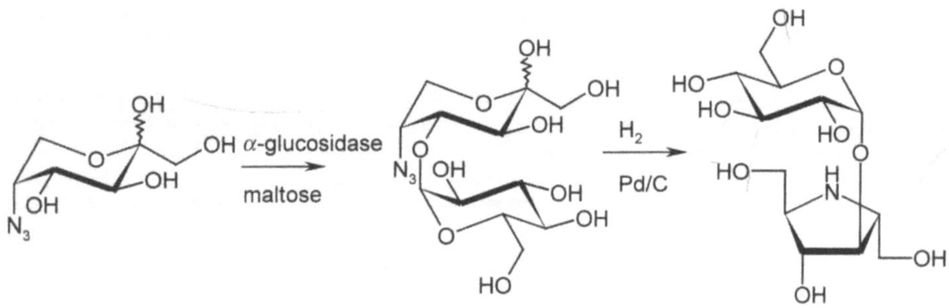

19a **104** **105**

Scheme 20

α-glucosidase from yeast in the presence of maltose as the donor. Glycosylation took place at positions O-1 and O-4; the isomers could be separated by chromatography on charcoal. Conventional hydrogenation of disaccharide **104** over Pd/C led to the O-4 glucosylated derivative **105**.

In the mid-nineteen-nineties, C-1 substituted derivatives of *DMDP* aroused interest. The first compounds prepared were the 1-deoxyfluoro-, 1-O-methyl, and 1-aminodeoxy derivatives of *DMDP* (**108**, **116**, and **118**). The 1-deoxyfluoro compound was easily accessible in two steps from known 5-azido-5,6-dideoxy-6-fluoro-D-glucofuranose (**106**, Scheme 21) by enzymatic isomerization to the

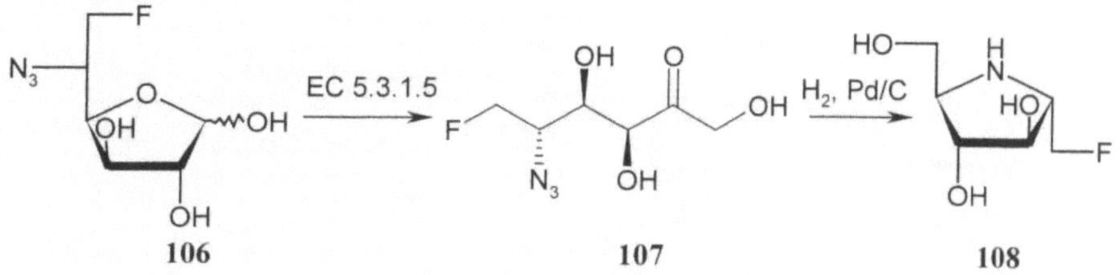

Scheme 21

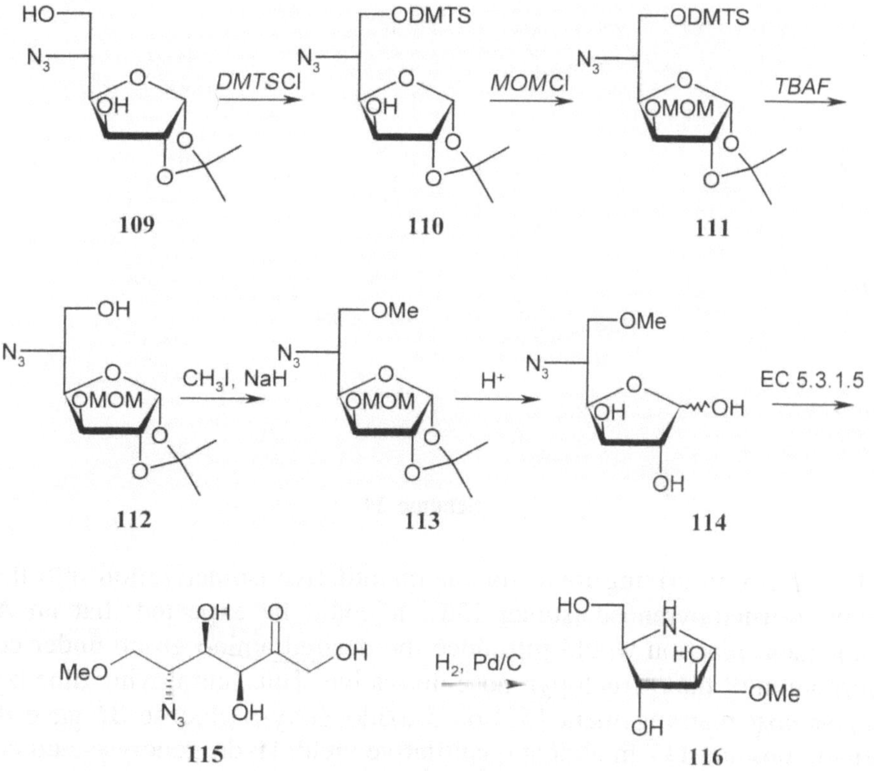

Scheme 22

open-chain ketose **107** employing glucose isomerase, followed by conventional intramolecular reductive amination leading to the 1-fluoro derivative **108** [54].

The 1-O-methyl derivative **116** [54] was prepared from 5-azido-5-deoxy-1,2-O-isopropylidene-α-*D*-glucofuranose (**109**, Scheme 22). Reaction with chlorodimethyl-(1,1,2-trimethylpropyl)-silane gave O-6 silyl ether **110**. O-Methoxymethylation afforded the fully protected sugar **111**. Subsequent desilylation of O-6 using tetrabutylammonium fluoride to obtain **112** and O-methylation with iodomethane in the presence of sodium hydride gave intermediate **113**. Acidic deprotection to **114** and enzymatic isomerization to the corresponding open-chain *D*-fructose derivative **115** followed by hydrogenation yielded **116**.

The 1-aminodeoxy compound **118** was obtained following a different approach [55] (Scheme 23). Bearing in mind that the release of ring strain in the 5-membered

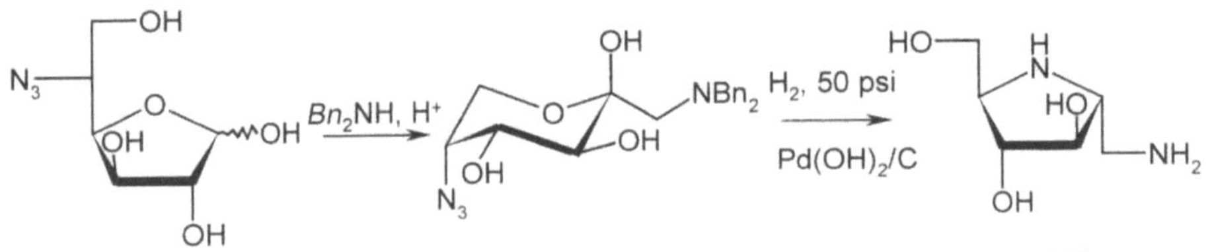

Scheme 23

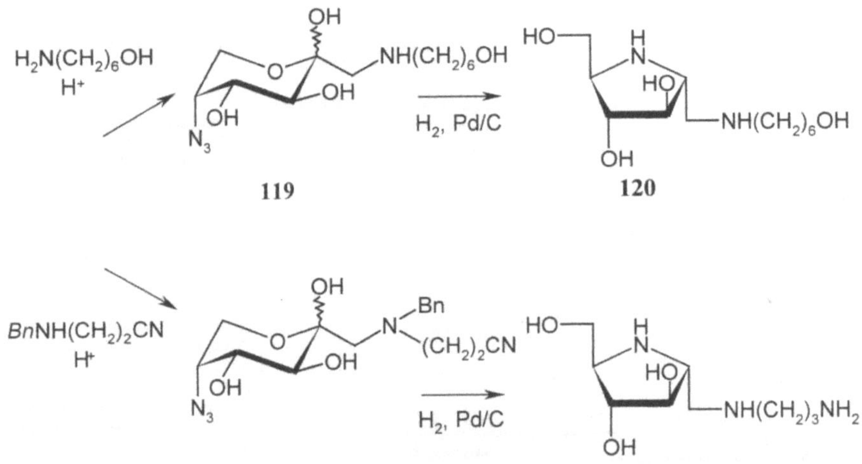

Scheme 24

ring **32** is a strong driving force for the quantitative isomerization into the corresponding D-fructopyranose isomer [56], it could be expected that an *Amadori* rearrangement reaction would introduce the desired amino group under concomitant formation of the D-fructopyranose derivative. Thus, employing dibenzylamine in an *Amadori* rearrangement [57] on 5-azidodeoxy-D-glucose **32** gave the rearrangement product **117** in almost quantitative yield. Hydrogenolysis and concomitant intramolecular reductive amination afforded **118**.

The *Amadori* rearrangement of 5-azidodeoxyglucose **32** was also performed employing 6-aminohexanol as well as 3-(N-phenyl)-amino propionitile to furnish the rearrangement products **119** and **121**, respectively (Scheme 24). Catalytic hydrogenation gave the 6-hydroxyhexyl derivative **120** and the triamine **122** [58].

118 was subsequently shown to react highly chemo- and regioselectively with acylating agents at the primary amine (Scheme 25); the 1-acetamido derivative **65** could easily be obtained by reaction of **118** with acetic anhydride in methanol, and the hydroxyhexanoyl, hexanoyl, and dodecanoyl derivatives **123**, **124**, and **125** resulted from the reaction with caprolactone, hexanoic anhydride, and dodecanoic anhydride, respectively [59].

To extend the series of hydroxyhexyl derivatives, the corresponding 1-O-hydroxyhexyl ether **127** was synthesized taking advantage of the same synthetic approach as reported for the methyl ether **116**, employing methoxymethyl-protected 6-hydroxybromohexane for the ether formation to obtain precursor **126** (Scheme 26).

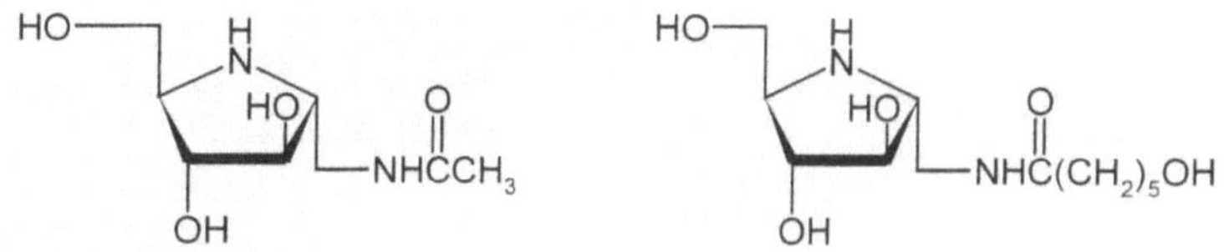

65 **123**

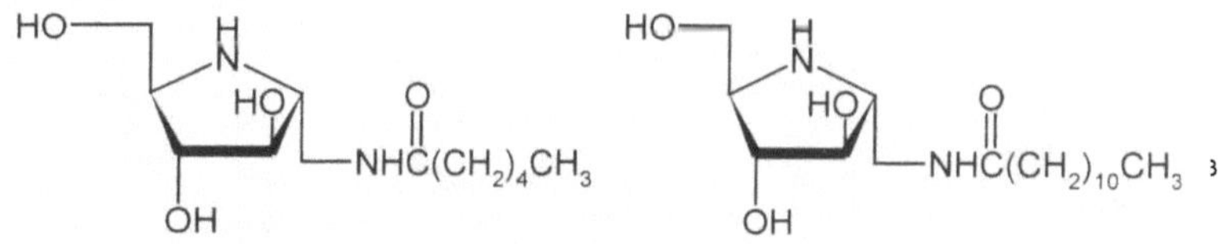

124 **125**

Scheme 25

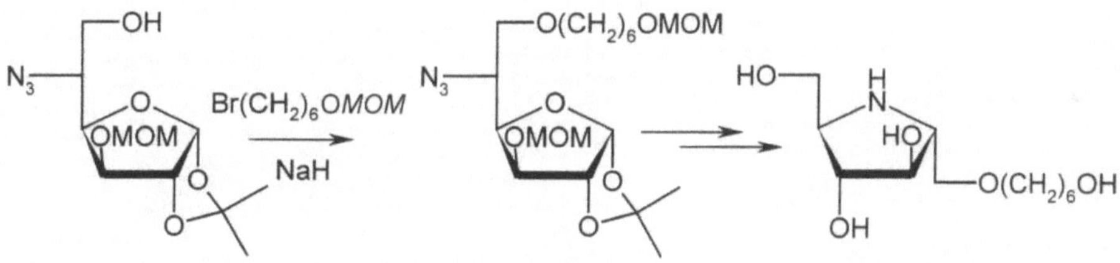

112 **126** **127**

Scheme 26

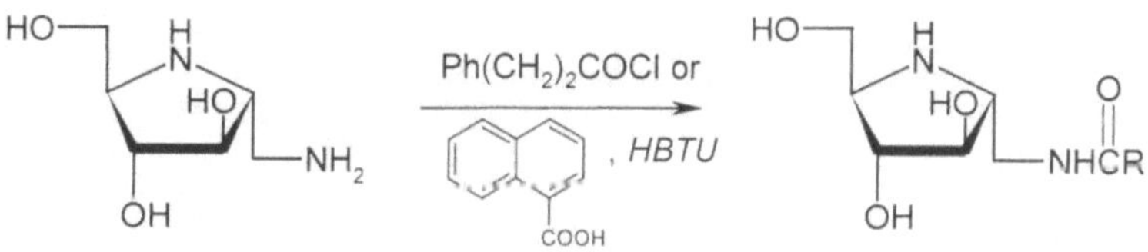

118

128: *R* = (CH$_2$)Ph
129: *R* = naphthyl

Scheme 27

This was converted to the desired inhibitor **127** following the conventional route [58]. Introduction of aromatic substituents at position 1 of **118** gave, for example, the phenylpropanoyl and naphthoyl derivatives **128** and **129** [59] (Scheme 27).

Other examples in this series are the phenyl (**130**), the 4- and the 3-(dimethylamino)-phenyl (**131** and **132**, respectively), and the coumarin derivative **134** (Scheme 28). These compounds were synthesized by coupling the primary amino group in **118** with the corresponding carboxylic acid moieties employing *HBTU* as the coupling reagent. Reaction of **118** with naphthylsulfonic chloride gave the corresponding sulfonamide derivative **133** [59].

Interestingly, all compounds bearing an aromatic substituent exhibited very good to excellent inhibitory activities with the β-glucosidase from *Agrobacterium sp.*

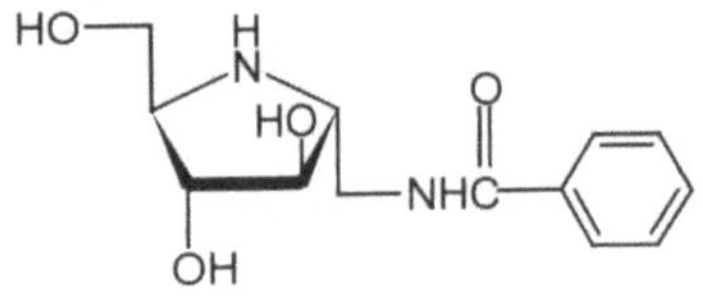

130

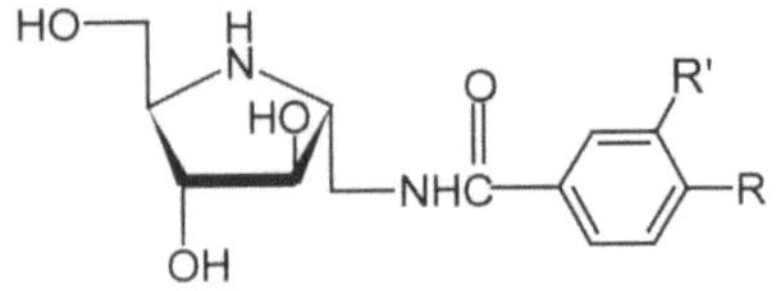

131: R = NMe_2, R' = H
132: R = H, R' = NMe_2

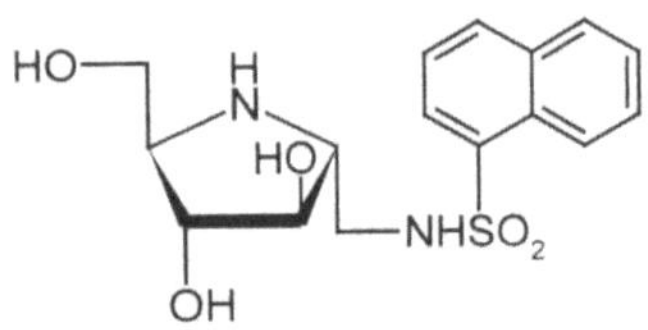

133

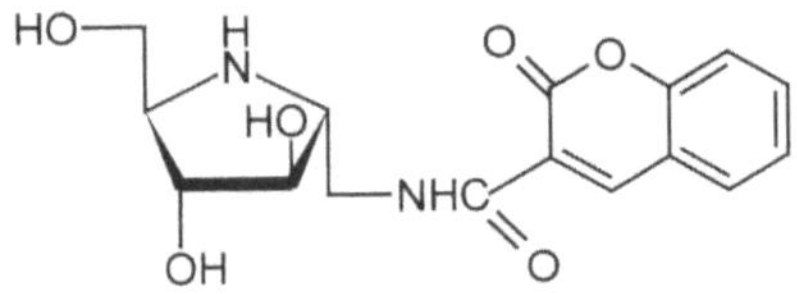

134

Scheme 28

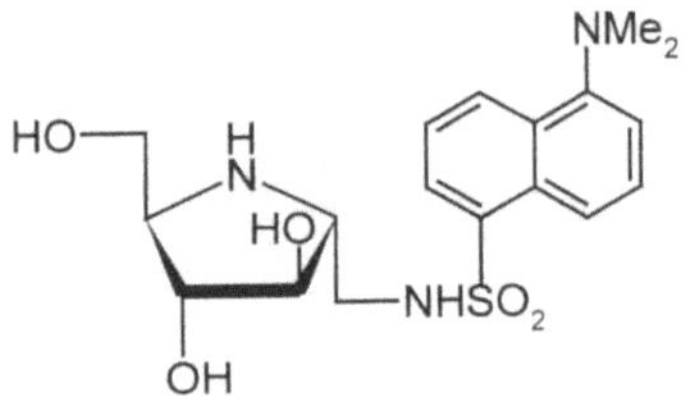

135

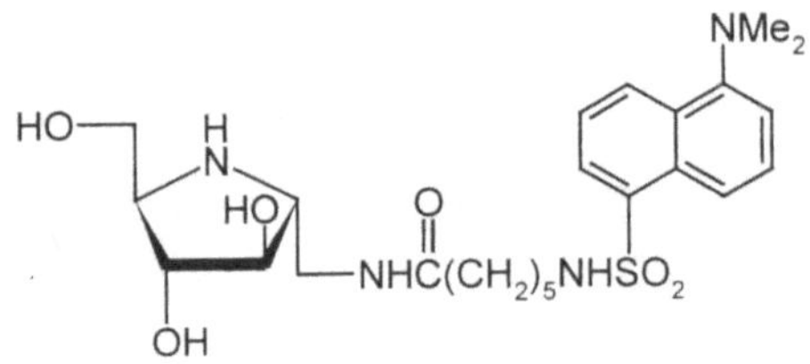

136

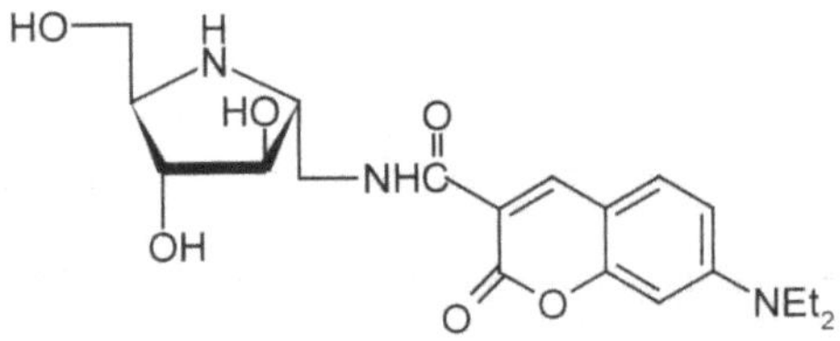

137

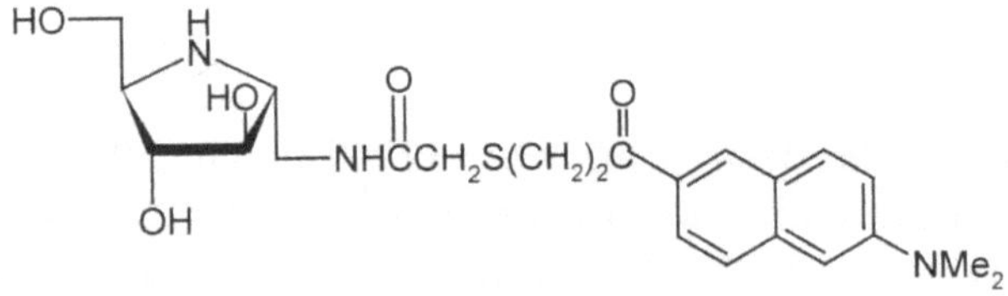

138

Scheme 29

with K_i values in the low micromolar to nanomolar range (Table 4). Thus, it was envisaged that derivatives of *DMDP* labelled with aromatic fluorescent tags might also be good inhibitors and, consequently, might be useful potential diagnostic compounds for monitoring enzyme–inhibitor interactions exploiting fluorescence spectrometry methods. N-Dansyl derivative **135** as well as a related compound featuring a C_6 spacer arm between the dye and the inhibitor (**136**) were synthesized [60] (Scheme 29). Their excellent inhibitory activities in the single-figure nanomolar range prompted the subsequent preparation of the diethylaminocoumarin and acrylodan derivatives **137** and **138**.

Biological Activities and Applications

DMDP has been shown to be a very interesting compound concerning its biological activities such as interaction with glycoprotein processing glucosidase I, anti-retroviral activity (including HIV), antifeedant properties against important pest insects, and trehalase, invertase, and D-fructofuranosidase inhibition. It also exhibits plant growth regulatory activity as well as anti-cancer properties.

Activity of DMDP and related compounds against various glycosidases

DMDP has been screened against a wide variety of glycosidases from several different sources as shown in Table 1. Compared to closely related compounds such as **2**, **3**, **5**, and **7**, *DMDP* shows to be a universal inhibitor of such enzymes.

A very interesting study was performed by *Fleet* and coworkers who investigated the inhibitory properties of *DMDP*, **2**, **3**, and **5** against mouse gut dissaccharidases. Inhibitors of such enzymes bear potential for the treatment of diabetes, obesity, and related metabolic disorders. Although these compounds failed to inhibit α-amylase, which to date is only inhibited by the pseudotetrasaccharide acarbose, these compounds have potential as research tools and can help to understand many aspects of carbohydrate metabolism [61].

A structural basis of inhibition of mammalian glycosidases has been deduced by *Asano* and coworkers when testing **1**, **2**, and **5** amongst other iminoalditols against several of these enzymes. *DMDP* was shown not to be an inhibitor of ER α-glucosidase II [22], whereas it is known to inhibit α-glucosidase I in cell cultures causing the accumulation of glycoproteins with high-mannose oligosaccharide structures and, thus, is a specific inhibitor of this enzyme [66].

The ability of *DMDP* and related iminoalditols to inhibit the hydrolysis of the glucoinolates sinigrin and progitin, which are thioglucosides present in the cruciferae having a variety of biological properties such as insect antifeedant activity, by inhibition of thioglucosidase has been shown by *Scofield et al.* [62]. Additionally, the protein processing glucosidase I from mung bean seedings was also shown to be inhibited by *DMDP* ($IC_{50} = 40\,\mu M$), **2** ($IC_{50} = 5$–$7\,\mu M$) and **3** ($IC_{50} = 2$–$3\,\mu M$) exhibited stronger activities [63].

Early investigations have shown that *DMDP* is an inhibitor of invertase (β-D-fructofuranosidase), an enzyme which is known to act by a mechanism analogous to that of glucosidases and that hydrolyses sucrose. *DMDP* showed 50% inhibition at about $1.5\,\mu M$ for the invertase at a *pH* above 6.5 [32]. *DMDP* also inhibits *PFP* (pyrophosphate-D-fructose-6-phosphate-1-phosphotransferase), an enzyme in plants, which is activated by fructose 2,6-diphosphate and catalyzes a reversible reaction at a crucial point in the synthesis of sucrose in young tissues and the utilization of this sugar in meristems. Because *DMDP* inhibits *PFP* to an extent of

48% in the forward direction and 52% in the reverse direction, it was expected to be useful in the bio-rational design of herbicides [41].

Anti-viral activity

Anti-HIV activity of *DMDP* is caused by inhibition of α-glucosidase I, an enzyme involved in the processing of N-linked olidosaccharides on glycoproteins. The inhibitory effect is weak when compared to the value of **2** and interestingly, contrasting **2**, N-alkylation does not lead to any increase in its potency [64]. Tested on influenza-infected MDCK cells, *DMDP* inhibited the processing of the viral hemagglutin by inhibiting α-glucosidase I [65] and inhibited at a concentrations of 250 μM α-glucosidase I to accumulate glycopeptides exhibiting Glc_3-Man_9-($Glac$NAc)$_2$ as the major oligosaccharide [66].

Effects on insects

Very early, *DMDP* has been found to be toxic to insects. The potential role as a plant protective chemical against a range of phytophagous insects was investigated by examining the effects on development, feeding behaviour, and functioning of taste receptors, and it could be shown that *DMDP* is an effective anti-feedant [67]. Furthermore, *DMDP* was found to be active against digestive enzymes from larvae of *Callosobruchus maculatus*, making it an interesting lead substance in the search for new insecticidal compounds [68]. IC_{50}-values with these enzymes were 0.6 micromolar against α-glucosidase and 340 micromolar against β-glucosidase, showing its lethal activity to the larvae of the bruchid beetle. The antifeedant activities against other insects such as the larvae of the lepidopterans *Spodoptera littoralis*, *Spodoptera frugiperda*, *Heliothis virescens*, and *Helicoverpa armigera* turned out to be consistently effective [69].

Several other insect species such as the orders *Orthoptera*, *Phasmida*, *Dictyoptera*, *Diptera*, and *Cleoptera* were screened for their inhibition of glycosidase activity by *DMDP* as well as other polyhydroxylated alkaloids. *DMDP* inhibited both α-glucosidase, including trehalase, and β-glucosidase activity strongly in these insects [70a]. Differentiation of glycosidase activity in some *Hemiptera* and *Lepidoptera* was also examined. The data obtained suggest that several separate enzymes or active sites are responsible for maltose and sucrose hydrolysis in aphids and *H. melpomone*, a species of larval *Lepidoptera* [70b]. Additionally, *DMDP* shows also potent antifeedant activity against important pest insects [71].

Activities of derivatives of DMDP

Related natural products bearing modifications at position 6 such as compounds **7** and **8** were found in various Hyacinthaceae and were tested against several glycosidases (Table 2). These studies were used for the explanation of the symptoms of poisoning live stock by bluebells [24].

Iminosugars **58** and its O-1-butyrate were believed to mimic the transition state stabilized by glycoprocessing enzymes and were tested against a few glycosidases

Table 3. Activities of **58** and **140**

Enzyme	*pH*	*DMDP*	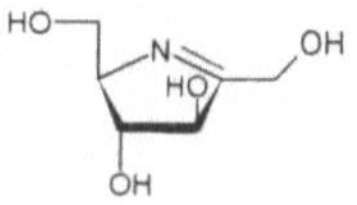 **58**	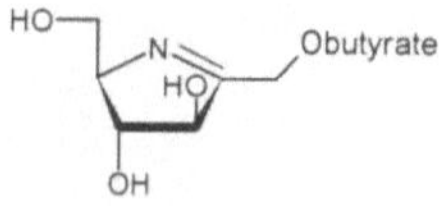	Ref.
β-Glucosidase (almonds)	6.8	$K_i = 7.8$	$K_i = 13 \pm 0.7$	NI	[45]
α-Glucosidase (brewer's yeast)	6.8	$K_i = 3.3$	$K_i = 2.6 \pm 0.2$	$K_i = 167 \pm 29$	[45]
β-Galactosidase (*E. coli*)	6.8	NI	$K_i = 276 \pm 51$	$K_i = 253 \pm 44$	[45]
α-Mannosidase (jack beans)	6.8	NI	$K_i = 17 \pm 1$	NI	[45]
α-Fucosidase (bovine epidimysis)	6.8	NI	$K_i = 381 \pm 71$	$K_i = 336 \pm 54$	[45]

(Table 3). Most notable was the inhibition of **58** against α-mannosidase, as *DMDP* itself does not show any activity against this enzyme at all [45].

Information about the active site of invertase (β-fructofuranosidase), the plant enzyme that splits sucrose into D-glucose and D-fructose, was obtained from the N-methyl and N,N-dimethyl derivatives **139** and **140** (Fig. 7) as well as with the N-(carboxypentyl) derivative **34** (K_i values of **139**, **140**, and **34** were 20, 130, and 50 μM, respectively, at $pH = 6$). The reduced biological activities of the N-alkyl derivatives as compared to parent compound **1** indicate steric clashes of the N-substituents with the active site. The *pH* dependence of the K_i values as well as the moderate impairment of inhibitory potency places yeast invertase among the small group of glycosidases which are strongly inhibited by cationic rather than basic glycon analogues [37]. Interestingly, *DMDP* had virtually no effect on the enzyme from the mammalian gut which catalyzes the same reaction [72].

The 1-acetamido derivative **65** was found to be a potent inhibitor of N-acetyl-glucosaminidases with K_i values of 9.8 and 1.9 μM for β-N-acetylglucosaminidase from bovine kidney and jack beans, respectively, at *pH* 6.5 [46] but exhibited only weak activity against α-glucosidase from *Saccharomyces sp.* ($K_i = 380\,\mu M$ at *pH* 7) [48] as would have been predictable from its structure.

Recently, **65** and its N-methyl derivative **79** were suggested as new drug leads for the therapy of osteroarthritis. This disease is characterized by a decreased concentration of glycosaminoglycans in articular cartilage, the degradation being mainly caused by hexosaminidases. This findings might open new applications of *DMDP* and its derivatives as drugs with chondroprotective activity [73].

Interestingly, the homologous N-acetamido compound **85** as well as the corresponding N-alkylated derivatives **83** and **84** showed activities against β-glucosidase from sweet almonds (K_i values of 2.2, 45, and 120 μM, respectively, at *pH* 5) [48]. The dimeric form of *DMDP* (**86**) exhibited K_i values of 53 μM against α-glucosidase from *Saccaromyces sp.* (*pH* 7) and 37 μM against β-glucosidase from sweet almonds (*pH* 5) [48].

Good glucosidase inhibitory activities were found for C-1 modified analogues of *DMDP*. The 1-deoxyfluoro, 1-O-methyl, and 1-aminodeoxy derivatives **108**,

Table 4. Activities of C-1 derivatives of *DMDP* against β-glucosidase *Agrobacterium sp.*, $pH = 7.0$

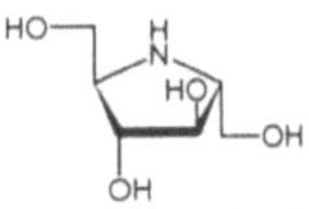 $K_i = 0.2$ [54] **1**	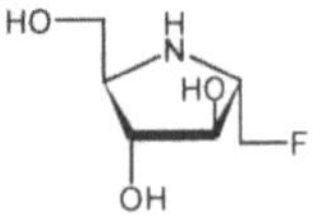 $K_i = 30$ [54] **108**	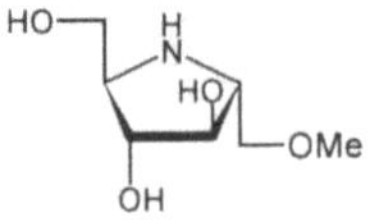 $K_i = 10$ [54] **116**	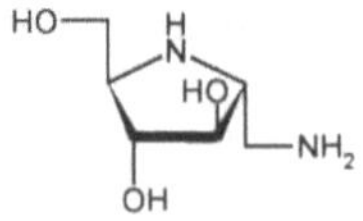 $K_i = 25$ [58] **118**
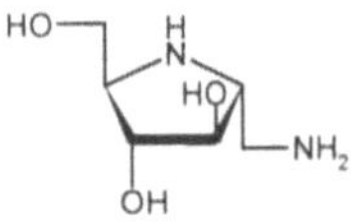 $K_i = 25$ [55] **118**	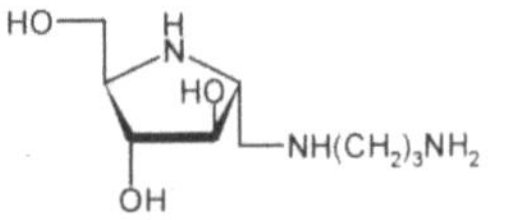 $K_i = 15$ [58] **122**	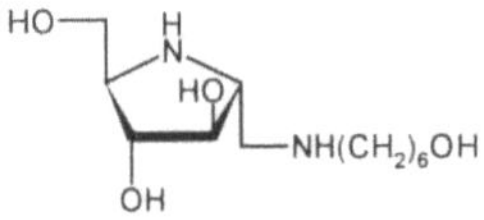 $K_i = 10$ [58] **120**	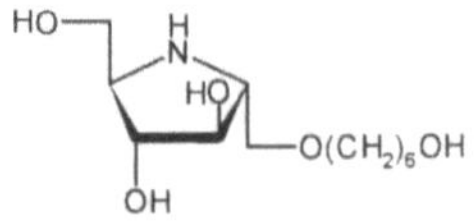 $K_i = 0.3$ [58] **127**
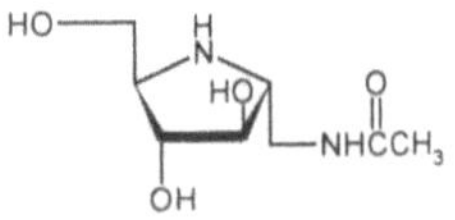 $K_i = 25$ [58] **65**	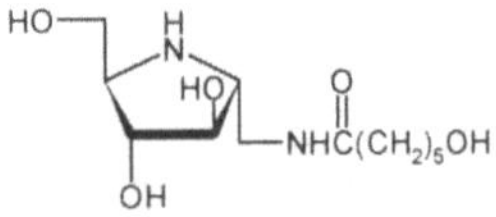 $K_i = 1.2$ [58] **123**	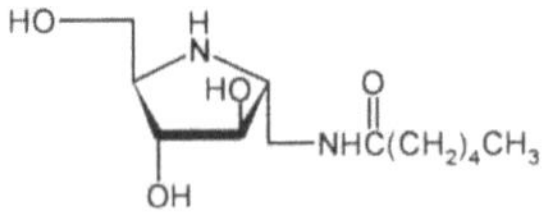 $K_i = 0.1$ [58] **124**	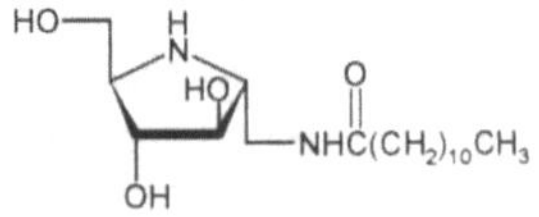 $K_i = 0.1$ [59] **125**
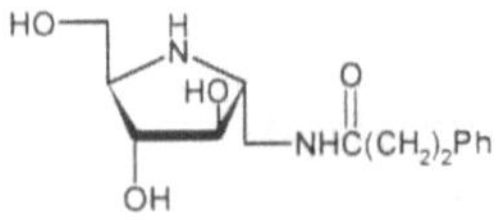 $K_i = 0.150$ [59] **128**	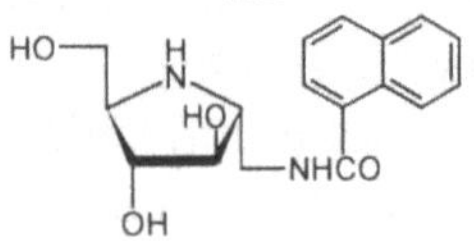 $K_i = 0.550$ [59] **129**	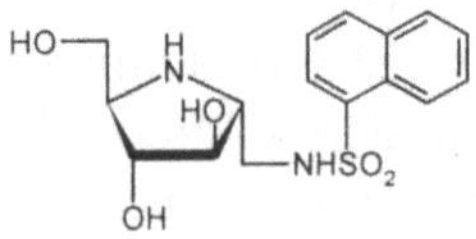 $K_i = 0.100$ [59] **133**	
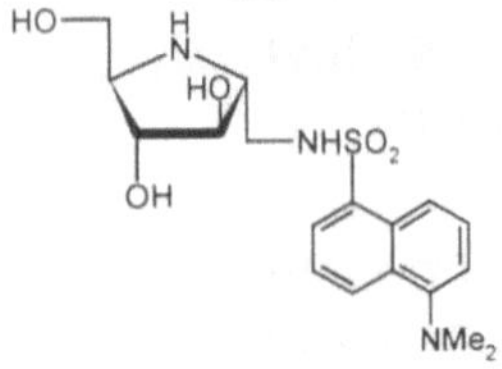 $K_i = 0.0024$ [60] **135**	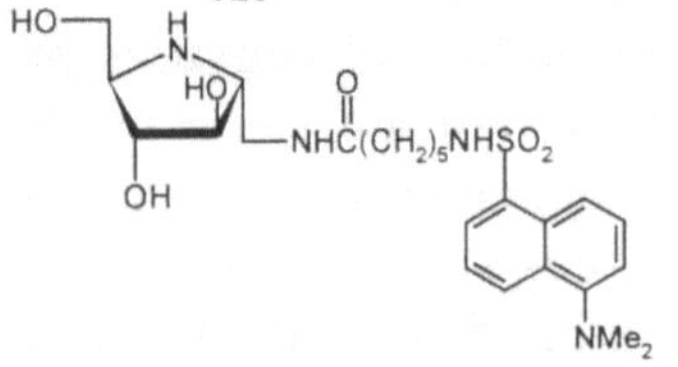 $K_i = 0.0021$ [60] **136**	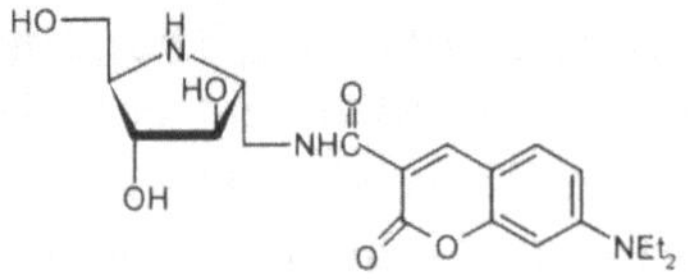 $K_i = 0.0012$ [59] **137**	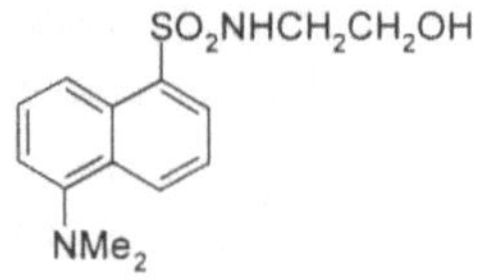 $K_i = 420$ [60]

116, and **118** exhibited moderate activities against β-glucosidase from *Argobacterium sp.* in the μM range (Table 4). Triamine **122** and the 1-N-hydroxyhexyl derivative **120** were similarly active. Surprisingly, all 1-N-acylamino derivatives such as **123**, **124**, and **125** showed K_i values comparable to that of **1**. Improved activities were detected with analogues bearing aromatic amides such as **128**, the corresponding sulfonamide **133**, and **129**. Fluorescently tagged compounds **135**, **136**, and **137** were found to be active in the low nanomolar range. This is noteworthy as only a few examples of related natural products and analogues, such as, for example, 1-deoxygalactonojirimycin (**141**, $K_i = 2\,\text{n}M$ with α-D-galactosidase from human placenta), 2-acetamido-1,2-dideoxygalactonijirimycin (**142**, $K_i = 1.2\,\text{n}M$ with N-Ac-β-D-glucosaminidase from jack beans), and 1-deoxyfuconojirimycin (**143** $K_i = 40\,\text{p}M$ with canine α-L-fucosidase) have been reported to show similar activities [8] (Fig. 8).

Exploitation of the fluorescence properties of compound **135** has shown that this inhibitor is bound as a 1:1-complex with the enzyme. The experiment was conducted by quenching the intrinsic tryptophane fluorescence of a defined amount

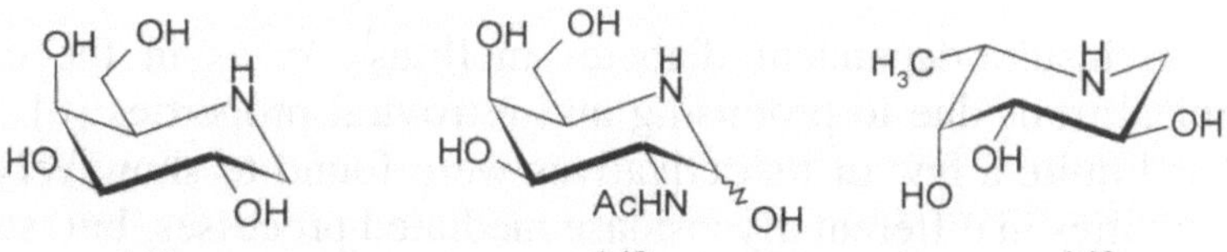

Fig. 8. Structures of compounds **141–143**

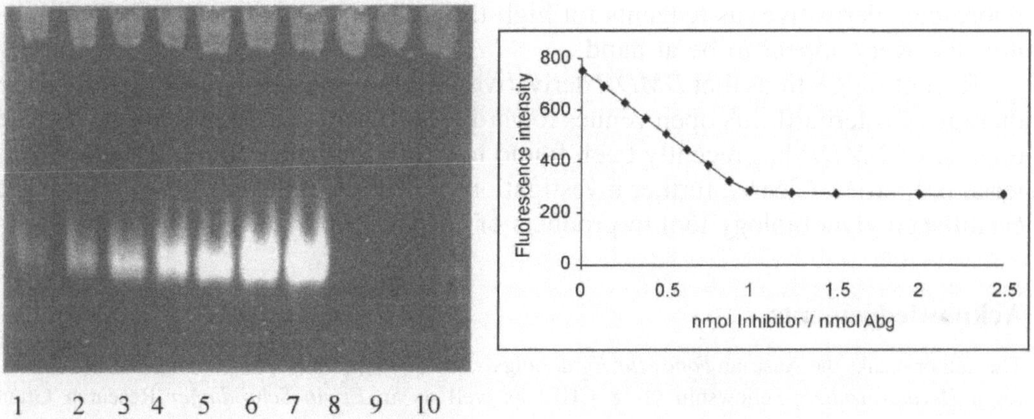

Fig. 9. Visualization of the fluoresent complex of **136** with *Agrobacterium sp.* β-glucosidase on native page and quenching of the intrinsic tryptophan fluorescence of *Agrobacterium sp.* β-glucosidase by titration with **136**

of *Agrobacterium sp.* β-glucosidase by titration with compound **135** (Fig. 9). The strong binding to the enzyme could be visualized very nicely on native PAGE gel with different amounts of enzyme incubated with **135** (Fig. 9, lanes 1–7), whereas other enzymes such as galactosidase, glucose isomerase, or albumin did not show binding to the inhibitor (lanes 8–10) [60].

From the studies mentioned above it can be concluded that the presence of aromatic substituents causes a dramatic increase of the inhibitory activities, resulting in K_i values in the nanomolar range. Consequently, it would be very interesting to test derivative **101** [51] as well as the corresponding allyl derivative [74].

Conclusions

Some well-known glycosidase inhibitors such as miglitol (**144**) and N-butyldeoxynojirimycin (**145**, Fig. 10) have been marketed for the treatment of non-insulin-

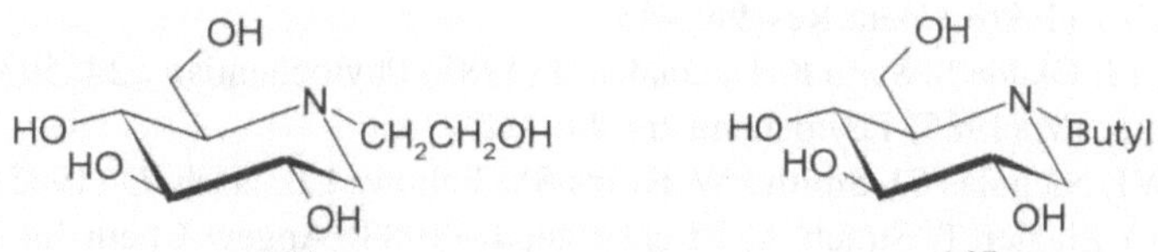

Fig. 10. Structures of compounds **144** and **145**

dependent and insulin-dependent diabetes mellitus, or, as in the case of **145**, undergo clinical trials due to promising anti-retroviral properties [8].

DMDP and quite a few of its derivatives were found to show very interesting biological activities in different glycosidase mediated processes, but, so far, despite many promising leads and fascinating biological properties, no commercial application has been reported. Nevertheless, compound **1** and some of its analogues have been shown to be powerful tools for glycobiology research. For example, purification of glycosidases by affinity chromatography employing immobilized **1** or fluorescent derivatives as reagents for high-throughput assays for enzyme or inhibitor discovery appear to be at hand.

Recent suggestions that *DMDP* derivatives might be new drug candidates for the therapy of osteroarthritis open venues towards pharmaceutical exploitation. A close relative of *DMDP* has recently been found to exhibit, albeit limited, antimycobacterial properties. Clearly, further investigations will show if the current gap from the established glycobiology tool to products of commercial value can be bridged.

Acknowledgements

The author thanks the Austrian *Fonds zur Förderung der wissenschaftlichen Forschung* (FWF), Vienna, for a *Hertha-Firnberg*-Fellowship (T-18 CHE) as well as an *Erwin Schrödinger* Research Grant (J 2092).

References

[1] Welter A, Jadot J, Dardenne G, Marlier M, Casimir J (1976) Phytochemistry **5**: 747
[2] Lamotte-Brasseur PJ, Dupont L, Dideberg O (1977) Acta Cryst **B33**: 409
[3] Paulesen H (1966) Angew Chem Int Ed Engl **5**: 405
[4] Liebig J, Wohler F (1837) Ann Pharmazie **21**: 96
[5] Legler G (1990) Adv Carbohydr Chem Biochem **48**: 319
[6] Sinnott ML (1990) Chem Rev **90**: 1171
[7] Withers SG, Namchuk M, Mosi R (1999) Potent Glycosidase Inhibitors: Transition State Mimics or Simple Fortuitous Binders? In: Stütz AE (ed) Iminosugars as Glycosidase Inhibitors. Wiley-VCH, Weinheim, p 188
[8] a) Stütz AE (ed) (1999) Iminosugars as Glycosidase Inhibitors. Wiley-VCH, Weinheim; b) Asano N, Nash RJ, Molyneux RJ, Fleet GWJ (2000) Tetrahedron Assymm **11**: 1645
[9] Paulsen H, Sangster I, Heyns K (1967) Chem Ber **100**: 802
[10] Schmist DD, Frommer W, Müller L, Truscheit E (1979) Naturwissenschaften **66**: 584
[11] Murao S, Miyata S (1980) Agric Biol Chem **44**: 219
[12] Hohenschutz LD, Bell EA, Jewess PJ, Leworthy DP, Pryce RJ, Arnold E, Clardy J (1981) Phytochemistry **20**: 811
[13] Molyneux J, Benson M, Wong RY, Tropea JE, Elbein AD (1988) J Nat Prod **51**: 1198
[14] a) Look GC, Fotsch CH, Wong C-H (1993) Acc Chem Res **26**: 182; b) Gijsen HJM, Qiao L, Fitz W, Wong C-H (1996) Chem Rev **96**: 443
[15] a) Furukawa J, Okuda S, Saito K, Hatanaka SI (1985) Phytochemistry **24**: 593; b) Nash RJ, Bell EA, Williams JM (1985) Phytochemistry **24**: 1620
[16] a) Fleet GWJ, Nicholas SJ, Smith PW, Evans SV, Fellows LE, Nash RJ (1985) Tetrahedron Lett **26**: 3127; b) Ziegler T, Straub A, Effenberger F (1988) Angew Chem Int Ed Engl **27**: 716; c) von der Osten CH, Sinskey AJ, Barbas CF, Pederson RL, Wang Y-F, Wong C-H (1989) J Am Chem Soc **111**: 3924

[17] Molyneux RJ, Pan YT, Tropea JE, Elbein AD, Lawyer CH, Hughes DJ, Fleet GWJ (1993) J Nat Prod **56**: 1356
[18] Asano N, Nishida M, Miyauchi M, Ikeda K, Yamamoto M, Kizu H, Kameda Y, Watson AA, Nash RJ, Fleet GWJ (2000) Phytochemistry **53**: 379
[19] a) Shibano M, Kitagawa S, Kusano G (1997) Chem Pharm Bull **45**: 505; b) Yoda H, Shimojo T, Takabe K (1999) Tetrahedron Lett **40**: 1335; c) Shibano M, Tsukamoto D, Fujimoto R, Masui Y, Sugimoto H, Kusano G (2000) Chem Pharm Bull **48**: 1281; d) Sprenger FK (Fitz) (2001) Diploma Thesis, Technical University, Graz
[20] Evans SV, Fellows LE, Shing TKM, Fleet GWJ (1985) Phytochemistry **24**: 1953
[21] a) Horn JM, Lees DC, Smith NG, Nash RJ, Fellows LE, Bell EA (1987) In: Labeyrie V, Fabres G, Lachaise D (eds) Proc 6th Symp Plant-Insect Relationships, p 394; b) Kite GC, Fellows LE, Fleet GWJ, Liu PS, Scofield AM, Smith NG (1988) Tetrahedron Lett **29**: 6483
[22] Asano N, Oseki K, Kizu H, Matsui K (1994) J Med Chem **37**: 3701
[23] Asano N, Nishida M, Kizu H, Matsui K (1997) J Nat Prod **60**: 98
[24] a) Watson AA, Nash RJ, Wormald MR, Harvey DJ, Dealler S, Lees E, Asano N, Kizu H, Kato A, Griffiths RC, Cairns AJ, Fleet GWJ (1997) Phytochemistry **46**: 255; b) Asano N, Kato A, Miyauchi M, Kizu H, Kameda Y, Watson AA, Nash RJ, Fleet GWJ (1998) J Nat Prod **61**: 625; c) Kato A, Adachi I, Miyauchi M, Ikeda K, Komae T, Kizu H, Kameda Y, Watson AA, Nash RJ, Wormald MR, Fleet GWJ, Asano N (1999) Carbohydr Res **316**: 95
[25] Ikeda K, Takahashi M, Nishida M, Miyauchi M, Kizu H, Kameda Y, Arisawa M, Watson AA, Nash RJ, Fleet GWJ, Asano N (2000) Carbohydr Red **323**: 73
[26] Kite GC, Sharp HJ, Hill PS, Boyce PC (1997) Biochem Syst Ecol **25**: 757
[27] Kim HS, Kim YH, Hong YS, Peak NS, Lee HS, Kim TH, Kim KW, Lee JJ (1999) Planta Medica **65**: 437
[28] Kite GC, Horn JM, Romeo JT, Fellows LE, Lees DC, Scofield AM, Smith NG (1990) Phytochemistry **29**: 103
[29] Kite GC, Fellows LE, Lees DC, Kitchen D, Monteith GB (1991) Biochem Syst Ecol **19**: 441
[30] Watanabe S, Kato H, Nagayama K, Abe H (1995) Biosci Biotech Biochem **59**: 936
[31] a) Fleet GWJ, Smith PW (1985) Tetrahedron Lett **26**: 1469; b) Fleet GWJ, Smith PW (1987) Tetrahedron **43**: 971
[32] Card PJ, Hitz WD (1985) J Org Chem **50**: 891
[33] a) Reitz AB, Baxter EW (1990) Tetrahedron Lett **31**: 6777; b) Baxter EW, Reitz AB (1994) J Org Chem **59**: 3175
[34] Liu KK-C, Kajimoto T, Chen L, Zhong Z, Ichikawa Y, Wong C-H (1991) J Org Chem **56**: 6280
[35] Hung RR, Straub JA, Whitesides GM (1991) J Org Chem **56**: 3849
[36] Duréault A, Portal M, Depezay JC (1991) Synlett **4**: 225
[37] Legler G, Korth A, Berger A, Ekhart C, Gradnig G, Stütz AE (1993) Carbohydr Res **250**: 67
[38] Fechter MH, Gradnig G, Berger A, Mirtl C, Schmid W, Stütz AE (1998) Carbohydr Res **309**: 367
[39] Legler G, Liedtke H (1995) Biol Chem Hoppe-Seyler **366**: 1117
[40] Hettkamp H, Legler G, Bause R (1984) Eur J Biochem **142**: 85
[41] Chorghade MS, Cseke CT, Liu PS (1994) Tetrahedron Asymm **11**: 2251
[42] Park KH, Yoon YJ, Lee SG (1994) Tetrahedron Lett **35**: 9737
[43] Ikota N (1995) Heterocycles **41**: 983
[44] a) Poitout L, Merrer YL, Depezay J-C (1996) Tetrahedron Lett **37**: 1609; b) Merrer YL, Poitout L, Depezay J-C, Dosbaa I, Geoffroy S, Foglietti M-J (1997) Bioorg Med Chem **5**: 519
[45] Takayama S, Martin R, Wu J, Laslo K, Siuzdak G, Wong C-W (1997) J Am Chem Soc **119**: 8146
[46] Takaoka Y, Kajimoto T, Wong C-H (1993) J Org Chem **58**: 4809
[47] Hiranuma S, Shimizu T, Nakata T, Kajimoto T, Wong C-H (1995) Tetrahedron Lett **36**: 8247
[48] Takebayashi M, Hiranuma S, Kanie Y, Kajimoto T, Kanie O, Wong C-H (1999) J Org Chem **64**: 5280
[49] Behr J-B, Evina CM, Phung N, Guillerm G (1997) J Chem Soc Perkin Trans 1, 1597

[50] Bouix C, Bisseret P, Eustache J (1998) Tetrahedron Lett **39**: 825
[51] Oliveria DF, Severino EA, Correia CRD (1999) Tetrahedron Lett **40**: 2083
[52] Scott DA, Krülle TM, Finn M, Nash RJ, Winters AL, Asano N, Butters TD, Fleet GWJ (1999) Tetrahedron Lett **40**: 7581
[53] Dax K, Ebner M, Peinsipp R, Stütz AE (1997) Tetrahedron Lett **38**: 225
[54] Andersen SM, Ebner M, Ekhart CW, Gradnig G, Legler G, Lundt I, Stütz AE, Withers SG, Wrodnigg TM (1997) Carbohydr Res **301**: 155
[55] Wrodnigg TM, Stütz AE, Withers SG (1997) Tetrahedron Lett **38**: 5463
[56] a) Berger A, de Raadt A, Gradnig G, Grasser M, Löw H, Stütz AE (1992) Tetrahedron Lett **33**: 7125; b) Stütz AE (2001) *D*-Xylose (*D*-Glucose) Isomerase and Related Enzymes in Carbohydrate Synthesis. Top Curr Chem 215; Stütz AE (ed) Glycoscience – Epimerization, Isomerization, Rearrangement Reactions of Carbohydrates. Springer, Heidelberg, p 77
[57] a) Hodge JE (1955) Adv. Carbohydr Chem **10**: 169 b) Wrodnigg TM, Eder B (2001) The Amadori and Heyns Rearrangements: Landmarks in the History of Carbohydrate Chemistry or Unrecognized Synthetic Opportunities? In: Stütz AE (ed) Glycoscience – Epimerization, Isomerization, Rearrangement Reactions of Carbohydrates. Springer, Heidelberg, p 115
[58] Wrodnigg TM, Gaderbauer W, Greimel P, Häusler H, Sprenger FK (Fitz), Stütz AE, Virgona C, Withers SG (2000) J Carbohydr Chem **19**: 975
[59] Wrodnigg TM, Withers SG, Stütz AE (2001) Bioorg Med Chem Lett **11**: 1063
[60] Hermetter A, Scholze H, Stütz AE, Withers SG, Wrodnigg TM (2001) Bioorg Med Chem Lett **11**: 1339
[61] Scofield AM, Fellows LE, Nash RJ, Fleet GWJ (1986) Life Science **39**: 645
[62] Scofield AM, Rossiter JT, Witham P, Kite GC, Nash RJ, Fellows LE (1990) Phytochemistry **29**: 107
[63] Szumlio T, Kaushal GP, Elbein AD (1986) Arch Biochem Biophys **247**: 261
[64] a) Tyms AS, Berrie EM, Ryder TA, Nash RJ, Hegarty MP, Taylor DL, Mobberley MA, Davis JM, Bell EA, Jeffries DJ, Taylor-Robinson D, Fellows LE (1987) Lancet **2**: 1025, b) Fleet GWJ, Karpas A, Dwek RA, Fellows LE, Tyms AS, Petursson S, Namgoong SK, Ramsden NG, Smith PW, Son JC, Wilson F, Witty DR, Jacob GS, Rademachern TW (1988) FEBS Letters **237**: 128, c) Asano N, Kizu H, Oseki K, Tomoaka E, Matsui K, Okamoto M, Baba M (1995) J Med Chem **38**: 2349
[65] a) Schwarz PM, Elbein AD (1985) J Biol Chem **260**: 14452
[66] Elbein AD, Mitchell M, Sanford BA, Fellows LE, Evans SV (1984) J Biol Chem **259**: 12409
[67] Blaney WM, Simmonds MSJ, Evans SV, Fellows LE (1984) Entomol Exp Appl **36**: 209
[68] Evans SV, Gatehouse AMR, Fellows LE (1985) Entomol Exp Appl **37**: 257
[69] Simmonds MSJ, Blaney WM, Fellows LE (1990) J Chem Ecol **16**: 3167
[70] a) Scofield AM, Witham P, Nash RJ, Kite GC, Fellows LE (1995) Comp Biochem Physiol **112A**: 187; b) Scofield AM, Witham P, Nash RJ, Kite GC, Fellows LE (1995) Comp Biochem Physiol **112A**: 197
[71] Fellows LE, Evans SV, Nash RJ, Bell EA (1986) Polyhydroxy-alkaloids as glycosidase inhibitors and their possible ecological role In: Green MB, Hedin PA (eds) Natural Resistance of Plants to Pests, Roles of Allelochemicals. American Chemical Society, Washington, DC, p 72
[72] Fellows LE (1987) Chem Brit 842
[73] Liu J, Shikham AR, Lotz MK, Wong C-H (2001) Chem Biol **8**: 701
[74] de Armas P, Garcia-Tellado F, Marrero-Tallado JJ (2000) Org Lett **2**: 313

Received November 14, 2001. Accepted November 19, 2001

Rearrangements in the Course of Fluorination by Diethylaminosulfur Trifluoride at C-2 of Glycopyranosides: Some New Parameters

Karl Dax*, **Martin Albert**, **David Hammond**, **Carina Illaszewicz**, **Thomas Purkarthofer**, **Martin Tscherner**, and **Hansjörg Weber**

Institut für Organische Chemie, Technische Universität Graz, A-8010 Graz, Österreich

Summary. Reaction of a series of 21 glycosides unprotected at O-2 and featuring various configurations with *DAST* (diethylaminosulfur trifluoride) was monitored by ^{19}F NMR spectroscopy. By means of the diacritical set of data (shift values and coupling constants) thus obtained for each product, identification of the operative mechanisms was possible. By correlation of these findings with stereochemical details from the structure of the educts, new parameters governing the choice of the reaction paths could be deduced. This evaluation led to the result that ring contraction after attack at C-2 of the ring oxygen and entry of the fluoride at C-1 is strongly favoured over all other possibilities. Exceptions are all derivatives of the manno series as well as all members of the *trans*-decalin type structure as present after 4,6-O-benzylidene acetal formation with educts having their ring substituents at C-4 and C-5 in diequatorial (*trans*) orientation. From these, α-D-mannopyranosides generally and of the *trans*-decalin types those with an additional 1,2-*trans*-configuration are prone to 1,2-aglycon migration and, again, entry of the fluoride at C-1. Additional pathways like alkoxy group migration, substitution under retention of configuration, or orthoester formation, are possible by participation of a suitably located neighbouring group at C-3 inasmuch as an alkoxy group interferes from an antiperiplanar orientation to the leaving group at C-2 and an acyloxy functionality attacks in a diequatorial relationship to the latter. The generally intended nucleophilic substitution by fluoride under inversion of configuration is of minor importance.

Keywords. Glycosides; Nucleophilic substitution; Fluorination; Rearrangements; Neighbour group participation; ^{19}F NMR spectroscopy; *DAST*.

Introduction

Within the carbohydrates, the so-called deoxyfluoro analogues which contain fluorine instead of a hydroxyl group are of special importance in biological studies [1]. Thus far, they have served, for example, as mechanistic probes to investigate receptor-substrate interactions, attachment sites of antibodies, specifity of enzymes, or metabolism and transport phenomena. For the stereoselective

---

* Corresponding author. E-mail: dax@orgc.tu-graz.ac.at

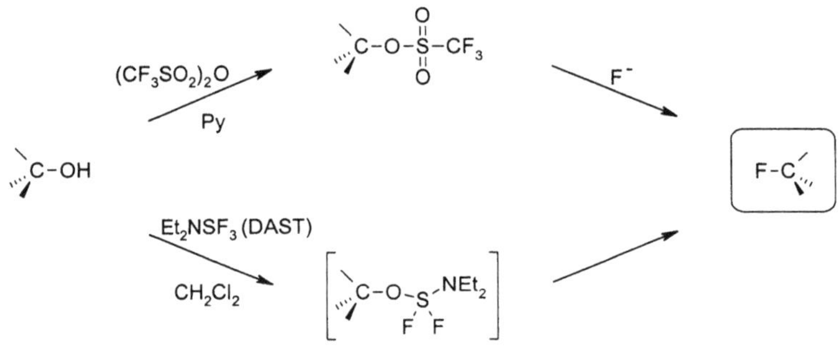

Scheme 1

chemical synthesis of these species, most frequently the principle of a nucleophilic displacement reaction with concomitant inversion of configuration (S_N2) is employed. This consists, as generally depicted in Scheme 1, either of the treatment of a sulfonate (trifluoromethanesulfonate, triflate) with fluoride anion or of the direct reaction of an unprotected alcohol with diethylaminosulfur trifluoride (*DAST*).

However, side reactions, many of them caused by neighbouring group (NG) participation, are often involved in such intended transformations [2–4]. Own unsuccessful efforts towards the synthesis of certain 2-deoxy-2-fluoro sugars applying the S_N2-strategy initiated attempts to categorize the set of hitherto observed reaction pathways in such nucleophilic displacement reactions for all non-glycosidic positions in carbohydrates [5]. From these, the structural prerequisites for the individual types of side reactions seemed evident:

i) Any vicinal alkoxy group oriented antiperiplanar to the leaving group (LG) (including that containing the ring oxygen) may, under inversion of configuration at the reaction centre, cause formation of an intermediate oxiranium ion (anchimeric assistance of the 'simple' type, Scheme 2). The oxiranium ion is opened, again under inversion of configuration, by the nucleophile at one of the carbons involved.

ii) A vicinal acyloxy group is able to interfere, with inversion of configuration, by attack of its carbonyl oxygen under formation of a cyclic acyloxonium ion ('complex' type, Scheme 3). The steric requirement is the diequatorial arrangement of both partners. The reaction is completed by attack of the external nucleophile at one of the carbons involved.

iii) Out of an antiperiplanar orientation relative to the LG, a hydrogen atom or an alkyl group (including parts of the sugar ring) may shift to the reaction centre with entry of the external nucleophile at the other end under inversion of configuration at both centres.

iv) As fluoride also may act as a base, in case of educts containing an axially oriented LG together with a vicinal *trans*-located hydrogen atom elimination (E) is often observed.

v) Fragmentation (F) may occur with the prerequisite that a well-stabilized (oxo)carbenium ion can be eliminated from the C-atom vicinal to that bearing the LG. This is usually the case with the anomeric centre in educts that contain the LG in equatorial position at C-3.

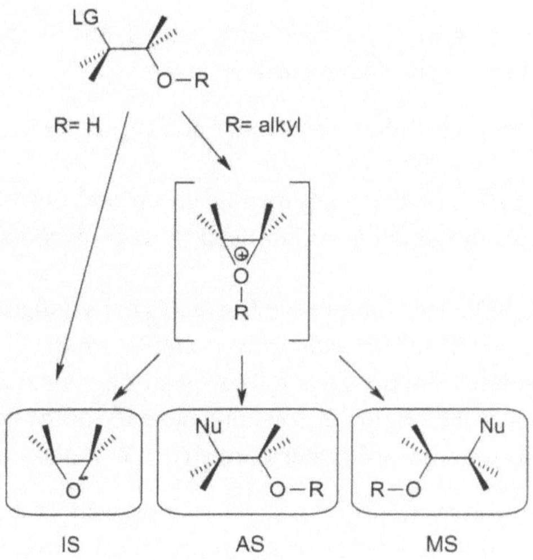

Scheme 2

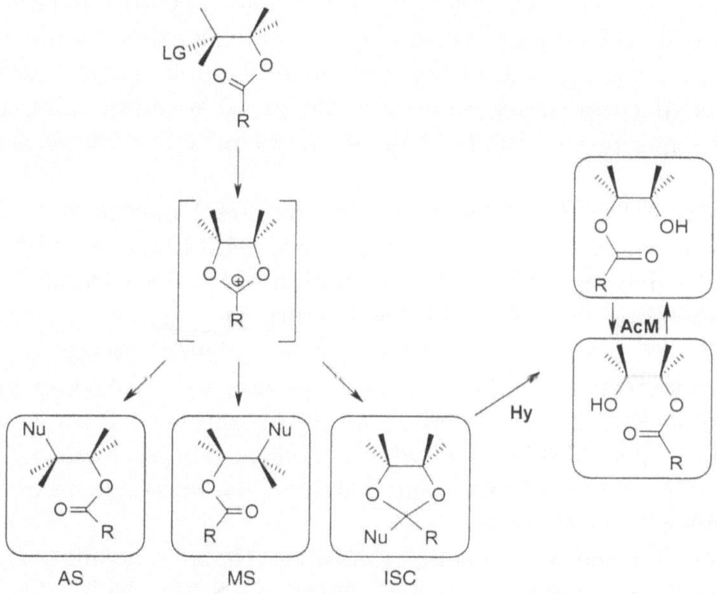

Scheme 3. Hy = hydrolysis, AcM = acyl migration

Generally it should be noted that mechanistic concertedness in the course of these reactions manifests itself by the stereochemical homogeneity of the product, whereas formation of structures effected by epimerization at one single centre points to the existence of intermediates; this is especially the case when the stabilized oxocarbenium ion containing C-1 is involved.

For the sake of clarity, within this account the following symbols [5] for the above cited mechanisms which involve neighbouring proup participation will be used:

AS: Anchimerically assisted substitution with retention of configuration at the position of the original activation

MS: Substitution by migration of the participating group and entry of the nucleophile (fluoride) at the position where the participating group departs with inversion of configuration at both centres

IS: Intramolecular nucleophilic substitution with inversion of configuration by a hetero atom of an NG under formation of a (small) heterocyclic ring ('simple' type)

ISC: Intramolecular substitution by the carbonyl oxygen of a vicinal acyl group under inversion of configuration at the position of the original activation and entry of the nucleophile at the carbon of the original carbonyl group to form an orthoester derivative ('complex' type)

In all instances, the position of activation in the sugar ring as well as the type of the participating group ('H' for hydrogen, 'NG' for any neighbouring group, 'rO' for ring oxygen) and its location in the educt are given in parentheses.

For the respective reactions at C-2 of glycopyranosides, the inventory of published data comprises the following reaction paths:

i) Straight S_N2-displacement is rather the exception than the rule; modest to good results are obtained with educts of β-D-manno- and β-D-gluco-configurations only. The disposition of the α-D-gluco-anomer to give this reaction is restricted to structures containing a 4,6-O-benzylidene protection (Table 1, entries 1–3).

ii) 2-Triflates of α-D-mannopyranosides are prone to elimination under formation of 2,3-unsaturated products upon treatment with fluoride anion (Table 1, entry 4).

iii) The corresponding *DAST*-reaction with the OH-2 unprotected α-D-mannopyranoside **1** causes 1,2-aglycon migration, MS(2,NG-1), with concomitant entry of the fluoride at C-1 under formation of the corresponding 2-O-methyl-D-glucopyranosyl fluorides **2** (Table 1, entry 5).

iv) Treatment of methyl 3-azido-4,6-O-benzylidene-3-deoxy-α-D-altropyranoside (**3**) with *DAST* (Table 1, entry 6) results in AS(2,NG-3 and/or NG-1) (substitution at C-2 with retention of configuration to form **4**) together with MS(2,NG-3) and MS(2,NG-1) which produce benzyl 2-azido-2,3-dideoxy-3-fluoro-α-D-glucopyranoside **5** and 3-azido-2-O-benzyl-3-deoxy-D-allopyranosyl fluorides **6**, respectively.

v) 2-Triflates of 6-deoxy-L-galactopyranosides (fucopyranosides, **7**), independent of the anomeric configuration, give ring contraction, MS(2,rO), to form 2,5-anhydro sugars **8** with inversion of configuration at C-2 and entry of fluorine at C-1 (Table 1, entry 7). This special type of side reaction is frequently observed with educts of galacto- and arabino-configurations when other nucleophiles than fluoride are applied.

In reactions of carbohydrate 2-triflates with other nucleophiles and in (mechanistically closely related) nitrous acid deaminations of 2-amino-2-deoxy sugars the following additional pathways have been observed:

i) Hydride shift, MS(2,H-3), occurred in the course of a diazotation reaction of methyl α-D-mannosaminide under formation of a 2-deoxy-3-ulose (Table 2, entry 1; this reaction formally corresponds to an elimination followed by *retro*-enolization). Additionally, aglycon migration took place.

ii) Alkyl shift, MS(2,C-4), was found with a nitrous acid deamination of methyl α- and β-D-glucosaminide to give a pentofuranose derivative with a formyl side

Table 1. Results from attempted fluorinations at C-2 of pyranosides (*TBAF* = tetrabutylammonium fluoride)

	Educt	Product(s)/Mechanism	Conditions	Yield/%	Ref.
1	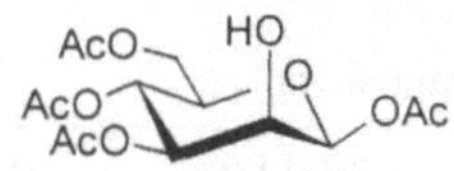	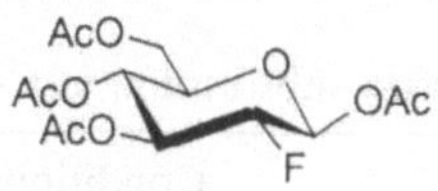 S_N2	*DAST* diglyme 100°C, 7 min	77	[6]
2	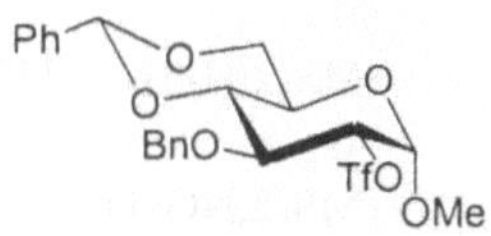	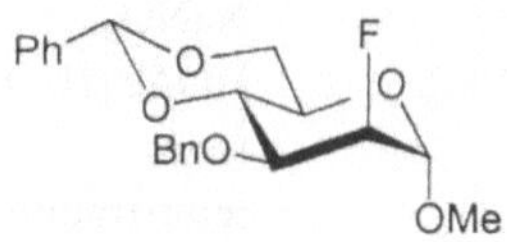 S_N2	*TBAF*, MeCN reflux, 2.5 h	30	[7]
3	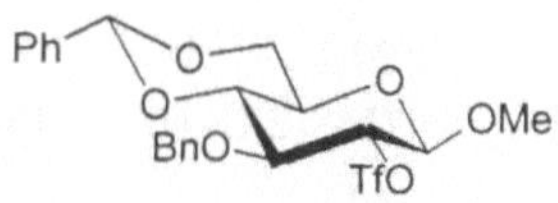	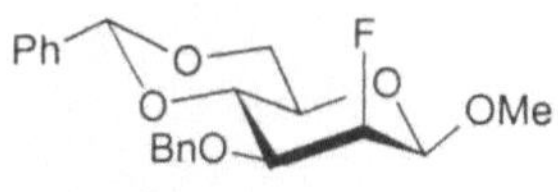 S_N2	*TBAF*, MeCN reflux, 80 min	77	[7]
4	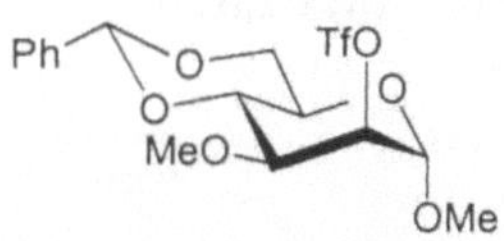	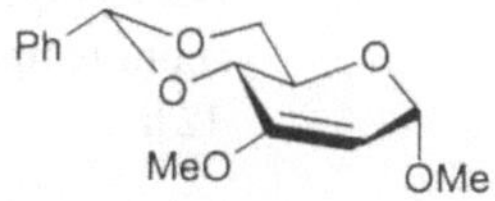 E	*TBAF*, MeCN reflux, 30 min	82	[8]
5	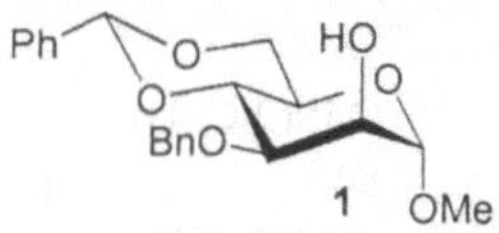 **1**	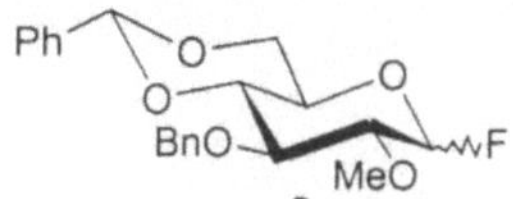 **2** MS(2,NG-1)	*DAST* diglyme 100°C, 30 min	76	[9]
		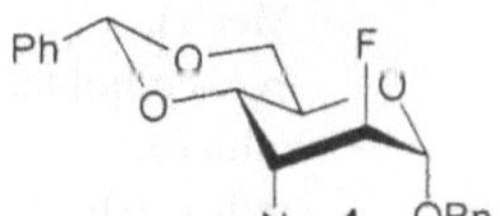 **4** AS(2,NG-1/3)		40 (AS)	
6	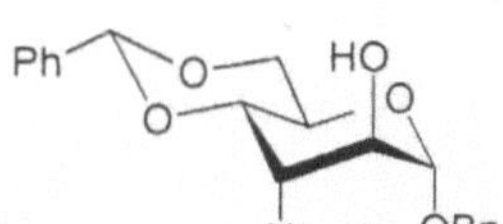 **3**	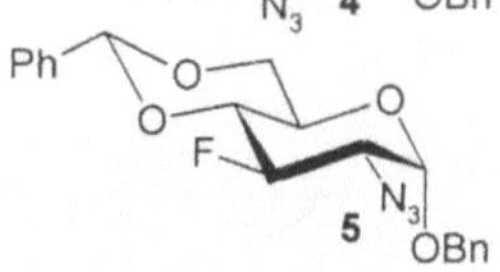 **5** MS(2,NG-3)	*DAST*, benzene reflux, 2 h	40 (MS(2,NG-3))	[10]
		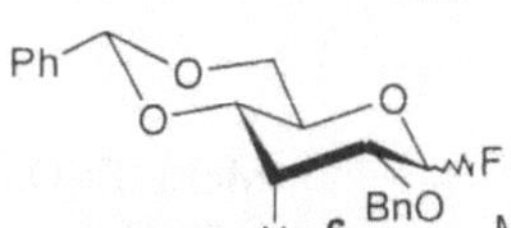 **6** MS(2,NG-1)		15 (MS(2,NG-1))	
7	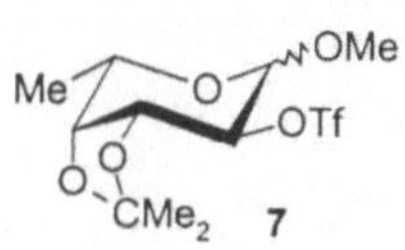 **7**	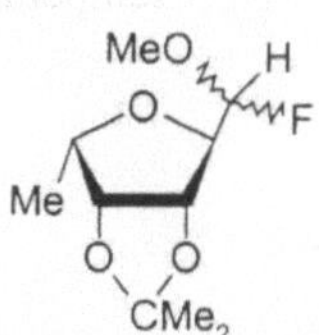 **8** MS(2,rO)	3HF · *TEA*, *TEA*, MeCN, room temperature, 16 h	60	[11]

chain (the former C-3) at C-2; this product was subject to epimerization at C-2 (Table 2, entries 2 and 3). The 'conventional' ring contraction, MS(2,rO) (compare Table 1, entry 7) was the main reaction.

iii) Orthoester formation, ISC(2,NG-3), resulted from a *trans*-diequatorial arrangement between the triflate and an acyloxy group at C-3 (Scheme 2, entry 4); the intermediate acyloxonium ion was partly hydrolyzed to a mixture of the respective 2- and 3-benzoate with inverted configuration at C-2.

Table 2. Further side reactions upon attempted nucleophilic substitution at C-2 of pyranosides

	Educt	Product(s)/Mechanism	Conditions	Yield/%	Ref.
1	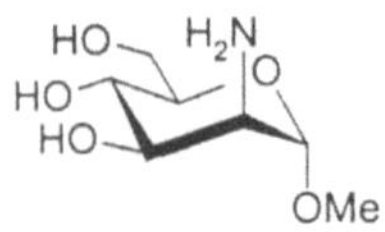	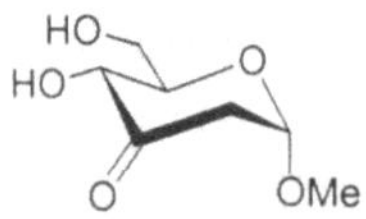 MS(2,H-3)	NaNO$_2$, HOAc/H$_2$O room temperature, 2 h	50 (MS(2,H-3)) 30 (MS(2,NG-1))	[12]
2	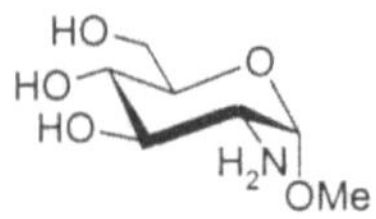	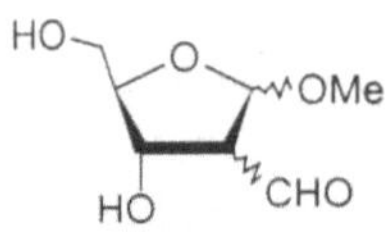 MS(2,C-4) + Hydrolysis	NaNO$_2$ ($pH = 3.5$) room temperature, 12 h	27 (MS(2,C-4)) 73 (MS(2,rO))	[13]
3	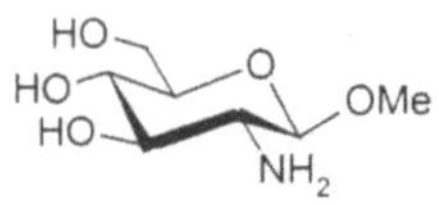			16 (MS(2,C-4)) 84 (MS(2,rO))	[13]
4	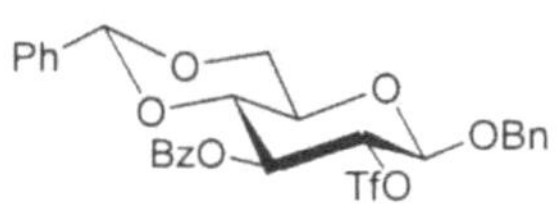	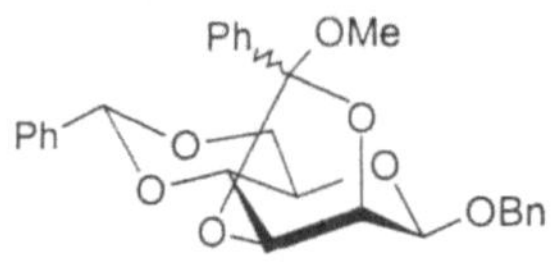 ISC	MeOH, *sym*-collidine, toluene, reflux, 6 h	90	[14]
5	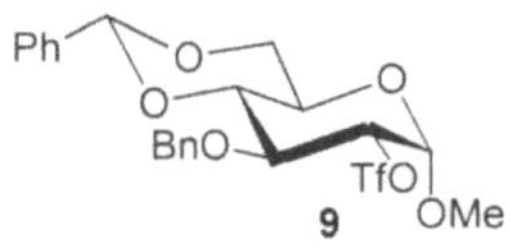 9	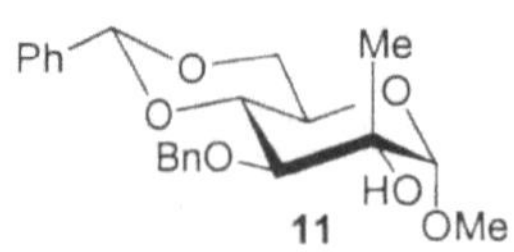 11	MeLi, Et$_2$O, $-50°C \rightarrow$ room temperature, 1 h	91	[15]
6	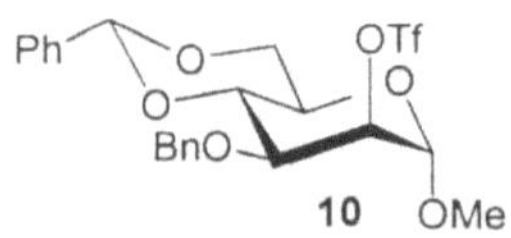 10			77	[15]

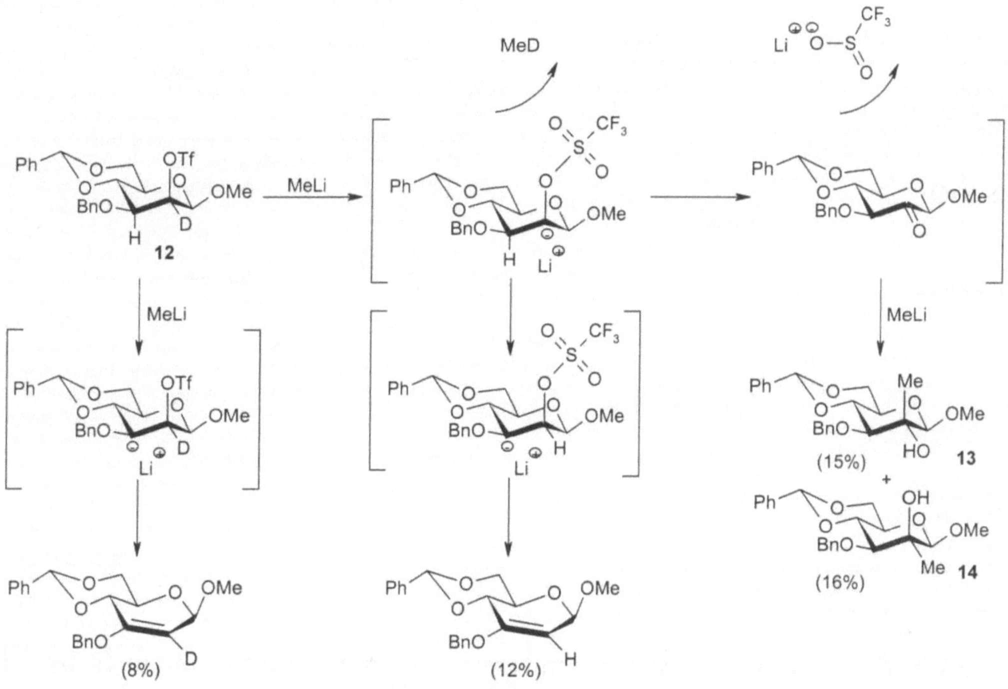

Scheme 4

iv) α-Hydrogen abstraction has recently been proven in experiments with deuterated educts by *El Nemr* and *Tsuchiya* [15] to be the initial step in the reaction of carbohydrate triflates with strong bases (alkyl lithium compounds). Thus, from the 2-triflates **9** and **10** of otherwise fully and identically protected α-D-gluco- and α-D-mannopyranoside, the same 2-C-alkyl derivative with D-gluco-configuration (**11**, Table 2, entries 5 and 6) was obtained. The whole set of products formed and the mechanisms involved for their formation is shown in Scheme 4 with the β-D-manno-educt **12**, where – besides unsaturated products – mixtures of 2-C-alkyl derivatives with D-gluco-(**13**) as well as D-manno-configuration (**14**) were formed due to the missing steric approach control in the final addition step.

Although this compilation sheds some light on the complex situation and even immediately allows the identification of a few seemingly contradictory results in the literature, a series of questions concerning the precise dependence on structural features has remained. Within this contribution we focus on the governing parameters for the most fascinating competitors, MS(2,rO) and MS(2,NG-1). Can these be made comprehensible when educts are used which – under the aspect of conformational flexibility – fulfill the stereochemical prerequisites for both possibilities?

As the collection of NMR data for all types of fluorinated reaction products obtained thus far in attempted syntheses by nucleophilic displacement reactions at C-2 had made available a set of diacritical ^{19}F NMR parameters (Table 3), the stage was set to conduct – under comparable conditions – a more

Table 3. Characteristic ^{19}F NMR data

Operative mechanism	Type of structure	δ/ppm	$^{2}J_{F,H}$/Hz
MS(2,rO)	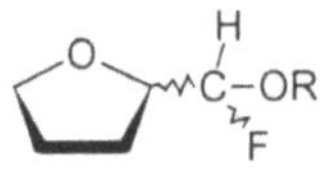	134–143	63–68
MS(2,NG-1)	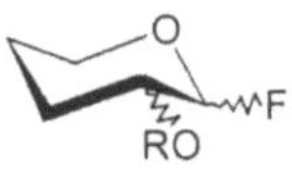	135–150	48–54
S_N2 or AS(2,NG-1/3)	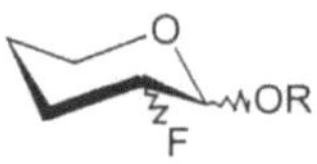	190–220	47–54
MS(2,NG-3)	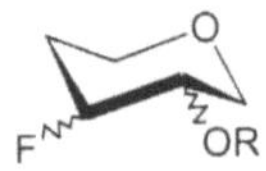		

comprehensive study simply by monitoring the crude reaction mixtures by ^{19}F NMR spectroscopy.

Results and Discussion

The first aim in these studies was to reexamine the reported (but by no means substantiated) 1,2-aglycon shifts, MS(2,NG-1), from *DAST*-treatment of 2-OH-unprotected educts of α-L-arabino- and β-D-galacto-configuration, respectively, containing a 3,4-O-isopropylidene group [16]. In contrast to this (and completely provable), the 2-sulfonates of similar compounds from the α-L- (Table 1, entry 7 [11]) as well as the β-D-galacto-series [17, 18] had been found to give ring contraction, MS(2,rO), on attempted nucleophilic displacements. When taking into account some conformational flexibility of these educt structures, the given 1,2-*trans*-orientation allows an antiperiplanar arrangement of LG and the respective migrating NG as the stereochemical prerequisite for both types of reactions. However, as can be seen from Table 4, entries 1 and 6, the *DAST*-treatment, under identical conditions as described [16] (in dichloromethane, 3 equivalents of *DAST*, room temperature), of benzyl 3,4-O-isopropylidene-α-D-arabinopyranoside (**15**, the enantiomer of the original educt [16]) as well as methyl 3,4-O-isopropylidene-6-O-trityl-β-D-galactopyranoside (**23**) caused MS(2,rO) only. In full agreement with previous findings [11, 17, 18] with sulfonate displacement reactions, the same pairs of ring-contracted products (but in differing individual ratios) were formed when the respective anomer (**16** or **22**; Table 4, entries 2 and 5) was reacted. MS(2,rO) was also exclusively observed when methyl 3,4-O-isopropylidene-β-D-arabinopyranoside (**18**, Table 4, entry 3) or its 2-triflate were treated with *DAST* and tetrabutylammonium fluoride, respectively. To elucidate any conformational barrier in the obviously clear-cut preference for MS(2,rO) over MS(2,NG-1), we turned to the (1,2-*trans*-configured) educts **20** (β-D-ribo) as well as **25** and **27** (α-D-altro) which – in comparison to the structures **15** and **23** – show inverted configurations

Table 4. Reactions with *DAST*

	Educt	Proposed structure of products	^{19}F NMR δ/ppm J/Hz	^{1}H NMR δ/ppm J/Hz	^{13}C NMR δ/ppm J/Hz	Operative mechanism
1	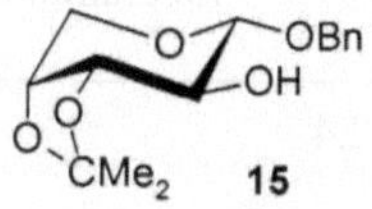 **15**	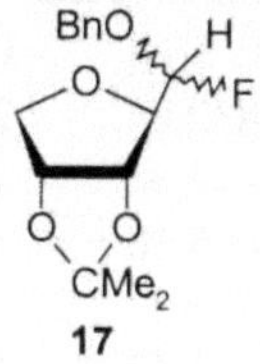 **17**	−140.0 67/26	H-1: 5.39 66.9/2.0	C-1: 111.2 219.8	MS(2,rO)
2	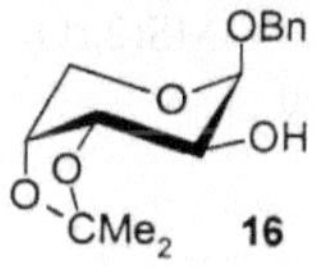 **16**		−134.2 63/5	H-1: 5.37 63.4/2.6	C-1: 109.3 220.7	
3	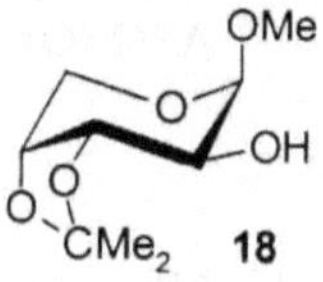 **18**	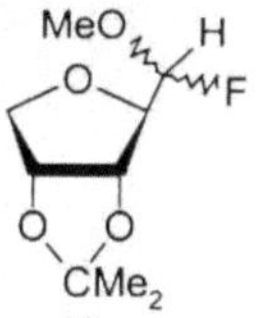 **19**	−141.3 67/24	H-1: 5.23 67.4/2.1	C-1: 114.2 222.9	MS(2,rO)
			−135.5 64/9	H-1: 5.18 63.6/2.7	C-1: 112.6 223.4	
4	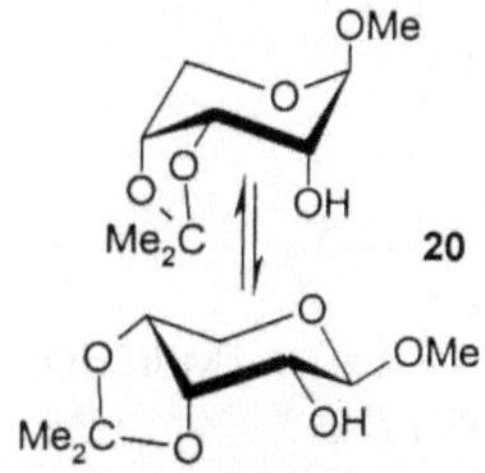 **20** ⇌ **21**	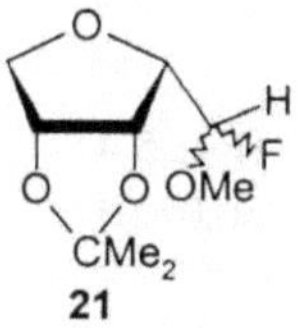 **21**	−138.6 67/6	H-1: 5.34 68.8/6.9	C-1: 112.8 220.6	MS(2,rO)
			−138.0 64/5	H-1: 5.42 62.8/7.0	C-1: 111.7 214.3	
5	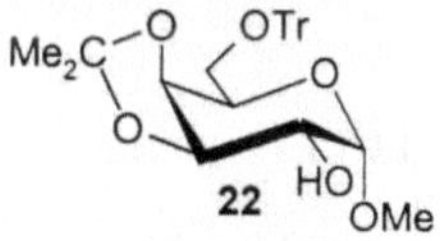 **22**	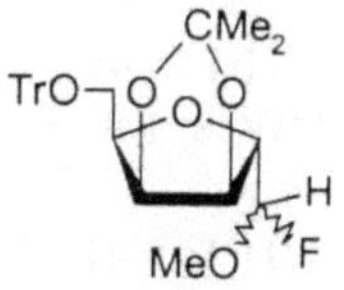 **24**	−140.8 67/24	H-1: 5.25 67.8/1.8	C-1: 113.4 220.5	MS(2,rO)
6	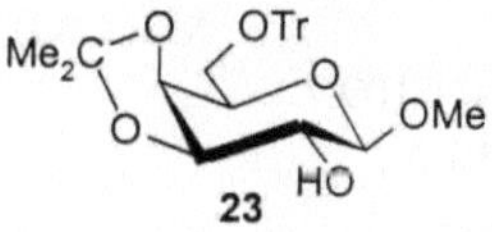 **23**		−135.4 64/6	H-1: 5.19 63.3/2.6	C-1: 111.7 220.4	
7	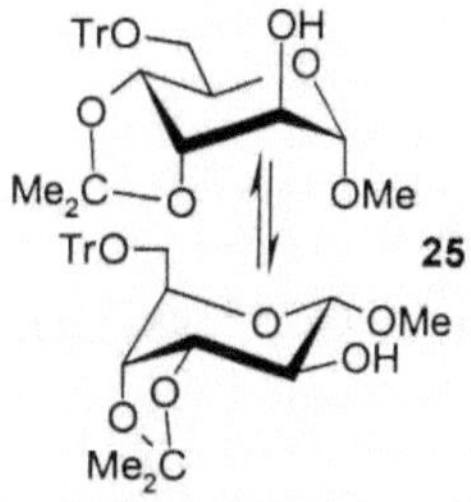 **25**	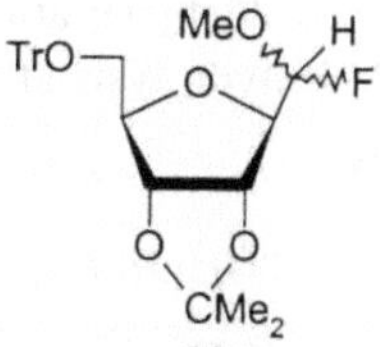 **26**	−142.7 67/15	H-1: 5.24 66.4/3.3	C-1: 111.8 220.7	MS(2,rO)
			−136.9 64/6	H-1: 5.19 64.6/4.1	C-1: 111.9 219.7	
8	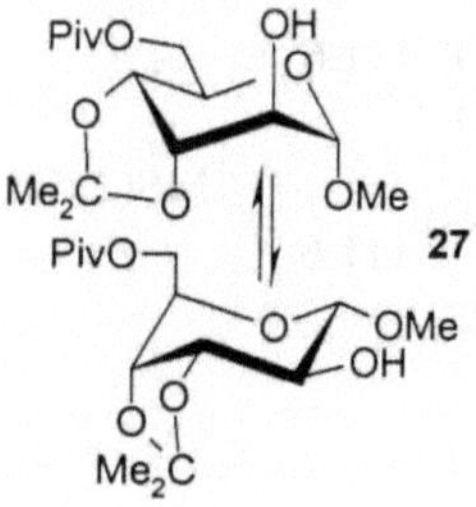 **27**	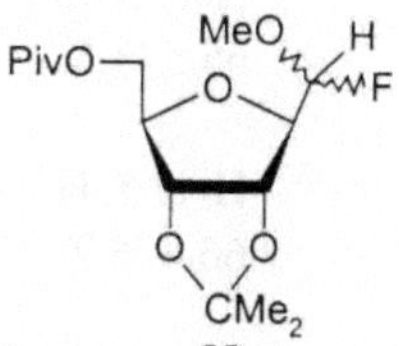 **28**	−143.3 67/18	H-1: 5.18 66.8/2.2	C-1: 111.6 220.9	MS(2,rO)
			−137.5 64/9	H-1: 5.13 64.2/3.1	C-1: 111.4 219.8	

(*continued*)

Table 4 *(continued)*

	Educt	Proposed structure of products	^{19}F NMR δ/ppm J/Hz	^{1}H NMR δ/ppm J/Hz	^{13}C NMR δ/ppm J/Hz	Operative mechanism
9	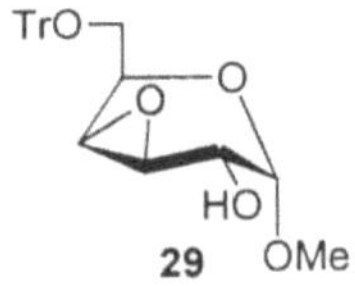 29	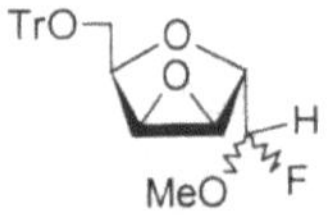 30	−142.4 67/21	H-1: 5.18 66.8/2.6	C-1: 112.4 222.1	MS(2,rO)
			−136.8 64/5	H-1: 5.16 63.7/3.1	not resolved	
		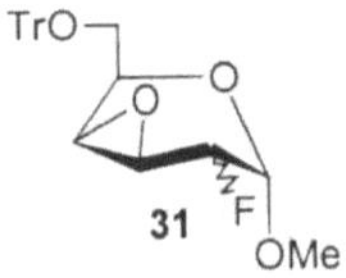 31	−196.5 47/12/7	H-1: 4.62 13.6 H-2: 4.60 45/3	C-1: 97.3 34.7 C-2: 85.2 176.8	S_N2 or AS(2,rO)
10	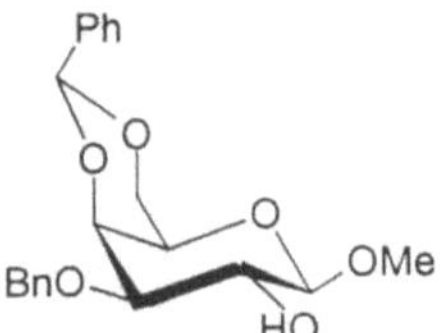 32	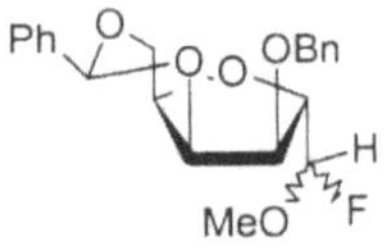 33	−142.3 66/14	H-1: 5.27 67/2	C-1: 112.3 219.2	MS(2,rO)
			−143.2 64/13	H-1: 5.26 63.7/1.8	C-1: 111.4 219.3	
11	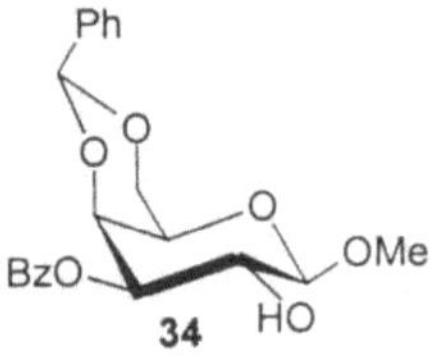 34	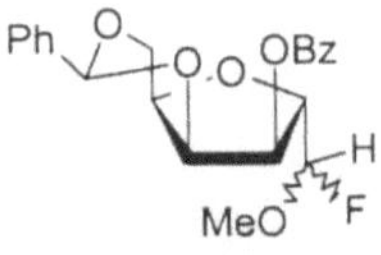 35	−141.9 66/16	H-1: 5.41 66.0/2.1	C-1: 112.5 221.5	MS(2,rO)
			−140.2 64/not resolved	H-1: 5.39 63.8/2.8	C-1: 111.5 220.3	
12	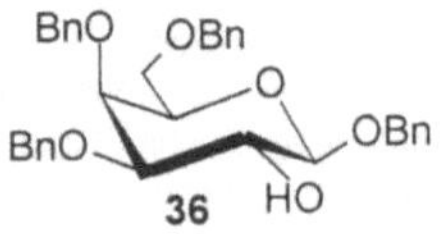 36	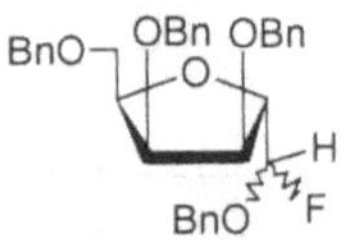 37	−141.9 67/18	H-1: 5.44 66.8/1.8	C-1: 110.2 221.7	MS(2,rO)
			−136.9 64/6	H-1: 5.41 63.7/2.4	C-1: 109.7 219.4	
13	 38	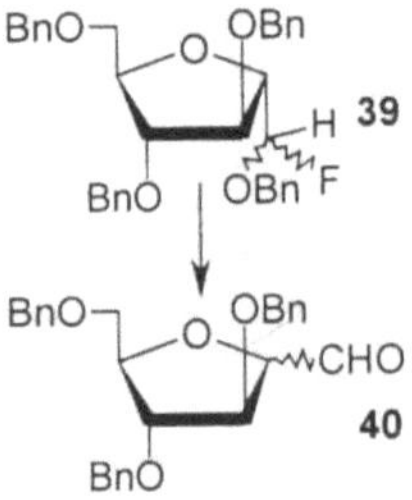 39, 40	−138.5 67/15	aldehyde, after hydrolysis at C-1		
			−134.7 67/9	H-1: 9.77	C-1: 202.9	MS(2,rO)
14	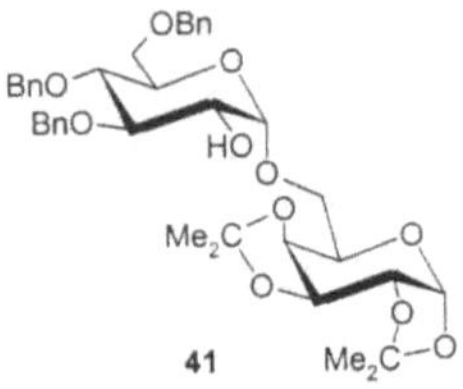 41	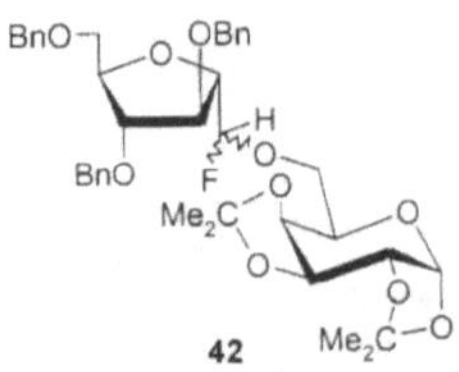 42	−135.9 67/15	H-1: 5.45 65.5/3.5	C-1: 111.6 221.7	MS(2,rO)
			−132.4 67/9	H-1: 5.41 66.4/6.2	C-1: 111.6 222	

(continued)

Table 4 *(continued)*

	Educt	Proposed structure of products	^{19}F NMR δ/ppm *J*/Hz	^{1}H NMR δ/ppm *J*/Hz	^{13}C NMR δ/ppm *J*/Hz	Operative mechanism
15	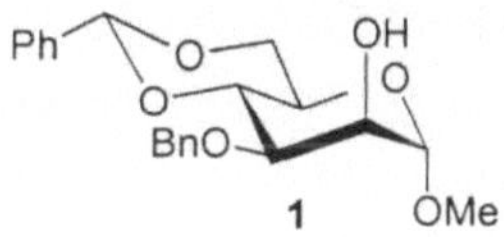 **1**	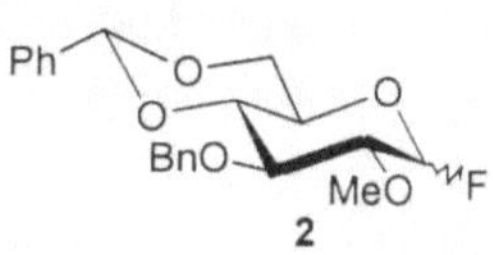 **2** Ref. [9]	α-anomer −148.1 52/27 β-anomer −135.6 55/12	α-anomer 5.70 52.0/2.9 β-anomer 5.29 53.5/6.2	α-anomer 105.6 229.7 β-anomer 109.5 218.2	MS(2,NG-1)
16	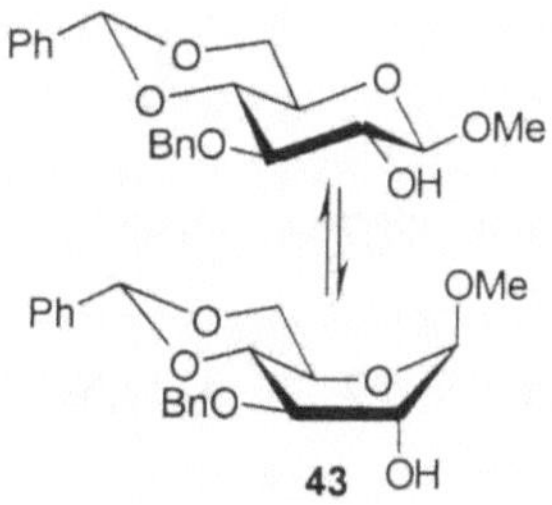 **43**	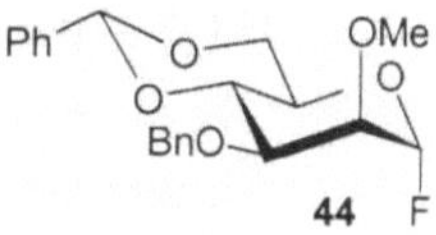 **44**	−139.2 49/not resolved	H-1: 5.50 49.5/1.8 H-2: 3.62 1.8/1.5	C-1: 106.7 224.7	MS(2,NG-1)
17	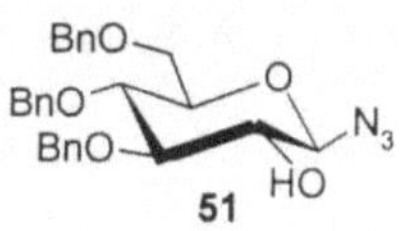 **51**	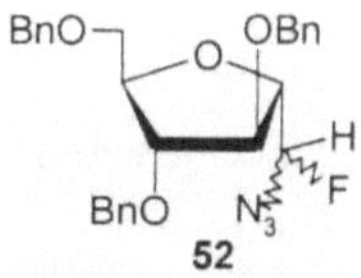 **52**	−148.4 56/11 −147.4 57/12	H-1: 5.54 55.8/5.6 H-1: 5.55 57/4	C-1: 99.9 217.4 C-1: 99.5 220	MS(2,rO)
		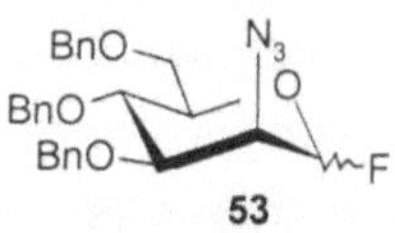 **53**	−135.7 50	not detected	not detected	MS(2,NG-1)
18	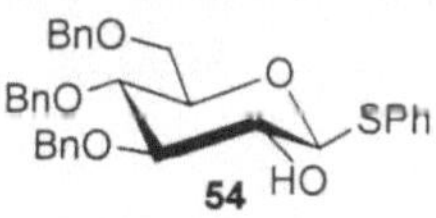 **54**	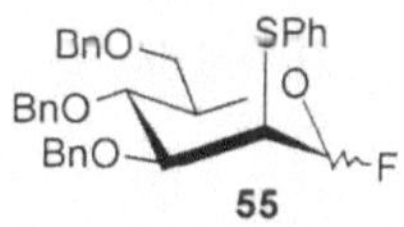 **55**	−126.0 52	H-1: 5.78 52.3/1.7	C-1: 108.4 225.7 C-2: 52.3 28.0	MS(2,NG-1)
19	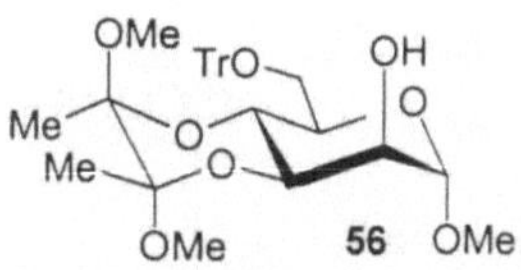 **56**	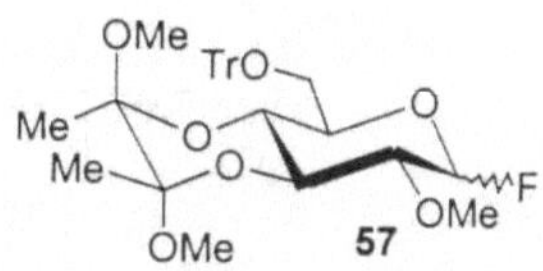 **57**	−147.8 53/23 −138.0 53/12	not detected	not detected	MS(2,NG-1)

(continued)

Table 4 *(continued)*

	Educt	Proposed structure of products	^{19}F NMR δ/ppm J/Hz	^{1}H NMR δ/ppm J/Hz	^{13}C NMR δ/ppm J/Hz	Operative mechanism
20	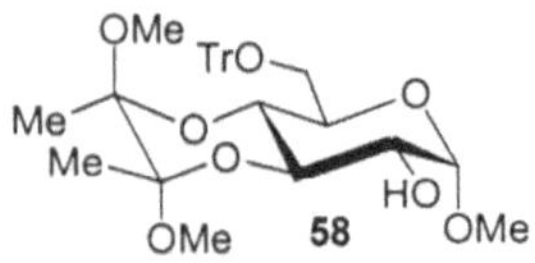 58	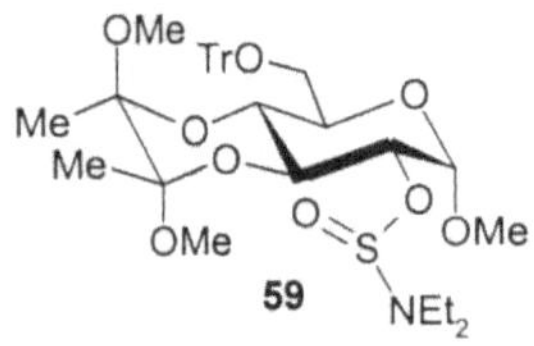 59	–	H-1: 4.91 3.6 H-1: 4.90 3.6	C-1: 98.8 C-1: 99.5	
		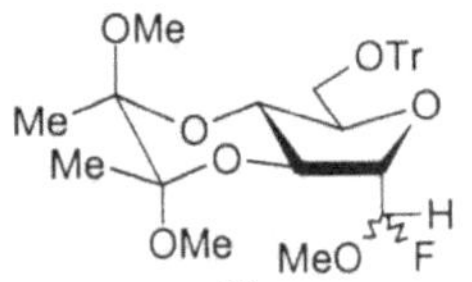 60	−142.5 66/10 −136.2 67/not resolved	not detected	not detected	MS(2,rO)
		61 unknown	−139.0 49/14	not detected	not detected	unknown
21	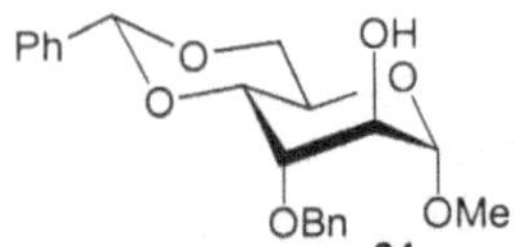 64	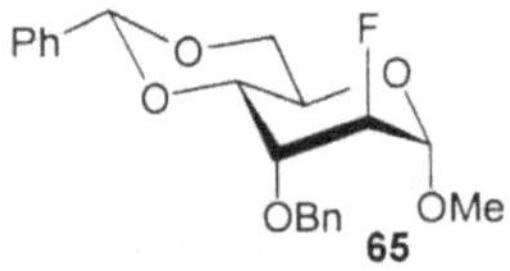 65	−192.5 43/10/10	not detected	C-1: 99.2 34.5	AS(2,NG-3/1)
		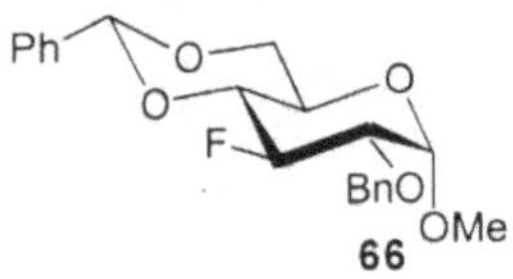 66	−199.0 54/10/10	not detected	C-1: 99.6 9.6	MS(2,NG-3)
22	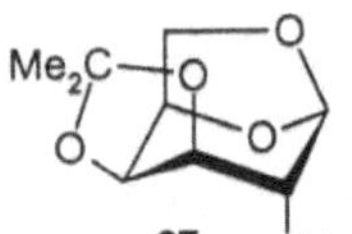 67	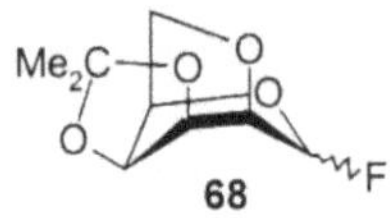 68	−122.1 67/6 −127.8 61/–	H-1: 5.85 65.9/1.5/1.5 H-1: 5.58 62.5 H-2: 4.00 (s)	C-1: 105.6 225.5 C-1: 106.3 227.8	MS(2,NG-1)
23	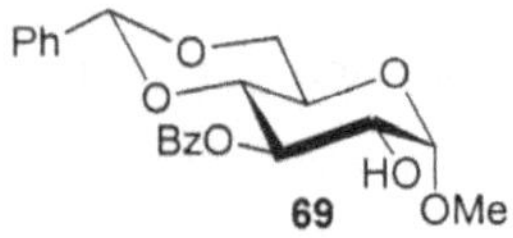 69	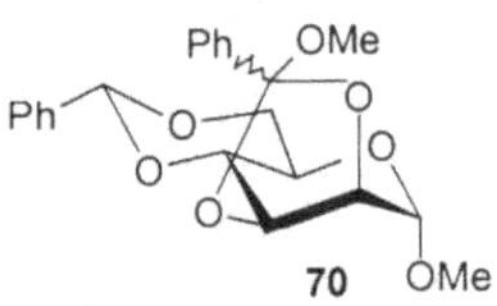 70	–	H-1: 5.06 (s) H-1: 5.02 (s)	C-1: 98.7 C-1: 98.9	ISC(2,NG-3)
		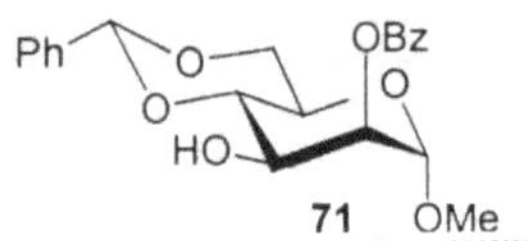 71	–	H-1: 4.85 (s)	C-1: 99.9	ISC(2,NG-3) + Hydrolysis

at C-1 as well as C-2. However, MS(2,rO) was found to be the sole reaction in all three cases (Table 4, entries 4, 7, and 8).

So far, in this series all educts contained a 3,4-*cis*-configuration and were, in this part of the molecule, also uniformly protected by the isopropylidene group. To test the influence of the latter, the conformationally more rigid methyl 3,4-anhydro-6-O-trityl-α-D-galactopyranoside (**29**) was studied; in this case (Table 4, entry 9), besides the ^{19}F NMR signals for the products **30** of the MS(2,rO)-reaction, that of **31** stemming from S_N2 or AS(2,rO) appeared (−196.4 ppm, 47/12/7 Hz). Subsequently, turning to the 3-O-protected derivatives of methyl 4,6-O-benzylidene-β-D-galactopyranoside, both benzyl ether **32** and benzoate **34** (Table 4, entries 10 and 11) were found to be subjects of MS(2,rO) exclusively. From the treatment of the 2-triflate of **34** with aliphatic alcohols in the presence of an organic base, formation of (the pair of) 2,3-orthoesters under inversion of configuration at C-2 has been claimed [14]. However, this reaction only had been proven with the β-D-gluco-analogue (see Table 2, entry 4). Thorough NMR measurements (private communication of the original authors) recently showed that the products are – as had been anticipated [5] – structural analogues of compounds **35** (Table 4, entry 11). Thus, even under these conditions, MS(2,rO) is preferred over ISC(2,NG-3).

To complete these extensive arabino/ribo- as well as galacto/altro-series and to start in the gluco/manno-range, a few not dioxolane ring protected derivatives of both were treated with *DAST*. Again (Table 4, entries 12 and 13), benzyl 3,4,6-tri-O-benzyl-β-D-galacto- (**36**) as well as the identically protected β-D-glucopyranoside (**38**) were subject to MS(2,rO). In the crude reaction mixture of **38** and *DAST*, a small proportion of benzyl fluoride (triplet at −206.6 ppm, 49 Hz) was also detected. During the standard isolation of the primary products, obviously hydrolysis at C-1 occurred under liberation of the aldehydo-2,5-anhydro-D-hexose derivative **40**. The disaccharidic educt **41**, consisting of an α-D-glucopyranosyl moiety with an unprotected OH-group at C-2 as donor and a 1,2:3,4-di-O-isopropylidene-α-D-galactopyranose as acceptor, was subject to this dominant type of side-reaction (Table 4, entry 14).

This MS(2,rO)-side reaction was also observed with 2-triflates of alkyl 3,4,6-tri-O-benzyl-α- and β-D-glucopyranosides under hydrolytic conditions (water/pyridine in *DMF*, 160°C, 5 min) and was used [19] for glycoside cleavage in the synthesis of chiral cyclopropylmethanol derivatives. Nevertheless, a parallel S_N2-reaction by fluoride was also observed upon treatment of the triflate of **38** with *TBAF* in *THF* [20].

In contrast to the findings described above, where MS(2,rO) was the generally dominating option, treatment of the corresponding derivatives from the α-D-manno-series containing an unprotected OH-2 with *DAST* is known [2, 3] to induce aglycon migration, MS(2,NG-1). This tendency is not restricted to 4,6-O-benzylidene protected educts (see Table 1, entry 5 [9]), but also dominates in cases where a cyclic acetal protection group is absent [16, 21, 22].

To validate our approach, methyl 3-O-benzyl-4,6-O-benzylidene-α-D-mannopyranoside (**1**) was subjected to reaction with *DAST* under our standard conditions, and the results ((MS(2,NG-1) as well as NMR data of product **2** (Table 4, entry 15)) were found to be in full agreement with the published ones from the *DAST*-reaction in diglyme at 100°C [9].

Table 5. Related reactions

	Educt	Product	NMR data δ/ppm, J/Hz	Yield/%	Ref.
1	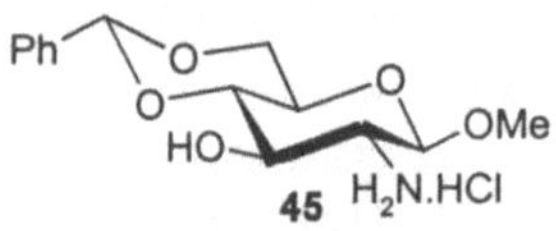 **45**	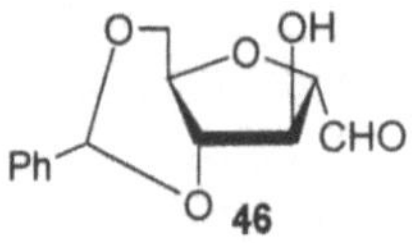 **46**	no data	50	[23]
2	 **47**	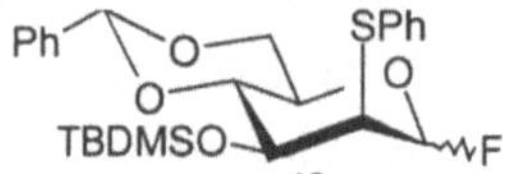 **48**	no data	88	[16]
3	 **49**	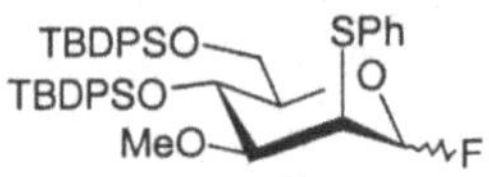 **50**	α-anomer: H-1: 5.34 (53.2/5.9) β-anomer: H-1: 5.54 (51.5/2.3)	86	[16]
4	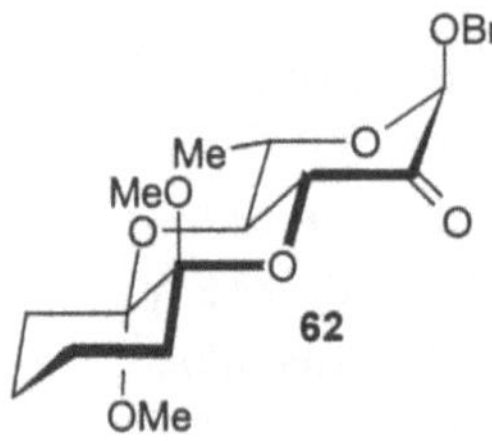 **62**	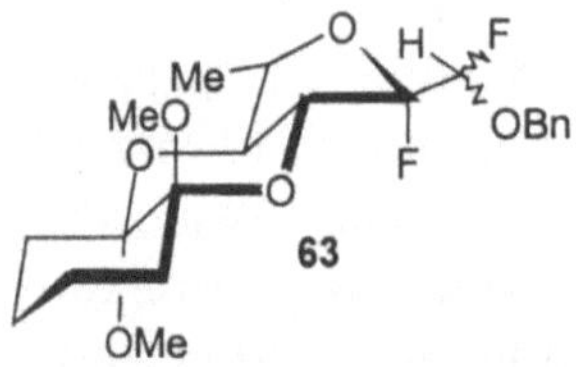 **63**	F-1: −144.3 (62.4/6.0) F-2: −118.9 (m) H-1: 5.32 (62.4/1.2) C-1: 106.2 (224.2/46.2) C-2: 110.3 (223.3/25.8)	19	[27]

The corresponding β-D-gluco-derivative **43** contains, in its $^4C_1$-conformation, an antiperiplaner orientation of the LG to the ring oxygen (rO) and a diequatorial one to the aglycon (NG-1). Due to 4,6-O-benzylidene protection it constitutes a rigid *trans*-decalin system. Thus far, for the displacement reactions of the corresponding triflates, S_N2 had been the choice [7]; the possible MS(2,rO), observed with educts devoid of this 'paralyzing' dioxane protection [19, 20], had never come into play[1]. However, there is a single report in the literature dealing with the nitrous acid deamination with methyl 2-amino-4,6-O-benzylidene-2-deoxy-β-D-glucopyranoside (**45**) of this *trans*-decalin type (Table 5, entry 1) [23] where from the elucidation of the product structure (that time by chemical means only) this MS(2,rO) had to be anticipated. In later experiments [24], isolation of the same product, 2,5-anhydro-4,6-O-benzylidene-D-mannose (**46**), from the corresponding benzyl glycoside could not be achieved (but its formation was not explicitly excluded).

In our case, against all expectations, the treatment of **43** with *DAST* caused aglycon migration, MS(2,NG-1), under formation of 3-O-benzyl-4,6-O-benzylidene-2-O-methyl-α-D-mannopyranosyl fluoride (**44**). For an explanation it can be argued that the centres involved in this rearrangement are located at the more flexible

[1] See Note Added in Proof at the end of this contribution

part of the *trans*-decalin system, thus possibly allowing *quasi*-diaxial arrangement of LG and NG-1 (as depicted in Table 4, entry 16). Furthermore it has been reported in the literature [16] that thioglycosides in the β-D-gluco- and β-D-galactopyranose series (as well β-D-galactopyranosyl azides) with unprotected OH-2, on treatment with *DAST*, give this MS(2,NG-1) in excellent yields (*e.g.* the phenyl thioglycosides **47** and **49** of the β-D-gluco-series shown in Table 5, entries 2 and 3).

To prove this exceptional migratory aptitude of the alkyl/arylthio as well as the azido group as compared to the aglycon of normal O-glycosides (*e.g.* **38**), azide **51** and thioglycoside **54** were treated with *DAST* under standard conditions. The results shown in Table 4, entries 17 and 18, give the impression that this report for the phenylthio group may be valid[1] (however, the given $J_{1,2}$ value of 5.9 Hz for the α-anomer of 2-phenylthio-mannopyranosyl fluoride **50** [16] cannot be imagined to be possible), whereas those from the azido-analogue could not be reproduced since a 10:1-preference for the MS(2,rO)-product **52** over that of the MS(2,NG-1)-reaction (**53**) was observed. As is evident from the literature [25], the geminal ^{19}F, ^{1}H couplings for derivatives containing an azido or arylthio substituent are by approximately 8–10 Hz smaller than the corresponding values in case of O-substitution.

The incorporation of the hexopyranose moiety into a *trans*-decalin ring system as in the 4,6-O-benzylidene protected educt **43** of the gluco-series obviously causes the exceptional reactivity order that the otherwise generally preferred MS(2,rO)[1] is 'prohibited', whereas (due to 1,2-*trans*-configuration) MS(2,NG-1) is 'allowed', even when a diequatorial orientation of LG and NG predominates. With the intention to bring the bridgeheads of the *trans*-annellation closer to the reaction centre, we chose the butane 2,3-*bis*-acetal (*BBA*) strategy [26] to simultaneously protect the diequatorial arranged OH-groups at C-3 and 4 in methyl α-D-manno- and -glucopyranoside. The corresponding educts, **56** and **58**, behaved as follows (Table 4, entries 19 and 20): The α-D-manno-isomer **56** solely gave the expected MS(2,NG-1)-products **57** (an identical result has very recently been published [27] with the corresponding cyclohexane diacetal (*CDA*) from the α-L-manno-series). However, the α-D-gluco-analogue **58**, under standard reaction conditions, led to at least eleven fluorinated products, most of them showing the characteristic ^{19}F NMR features of ring-contracted structures. Only when the reaction rate was drastically reduced (by conducting the treatment with *DAST* at -20°C in the presence of an excess of pyridine), formation of one single product could be observed by TLC monitoring. After isolation, it did not contain fluorine but a diethylamino moiety. From the similarity of the NMR spectra to those of the starting material **58** (except chemical shift values for H-2 and H-3), its structure was anticipated to be that of the two (diastereomeric) 2-O-diethylaminosulfinyl derivatives **59** (formed by hydrolysis of both S–F-bonds in the intermediate of the *DAST*-reaction, see Scheme 1). The distinct results presented in Table 4, entry 20, were then obtained after prolonged reaction time under these conditions. MS(2,rO) to give **60** clearly dominated (compare results from the *DAST*-treatment of a structurally related 2-ulose **62** bearing a 3,4-*CDA* protection [27], where ring contraction was observed; Table 5, entry 4). For the additionally observed signal ($\delta = -139.0$ ppm, $J = 49$ and 14 Hz), which corresponds to a glycosyl fluoride, presently no structure can be suggested (**61**). However, it proved, by the admixture of the reaction products **57**, not to be identical with the β-anomer of 2-O-methyl-glucopyranosyl fluoride **57** ($\delta = -138.0$ ppm, $J = 53$ and 12 Hz). The latter was formed in the *DAST*-reaction of **56** (Table 4, entry 19), but its generation from **58** would have also been conceivable [24] as the opening of the intermediate oxiranium ion in the MS(2,rO)-mechanism by the neighbouring methoxy group from C-1 under overall retention of configuration.

Furthermore, two other educts were investigated, which both have a relatively rigid conformation as well as an axial OH-group at C-2. In the first case, with methyl 3-O-benzyl-4,6-O-benzylidene-α-D-altropyranoside (**64**), two antiperiplanar oriented alkoxy groups are present in vicinal positions to the LG, whereas in the second example, with 1,6-anhydro-3,4-O-isopropylidene-β-D-galactopyranose (**67**), the oxygen atom of the 1,6-anhydro ring occupies the antiperiplanar position. The results obtained with **64** (Table 4, entry 21) parallel those from the corresponding 3-azido-3-deoxy analogue **3** depicted in Table 1, entry 6, except that aglycon migration (MS(2,NG-1)) having led to compound **6** was not observed in this case. In comparison to the MS(2,rO) reactions found with the educts **25** and **27**, both of α-D-altro-configuration as well, the diacritical role of the protecting group with respect to the necessary conformational interconversion becomes evident. As expected, the *DAST*-treatment of **67** resulted in migrative substitution at C-2 by the oxygen of the 1,6-anhydro ring with inversion of configuration and entry of fluoride at C-1 (MS(2,NG-1)) to yield 2,6-anhydro-D-talopyranosyl fluorides **68**. Noteworthy, the ^{19}F NMR resonances are located at higher fields and show the same magnitude of geminal couplings as characteristic for MS(2,rO)-products. These results are in good agreement with those reported from the *DAST* reaction of the unprotected 1,6-anhydro-β-D-galactopyranose [28].

Finally, to test the S_N2- *vs.* the ISC(2,NG-3)-option, methyl 3-O-benzoyl-4,6-O-benzylidene-α-D-glucopyranoside (**69**) was reacted with *DAST* (Table 4, entry 23). Although TLC monitoring showed product formation, no discrete resonance signal could be observed by ^{19}F NMR. After work-up, the products were found to be those of ISC (**70**) and subsequent hydrolysis (**71**) as well as (not shown in entry 23) the 2-O-diethylaminosulfinyl derivative of educt **69**.

In conclusion, the following ‘rules’, additional to those mentioned in the introduction, can be derived from the results described herein on the treatment of alkyl hexopyranosides containing an unprotected OH-group at C-2 with *DAST* in dichloromethane at room temperature:

i) MS(2,rO), the ring contraction under attack of the ring oxygen, is generally preferred over all other possibilities, even when the predominating conformation does not contain OH-2 in the (*quasi*-essential) equatorial position. This is especially valid in the class of pento- and hexopyranosides which contain a (*cis*)-3,4-O-isopropylidene protecting group. Exceptions from this type of reaction are all educts of the manno-series as well as those 4,6-O-benzylidenated hexopyranosides which are members of the *trans*-decalin family (derivatives of the gluco-, allo-, and altro-series as they contain their ring substituents at C-4 and C-5 in (*trans*)-diequatorial orientation)[1].

ii) MS(2,NG-1), the so-called aglycon migration, seems to be the ‘opposite’ alternative to MS(2,rO) as they, with O-glycosidic educts, have never been observed together in the same reaction. Independently of the protecting groups present, MS(2,NG-1) finds its stage set for single action in all derivatives of α-D-mannopyranoside.

Table 6. ^{13}C NMR data (δ/ppm, J/Hz); assignments marked with an asterisk may be interchanged

	C-1	C-2	C-3	C-4	C-5	C-6	Others
17a	111.2	84.8	80.9	81.5	74.7	–	71.7 (*Bn*), 112.5/26.6/25.0 (CMe_2)
	219.8	21.0	3.6				
17b	109.3	84.3	80.8	81.4	74.7	–	71.3 (*Bn*), 112.5/26.6/25.0 (CMe_2)
	220.7	29.4					
19a	114.2	85.3	81.4	82.1	75.2	–	57.7 (OMe), 113.4/26.8/25.1 (CMe_2)
	222.9	21.5	3.5				
19b	112.6	84.9	81.3	82.0	75.2	–	57.8 (OMe), 113.6/26.8/25.1 (CMe_2)
	223.4	29.5					
21a	112.8	82.7	81.3	81.5	74.1	–	57.9 (OMe), 113.5/26.2/24.9 (CMe_2)
	220.6	22.2	6.2				
21b	111.7	82.2	80.8	81.5	74.2	–	57.8 (OMe), 113.5/26.3/25.0 (CMe_2)
	214.3	35.8					
26a	111.8	84.7	80.8	82.4	84.8	64.4	57.5 (OMe), 114.1/27.6/25.7 (CMe_2),
	220.0	24.9	3.5				87.0 (CPh_3)
26b	111.9	84.9	81.1	82.8	85.2	64.8	57.5 (OMe), 113.8/27.6/25.7 (CMe_2),
	219.7	26.6	2.3				87.0 (CPh_3)
28a	111.6	84.5	80.5	81.9	83.3	63.9	57.6 (OMe), 114.4/27.5/25.6 (CMe_2),
	220.9	24.0	3.8				178.4/39.0/27.4 (*Piv*)
28b	111.4	85.1	80.7	82.3	83.7	64.3	57.6 (OMe), 114.1/27.5/25.6 (CMe_2),
	220.7	29.4	1.5				178.4/39.0/27.4 (*Piv*)
31	97.3	85.2	51.0	48.6	65.3	63.3	55.4 (OMe), 87.0 (CPh_3)
	34.7	176.8	24.4				
33a	111.4	80.6	79.2	74.3*	73.3*	67.6*	57.5 (OMe), 72.2* (*Bn*), 99.0 (CHPh)
	219.3	25.0	[n.r.]				
33b	112.3	80.7	79.2	74.5*	73.3*	67.5*	57.4 (OMe), 72.2* (*Bn*), 99.1 (CHPh)
	219.2	22.0	[n.r.]				
35a	111.2	79.8	74.4	74.8*	73.6*	67.3*	57.5 (OMe), 166.2 (*Bz*), 98.7 (CHPh)
	221.3	23.3	4.0				
35b	111.5	79.7	74.1	74.7*	73.6*	67.3*	57.6 (OMe), 166.2 (*Bz*), 98.7 (CHPh)
	220.3	27.7	1.3				
2α	105.6	81.5	78.2	81.2	64.7	68.8	60.2 (OMe), 75.4 (*Bn*), 101.6 (CHPh)
	229.7	24.1			3.8		
2β	109.5	83.3	79.6	80.7	65.6	68.8	60.5 (OMe), 74.7 (*Bn*), 101.5 (CHPh)
	218.2	24.1	8.8				
44	106.7	78.3	75.6	78.5	66.3	68.6	60.9 (OMe), 73.7 (*Bn*), 101.9 (CHPh)
	224.7	35.6	2.7		2.7		
55	108.4	52.3	77.8*	75.3	74.2*	68.6	74.2/73.5/71.9 (*Bn*)
	225.7	28.0	2.7		3.7		
58	99.9	69.5*	70.2*	71.0*	66.4	62.0	55.3 (OMe), 99.7/48.4/48.2/18.0/ 17.8 (*BBA*), 86.6 (CPh_3)
59a	98.8	69.9	68.2	68.7	66.8	61.9	54.9 (OMe), 99.6/48.1/48.0/17.6/ 17.5 (*BBA*), 86.3 (CPh_3), 36.7/36.6/ 14.0/13.9 (NEt_2)
59b	99.5	71.9	68.1	66.7	68.7	61.7	

(*continued*)

Table 6 (*continued*)

	C-1	C-2	C-3	C-4	C-5	C-6	Others
65	99.2	88.2	72.2	71.1	58.7	69.5	56.1 (OMe), 73.6 (*Bn*), 102.6 (CHPh)
	34.5	174.2	27.2	1.9			
66	99.6	80.0	91.4	77.2	61.8	69.1	55.7 (OMe), 73.8 (*Bn*), 101.9 (CHPh)
	9.6	16.9	187.2	~15	~5		
68a	105.6	65.3	70.5	69.9	69.3	60.8	110.9/26.2/24.5 (CMe_2)
	225.5	21.4	4.2		1.5		
68b	106.3	68.4	73.5	70.5	68.7	63.0	113.0/26.2/25.1 (CMe_2)
	227.8	19.5	9.2	6.8	2.0		
70a	98.7	76.7	75.9	79.7	60.2	69.1	55.6 (OMe), 102.0 (CHPh), 121.0/51.2 (orthoester)
70b	98.9	77.0	75.2	79.5	60.5	69.1	55.6 (OMe), 102.2 (CHPh), 122.1/51.6 (orthoester)
71	99.9	72.8	67.7	79.4	63.6	69.0	55.4 (OMe), 102.5 (CHPh), 166.4 (*Bz*)

iii) MS(2,NG-1) is also operative in pyranoid educts with a 4,6-O-benzylidene protection as soon as a 1,2-*trans*- as well as 4,5-*trans*-orientation of their ring substituents is given (α-D-altro and β-D-gluco[1]; β-D-allo was not probed). In the case of 4,6-O-benzylidene-α-D-altropyranosides, MS(2,NG-3), the migrative substitution ('simple' case) by the NG from C-3, as well as AS(2,NG-1 and/or 3), the anchimerically assisted substitution under retention of configuration, become significant as well.

iv) ISC(2,NG-3), the participation of the 'complex' type (out of a diequatorial arrangement to the LG) by a neighbouring acyloxy group, ranks behind a possible MS(2,rO) but before S_N2.

v) S_N2, the usually intended nucleophilic substitution under inversion of configuration, is the reaction of choice with β-D-mannopyranosides only.

vi) E, elimination, was never observed, although many educts fulfilled the stereochemical prerequisite of an antiperiplanar orientation between the LG and a vicinal hydrogen atom at C-3 or C-1.

Synthesis of Educts

Many of the educts used in this study were taken either directly or in form of essential precursors from the carbohydrate reference collection of this institute or were gifts of Chemprosa Company, Lannach (Austria). The eventually necessary final transformations on donated precursors as well as the syntheses of the remaining educts were conducted according to literature procedures. The regioselective introduction of the butane *bis*-acetal protection in educt **58** was achieved starting from the known methyl 2-benzoyl-α-D-glucopyranoside. All compounds gave satisfactory ^{1}H and ^{13}C NMR spectra, but were not further characterized.

Table 7. ^{1}H NMR data (δ/ppm, J/Hz)

	H-1	H-2	H-3	H-4	H-5a	H-5b/6a	H-6b	Others
17a	5.39 66.9/2.0	4.13 25.7/2.0/1.5	4.81 not resolved	4.96 6.4/1.5/1.5	3.98–4.03 not resolved		–	4.92 (12.0)/4.67 (12.0/1.5) (*Bn*), 1.50/1.34 (CMe_2)
17b	5.37 63.4/2.6	4.20 not resolved	4.81 not resolved	4.96 6.4/1.5/1.5	3.98–4.03 not resolved		–	4.92 (12.0)/4.65 (12.0/1.5) (*Bn*), 1.50/1.34 (CMe_2)
19a	5.24 67.4/2.0	4.09 25.0/2.0/1.5	4.92 5.7/1.5	4.81 5.7/2.8/2.8	4.02 2.8/2.8		–	3.57 (1.2) (OMe), 1.52/ 1.33 (CMe_2)
19b	5.18 63.6/2.6	4.15 7.0/2.6/1.0	4.92 6.2/1.0	4.81 6.2/2.7/2.7	3.99 2.7/2.7		–	3.55 (1.3) (OMe), 1.52/ 1.33 (CMe_2)
24a	5.25 67.8/1.8	4.05 24.4/not resolved	4.93 6.2/not resolved	4.76 5.7/4.4	4.2 not resolved	3.40 9.2/5.7	3.32 9.2/6.6	3.55 (OMe), 1.43/1.32 (CMe_2)
28a	5.18 66.8/2.2	4.1 not resolved	4.78 6.6/3.5	4.3 5.7/3.5	4.0–4.25 not resolved			3.56 (OMe), 1.51/1.32 (CMe_2), 1.19 (*Piv*)
35a	5.41 66.0/2.1	4.63 16/9/2	5.59 8.3/4.3	4.84 not resolved	4.09 not resolved	4.43 12.2	4.12 12.2/1.5	3.61 (OMe), 5.50 (CHPh)
35b	5.39 63.8/2.8	4.68 7.8/2.8	5.64 7.8/4.2	4.84 not resolved	4.09 not resolved	4.43 12.2	4.12 12.2/1.5	3.59 (OMe), 5.50 (CHPh)
58	4.91 3.6	3.5–3.95 not resolved				3.40 11.4	3.14 11.4/5	3.49 (OMe), 3.29/3.02/ 1.33/1.15 (*BBA*)
59a	4.91 3.6	4.29 10.3/3.6	4.09 10.3/9.8	3.76 10.3/9.8	3.9 not resolved	3.1–3.4 not resolved		3.47 (OMe), 3.29/3.02/ 1.33/1.15 (*BBA*)
59b	4.90 3.6	4.18 10.3/3.6	4.06 10.3/9.8	3.83 10.3/9.8	3.9 not resolved	3.1–3.4 not resolved		3.45 (OMe), 3.29/3.02/ 1.33/1.15 (*BBA*)
68a	5.85 65.9/1.5/1.5	4.15 not resolved	4.47 7.8	4.44 7.8	4.14 not resolved	4.07 9.5/6.4	3.68 9.5/1.5	1.58/1.38 (CMe_2)
68b	5.58 62.5	4.00 –	4.22 7.8	4.39 7.8/4.4	4.14–4.20 not resolved			1.58/1.39 (CMe_2)

Characterization of products

As in most cases the formed mixtures of diastereomers were found to be hardly separable, unequivocal identification of individual ^{1}H and ^{13}C NMR signals was not possible in all instances; this was especially the case with proton resonance spectra. Data essential for the structure assignments are included in Table 4; further data are collected in Tables 6 and 7 (signals corresponding to aromatic moieties of protecting groups are omitted). The configuration at C-1 within a pair of diastereomers formed by MS(2,rO) can be deduced from the set of couplings observed with the nuclei F, H-1, and H-2 [11], but this was not done with the products obtained here. They are designated simply as isomers 'a' or 'b'; this holds also for the pair of diastereoisomers in compound **59**.

Experimental

General

NMR spectra were recorded at 300.13 or 200 MHz (^{1}H), 75.47 or 50.29 MHz (^{13}C), and 282.4 MHz (^{19}F) using a Bruker MSL 300 and a Varian Gemini 200 apparatus, respectively, except for the elucidation of **59** where a Varian Inova 500 spectrometer was employed. As reference standards, tetramethylsilane (^{1}H and ^{13}C NMR) and trichlorofluoromethane (^{19}F NMR) were used. TLC was performed on silica gel 60 F_{254} precoated aluminum plates (Merck 5554) with detection by charring after spraying with vanillin/H_2SO_4 (1%). For column chromatography, silica gel 60, 230–400 mesh (Merck 9385), was employed.

General procedure for the DAST-reaction

Analytical Scale: To a cold (0°C) solution of 0.1 mmol educt in 1 cm^3 absolute CH_2Cl_2, 0.04–0.05 cm^3 *DAST* (0.30–0.38 mmol) were added under stirring. After 10 min the cooling bath was removed, and the mixture was stirred at room temperature for 5 h under monitoring by TLC (generally, the reaction products are faster moving than the educt). For the ^{19}F NMR measurements, 0.4 cm^3 of the reaction mixture were transferred into the NMR tube and diluted with $CDCl_3$.

Preparative Scale: The reaction conditions were the same as used in the analytical scale. The amount of educt, solvent, and reagent was ten- to fifteen-fold. The reaction time was extended to 15–24 h. Work-up started with dilution of the reaction mixture with CH_2Cl_2 and quenching of the excess reagent by addition of an excess of MeOH. After 15 to 20 min the mixture was then extracted with H_2O, followed by aqueous $NaHCO_3$ and again H_2O. After drying with Na_2SO_4 the solvent was evaporated, and the residue was chromatographically purified on silica gel.

Note Added in Proof

After having submitted the manuscript we became aware of a recent paper dealing with the *DAST* reaction of glycosides and thioglycosides of 3-deoxy-3-C-methyl-3-nitro-D/L-gluco- as well as -mannopyranoses [29]. These educt structures, in the same way as ours, are subject to either ring contraction (MS(2,rO)) or aglycon migration (AS(2,NG-1)) and show, as is to be expected, a higher degree of reactivity in displacement reactions at C-2. As the presented generalizations concerning the diacritic structural parameters in the choice of the side reaction were in contrast to those we had deduced from our findings, a thorough inspection of the reported ^{1}H and ^{13}C NMR data was carried out in order to re-examine the elucidation of product structures. From the value of the geminal ^{19}F,

^{1}H couplings ($^2J_{F,H}$) and other diacritical data (*e.g.* chemical shifts of H-2 and C-2, respectively) it has to be concluded that the structure appointments are wrong in 8 cases. According to our interpretation of the data, the methyl α-L-manno derivative as well as all thioglycosides (α- and β-anomers from the D/L-gluco series) were subject to AS(2,NG-1), whereas all O-glycosides with α- or β-D/L-gluco configuration, including those containing a 4,6-O-benzylidene protection, gave MS(2,rO) only. This is in full agreement with parameters we have deduced herein from our results, except that in the (more reactive) 3-deoxy-3-C-methyl-3-nitro-glucopyranose case, a 4,6-O-benzylidene protection does no more cause prevention of the ring contractive MS(2,rO)-reaction.

Acknowledgements

We appreciate financial support by the *Fonds zur Förderung der wissenschaftlichen Forschung*, Vienna (P-11021 OECH) and thank Chemprosa, Lannach (Austria) for the gift of educts for this investigation. Student coworkers in the synthesis of educts as well as their treatment with *DAST* have been: *R. Peinsipp, J. Wolfgang, G. Koller, B. J. Paul, F. Burghart, C. Gruber, W. Haas, M. Grininger, R. Harrer, W. Kroutil, S. Leitgeb, M. Waldhuber, D. Kober, T. Sovic*, and *J. Steinreiber*. They are all thanked for their valuable help in these laborious studies.

References

[1] For an overview see: Liebmann JF, Greenberg A, Dolbier WR (eds) (1988) Fluorine-Containing Molecules, Structure, Reactivity, Synthesis and Applications. VCH, Weinheim
[2] Detailed data are collected in: Tsuchiya T (1990) Adv Carbohydr Chem Biochem **48**: 91
[3] Dax K, Albert M, Ortner J, Paul BJ (2000) Carbohydr Res **327**: 47
[4] Dax K, Albert M, Ortner J, Paul BJ (1999) Curr Org Chem **3**: 287
[5] Dax K, Albert M (2001) In: Stütz AE (ed) Glycoscience: Epimerisation, Isomerisation and Rearrangement Reactions in Carbohydrates. Springer, Berlin, p 193
[6] Kovac P (1986) Carbohydr Res **153**: 168
[7] Haradahira T, Maeda M, Omae H, Yano Y, Kojima M (1984) Chem Pharm Bull **32**: 4758
[8] Haradahira T, Maeda M, Kai Y, Omae H, Kojima M (1985) Chem Pharm Bull **33**: 165
[9] Kovac P, Yeh HJC, Jung GL, Glaudemans CPJ (1986) J Carbohydr Chem **5**: 497
[10] Castillon S, Dessinges A, Faghih R, Lukacs G, Olesker A, Thang TT (1985) J Org Chem **50**: 4913
[11] Baer H, Hernandez Mateo F, Siemsen L (1989) Carbohydr Res **187**: 67
[12] Llewellyn JW, Williams JM (1973) J Chem Soc Perkin Trans I, 1997
[13] Erbing C, Lindberg B, Svensson S (1973) Acta Chem Scand **27**: 3699
[14] Ivanova IA, Nikolaev AV (1998) J Chem Soc Perkin Trans I, 3093
[15] a) El Nemr A, Tsuchiya T (2001) Carbohydr Res **330**: 205; b) El Nemr A, Tsuchiya T (1997) Carbohydr Res **303**: 267; c) El Nemr A, Tsuchiya T (1997) Carbohydr Res **301**: 77; d) El Nemr A, Tsuchiya T, Kobayashi Y (1996) Carbohydr Res **293**: 31; e) El Nemr A, Tsuchiya T (1995) Tetrahedron Lett **36**: 7665
[16] Nicolaou KC, Ladduwahetty T, Randall JL, Chucholowski A (1986) J Am Chem Soc **108**: 2466
[17] El Sayed Ahmed FM, David S, Vatele J-M (1986) Carbohydr Res **155**: 19
[18] Fleet GWJ, Seymour LC (1987) Tetrahedron Lett **28**: 3015
[19] a) Charette AB, Cote B, Marcoux J-F (1991) J Am Chem Soc **113**: 8166; b) Charette AB, Cote B (1993) J Org Chem **58**: 933
[20] Ogawa T, Takahashi Y (1983) J Carbohydr Chem **2**: 461
[21] Street IP, Withers SG (1986) Can J Chem **64**: 1400
[22] a) Bliard C, Herczegh P, Olesker A, Lukacs G (1989) J Carbohydr Chem **8**: 103; b) Bliard C, Cabrera Escribano F, Lukacs G, Olesker A, Sarda P (1987) J Chem Soc Chem Commun: 368

[23] Akiya S, Osawa T (1959) Chem Pharm Bull **7**: 277; see also Irvine JC, Hynd OA (1914) J Chem Soc 698
[24] Chan W-P, Gross PH (1980) J Org Chem **45**: 1369
[25] a) Robins MJ, Wnuk SF, Mullah KB, Dalley NK (1994) J Org Chem **59**: 544; b) Fuchigami T, Konno A, Nakagawa K, Shimojo M (1994) J Org Chem **59**: 5938; c) Takeuchi Y, Asahina M, Hori K, Koizumi T (1988) J Chem Soc Perkin Trans I, 1149
[26] Montchamp J-L, Tian F, Hart ME, Frost JW (1996) J Org Chem **61**: 3897
[27] Aghmiz ML, Diaz Y, Jana GH, Matheu MI, Echarri R, Castillon S, Jimeno ML (2001) Tetrahedron **57**: 6733
[28] Baillargeon DJ, Reddy GS (1986) Carbohydr Res **154**: 275
[29] Borrachero P, Cabrera-Escribano F, Carmona AT, Gomez-Guillen M (2000) Tetrahedron Asymm **11**: 2927

Received November 26, 2001. Accepted November 28, 2001

Allyl Protecting Group Mediated Intramolecular Aglycon Delivery (IAD) of Glycosyl Fluorides

Ian Cumpstey[1], **Antony J. Fairbanks**[1,*], and **Alison J. Redgrave**[2]

[1] Dyson Perrins Laboratory, University of Oxford, Oxford OX1 3QY, United Kingdom
[2] GlaxoSmithKline, Medicines Research Centre, Stevenage SG1 2NY, United Kingdom

Summary. Stereospecific 1,2-*cis*-glycosylation of 2-O-allyl protected glucosyl and mannosyl fluorides can be achieved *via* a sequence of allyl isomerization, N-iodosuccinimide mediated tethering, and intramolecular aglycon delivery (IAD). Fluoride is advantageous as an anomeric leaving group since extended reaction times can be employed to tether hindered aglycon alcohols without competitive anomeric activation. Tin(II) chloride mediated intramolecular glycosylation furnishes the desired α-glucosides and β-mannosides in an entirely stereoselective manner.

Keywords. Carbohydrates; Glycosides; Glycosylation; Stereoselectivity; Intramolecular aglycon delivery (IAD).

Introduction

Intramolecular aglycon delivery (IAD) is an attractive synthetic approach for the formation of 1,2-*cis*-glycosidic linkages. The technique of temporarily tethering together glycosyl donor and glycosyl acceptor *via* the 2-hydroxyl group of the glycosyl donor was originally pioneered for the synthesis of β-mannosides by *Hindsgaul* [1] and *Stork* [2] and more recently to greater effect by *Ogawa* [3]. The principle of the approach is that the glycosylation reaction, which ensues after tethering, must be intramolecular in nature. This therefore enforces the formation of a 1,2-*cis*-glycosidic linkage since the nucleophile must be delivered to the anomeric centre formally *syn* to the 2-hydroxyl group.

As part of our ongoing interest in the synthesis of oligosaccharides containing 1,2-*cis*-linkages we recently reported the development of an N-iodosuccinimide (*NIS*) mediated one-pot modification of the original *Hindsgaul* methodology [4, 5]. Subsequently we have developed 2-O-allyl protected thioglycosides as glycosyl donors for IAD [6]. Herein, an enol ether at position 2 of the glycosyl donor is accessed in near quantitative yield by isomerization of a 2-O-allyl protecting group. Subsequent tethering and intramolecular glycosylation then yield the 1,2-*cis*-glycoside. However, whereas good yields were obtained for *NIS* mediated

* Corresponding author. E-mail: antony.fairbanks@chemistry.ox.ac.uk

tethering of thioglycoside donors with a variety of simple aglycon alcohols, in the cases of more hindered carbohydrate alcohols where tethering occurred more slowly, limitations arose due to competitive activation of the anomeric leaving group. The use of an orthogonal anomeric leaving group which cannot be activated during tethering could provide a solution to this problem. Reported herein are full details of investigations into the use of glycosyl fluorides as donors for allyl mediated IAD [7].

Results and Discussion

As substrates for IAD investigations the manno (**1**) and gluco glycosyl (**2**) fluorides bearing allyl protection of the 2-hydroxyl were prepared from their respective orthoesters. Thus, *DAST* opening of **3** [8] and **4** [9] followed by an *n*-propylamine mediated deacetylation of acetates **5** and **6** gave alcohols **7** and **8** [10]. Finally, reprotection of the 2-hydroxyl group by treatment with allyl bromide and sodium hydride in *DMF* furnished the desired donors **1** and **2** in excellent overall yields (**1**: 78% yield over three steps from **3**; **2**: 84% yield over three steps from **4**; Scheme 1).

The extremely efficient allyl isomerization following *Boons*' protocol [11] proceeded as smoothly as had been previously observed in the cases of 2-O-allyl protected thioglycosides [6]; thus, enol ethers **9** and **10** were furnished in excellent yields (96% and 98%, respectively; Scheme 2).

Studies then turned to an investigation of tethering reactions of **9** and **10** with a variety of aglycon alcohols. Previous work had demonstrated that N-iodosuccinimide (*NIS*) was a suitable reagent for this type of tethering reaction, whereas attempted acid catalyzed tethering had proved unsuccessful. Therefore, *NIS*-mediated tethering of the manno vinyl ether **9** was carried out with a variety of alcohols **11a–f,h** under our established reaction conditions [12]. Pleasingly, the desired mixed acetals **12a–f,h** were obtained in good to excellent yields (Table 1). The tethering yields for simple non-carbohydrate and primary carbohydrate alcohols are excellent. Also notable are the cases of the secondary carbohydrate alcohols **11e** and **11f** and the steroid **11h**, whereby longer reaction times and thus

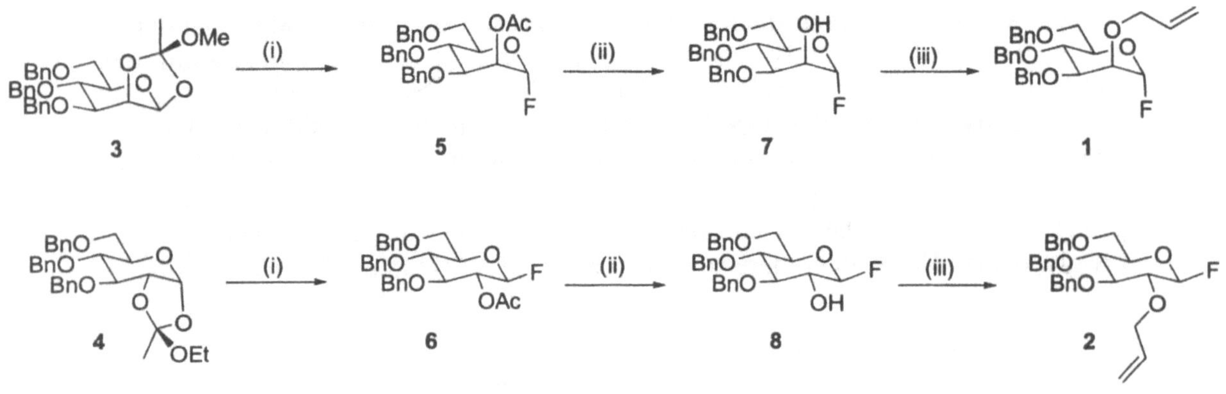

Scheme 1. (i) *DAST*, *DCM*, 0°C; **5**: 98%, **6**: 84%; (ii) *n*-PrNH$_2$, MeOH, *THF*, 45°C; **7**: 83%, **8**: 91% from **4**; (iii) allyl bromide, NaH, *DMF*, 0°C; **1**: 96%, **2**: 93%

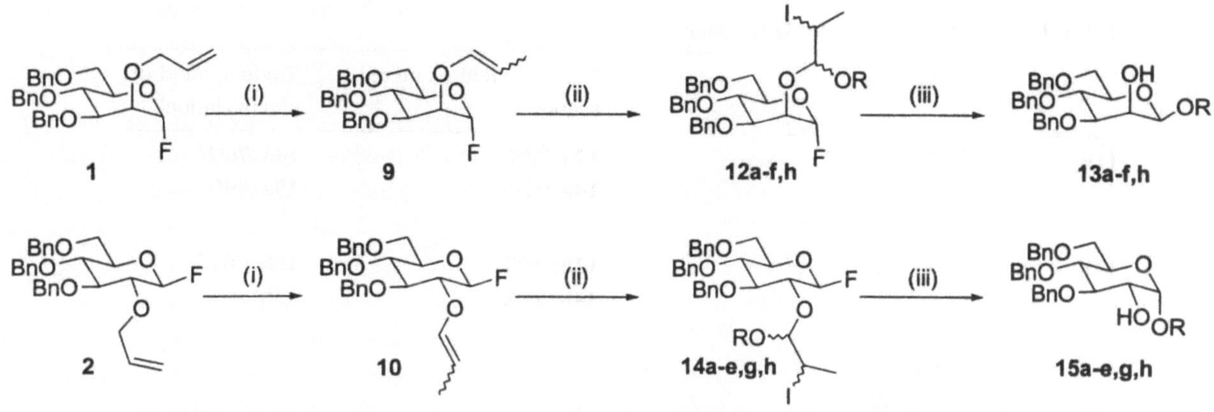

Scheme 2. (i) *Wilkinson*'s catalyst, BuLi, *THF*, 70°C; **9**: 96% from **1**, **10**: 98% from **2**; (ii) *R*OH **13a–h**, *NIS*, *DCE*, −40°C to r.t., 4 Å sieves; (iii) AgO*Tf*, *DTBMP*, $SnCl_2$, *DCE* or MeCN, 50°C; then *TFA*, H_2O or *NIS*, H_2O

more efficient tethering were possible than during previous studies on thioglycosides where competitive anomeric activation occurred. However, it should be noted that in the case of the more hindered alcohols **11d–f,h** an increasing amount of the succinimide trapped material **16** (Fig. 1) was also isolated. Formation of **16** occurs by competitive nucleophilic attack by succinimide rather than the desired alcohol, and therefore indicates a potential limitation of the use of *NIS* as tethering reagent in these cases.

In the gluco series, tethering of enol ether **10** with alcohols **11a–e,g,h** proceeded likewise to give the mixed acetals **14a–e,g,h**. Whereas excellent yields were obtained for alcohols **11a–c**, slightly lower yields were observed than in the manno series for some of the more bulky carbohydrates **11d,e**. Overall it is clear that in general tethering is marginally less efficient in the gluco series. The steroid **11h** was only partially soluble in dichloroethane (*DCE*), but a change of the solvent to *THF* produced an acceptable 68% yield of mixed acetals **14h**. Again some succinimide trapped material, in the gluco series compound **17**, was observed in the cases of the more hindered alcohols.

Attention then turned to activation of the glycosyl fluorides and intramolecular glycosylation. Although activation of glycosyl fluorides can be carried out by the use of any number of different *Lewis* acids [13], in order to ensure complete stereoselectivity it is crucial that the mixed acetal tether, which itself could be susceptible to *Lewis* acid catalyzed cleavage, remains intact so that IAD can operate. Therefore, a series of several different reaction conditions often employed for the activation of glycosyl fluorides were investigated, including *Cp*$_2$HfCl$_2$/AgClO$_4$/*DTBMP* [14], $BF_3 \cdot OEt_2$ [15], and $BF_3 \cdot OEt_2$/2,6-di-*tert*-butyl-4-methylpyridine (*DTBMP*). Whereas some of the desired products were usually observed, in these cases product formation was either accompanied by hydrolysis of the tether to various extents, the formation of α/β mixtures of products (presumably arising from intermolecular glycosylation following the hydrolysis of the tether), or occasionally the formation of an intractable mixture of products. Optimum results were obtained using the *Mukaiyama* activation

Table 1. Tethering and glycosylation yields

	Alcohol	Product/yield of mixed acetals[a]	Product/yield of glycosylation[c]
11a	MeOH	**12a**/97% **14a**/99%	**13a**/76%[e] **15a**/89%[e]
11b	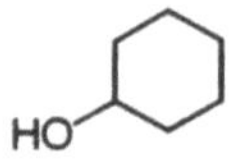	**12b**/99% **14b**/99%	**13b**/70%[e] **15b**/66%[e]
11c	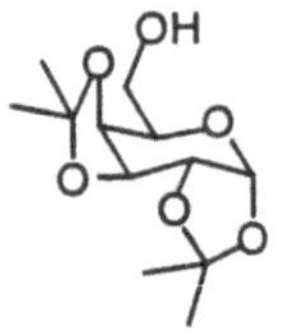	**12c**/98% **14c**/91%	**13c**/61%[d,f] **15c**/63%[d]
11d	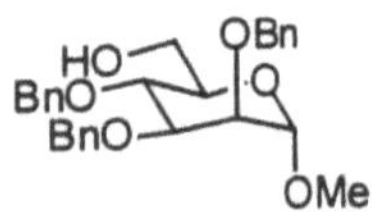	**12d**/80% **14d**/79%	**13d**/49%,[e] 75%[d] **15d**/44%,[e] 46%[d]
11e	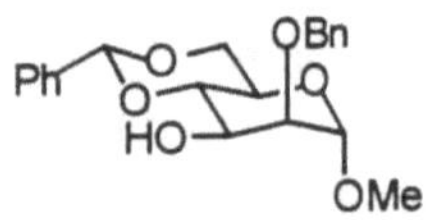	**12e**/83% **14e**/52%	**13e**/55%[d,f] **15e**/66%[d,g]
11f	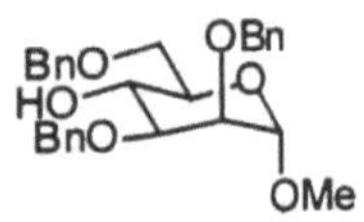	**12f**/37%	**13f**/50%[d]
11g	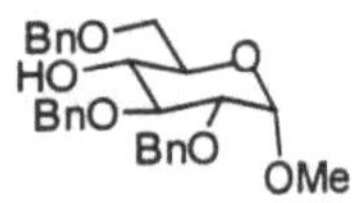	**14g**/39%	**15g**/45%[d]
11h	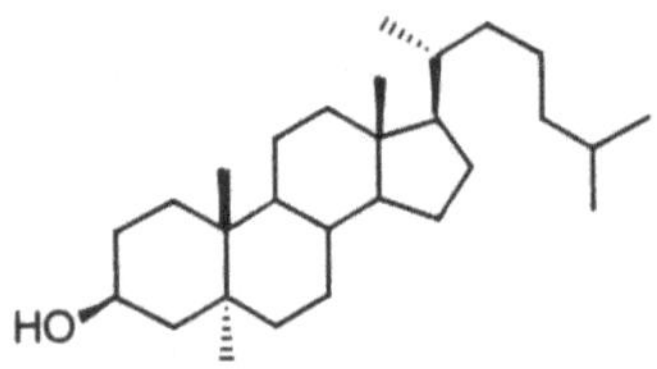	**12h**/72% **14h**/68%[b]	**13h**/59%[d] **15h**/87%[d]

[a] Isolated yields; reactions carried out in *DCE* except where otherwise stated; [b] reaction carried out in *THF*; [c] isolated yields, acid treatment with *TFA* except where otherwise stated; [d] reaction carried out in *DCE*; [e] reaction carried out in acetonitrile; [f] no acid treatment; [g] treatment with *NIS*, H_2O

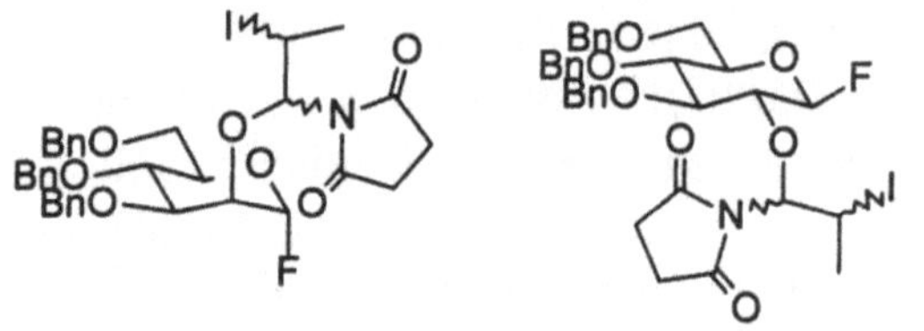

16 **17**

Fig. 1. Succinimide trapped materials

conditions [16], *i.e.* tin(II) chloride in combination with silver triflate, addition of the hindered base *DTBMP*, and heating to 50°C so that the reaction proceeded at an efficient rate. Under these conditions all mixed acetals furnished the desired β-mannosides **13a–f,h** and α-glucosides **15a–e,g,h** stereospecifically as the pure 1,2-*cis*-anomers in all cases in good to excellent yields (Table 1).

Two features of the intramolecular glycosylation reaction are worth mentioning. The first is that the efficiency of glycosylation is solvent dependent. For example, it was observed that whereas the glycosylation of the manno mixed acetals **12c** was slow in acetonitrile and occurred at the same time as partial hydrolysis of the tether, an otherwise identical experiment employing *DCE* as solvent proceeded very quickly (*ca.* 2 h) and without hydrolysis. Subsequent glycosylation reactions were therefore carried out with *DCE* as the solvent.

The second important feature worthy of special consideration is the potential formation of side-products by nucleophilic trapping of the oxonium ion produced subsequent to the intramolecular glycosylation reaction. Indeed the question as to the possible fate of this type of oxonium ion has recently arisen in the *Ogawa/Ito para*-methoxybenzyl (*PMB*) IAD system [17]. In our earlier work on thioglycosides [5], byproducts were observed after glycosylation that were identified as mixed acetals such as **18** (Fig. 2). These presumably arose from trapping of the oxonium ion produced subsequent to the glycosylation reaction by an external alcohol acting as a nucleophile. In the thioglycoside cases, treatment of the crude reaction mixture with trifluoroacetic acid (*TFA*) during work-up resulted in the hydrolysis of any such mixed acetals and in an increased the yield of the desired 1,2-*cis*-glycoside.

Analysis of the intramolecular glycosylation reactions of the tethered glycosyl fluorides by TLC revealed that varying amounts of byproducts were also sometimes produced. Interestingly, the formation of these byproducts was generally more prevalent in the gluco rather than the manno series. Treatment of these materials with *TFA* following the standard procedure developed earlier always yielded further 1,2-*cis*-glycoside product, but it was notable that by-product hydrolysis was much faster than had previously been observed in the thioglycoside work. Therefore, in the case of alcohol **11e**, which possesses an acid sensitive protecting group which could be cleaved by acid treatment, rather than adopting this acid work-up procedure in order to optimize the yield of disaccharide product, the identity of this by-product was investigated. Rather surprisingly, the minor byproducts were identified as the enol ethers **19** [18]. Treatment of these enol ethers **19** with *NIS*/H_2O proved an excellent alternative to the standard acidic work-up procedure since it rapidly resulted in the formation of the desired glycoside **15e**, without any cleavage of the acid sensitive 4,6-benzylidene protecting group (Scheme 3).

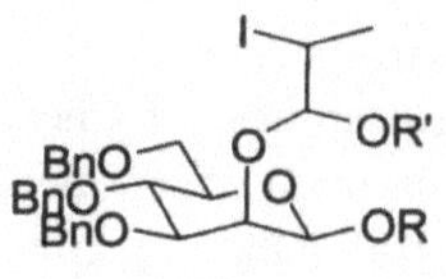

18

Fig. 2. Trapping after glycosylation

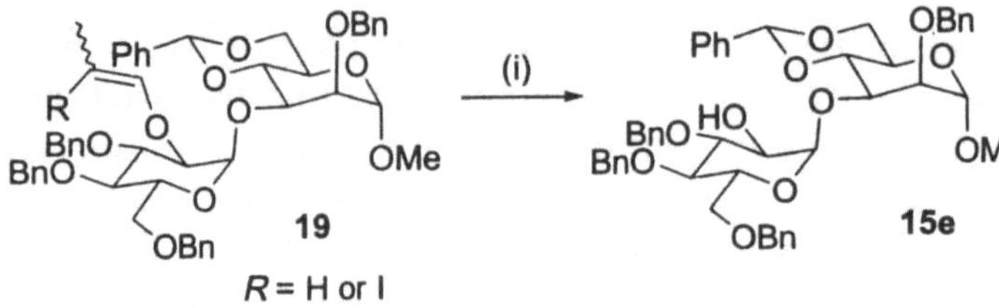

Scheme 3. (i) *NIS*, *THF*, H_2O; then Et_3N, ~70%

In summary it has been demonstrated that glycosyl fluorides are excellent glycosyl donors for the allyl mediated IAD approach to 1,2-*cis*-glycosides, and that tethering and stereospecific intramolecular glycosylation may be achieved for a variety of primary and secondary carbohydrate alcohols. In particular, the use of glycosyl fluorides is advantageous in that tethering efficiency can be increased in the case of bulky secondary carbohydrate alcohols by the use of extended reaction times. Comparison with alternative methods of IAD indicates that allyl mediated IAD is superior to the original approaches of *Stork* and *Hindsgaul* in terms of both simplicity of application and yield and can therefore be considered as complementary to the *Ogawa PMB* approach. Although tethering is more efficient in the *PMB* system, the allyl system has the advantages that there is no requirement for cyclic 4,6-protection of the glycosyl donor in order to produce good yields for the glycosylation step and that the technique is also applicable for the formation of α-gluco linkages. The current disadvantages of allyl IAD are competitive trapping by succinimide during attempted tethering with the most hindered alcohol acceptors and the formation of mixed acetal or enol ether side products by further reaction of the oxonium ion produced after intramolecular glycosylation has occurred. Further investigations into the use of allyl derived vinyl ethers for 1,2-*cis*-glycosylation procedures are currently in progress, and the results will be reported in due course.

Experimental

General

Melting points were recorded on a Kofler hot block and are uncorrected. Proton nuclear magnetic resonance spectra were recorded on Bruker DPX 400 (400 MHz) or Bruker AMX 500 (500 MHz) spectrometers. Carbon nuclear magnetic resonance spectra were recorded on Bruker AC 200 (50.3 MHz), Bruker DPX 400 (100.6 MHz), or Bruker AMX 500 (125.7 MHz) spectrometers. Multiplicities were assigned using the DEPT sequence. All chemical shifts are quoted on the δ-scale in ppm. Infrared spectra were recorded on a Perkin-Elmer 150 *Fourier* transform spectrophotometer. Low resolution mass spectra were recorded on a Micromass Platform 1 APCI using atmospheric pressure chemical ionisation (APCI). High resolution mass spectra (electrospray) were performed on a Waters 2790-Micromass LCT electrospray ionization mass spectrometer or by the EPSRC Mass Spectrometry Service Centre, Department of Chemistry, University of Wales, Swansea, on a MAT900 XLT electrospray ionization mass spectrometer. Optical rotations were measured on a Perkin-Elmer 241 polarimeter with a path length of 1 dm. Concentrations are given in $g/100\,cm^3$. Microanalyses were performed by the microanalytical services of Elemental Microanalysis Ltd, Devon; the results agreed with the calculated values within experimental error. Thin layer chromatography (TLC) was carried out on Merck Kieselgel 0.22–0.25 mm thickness glass-backed sheets pre-coated with $60F_{254}$ silica. Plates were developed using 5% w/v ammonium molybdate in 2 *M* H_2SO_4. Flash column chromatography was carried out using Sorbsil C60 40/60 silica. Solvents and reagents were dried and purified before use according to standard procedures; MeOH was distilled from NaH, CH_3CN and CH_2Cl_2 were distilled from CaH_2, and *THF* was distilled from a solution of sodium benzophenone ketyl immediately before use. Anhydrous *DMF* and anhydrous dichloroethane (*DCE*) were purchased from Aldrich. Petrol refers to the fraction of light petroleum ether boiling in the range of 40–60°C. All procedures requiring anhydrous conditions were performed under an Ar atmosphere in glassware which was flame-dried before use.

2-O-Allyl-3,4,6-tri-O-benzyl-α-D-mannopyranosyl fluoride (**1**; $C_{30}H_{33}FO_5$)

Alcohol **7** (960 mg, 2.12 mmol) was dissolved in anhydrous *DMF* (5 cm^3) and cooled to 0°C. Allyl bromide (0.36 cm^3, 4.24 mmol) was added; then, NaH (60% in mineral oil, 170 mg, 4.24 mmol) was added portionwise. After 1 h, TLC (ethyl acetate:petrol = 1:3) indicated formation of a single product ($R_f = 0.5$) and complete consumption of starting material ($R_f = 0.2$). The reaction mixture was quenched with 100 cm^3 H_2O and extracted with 2 × 100 cm^3 CH_2Cl_2. The combined organic extracts were dried ($MgSO_4$), filtered, and concentrated *in vacuo*. The residue was purified by flash column chromatography (petrol:ethyl acetate = 5:1) to afford allylated glycosyl fluoride **1** (1.01 g, 96%) as a colourless oil.

$[\alpha]_D^{27} = +35.2$ ($c = 1.4$, $CHCl_3$); ^{1}H NMR (400 MHz, $CDCl_3$): $\delta = 3.71$ (1H, dd, $J_{5,6} = 1.9$ Hz, $J_{6,6'} = 10.9$ Hz, H-6), 3.67 (1H, dd, $J_{5,6'} = 4.5$ Hz, H-6′), 3.84–3.85 (1H, m, H-2), 3.86–3.98 (2H, m, H-3, H-5), 4.02 (1H, at, $J = 9.5$ Hz, H-4), 4.17 (1H, dd, $J_{gem} = 13.1$ Hz, $J = 6.0$ Hz, OC*H*H′), 4.25 (1H, dd, $J = 5.7$ Hz, OCH*H′*), 4.51, 4.88 (2H, ABq, J_{AB} 10.8 Hz, $PhCH_2$), 4.54, 4.68 (2H, ABq, $J_{AB} = 12.3$ Hz, $PhCH_2$), 4.73 (2H, s, $PhCH_2$), 5.24 (1H, dd, $J_{gem} = 1.5$ Hz, $J_Z = 10.3$ Hz, $OCH_2CH{=}CH_EH_Z$), 5.31 (1H, dd, $J_E = 17.2$ Hz, $OCH_2CH{=}CH_EH_Z$), 5.64 (1H, dd, $J_{1,2} = 1.8$ Hz, $J_{1,F} = 50.6$ Hz, H-1), 5.93 (1H, ddat, $OCH_2CH{=}CH_2$), 7.16–7.41 (15H, m, Ar–H); ppm; ^{13}C NMR ($CDCl_3$, 100 MHz): $\delta = 68.5$ (t, C-6), 72.6, 72.6 (2 × t, $PhCH_2$, $OCH_2CH{=}CH_2$), 73.3 (dd, $^2J_{C\text{-}2,F} =$ 35.3 Hz, C-2), 73.4, 75.2 (2 × t, 2 × $PhCH_2$), 73.9 (d, C-4), 74.0 (d, C-5), 79.0 (d, C-3), 106.4 (dd, $^1J_{C\text{-}1,H\text{-}1} = 182.2$ Hz, $^1J_{C\text{-}1,F} = 222.6$ Hz, C-1), 118.1 (t, $OCH_2CH{=}CH_2$), 127.6, 127.7, 127.8, 127.9, 127.9, 128.1, 128.1, 128.3, 128.3, 128.5, 128.7 (11 × d, Ar–CH), 134.5 (d, $OCH_2CH{=}CH_2$), 138.0, 138.0, 138.1 (3 × s, Ar–C); ppm; MS ($APCI^+$): $m/z = 515$ ($M + Na^+$, 100%); HRMS: calcd. for $C_{30}H_{37}NO_5F$ (MNH_4^+): 510.2656, found: 510.2657.

2-O-Allyl-3,4,6-tri-O-benzyl-β-D-glucopyranosyl fluoride (**2**; $C_{30}H_{33}FO_5$)

Alcohol **8** (3.94 g, 8.70 mmol) was dissolved in anhydrous *DMF* (10 cm^3) and cooled to 0°C. Allyl bromide (1.5 cm^3, 17.4 mmol) was added; then, NaH (60% in mineral oil, 696 mg, 17.4 mmol) was added portionwise. After 1 h, TLC (ethyl acetate:petrol = 1:3) indicated formation of a single product ($R_f = 0.4$) and complete consumption of starting material ($R_f = 0.2$). The reaction was quenched with MeOH (6 cm^3) and concentrated *in vacuo*. The residue was partitioned between CH_2Cl_2 (2 × 100 cm^3) and H_2O (100 cm^3). The combined organic layers were dried ($MgSO_4$), filtered, and concentrated *in vacuo*. The residue was purified by flash column chromatography (petrol:ethyl acetate = 6:1) to afford allylated glycosyl fluoride **2** (3.99 g, 93%) as a colourless oil.

$[\alpha]_D^{22} = +9.2$ ($c = 0.5$, $CHCl_3$); ^{1}H NMR (400 MHz, $CDCl_3$): $\delta = 3.49$ (1H, ddd, $J_{1,2} = 6.9$ Hz, $J_{2,3} = 8.6$ Hz, $J_{2,F} = 12.3$ Hz, H-2), 3.59 (1H, ddd, $J_{4,5} = 9.5$ Hz, $J_{5,6} = 2.6$ Hz, $J_{5,6'} = 3.7$ Hz, H-5), 3.64 (1H, at, $J = 8.7$ Hz, H-3), 3.69–3.77 (3H, m, H-4, H-6, H-6′), 4.20 (1H, dd, $J_{gem} = 12.4$ Hz, $J = 5.8$ Hz, OC*H*H′), 4.34 (1H, dd, $J = 5.7$ Hz, OCH*H′*), 4.54, 4.93 (2H, ABq, $J_{AB} = 10.8$ Hz, $PhCH_2$), 4.55, 4.64 (2H, ABq, $J_{AB} = 12.2$ Hz, $PhCH_2$), 4.80, 4.93 (2H, ABq, $J_{AB} = 10.8$ Hz, $PhCH_2$), 5.21 (1H, dd, $J_{1,F} = 52.8$ Hz, H-1), 5.22 (1H, dd, $J_{gem} = 1.6$ Hz, $J_Z = 10.3$ Hz, $OCH_2CH{=}CH_EH_Z$), 5.32 (1H, dd, $J_E = 17.3$ Hz, $OCH_2CH{=}CH_EH_Z$), 5.95 (1H, ddt, $OCH_2CH{=}CH_2$), 7.14–7.38 (15H, m,

Ar–H) ppm; ^{13}C NMR δ_C ($CDCl_3$, 100 MHz): δ = 68.3, 73.3, 73.5, 75.0, 75.5 (5 × t, 3 × $PhCH_2$, $OCH_2CH=CH_2$, C-6), 74.7, 76.8, 83.4 (3 × d, C-3, C-4, C-5), 81.2 (dd, $^2J_{C\text{-}2,F}$ = 21.5 Hz, C-2), 109.7 (dd, $^1J_{C\text{-}1,F}$ = 215.5 Hz, C-1), 117.6 (t, $OCH_2CH=CH_2$), 127.7, 127.8, 127.8, 127.9, 127.9, 128.4, 128.4 (7 × d, Ar–CH), 134.3 (d, $OCH_2CH=CH_2$), 137.8, 137.8, 138.2 (3 × s, Ar–C) ppm; MS (APCI$^+$): m/z = 515 (M + Na$^+$, 10%).

2-O-Acetyl-3,4,6-tri-O-benzyl-α-D-mannopyranosyl fluoride (**5**; $C_{29}H_{31}FO_6$)

Orthoester **3** (16.7 g, 32.8 mmol) was dissolved in 80 cm^3 freshly distilled CH_2Cl_2. The mixture was cooled to 0°C, and (diethylamino)-sulfur trifluoride (*DAST*, 5.1 cm^3, 37.7 mmol) was added. The reaction mixture was stirred at 0°C. After 7 h, TLC (petrol:ethyl acetate = 3:1) indicated formation of a product (R_f = 0.5) and complete consumption of starting material (R_f = 0.3). The reaction was diluted with diethyl ether (200 cm^3) and then washed with H_2O (2 × 200 cm^3). The combined organics were dried ($MgSO_4$), filtered, and concentrated *in vacuo*. The residue was purified by flash column chromatography (petrol:ethyl acetate = 5:1, 1% Et_3N) to afford acetylated glycosyl fluoride **5** (15.8 g, 98%) as a pale yellow oil.

$[\alpha]_D^{23}$ = +13.3 (c = 1.0, $CHCl_3$); IR (thin film): ν = 1751 (s, C=O) cm^{-1}; ^{1}H NMR (400 MHz, $CDCl_3$): δ = 2.17 (3H, s, CH_3), 3.72 (1H, d, $J_{6,6'}$ = 10.9 Hz, H-6), 3.82 (1H, dd, $J_{5,6'}$ = 2.7 Hz, H-6′), 3.97–3.99 (3H, m, H-3, H-4, H-5), 4.49, 4.87 (2H, ABq, J_{AB} = 10.7 Hz, $PhCH_2$), 4.52, 4.69 (2H, ABq, J_{AB} = 12.1 Hz, $PhCH_2$), 4.57, 4.72 (2H, ABq, J_{AB} = 10.9 Hz, $PhCH_2$), 5.48 (1H, br s, H-2), 5.62 (1H, dd, $J_{1,2}$ = 1.9 Hz, $J_{1,F}$ = 49.1 Hz, H-1), 7.15–7.38 (15H, m, Ar–H) ppm; ^{13}C NMR ($CDCl_3$, 100 MHz): δ = 21.0 (q, CH_3), 67.0 (dd, $^2J_{C\text{-}2,F}$ = 39.9 Hz, C-2), 68.2 (t, C-6), 72.1, 73.5, 75.3 (3 × t, 3 × $PhCH_2$), 73.3 (d, C-4), 73.8 (dd, $^3J_{C\text{-}5,F}$ = 2.0 Hz, C-5), 77.2 (d, C-3), 105.5 (dd, $^1J_{C\text{-}1,H}$ = 182.6 Hz, $^1J_{C\text{-}1,F}$ = 221.0 Hz, C-1), 127.8, 127.8, 127.9, 128.0, 128.1, 128.4, 128.5 (7 × d, Ar–CH), 137.6, 138.0, 138.1 (3 × s, Ar–C), 170.1 (s, C=O) ppm; MS (APCI$^+$) = 517 (M + Na$^+$, 18%); HRMS: calcd. for $C_{29}H_{35}NO_6F$ (MNH_4^+): 512.2448, found: 512.2451.

2-O-Acetyl-3,4,6-tri-O-benzyl-β-D-glucopyranosyl fluoride (**6**; $C_{29}H_{31}FO_6$)

Orthoester **4** (1.0 g, 1.92 mmol) was dissolved in 8 cm^3 freshly distilled CH_2Cl_2. The mixture was cooled to 0°C, and *DAST* (0.33 cm^3, 2.50 mmol) was added. The reaction mixture was stirred at 0°C. After 3.5 h, TLC (CH_2Cl_2: ether = 14:1) indicated formation of a product (R_f = 0.8) and complete consumption of starting material (R_f = 0.7). The reaction was quenched with ice-water (200 cm^3) and extracted with diethyl ether (2 × 200 cm^3). The combined organics were dried ($MgSO_4$), filtered, and concentrated *in vacuo*. The residue was purified by flash column chromatography (petrol:ethyl acetate = 4:1) to afford acetylated glycosyl fluoride **6** (796 mg, 84%) as a colourless oil.

$[\alpha]_D^{22}$ = +15.3 (c = 1.25, $CHCl_3$); IR (thin film): ν = 1750 (s, C=O) cm^{-1}; ^{1}H NMR (400 MHz, $CDCl_3$): δ = 2.02 (3H, s, CH_3), 3.67–3.75 (4H, m, H-3, H-5, H-6, H-6′), 3.86 (1H, at, J = 9.0 Hz, H-4), 4.56, 4.78 (2H, ABq, J_{AB} = 11.2 Hz, $PhCH_2$), 4.56, 4.63 (2H, ABq, J_{AB} = 12.1 Hz, $PhCH_2$), 4.71, 4.78 (2H, ABq, J_{AB} = 11.6 Hz, $PhCH_2$), 5.09–5.15 (1H, m, H-2), 5.27 (1H, dd, $J_{1,2}$ = 6.1 Hz, $J_{1,F}$ = 52.9 Hz, H-1), 7.16–7.36 (15H, m, Ar–H) ppm; ^{13}C NMR ($CDCl_3$, 100 MHz): δ = 20.8 (q, CH_3), 68.4, 73.6, 74.4, 74.9 (4 × t, C-6, 3 × $PhCH_2$), 72.5 (dd, C-2), 75.0, 81.5, 81.6 (3 × d, C-3, C-4, C-5), 106.7 (dd, C-1), 127.7, 127.8, 127.8, 127.9, 128.0, 128.4, 128.4, 128.9 (8 × d, Ar–CH), 137.6, 137.8 (2 × s, Ar–C), 169.4 (s, C=O) ppm; ^{19}F NMR ($CDCl_3$, 376 MHz): δ = −136.8 (dd, $J_{1,F}$ = 52.9 Hz, $J_{2,F}$ = 10.2 Hz) ppm; MS (APCI$^+$): m/z = 533 (M + K$^+$, 100%), 517 (M + Na$^+$, 15%), 475 (M-F, 5%).

3,4,6-Tri-O-benzyl-α-D-mannopyranosyl fluoride (**7**; $C_{27}H_{29}FO_5$)

Acetate **5** (2.0 g, 4.04 mmol) was dissolved in MeOH (5 cm^3), *THF* (10 cm^3), and *n*-propylamine (5 cm^3), and the reaction mixture was stirred at 45°C. After 15 h, TLC (ethyl acetate:petrol = 1:3) indicated formation of a product ($R_f = 0.2$) and complete consumption of starting material ($R_f = 0.4$). The mixture was concentrated *in vacuo* and purified by flash column chromatography (petrol:ethyl acetate = 3:1, 1% Et_3N) to afford alcohol **7** (1.56 g, 83%) as a colourless oil.

$[\alpha]_D^{23} = +34.2$ ($c = 0.5$, $CHCl_3$); ^{1}H NMR (400 MHz, $CDCl_3$): $\delta = 3.70$ (1H, dd, $J_{5,6} = 1.6$ Hz, $J_{6,6'} = 10.8$ Hz, H-6), 3.78 (1H, dd, $J_{5,6'} = 3.5$ Hz, H-6′), 3.87–3.90 (1H, m, H-3), 3.94–3.96 (2H, m, H-4, H-5), 4.12–4.13 (1H, m, H-2), 4.54, 4.67 (2H, ABq, $J_{AB} = 12.2$ Hz, $PhCH_2$), 4.54, 4.83 (2H, ABq, $J_{AB} = 10.8$ Hz, $PhCH_2$), 4.71, 4.74 (2H, ABq, $J_{AB} = 11.4$ Hz, $PhCH_2$), 5.68 (1H, dd, $J_{1,2} = 1.7$ Hz, $J_{1,F} = 49.5$ Hz, H-1), 7.17–7.37 (15H, m, Ar–H) ppm; ^{13}C NMR ($CDCl_3$, 100 MHz): $\delta = 67.1$ (dd, $^2J_{C\text{-}2,F} = 39.0$ Hz, C-2), 68.2 (d, C-6), 72.4, 73.5, 75.2 (3 × t, 3 × $PhCH_2$), 73.2 (d, C-4), 73.4 (dd, $^3J_{C\text{-}5,F} = 2.7$ Hz, C-5), 79.1 (dd, $^3J_{C\text{-}3,F} = 1.8$ Hz, C-3), 107.2 (dd, $^1J_{C\text{-}1,H\text{-}1} = 181.9$ Hz, $^1J_{C\text{-}1,F} = 217.7$ Hz, C-1), 127.7, 127.8, 127.9, 127.9, 128.0, 128.1, 128.4, 128.6 (8 × d, Ar–CH), 137.6, 137.9, 138.0 (3 × s, Ar–C) ppm.

3,4,6-Tri-O-benzyl-β-D-glucopyranosyl fluoride (**8**; $C_{27}H_{29}FO_5$)

Orthoester **4** (5.00 g, 9.62 mmol) was dissolved in 20 cm^3 freshly distilled CH_2Cl_2 in a flame-dried flask, cooled to 0°C under Ar, and *DAST* (1.65 cm^3, 12.5 mmol) was added. After 7.5 h, TLC (ethyl acetate:petrol = 1:3) indicated formation of a product ($R_f = 0.45$) and complete consumption of starting material ($R_f = 0.5$). The reaction was quenched with water (200 cm^3) and extracted with diethyl ether (2 × 150 cm^3). The organic extracts were dried ($MgSO_4$), filtered, and concentrated *in vacuo*. The residue was dissolved in *THF* (10 cm^3), MeOH (5 cm^3), and *n*-propylamine (5 cm^3) and stirred at 45°C. After 16.5 h, TLC (ethyl acetate:petrol = 1:3) indicated formation of a single product ($R_f = 0.2$). The mixture was concentrated *in vacuo*. The residue was purified by flash column chromatography (petrol:ethyl acetate = 4:1) to afford alcohol **8** (3.94 g, 91%) as a white crystalline solid.

M.p.: 84°C (diethyl ether/petrol; Ref. [10]: 80–82°C); $[\alpha]_D^{22} = +31.5$ ($c = 0.55$ in $CHCl_3$; Ref. [10]: $[\alpha]_D^{22} = +25.2$); ^{1}H NMR (400 MHz, $CDCl_3$): $\delta = 3.59$–3.80 (6H, m, H-2, H-3, H-4, H-5, H-6, H-6′), 4.54, 4.62 (2H, ABq, $J_{AB} = 12.0$ Hz, $PhCH_2$), 4.57, 4.78 (2H, ABq, $J_{AB} = 11.1$ Hz, $PhCH_2$), 4.81, 4.85 (2H, ABq, $J_{AB} = 11.5$ Hz, $PhCH_2$), 5.17 (1H, d, $J_{1,2} = 6.5$ Hz, $J_{1,F} = 53.0$ Hz, H-1), 7.17–7.40 (15H, m, Ar–H) ppm.

2-O-Prop-1′-enyl-3,4,6-tri-O-benzyl-α-D-mannopyranosyl fluoride (**9**; $C_{30}H_{33}FO_5$)

Wilkinson's catalyst (185 mg, 0.20 mmol) was dissolved in 4 cm^3 freshly distilled *THF*, and the solution was degassed. *n*-Butyllithium (1.6 *M* solution in hexanes, 0.19 cm^3, 0.30 mmol) was added, and the mixture was stirred for 1 h. Allylated mannosyl fluoride **1** (975 mg, 1.98 mmol) was dissolved in 4 cm^3 freshly distilled *THF* and heated to 70°C. The catalyst solution was added *via* a cannular under Ar. After 1 h, TLC (ethyl acetate:petrol = 1:3) indicated formation of a single product ($R_f = 0.51$) and complete consumption of starting material ($R_f = 0.49$). The reaction was allowed to cool, diluted with CH_2Cl_2 (20 cm^3), and concentrated *in vacuo*. The residue was purified by flash column chromatography (petrol:ethyl acetate = 6:1, 1% Et_3N) to afford vinyl ethers **9** (934 mg, 96%) as a pale yellow oil.

Partial data: ^{1}H NMR (400 MHz, $CDCl_3$): *Z*:*E* = 2.7:1; *E*-isomer: $\delta = 1.55$ (3H, dd, $J = 1.7$ Hz, $J = 6.8$ Hz, OCH=CHC*H*$_3$), 5.02 (1H, dq, $J_E = 12.4$ Hz, OCH=C*H*CH$_3$), 5.66 (1H, dd, $J_{1,2} = 1.9$ Hz, $J_{1,F} = 50.2$ Hz, H-1), 6.12 (1H, dq, OC*H*=CHCH$_3$) ppm; *Z*-isomer: $\delta = 1.63$ (3H, dd, $J = 1.6$ Hz, $J = 6.9$ Hz, OCH=CHC*H*$_3$), 5.62 (1H, dd, $J_{1,2} = 2.0$ Hz, $J_{1,F} = 50.5$ Hz, H-1), 5.97 (1H, dq, $J_Z = 6.0$ Hz, OC*H*=CHCH$_3$) ppm; ^{13}C NMR ($CDCl_3$, 100 MHz): *E*-isomer: $\delta = 12.3$ (q, OCH=CHC*H*$_3$), 102.4 (d, OCH=*C*HCH$_3$), 105.6 (dd, $^1J_{C\text{-}1,F} = 221.7$ Hz, C-1), 145.5 (d, O*C*H=CHCH$_3$) ppm; *Z*-isomer: $\delta = 9.4$ (q, OCH=CH*C*H$_3$), 104.2 (d, OCH=*C*HCH$_3$), 106.1 (dd, $^1J_{C\text{-}1,F} = 220.7$ Hz, C-1), 144.5 (d, O*C*H=CHCH$_3$) ppm; MS (APCI$^+$): $m/z = 515$ (M + Na$^+$, 100%); HRMS: calcd. for $C_{30}H_{37}NO_5F$ (MNH_4^+): 510.2656, found: 510.2663.

2-O-Prop-1′-enyl-3,4,6-tri-O-benzyl-β-D-glucopyranosyl fluoride (**10**; $C_{30}H_{33}FO_5$)

Wilkinson's catalyst (390 mg, 0.42 mmol) was dissolved in 8 cm^3 freshly distilled *THF*, and the solution was degassed. *n*-Butyl lithium (1.6 *M* solution in hexanes, 0.4 cm^3, 0.63 mmol) was added, and the mixture was stirred for 1 h. Allylated glucosyl fluoride **2** (2.06 g, 4.19 mmol) was dissolved in 8 cm^3 freshly distilled *THF* and heated to 70°C. The catalyst solution was added *via* a cannular under Ar. After 45 min, the reaction was allowed to cool, diluted with 20 cm^3 CH_2Cl_2, and concentrated *in vacuo*. The residue was purified by flash column chromatography (petrol:ethyl acetate = 5:1, 1% Et_3N) to afford vinyl ethers **10** (2.02 g, 98%) as a pale yellow oil.

Partial data: ^{1}H NMR (400 MHz, $CDCl_3$): *E*:*Z* = 1:2.3; *E*-isomer: δ = 1.58 (3H, dd, J = 1.6 Hz, J = 7.0 Hz, OCH=CHC*H*$_3$), 5.08 (1H, dq, J_E = 12.2 Hz, OCH=C*H*CH$_3$), 5.23 (1H, dd, $J_{1,2}$ = 6.3 Hz, $J_{1,F}$ = 52.5 Hz, H-1), 6.23 (1H, d, OC*H*=CHCH$_3$) ppm; *Z*-isomer: δ = 1.65 (3H, dd, J = 1.7 Hz, J = 6.8 Hz, OCH=CHC*H*$_3$), 4.51 (1H, aquin, J = 6.6 Hz, OCH=C*H*CH$_3$), 5.25 (1H, m, $J_{1,2}$ = 6.4 Hz, $J_{1,F}$ = 52.7 Hz, H-1), 6.16 (1H, dd, J_Z = 6.1 Hz, OC*H*=CHCH$_3$) ppm; ^{13}C NMR ($CDCl_3$, 100 MHz): *E*-isomer: δ = 12.2 (q, OCH=CH*C*H$_3$), 102.1 (d, OCH=*C*HCH$_3$), 108.4 (dd, $^1J_{C\text{-}1,F}$ = 217.0 Hz, C-1), 146.4 (d, O*C*H=CHCH$_3$) ppm; *Z*-isomer: δ = 9.3 (q, OCH=CH*C*H$_3$), 101.5 (d, OCH=*C*HCH$_3$), 108.5 (dd, $^1J_{C\text{-}1,F}$ = 213.3 Hz, C-1), 145.6 (d, O*C*H=CHCH$_3$) ppm; MS (ES$^+$): m/z = 510 (M + NH_4^+, 100%); HRMS: calcd. for $C_{30}H_{37}NO_5F$ (MNH_4^+): 510.2656, found: 510.2663.

General procedure for NIS mediated tethering

NIS (3 equiv.) and powdered molecular sieves (4 Å) were stirred in 1 cm^3 dry *DCE* under Ar at −40°C. The alcohol (1.5–3 equiv.) was added *via* a cannular under Ar in 1.5 cm^3 dry *DCE*. The vinyl ether (0.1 mmol) was added *via* a cannular under Ar in 1.5 cm^3 dry *DCE*, and the mixture was allowed to warm slowly to room temperature. After 1–24 h, the mixture was diluted with *DCM*, washed with aqueous $Na_2S_2O_3$ solution, dried, filtered and concentrated *in vacuo*. The resulting residue was purified by flash column chromatography to give the mixed acetals.

2-O-(2-Iodo-1-methoxypropyl)-3,4,6-tri-O-benzyl-α-D-mannopyranosyl fluoride (**12a**; $C_{31}H_{36}FIO_6$)

Vinyl ethers **9** (80 mg, 0.16 mmol), *NIS* (109 mg, 0.49 mmol), and **11a** (0.010 ml, 0.24 mmol) gave mixed acetals **12a** (102 mg, 97%) as a colourless oil; MS (APCI$^+$): m/z = 673 (M + Na$^+$, 15%).

2-O-(1-Cyclohexyloxy-2-iodopropyl)-3,4,6-tri-O-benzyl-α-D-mannopyranosyl fluoride (**12b**; $C_{36}H_{44}FIO_6$)

Vinyl ethers **9** (80 mg, 0.16 mmol), *NIS* (109 mg, 0.49 mmol), and **11b** (0.034 cm^3, 0.33 mmol) gave mixed acetals **12b** (115 mg, 99%) as a colourless oil; MS (APCI$^+$): m/z = 741 (M + Na$^+$, 5%).

2-O-(1-(1,2:3,4-di-O-isopropylidene-α-D-galactopyranos-6-O-yl)-2-iodopropyl)-3,4,6-tri-O-benzyl-α-D-mannopyranosyl fluoride (**12c**; $C_{42}H_{52}FIO_{11}$)

Vinyl ethers **9** (178 mg, 0.36 mmol), *NIS* (243 mg, 1.1 mmol), and **11c** (212 mg, 0.82 mmol) gave mixed acetals **12c** (311 mg, 98%) as a colourless oil; MS (ES$^+$): m/z = 917 (M + K$^+$, 25%), 901 (M + Na$^+$, 100%), 896 (M + NH_4^+, 32%).

2-O-(2-Iodo-1-(methyl 2,3,4-tri-O-benzyl-α-D-mannopyranosid-6-O-yl)-propyl)-3,4,6-tri-O-benzyl-α-D-mannopyranosyl fluoride (**12d**; $C_{58}H_{64}FIO_{11}$)

Vinyl ethers **9** (80 mg, 0.16 mmol), *NIS* (109 mg, 0.49 mmol), and **11d** (151 mg, 0.33 mmol) gave mixed acetals **12d** (140 mg, 80%) as a colourless oil; MS (APCI$^-$): m/z = 1117 (M + Cl$^-$, 92%).

2-O-(2-Iodo-1-(methyl 2-O-benzyl-4,6-O-benzylidene-α-D-mannopyranosid-3-O-yl)-propyl)-3,4,6-tri-O-benzyl-α-D-mannopyranosyl fluoride (**12e**; $C_{51}H_{56}FIO_{11}$)

Vinyl ethers **9** (80 mg, 0.16 mmol), *NIS* (109 mg, 0.49 mmol), and **11e** (121 mg, 0.33 mmol) gave mixed acetals **12e** (134 mg, 83%) as a colourless oil; MS (APCI$^-$): $m/z = 1025$ (M + Cl$^-$, 15%).

2-O-(2-Iodo-1-(methyl 2,3,6-tri-O-benzyl-α-D-mannopyranosid-4-O-yl)-propyl)-3,4,6-tri-O-benzyl-α-D-mannopyranosyl fluoride (**12f**; $C_{58}H_{64}FIO_{11}$)

Vinyl ethers **9** (80 mg, 0.16 mmol), *NIS* (109 mg, 0.49 mmol), and **11f** (226 mg, 0.49 mmol) gave mixed acetals **12f** (65 mg, 37%) as a colourless oil; MS (APCI$^-$): $m/z = 1117$ (M + Cl$^-$, 8%).

2-O-(1-(Cholestan-3-β-yloxy)-2-iodopropyl)-3,4,6-tri-O-benzyl-α-D-mannopyranosyl fluoride (**12h**; $C_{57}H_{80}FIO_6$)

Vinyl ethers **9** (80 mg, 0.16 mmol), *NIS* (109 mg, 0.49 mmol), and **11h** (136 mg, 0.33 mmol) gave mixed acetals **12h** (118 mg, 72%) as a colourless oil; MS (APCI$^-$): $m/z = 1041$ (M + Cl$^-$, 18%).

General procedure for intramolecular glycosylation

2,6-Di-*tert*-butyl-4-methylpyridine (*DTBMP*, 2 equiv.), silver triflate (2 equiv.), anhydrous tin(II) chloride (2 equiv.), and powdered molecular sieves (4 Å) were stirred in 1 cm^3 dry *DCE* under Ar at 50°C. The mixed acetals (0.1 mmol) were added *via* a cannular under argon in 3 cm^3 dry *DCE*, and the reaction was stirred at 50°C until TLC indicated disappearance of starting material. *TFA* (2 cm^3) and H_2O (1 cm^3) were added, and the solution was stirred for further 30 min. Diethyl ether was added, and the mixture filtered through Celite®, washed with saturated aqueous $NaHCO_3$ and brine, dried ($MgSO_4$), filtered, and concentrated *in vacuo*. The resulting residue was purified by flash column chromatography to give the pure 1,2-*cis*-glycoside.

Methyl 3,4,6-tri-O-benzyl-β-D-mannopyranoside (**13a**; $C_{28}H_{32}O_6$)

Mixed acetals **12a** (102 mg, 0.157 mmol), 2,6-di-*tert*-butyl-4-methylpyridine (64 mg, 0.314 mmol), silver trifluoromethanesulfonate (81 mg, 0.314 mmol), and anhydrous tin(II) chloride (60 mg, 0.314 mmol) gave β-mannoside **13a** (55 mg, 76%) as a colourless oil.

$[\alpha]_D^{24} = -13$ ($c = 0.75$, $CHCl_3$; Ref. [4]: $[\alpha]_D = -13$ ($c = 1.0$, $CHCl_3$)); ^{1}H NMR (500 MHz, $CDCl_3$): $\delta = 3.47$ (1H, ddd, $J_{4,5} = 9.5$ Hz, $J_{5,6} = 5.0$ Hz, $J_{5,6'} = 1.3$ Hz, H-5), 3.57 (3H, s, OCH_3), 3.57–3.60 (1H, m, H-3), 3.74 (1H, dd, $J_{6,6'} = 10.8$ Hz, H-6), 3.80 (1H, dd, H-6′), 3.88 (1H, at, $J = 9.4$ Hz, H-4), 4.11 (1H, d, $J_{2,3} = 3.0$ Hz, H-2), 4.35 (1H, s, H-1), 4.55, 4.90 (2H, ABq, $J_{AB} = 10.8$ Hz, $PhCH_2$), 4.58, 4.65 (2H, ABq, $J_{AB} = 12.3$ Hz, $PhCH_2$), 4.69, 4.78 (2H, ABq, $J_{AB} = 11.7$ Hz, $PhCH_2$), 7.22–7.40 (15H, m, Ar–H) ppm; ^{13}C NMR (125 MHz, $CDCl_3$): $^1J_{C\text{-}1,H\text{-}1} = 156.7$ Hz.

Cyclohexyl 3,4,6-tri-O-benzyl-β-D-mannopyranoside (**13b**; $C_{33}H_{40}O_6$)

Mixed acetals **12b** (112 mg, 0.156 mmol), 2,6-di-*tert*-butyl-4-methylpyridine (64 mg, 0.312 mmol), silver trifluoromethanesulfonate (80 mg, 0.312 mmol), and anhydrous tin(II) chloride (59 mg, 0.312 mmol) gave β-mannoside **13b** (50 mg, 60% plus a further 8 mg (10%) by subsequent acid treatment of the combined minor products; overall yield: 58 mg, 70%) as a colourless oil.

$[\alpha]_D^{24} = -20$ ($c = 0.4$, $CHCl_3$; Ref. [4]: $[\alpha]_D = -19$ ($c = 1.6$, $CHCl_3$)); ^{1}H NMR (500 MHz, $CDCl_3$): $\delta = 1.20$–2.02 (10H, m, cyclohexyl-H), 2.50 (1H, br s, OH-2), 3.43–3.45 (1H, m, H-5), 3.58 (1H, br d, $J_{3,4} = 8.6$ Hz, H-3), 3.61–3.81 (3H, m, H-6, H-6′, OCH), 3.86 (1H, at, $J = 9.3$ Hz, H-4),

4.08 (1H, br s, H-2), 4.56 (1H, s, H-1), 4.58, 4.63 (2H, ABq, $J_{AB} = 12.1$ Hz, $PhCH_2$), 4.58, 4.91 (2H, ABq, $J_{AB} = 10.7$ Hz, $PhCH_2$), 4.69, 4.80 (2H, ABq, $J_{AB} = 11.7$ Hz, $PhCH_2$), 7.23–7.41 (15H, m, Ar–H) ppm; ^{13}C NMR (125 MHz, $CDCl_3$): $^1J_{C\text{-}1,H\text{-}1} = 155.3$ Hz.

3,4,6-Tri-O-benzyl-β-D-mannopyranosyl-(1 → 6)-1,2:3,4-di-O-isopropylidene-α-D-galactopyranose (**13c**; $C_{39}H_{48}O_{11}$)

Mixed acetals **12c** (124 mg, 0.14 mmol), 2,6-di-*tert*-butyl-4-methylpyridine (57 mg, 0.28 mmol), silver trifluoromethanesulfonate (72 mg, 0.28 mmol), and anhydrous tin(II) chloride (53 mg, 0.28 mmol) gave β-manno disaccharide **13c** (60 mg, 61%) as a colourless oil.

$[\alpha]_D^{22} = -50.7$ ($c = 0.55$, $CHCl_3$; Ref. [4]: $[\alpha]_D = -53$ ($c = 1.3$, $CHCl_3$)); ^{1}H NMR (400 MHz, $CDCl_3$): $\delta = 1.32$, 1.34, 1.44, 1.54 (12H, 4 × s, 4 × CH_3), 3.42 (1H, ddd, $J_{4,5} = 9.8$ Hz, $J_{5,6} = 2.2$ Hz, $J_{5,6'} = 4.6$ Hz, H-5$_b$), 3.56 (1H, dd, $J_{2,3} = 3.1$ Hz, $J_{3,4} = 9.2$ Hz, H-3$_b$), 3.70–3.78 (3H, m, H-6$_a$, H-6$_b$, H-6$'_b$), 3.92 (1H, at, $J = 9.5$ Hz, H-4$_b$), 4.03 (1H, dat, $J = 2.2$ Hz, $J_{5,6} = 7.8$ Hz, H-5$_a$), 4.13 (1H, dd, $J_{5,6'} = 2.7$ Hz, $J_{6,6'} = 11.2$ Hz, H-6$'_a$), 4.22 (1H, dd, $J_{3,4} = 7.9$ Hz, $J_{4,5} = 1.9$ Hz, H-4$_a$), 4.22 (1H, d, H-2$_b$), 4.32 (1H, dd, $J_{1,2} = 5.0$ Hz, $J_{2,3} = 2.4$ Hz, H-2$_a$), 4.5 (1H, s, H-1$_b$), 4.52, 4.90 (2H, ABq, J_{AB} 10.8 Hz, $PhCH_2$), 4.54, 4.64 (2H, ABq, $J_{AB} = 12.1$ Hz, $PhCH_2$), 4.60 (1H, dd, H-3$_a$), 4.66, 4.78 (2H, ABq, $J_{AB} = 12.0$ Hz, $PhCH_2$), 5.55 (1H, d, H-1$_a$), 7.18–7.39 (15H, m, Ar–H) ppm; ^{13}C NMR (125 MHz, $CDCl_3$): $^1J_{C\text{-}1b,H\text{-}1b} = 160.0$ Hz.

Methyl 3,4,6-tri-O-benzyl-β-D-mannopyranosyl-(1 → 6)-2,3,4-tri-O-benzyl-α-D-mannopyranoside (**13d**; $C_{55}H_{60}O_{11}$)

Mixed acetals **12d** (136 mg, 0.126 mmol), 2,6-di-*tert*-butyl-4-methylpyridine (52 mg, 0.25 mmol), silver trifluoromethanesulfonate (65 mg, 0.312 mmol), and anhydrous tin(II) chloride (48 mg, 0.312 mmol) gave β-manno disaccharide **13d** (125 mg, 75%) as a colourless oil.

$[\alpha]_D^{24} = +16.0$ ($c = 1.1$ in $CHCl_3$; Ref. [2]: $[\alpha]_D = +16.5$ ($c = 1.0$, $CHCl_3$)); ^{1}H NMR (500 MHz, $CDCl_3$): $\delta = 3.30$ (3H, s, CH_3), 3.40 (1H, ddd, $J_{4,5} = 9.7$ Hz, $J_{5,6} = 4.7$ Hz, $J_{5,6'} = 2.4$ Hz, H-5$_b$), 3.52 (1H, dd, $J_{2,3} = 3.0$ Hz, $J_{3,4} = 9.2$ Hz, H-3$_b$), 3.70–3.76 (3H, m, H-6$_b$, H-6$'_b$, H-6$_a$), 3.78–3.81 (3H, m, H-2$_a$, H-4$_a$, H-5$_a$), 3.90–3.92 (1H, m, H-3$_a$), 3.92 (1H, at, $J = 9.4$ Hz, H-4$_b$), 4.06 (1H, d, H-2$_b$), 4.23 (1H, dd, $J_{6,6'} = 10.0$ Hz, H-6$'_a$), 4.37 (1H, s, H-1$_b$), 4.56, 4.93 (2H, ABq, $J_{AB} = 11.4$ Hz, $PhCH_2$), 4.56, 4.63 (2H, ABq, $J_{AB} = 13.0$ Hz, $PhCH_2$), 4.57, 4.91 (2H, ABq, $J_{AB} = 11.0$ Hz, $PhCH_2$), 4.62 (2H, s, $PhCH_2$), 4.69, 4.79 (2H, ABq, $J_{AB} = 12.3$ Hz, $PhCH_2$), 4.71, 4.76 (2H, ABq, $J_{AB} = 12.7$ Hz, $PhCH_2$), 4.72 (1H, d, $J_{1,2} = 1.7$ Hz, H-1$_a$), 7.22–7.41 (30H, m, Ar–H) ppm; ^{13}C NMR (125 MHz, $CDCl_3$): $^1J_{C\text{-}1b,H\text{-}1b} = 157.9$ Hz.

Methyl 3,4,6-tri-O-benzyl-β-D-mannopyranosyl-(1 → 3)-2-O-benzyl-4,6-O-benzylidene-α-D-mannopyranoside (**13e**; $C_{48}H_{52}O_{11}$)

Mixed acetals **12e** (126 mg, 0.127 mmol), 2,6-di-*tert*-butyl-4-methylpyridine (52 mg, 0.25 mmol), silver trifluoromethanesulfonate (65 mg, 0.25 mmol), and anhydrous tin(II) chloride (48 mg, 0.25 mmol) gave β-manno disaccharide **13e** (56 mg, 55%) as a colourless oil.

$[\alpha]_D^{22} = +5.1$ ($c = 0.35$, $CHCl_3$); IR (thin film): $\nu = 3479$ (br, OH) cm^{-1}; ^{1}H NMR (400 MHz, $CDCl_3$): $\delta = 3.27$ (1H, dt, $J_{4,5} = 9.6$ Hz, $J_{5,6} = 3.5$ Hz, H-5$_b$), 3.32 (3H, s, OCH_3), 3.39 (1H, dd, $J_{2,3} = 3.1$ Hz, $J_{3,4} = 9.0$ Hz, H-3$_b$), 3.70 (2H, d, H-6$_b$, H-6$'_b$), 3.79–3.87 (3H, m, H-2$_a$, H-5$_a$, H-6$_a$), 3.89 (1H, at, $J = 9.1$ Hz, H-4$_b$), 3.94 (1H, d, H-2$_b$), 4.19 (1H, at, $J = 9.7$ Hz, H-4$_a$), 4.26 (1H, s, H-1$_b$), 4.28 (1H, dd, $J_{5,6} = 4.2$ Hz, $J_{6,6'} = 9.6$ Hz, H-6$'_a$), 4.49, 4.58 (2H, ABq, $J_{AB} = 12.0$ Hz, $PhCH_2$), 4.49 (1H, dd, $J_{2,3} = 3.3$ Hz, $J_{3,4} = 10.2$ Hz, H-3$_a$), 4.55, 4.87 (2H, ABq, $J_{AB} = 10.8$ Hz, $PhCH_2$), 4.62, 4.77 (2H, ABq, $J_{AB} = 12.2$ Hz, $PhCH_2$), 4.63, 4.73 (2H, ABq, $J_{AB} = 12.2$ Hz, $PhCH_2$), 4.77 (1H, d, $J_{1,2} = 1.4$ Hz, H-1$_a$), 5.61 (1H, s, PhCH), 7.21–7.49 (25H, m, Ar–H) ppm; ^{13}C NMR ($CDCl_3$,

100 MHz): $\delta = 54.9$ (q, OCH_3), 64.0, 68.4, 71.8, 74.0, 75.1, 75.9, 77.2, 80.8 (8 × d, C-2_b, C-3_b, C-4_b, C-5_b, C-2_a, C-3_a, C-4_a, C-5_a), 68.8, 68.9, 71.1, 73.0, 73.5, 75.1 (6 × t, C-6_b, C-6_a, 4 × $PhCH_2$), 96.4 (d, $^1J_{C\text{-}1,H\text{-}1} = 158.9$ Hz, C-1_b), 99.7 (d, $^1J_{C\text{-}1,H\text{-}1} = 168.6$ Hz, C-1_a), 101.8 (d, PhCH), 126.0, 127.4, 127.6, 127.7, 127.7, 127.8, 128.0, 128.0, 128.1, 128.2, 128.2, 128.3, 128.4, 128.4, 129.0 (15 × d, Ar–CH), 137.3, 137.6, 138.1, 138.3, 138.5 (5 × s, Ar–C); MS (APCI$^-$): $m/z = 839$ (M + Cl^-, 100%); MS (ES$^+$): $m/z = 822$ (M + NH_4^+, 100%); HRMS: calcd. for $C_{48}H_{56}NO_{11}$ (MNH_4^+): 822.3853, found: 822.3864.

Methyl 3,4,6-tri-O-benzyl-β-D-mannopyranosyl-(1 → 4)-2,3,6-tri-O-benzyl-α-D-mannopyranoside (**13f**; $C_{55}H_{60}O_{11}$)

Mixed acetals **12f** (65 mg, 0.06 mmol), 2,6-di-*tert*-butyl-4-methylpyridine (25 mg, 0.12 mmol), silver trifluoromethanesulfonate (31 mg, 0.12 mmol), and anhydrous tin(II) chloride (23 mg, 0.12 mmol) gave β-manno disaccharide **13f** (27 mg, 50%) as a colourless oil.

$[\alpha]_D^{22} = +21.4$ ($c = 0.35$, $CHCl_3$); IR (thin film): $\nu = 3416$ (br, OH) cm^{-1}; ^{1}H NMR (400 MHz, $CDCl_3$): $\delta = 3.29$ (1H, ddd, $J_{4,5} = 9.7$ Hz, $J_{5,6} = 2.2$ Hz, $J_{5,6'} = 4.5$ Hz, H-5_b), 3.34 (3H, s, OCH_3), 3.39 (1H, dd, $J_{2,3} = 3.3$ Hz, $J_{3,4} = 9.2$ Hz, H-3_b), 3.62 (1H, dd, $J_{6,6'} = 11.2$ Hz, H-6_b), 3.65 (1H, dd, H-$6'_b$), 3.75–3.80 (3H, m, H-2_a, H-5_a, H-6_a), 3.86 (1H, at, $J = 9.5$ Hz, H-4_b), 3.88 (1H, dd, $J_{5,6'} = 3.7$ Hz, $J_{6,6'} = 11.0$ Hz, H-$6'_a$), 3.97 (1H, dd, $J_{2,3} = 3.2$ Hz, H-3_a), 4.01 (1H, d, H-2_b), 4.32 (1H, at, $J = 9.2$ Hz, H-4_a), 4.43, 4.56 (2H, ABq, $J_{AB} = 12.1$ Hz, $PhCH_2$), 4.48, 4.54 (2H, ABq, $J_{AB} = 12.0$ Hz, $PhCH_2$), 4.51, 4.86 (2H, ABq, $J_{AB} = 11.1$ Hz, $PhCH_2$), 4.57, 4.71 (2H, ABq, $J_{AB} = 11.7$ Hz, $PhCH_2$), 4.59, 4.70 (2H, ABq, $J_{AB} = 11.8$ Hz, $PhCH_2$), 4.68, 4.72 (2H, ABq, $J_{AB} = 12.9$ Hz, $PhCH_2$), 4.69 (1H, s, H-1_b), 4.77 (1H, d, $J_{1,2} = 1.9$ Hz, H-1_a), 7.19–7.36 (30H, m, Ar–H) ppm; ^{13}C NMR ($CDCl_3$, 100 MHz): $\delta = 54.7$ (q, OCH_3), 67.4 (d, C-2_b), 69.0 (t, C-6_b), 69.4 (t, C-6_a), 70.9 (d, C-5_a), 70.8, 72.2, 72.5, 73.2, 73.3, 75.0 (6 × t, 6 × $PhCH_2$), 72.9 (d, C-4_a), 73.8 (d, C-4_b), 74.8 (d, C-2_a), 75.5 (d, C-5_b), 79.0 (d, C-3_a), 81.8 (d, C-3_b), 98.9 (d, $^1J_{C\text{-}1,H\text{-}1} = 172.5$ Hz, C-1_a), 100.0 (d, $^1J_{C\text{-}1,H\text{-}1} = 161.5$ Hz, C-1_b), 127.3, 127.4, 127.4, 127.5, 127.5, 127.5, 127.5, 127.6, 127.7, 127.9, 128.0, 128.1, 128.2, 128.2 (14 × d, Ar–CH), 137.9, 138.1, 138.2, 138.2, 138.3, 138.4 (6 × s, Ar–C) ppm; MS (APCI$^-$): $m/z = 931$ (M + Cl^-, 100%); MS (ES$^+$): $m/z = 914$ (M + NH_4^+, 100%); HRMS: calcd. for $C_{55}H_{64}NO_{11}$ (MNH_4^+): 914.4479, found: 914.4473.

Cholestan-3′-β-yl 3,4,6-tri-O-benzyl-β-D-mannopyranoside (**13h**; $C_{54}H_{76}O_6$)

Mixed acetals **12h** (121 mg, 0.12 mmol), 2,6-di-*tert*-butyl-4-methylpyridine (49 mg, 0.24 mmol), silver trifluoromethanesulfonate (61 mg, 0.24 mmol), and anhydrous tin(II) chloride (46 mg, 0.24 mmol) gave β-mannoside **13h** (58 mg, 59%) as colourless crystals.

M.p.: 100–102°C (EtOH); $[\alpha]_D^{24} = -4.9$ ($c = 0.75$, $CHCl_3$); IR (thin film): $\nu = 3435$ (br, OH) cm^{-1}; ^{1}H NMR (400 MHz, $CDCl_3$): $\delta = 0.58$–2.00 (31H, m, steroid CH, CH_2), 0.66, 0.81 (6H, 2 × s, 2 × CH_3), 0.87 (3H, d, $J = 6.6$ Hz, CH_3), 0.88 (3H, d, $J = 6.8$ Hz, CH_3), 0.91 (3H, d, $J = 6.5$ Hz, CH_3), 2.11 (1H, br s, OH-2), 3.43 (1H, ddd, $J_{4,5} = 9.7$ Hz, $J_{5,6} = 5.6$ Hz, $J_{5,6'} = 1.9$ Hz, H-5), 3.57 (1H, dd, $J_{2,3} = 3.2$ Hz, $J_{3,4} = 9.1$ Hz, H-3), 3.69 (1H, dd, $J_{6,6'} = 10.7$ Hz, H-6), 3.70–3.77 (1H, m, steroid OCH), 3.80 (1H, dd, H-6′), 3.84 (1H, at, $J = 9.3$ Hz, H-4), 4.06 (1H, d, H-2), 4.56, 4.91 (2H, ABq, $J_{AB} = 10.9$ Hz, $PhCH_2$), 4.57 (1H, s, H-1), 4.58, 4.63 (2H, ABq, $J_{AB} = 12.1$ Hz, $PhCH_2$), 4.68, 4.79 (2H, ABq, $J_{AB} = 11.9$ Hz, $PhCH_2$), 7.21–7.40 (15H, m, Ar–H) ppm; ^{13}C NMR ($CDCl_3$, 100 MHz): $\delta = 12.0$, 12.2, 18.7, 22.5, 22.8 (5 × q, 5 × steroid CH_3), 21.2, 23.8, 24.2, 28.2, 28.8, 29.2, 32.1, 34.2, 36.2, 37.0, 39.5, 40.0 (12 × t, 12 × steroid CH_2), 28.0, 35.6, 35.8, 44.6, 54.4, 56.2, 56.5 (7 × d, 7 × steroid CH), 35.4, 42.6 (2 × s, 2 × steroid C), 68.8 (d, C-2), 69.4 (t, C-6), 71.3, 73.4, 75.2 (3 × t, 3 × $PhCH_2$), 74.3 (d, C-4), 75.1 (d, C-5), 77.5 (d, steroid OCH), 81.7 (d, C-3), 97.2 (d, $^1J_{C\text{-}1,H\text{-}1} = 155.6$ Hz, C-1), 127.5, 127.7, 127.8, 127.9, 128.1, 128.3, 128.4, 128.5 (8 × d, Ar–CH), 137.9, 138.2, 138.3 (3 × s, Ar–C) ppm.

2-O-(2-Iodo-1-methoxypropyl)-3,4,6-tri-O-benzyl-β-D-glucopyranosyl fluoride (**14a**; $C_{31}H_{36}FIO_6$)

Vinyl ethers **10** (300 mg, 0.61 mmol), *NIS* (410 mg, 1.83 mmol), and **11a** (0.037 cm^3, 0.92 mmol) gave mixed acetals **14a** (399 mg, >99%) as a colourless oil; MS (APCI$^+$): $m/z = 673$ (M + Na$^+$, 15%).

2-O-(1-Cyclohexyloxy-2-iodopropyl)-3,4,6-tri-O-benzyl-β-D-glucopyranosyl fluoride (**14b**; $C_{36}H_{44}FIO_6$)

Vinyl ethers **10** (80 mg, 0.16 mmol), *NIS* (109 mg, 0.49 mmol), and **11b** (0.034 cm^3, 0.33 mmol) gave mixed acetals **14b** (116 mg, 99%) as a colourless oil; MS (APCI$^+$): $m/z = 741$ (M + Na$^+$, 5%).

2-O-(1-(1,2:3,4-Di-O-isopropylidene-D-galactopyranos-6-O-yl)-2-iodopropyl)-3,4,6-tri-O-benzyl-β-D-glucopyranosyl fluoride (**14c**; $C_{42}H_{52}FIO_{11}$)

Vinyl ethers **10** (80 mg, 0.36 mmol), *NIS* (109 mg, 0.49 mmol), **11c** (84 mg, 0.82 mmol) gave mixed acetals **14c** (130 mg, 91%) as a colourless oil; MS (APCI$^-$): $m/z = 913$ (M + Cl$^-$, 4%); MS (APCI$^+$): $m/z = 901$ (M + Na$^+$, 5%).

2-O-(2-Iodo-1-(methyl 2,3,4-tri-O-benzyl-α-D-mannopyranosid-6-O-yl)propyl)-3,4,6-tri-O-benzyl-β-D-glucopyranosyl fluoride (**14d**; $C_{58}H_{64}FIO_{11}$)

Vinyl ethers **10** (80 mg, 0.16 mmol), *NIS* (109 mg, 0.49 mmol), and **11d** (151 mg, 0.33 mmol) gave mixed acetals **14d** (139 mg, 79%) as a colourless oil; MS (APCI$^-$): $m/z = 1117$ (M + Cl$^-$, 92%).

2-O-(2-Iodo-1-(methyl 2-O-benzyl-4,6-O-benzylidene-α-D-mannopyranosid-3-O-yl)propyl)-3,4,6-tri-O-benzyl-β-D-glucopyranosyl fluoride (**14e**; $C_{51}H_{56}FIO_{11}$)

Vinyl ethers **10** (80 mg, 0.16 mmol), *NIS* (109 mg, 0.49 mmol), and **11e** (121 mg, 0.33 mmol) gave mixed acetals **14e** (83 mg, 52%) as a colourless oil; MS (ES$^+$): $m/z = 1029$ (M + K$^+$, 74%), 1013 (M + Na$^+$, 100%), 1008 (M + NH_4^+, 15%).

2-O-(2-Iodo-1-(methyl 2,3,6-tri-O-benzyl-α-D-glucopyranosid-4-O-yl)propyl)-3,4,6-tri-O-benzyl-β-D-glucopyranosyl fluoride (**14g**; $C_{58}H_{65}FIO_{11}$)

Vinyl ethers **10** (80 mg, 0.16 mmol), *NIS* (109 mg, 0.49 mmol), and **11g** (221 mg, 0.48 mmol) gave mixed acetals **14g** (69 mg, 39%) as a colourless oil; MS (ES$^+$): $m/z = 1121$ (M + K$^+$, 22%), 1105 (M + Na$^+$, 100%), 1100 (M + NH_4^+, 21%).

2-O-(1-(Cholestan-3-β-yloxy)-2-iodopropyl)-3,4,6-tri-O-benzyl-β-D-glucopyranosyl fluoride (**14h**; $C_{57}H_{80}FIO_6$)

Vinyl ethers **10** (71 mg, 0.13 mmol), *NIS* (103 mg, 0.40 mmol), and **11h** (84 mg, 0.82 mmol) in *THF* (3 cm^3) gave mixed acetals **14h** (99 mg, 68%) as a colourless oil; MS (ES$^+$): $m/z = 1045$ (M + K$^+$, 100%), 1029 (M + Na$^+$, 90%).

Methyl 3,4,6-tri-O-benzyl-α-D-glucopyranoside (**15a**; $C_{28}H_{32}O_6$)

Mixed acetals **14a** (77 mg, 0.118 mmol), 2,6-di-*tert*-butyl-4-methylpyridine (48 mg, 0.236 mmol), silver trifluoromethanesulfonate (61 mg, 0.236 mmol), and anhydrous tin(II) chloride (45 mg,

0.236 mmol) gave α-methyl glucoside **15a** (49 mg, 89%) as a white solid which was recrystallized from diethyl ether/petrol.

M.p.: 78–81°C (diethyl ether/petrol; Ref. [4]: 83–84°C); $[\alpha]_D^{22} = +70.8$ ($c = 1.2$, $CHCl_3$; Ref. [4]: $[\alpha]_D = +90$ ($c = 1.1$, $CHCl_3$)); ^{1}H NMR (400 MHz, $CDCl_3$): $\delta = 2.11$ (1H, d, $J_{OH,2} = 7.4$ Hz, OH-2), 3.40 (3H, s, OCH_3), 3.60–3.76 (6H, m, H-2, H-3, H-4, H-5, H-6, H-6′), 4.47, 4.79 (2H, ABq, $J_{AB} = 10.7$ Hz, $PhCH_2$), 4.49, 4.62 (2H, ABq, $J_{AB} = 12.1$ Hz, $PhCH_2$), 4.78 (1H, d, $J_{1,2} = 3.2$ Hz, H-1), 4.83, 4.88 (2H, ABq, $J_{AB} = 11.2$ Hz, $PhCH_2$), 7.11–7.36 (15H, m, Ar–H) ppm.

Cyclohexyl 3,4,6-tri-O-benzyl-α-D-glucopyranoside (**15b**; $C_{33}H_{40}O_6$)

Mixed acetals **14b** (77 mg, 0.118 mmol), 2,6-di-*tert*-butyl-4-methylpyridine (65 mg, 0.318 mmol), silver trifluoromethanesulfonate (82 mg, 0.318 mmol), and anhydrous tin(II) chloride (60 mg, 0.318 mmol) gave α-glucoside **15b** (56 mg, 66%) as a white solid which was recrystallized from diethyl ether/petrol.

M.p.: 92–94°C (diethyl ether/petrol); $[\alpha]_D^{22} = +90.0$ ($c = 0.3$, $CHCl_3$; Ref. [4]: $[\alpha]_D = +60$ ($c = 0.6$, $CHCl_3$)); ^{1}H NMR (400 MHz, $CDCl_3$): $\delta = 1.24$–1.92 (10H, m, cyclohexyl-H), 2.06 (1H, d, $J = 2.7$ Hz, OH-2), 3.61–3.75 (5H, m, H-2, H-3, H-4, H-6, cyclohexyl OCH), 3.77 (1H, dd, $J_{5,6'} = 3.9$ Hz, $J_{6,6'} = 10.7$ Hz, H-6′), 3.87–3.90 (1H, m, H-5), 4.49, 4.83 (2H, ABq, $J_{AB} = 10.7$ Hz, $PhCH_2$), 4.51, 4.65 (2H, ABq, $J_{AB} = 12.0$ Hz, $PhCH_2$), 4.84, 4.99 (2H, ABq, $J_{AB} = 11.0$ Hz, $PhCH_2$), 5.03 (1H, d, $J_{1,2} = 3.5$ Hz, H-1), 7.14–7.42 (15H, m, Ar–H) ppm.

3,4,6-Tri-O-benzyl-α-D-glucopyranosyl-(1 → 6)-1,2:3,4-di-O-isopropylidene-D-galactopyranose (**15c**; $C_{39}H_{48}O_{11}$)

Mixed acetals **14c** (124 mg, 0.14 mmol), 2,6-di-*tert*-butyl-4-methylpyridine (58 mg, 0.28 mmol), silver trifluoromethanesulfonate (73 mg, 0.28 mmol), and anhydrous tin(II) chloride (54 mg, 0.28 mmol) gave α-gluco disaccharide **15c** (62 mg, 63%) as a colourless oil.

$[\alpha]_D^{22} = +27.7$ ($c = 0.35$, $CHCl_3$; Ref. [4]: $[\alpha]_D = +29$ ($c = 0.9$, $CHCl_3$)); ^{1}H NMR (400 MHz, $CDCl_3$): $\delta = 1.34$, 1.35, 1.45, 1.54 (12H, 4 × s, 4 × CH_3), 3.63–3.79 (6H, m, H-2_b, H-3_b, H-4_b, H-6_b, H-$6'_b$, H-6_a), 3.85 (1H, ddd, $J_{4,5} = 9.9$ Hz, $J_{5,6} = 2.2$ Hz, $J_{5,6'} = 3.0$ Hz, H-5_b), 3.91 (1H, dd, $J_{5,6'} = 6.7$ Hz, $J_{6,6'} = 10.2$ Hz, H-$6'_a$), 4.00 (1H, atd, $J = 6.7$ Hz, $J_{4,5} = 1.9$ Hz, H-5_a), 4.25 (1H, dd, $J_{3,4} = 7.8$ Hz, H-4_a), 4.34 (1H, dd, $J_{1,2} = 4.9$ Hz, $J_{2,3} = 2.2$ Hz, H-2_a), 4.49, 4.83 (2H, ABq, $J_{AB} = 10.3$ Hz, $PhCH_2$), 4.50, 4.64 (2H, ABq, $J_{AB} = 11.7$ Hz, $PhCH_2$), 4.63 (1H, dd, H-3_a), 4.82, 4.98 (2H, ABq, $J_{AB} = 10.7$ Hz, $PhCH_2$), 4.93 (1H, d, H-1_b), 5.53 (1H, d, H-1_a), 7.12–7.41 (15H, m, Ar–H) ppm.

Methyl 3,4,6-tri-O-benzyl-α-D-glucopyranosyl-(1 → 6)-3,4,6-tri-O-benzyl-α-D-mannopyranoside (**15d**; $C_{55}H_{60}O_{11}$)

Mixed acetals **14d** (49 mg, 0.045 mmol), 2,6-di-*tert*-butyl-4-methylpyridine (19 mg, 0.09 mmol), silver trifluoromethanesulfonate (23 mg, 0.09 mmol), and anhydrous tin(II) chloride (17 mg, 0.09 mmol) gave α-gluco disaccharide **15d** (18 mg, 44%) as a colourless oil.

$[\alpha]_D^{22} = +65.5$ ($c = 1.1$, $CHCl_3$); ^{1}H NMR (400 MHz, $CDCl_3$): $\delta = 3.30$ (3H, s, OCH_3), 3.59–3.80 (8H, m, H-2_b, H-3_b, H-4_b, H-5_b, H-6_b, H-$6'_b$, H-5_a, H-6_a), 3.79 (1H, dd, $J_{1,2} = 2.0$ Hz, $J_{2,3} = 3.0$ Hz, H-2_a), 3.91 (1H, dd, $J_{3,4} = 9.4$ Hz, H-3_a), 4.04–4.10 (2H, m, H-4_a, H-$6'_a$), 4.47, 4.84 (2H, ABq, $J_{AB} = 10.8$ Hz, $PhCH_2$), 4.48, 4.61 (2H, ABq, $J_{AB} = 12.2$ Hz, $PhCH_2$), 4.62 (2H, s, $PhCH_2$), 4.62, 4.98 (2H, ABq, $J_{AB} = 11.1$ Hz, $PhCH_2$), 4.70 (1H, d, H-1_a), 4.71, 4.92 (2H, ABq, $J_{AB} = 11.1$ Hz, $PhCH_2$), 4.72 (2H, s, $PhCH_2$), 5.00 (1H, d, $J_{1,2} = 2.9$ Hz, H-1_b), 7.14–7.39 (30H, m, Ar–H) ppm; ^{13}C NMR ($CDCl_3$, 100 MHz): $\delta = 54.9$ (q, CH_3), 67.7, 68.4, 72.0, 72.8, 73.4, 75.0, 75.1, 75.2 (8 × t, 6 × $PhCH_2$, C-6_b, C-6_a), 70.7, 71.3, 73.8, 74.1, 74.2, 77.0, 79.9, 83.3 (8 × d, C-2_b, C-3_b, C-4_b, C-5_b, C-2_a, C-3_a, C-4_a, C-5_a), 98.9, 100.6 (2 × d, C-1_b, C-1_a), 127.4, 127.6, 127.6, 127.7, 127.8, 127.8, 127.9, 127.9, 128.0, 128.1, 128.3, 128.3, 128.4 (13 × d, Ar–CH), 138.0, 138.2, 138.3, 138.4, 138.9

(5 × s, Ar–C) ppm; MS (APCI$^+$): $m/z = 919$ (M + Na$^+$, 55%); MS (APCI$^-$): $m/z = 931$ (M + Cl$^-$, 10%); HRMS: calcd. for $C_{55}H_{61}O_{11}$ (MH$^+$): 897.4214, found: 897.4229.

Methyl 3,4,6-tri-O-benzyl-α-D-glucopyranosyl-(1 → 3)-2-O-benzyl-4,6-O-benzylidene-α-D-mannopyranoside (**15e**; $C_{48}H_{52}O_{11}$)

Mixed acetals **14e** (77 mg, 0.078 mmol), 2,6-di-*tert*-butyl-4-methylpyridine (32 mg, 0.16 mmol), silver trifluoromethanesulfonate (40 mg, 0.16 mmol), and anhydrous tin(II) chloride (30 mg, 0.16 mmol) gave α-gluco disaccharide **15e** (23 mg, 37%) as white crystals.

M.p.: 145–147°C (diethyl ether/petrol); $[\alpha]_D^{22} = +42.4$ ($c = 0.25$, $CHCl_3$); IR (thin film): $\nu = 3462$ (br, OH) cm^{-1}; ^{1}H NMR (400 MHz, $CDCl_3$): $\delta = 2.65$ (1H, d, $J_{OH,2} = 9.8$ Hz, OH-2), 3.34 (3H, s, OCH_3), 3.56 (1H, at, $J = 9.2$ Hz, H-4$_b$), 3.65–3.68 (2H, m, H-6$_b$, H-6$'_b$), 3.71 (1H, atd, $J_{1,2} = 3.4$ Hz, $J = 9.6$ Hz, H-2$_b$), 3.75 (1H, at, $J = 9.3$ Hz, H-3$_b$), 3.81–3.88 (3H, m, H-2$_a$, H-5$_a$, H-6$_a$), 3.90–3.94 (1H, m, H-5$_b$), 4.22–4.24 (2H, m, H-4$_a$, H-6$'_a$), 4.28 (1H, dd, $J_{2,3} = 3.7$ Hz, $J_{3,4} = 9.2$ Hz, H-3$_a$), 4.47, 4.85 (2H, ABq, $J_{AB} = 11.0$ Hz, $PhCH_2$), 4.48, 4.57 (2H, ABq, $J_{AB} = 12.0$ Hz, $PhCH_2$), 4.66, 4.76 (2H, ABq, $J_{AB} = 12.0$ Hz, $PhCH_2$), 4.69, 4.82 (2H, ABq, $J_{AB} = 10.9$ Hz, $PhCH_2$), 4.69 (1H, d, $J_{1,2} = 1.4$ Hz, H-1$_a$), 5.21 (1H, d, H-1$_b$), 5.63 (1H, s, PhCH), 7.14–7.49 (25H, m, Ar–H) ppm; ^{13}C NMR ($CDCl_3$, 100 MHz): $\delta = 54.9$ (q, OCH_3), 63.9, 71.5, 73.8, 76.1, 77.0, 77.6, 78.2, 83.4 (8 × d, C-2$_a$, C-3$_a$, C-4$_a$, C-5$_a$, C-2$_b$, C-3$_b$, C-4$_b$, C-5$_b$), 68.7, 68.8, 73.4, 73.5, 74.8, 75.2 (6 × t, C-6$_a$, C-6$_b$, 4 × $PhCH_2$), 100.0, 100.8, 101.8 (3 × d, C-1$_a$, C-1$_b$, PhCH), 125.6, 127.5, 127.6, 127.6, 127.7, 127.8, 127.8, 127.9, 128.0, 128.2, 128.3, 128.3, 128.3, 128.5, 129.1 (15 × d, Ar–CH), 137.0, 137.8, 137.9, 138.4, 138.7 (5 × s, Ar–C) ppm; MS (ES$^+$): $m/z = 843$ (M + K$^+$, 50%), 827 (M + Na$^+$, 100%); HRMS: calcd. for $C_{48}H_{56}NO_{11}$ (MNH_4^+): 822.3853, found: 822.3842.

Methyl 3,4,6-tri-O-benzyl-α-D-glucopyranosyl-(1 → 4)-2,3,6-tri-O-benzyl-α-D-glucopyranoside (**15g**; $C_{55}H_{60}O_{11}$)

Mixed acetals **14g** (65 mg, 0.06 mmol), 2,6-di-*tert*-butyl-4-methylpyridine (25 mg, 0.12 mmol), silver trifluoromethanesulfonate (31 mg, 0.12 mmol), and anhydrous tin(II) chloride (23 mg, 0.12 mmol) gave α-gluco disaccharide **15g** (24 mg, 45%) as a colourless oil.

$[\alpha]_D^{22} = +35.1$ ($c = 1.05$, $CHCl_3$); IR (thin film): $\nu = 3418$ (br, OH) cm^{-1}; ^{1}H NMR (400 MHz, $CDCl_3$): $\delta = 3.38$ (3H, s, OCH_3), 3.45 (1H, br d, $J_{OH,2} = 9.1$ Hz, OH-2$_b$), 3.55 (1H, at, $J = 9.1$ Hz, H-3$_b$), 3.55 (1H, dd, $J_{5,6} = 1.8$ Hz, $J_{6,6'} = 10.6$ Hz, H-6$_a$), 3.61–3.66 (4H, m, H-2$_b$, H-2$_a$, H-3$_a$, H-6$'_a$), 3.68 (1H, dd, $J_{5,6} = 1.9$ Hz, $J_{6,6'} = 11.0$ Hz, H-6$_b$), 3.69–3.71 (1H, m, H-5$_b$), 3.86–3.90 (1H, m, H-5$_a$), 3.88 (1H, at, $J = 9.5$ Hz, H-4$_b$), 3.93 (1H, dd, $J_{5,6'} = 3.9$ Hz, H-6$'_b$), 3.99 (1H, at, $J = 9.3$ Hz, H-4$_a$), 4.44, 4.57 (2H, ABq, $J_{AB} = 12.2$ Hz, $PhCH_2$), 4.49, 4.83 (2H, ABq, $J_{AB} = 10.6$ Hz, $PhCH_2$), 4.49, 4.59 (2H, ABq, $J_{AB} = 11.6$ Hz, $PhCH_2$), 4.57, 4.69 (2H, ABq, $J_{AB} = 11.0$ Hz, $PhCH_2$), 4.59, 4.70 (2H, ABq, $J_{AB} = 12.0$ Hz, $PhCH_2$), 4.61 (1H, d, $J_{1,2} = 3.6$ Hz, H-1$_b$), 4.79, 5.18 (2H, ABq, $J_{AB} = 10.6$ Hz, $PhCH_2$), 5.19 (1H, d, $J_{1,2} = 3.1$ Hz, H-1$_a$), 7.16–7.36 (30H, m, Ar–H) ppm; ^{13}C NMR ($CDCl_3$, 100 MHz): $\delta = 55.2$ (q, OCH_3), 68.4 (t, C-6$_b$), 68.7 (t, C-6$_a$), 70.0 (d, C-5$_b$), 71.5 (d, C-5$_a$), 73.0, 73.2, 73.3, 74.9, 75.1, 75.2 (6 × t, 6 × $PhCH_2$), 74.0, 80.2, 83.6 (3 × d, C-2$_b$, C-2$_a$, C-3$_a$), 76.9 (d, C-3$_b$), 77.6 (d, C-4$_b$), 80.5 (d, C-4$_a$), 97.7 (d, C-1$_b$), 101.2 (d, C-1$_a$), 127.3, 127.4, 127.5, 127.6, 127.7, 127.8, 127.8, 128.0, 128.1, 128.2, 128.2, 128.3, 128.4 (13 × d, Ar–CH), 137.5, 137.7, 137.7, 137.9, 138.2, 138.7 (6 × s, Ar–C) ppm; MS (ES$^+$): $m/z = 935$ (M + K$^+$, 14%), 919 (M + Na$^+$, 100%), 916 (2M + K$^+$ + H$^+$, 21%), 914 (M + NH_4^+, 14%); HRMS: calcd. for $C_{55}H_{64}NO_{11}$ (MNH_4^+): 914.4479, found: 914.4476.

Cholestan-3′-β-yl 3,4,6-tri-O-benzyl-α-D-glucopyranoside (**15h**; $C_{54}H_{76}O_6$)

Mixed acetals **14h** (41 mg, 0.04 mmol), 2,6-di-*tert*-butyl-4-methylpyridine (16 mg, 0.08 mmol), silver trifluoromethanesulfonate (21 mg, 0.08 mmol), and anhydrous tin(II) chloride (15 mg, 0.08 mmol) gave α-glucoside **15h** (29 mg, 87%) as white crystals.

M.p.: 108–109°C; $[\alpha]_D^{24} = +98.5$ ($c = 0.20$, $CHCl_3$); IR (thin film): $\nu = 3487$ (br, OH) cm^{-1}; ^1H NMR (400 MHz, $CDCl_3$): $\delta = 0.59$–1.99 (31H, m, steroid CH, CH_2), 0.66, 0.81 (6H, 2 × s, 2 × CH_3), 0.88 (3H, d, $J = 6.5$ Hz, CH_3), 0.89 (3H, d, $J = 6.5$ Hz, CH_3), 0.91 (3H, d, $J = 6.6$ Hz, CH_3), 2.10 (1H, br d, $J_{OH,2} = 8.0$ Hz, OH-2), 3.57–3.79 (4H, m, H-2, H-3, H-4, OCH), 3.68 (1H, dd, $J_{5,6} = 2.0$ Hz, $J_{6,6'} = 10.5$ Hz, H-6), 3.77 (1H, dd, $J_{5,6'} = 4.0$ Hz, H-6′), 3.89–3.92 (1H, m, H-5), 4.49, 4.84 (2H, ABq, $J_{AB} = 10.7$ Hz, $PhCH_2$), 4.51, 4.65 (2H, ABq, $J_{AB} = 11.9$ Hz, $PhCH_2$), 4.84, 4.99 (2H, ABq, $J_{AB} = 11.2$ Hz, $PhCH_2$), 5.04 (1H, d, $J_{1,2} = 3.6$ Hz, H-1), 7.14–7.41 (15H, m, Ar–H) ppm; ^{13}C NMR ($CDCl_3$, 100 MHz): $\delta = 12.0$, 12.3, 18.6, 22.6, 22.8 (5 × q, 5 × steroid CH_3), 21.2, 23.8, 24.2, 27.8, 28.2, 28.7, 32.0, 36.0, 36.1, 36.8, 39.5, 40.0 (12 × t, 12 × steroid CH_2), 28.0, 35.5, 35.8, 44.9, 54.2, 56.2, 56.4 (7 × d, 7 × steroid CH), 35.4, 42.6 (2 × s, 2 × steroid C), 68.6, 73.4, 75.0, 75.3 (4 × t, 3 × $PhCH_2$, C-6), 70.4, 73.0, 77.3, 77.3, 83.8 (5 × d, C-2, C-3, C-4, C-5, steroid OCH), 96.9 (d, C-1), 127.6, 127.7, 127.7, 127.9, 127.9, 128.3, 128.3 (7 × d, Ar–CH), 137.9, 138.2, 138.8 (3 × s, Ar–C) ppm.

Hydrolsis of vinyl ethers **19**

Vinyl ethers **19** (30 mg, 0.04 mmol) were dissolved in 2 cm^3 *THF* and 0.2 cm^3 H_2O. *NIS* (9 mg, 0.04 mmol) was added to the stirred mixture at room temperature. After 50 min, the mixture was partitioned between 30 cm^3 $CHCl_2$ and 30 cm^3 of a 10% aqueous solution of $Na_2S_2O_3$. Triethylamine (2 cm^3) was added to the organic layer which was then concentrated *in vacuo* and purified by flash column chromatography (petrol:ethyl acetate = 3:1) to give α-gluco disaccharide **15e** (18 mg, ~70%) as a white solid, identical to that described previously.

Data for succinimide trapped material: 2-O-(2-Iodo-1-(succinimid-N-yl)-propyl)-3,4,6-tri-O-benzyl-α-D-mannopyranosyl fluoride (**16**; $C_{34}H_{37}FINO_7$)

Isolated in varying amounts during manno tethering reactions as single diastereomer as a colourless oil; $[\alpha]_D^{22} = +60.0$ ($c = 0.75$, $CHCl_3$); IR (thin film): $\nu = 1709$ (s, C=O) cm^{-1}; ^1H NMR (400 MHz, $CDCl_3$): $\delta = 1.98$ (3H, d, $J = 6.8$ Hz, CH_3), 2.76 (4H, s, $COCH_2CH_2CO$), 3.66–3.73 (2H, m, H-6, H-6′), 3.90 (3H, br s, H-3, H-4, H-5), 3.99 (1H, br s, H-2), 4.51, 4.62 (2H, ABq, $J_{AB} = 12.2$ Hz, $PhCH_2$), 4.51, 4.82 (2H, ABq, $J_{AB} = 11.0$ Hz, $PhCH_2$), 4.68, 4.90 (2H, ABq, $J_{AB} = 11.0$ Hz, $PhCH_2$), 5.12 (1H, dq, $J = 10.2$ Hz, CHI), 5.39 (1H, dd, $J_{1,2} = 2.0$ Hz, $J_{1,F} = 50.5$ Hz, H-1), 5.72 (1H, d, CHN), 7.15–7.47 (15H, m, Ar–H) ppm; ^{13}C NMR ($CDCl_3$, 100 MHz): $\delta = 23.2$, 25.0 (d, q, CH_3, CHI), 27.9 (t, $COCH_2CH_2CO$), 68.4, 73.0, 73.4, 75.1 (4 × t, C-6, 3 × $PhCH_2$), 72.7 (dd, $^2J_{C\text{-}2,F} = 6.6$ Hz, C-2), 73.6, 74.2, 78.9 (3 × d, C-3, C-4, C-5), 86.4 (d, CHN), 106.3 (dd, $^1J_{C\text{-}1,F} = 221.0$ Hz, C-1), 127.6, 127.8, 127.8, 128.0, 128.0, 128.3, 128.4, 128.5, 128.6 (9 × d, Ar–CH), 137.5, 137.8, 138.0 (3 × s, Ar–C) ppm; MS (ES^+): $m/z = 735$ ($M + NH_4^+$, 100%); HRMS: calcd. for $C_{34}H_{41}N_2O_7FI$ (MNH_4^+): 735.1943, found: 725.1949.

Acknowledgements

We gratefully acknowledge financial support from the EPSRC and GlaxoSmithKline as well as the use of the EPSRC National Mass Spectrometry Service at Swansea and the Chemical Database Service at Daresbury.

References

[1] Barresi F, Hindsgaul O (1994) Can J Chem **72**: 1447; Barresi F, Hindsgaul O (1992) Synlett 759; Barresi F, Hindsgaul O (1991) J Am Chem Soc **113**: 9377
[2] Stork G, Kim G (1992) J Am Chem Soc **114**: 1087; Stork G, La Clair JJ (1996) J Am Chem Soc **118**: 247

[3] Lergenmuller M, Nukada T, Kuramochi K, Dan A, Ogawa T, Ito Y (1999) Eur J Org Chem 1367; Dan A, Ito Y, Ogawa T (1995) J Org Chem **60**: 4680; Ito Y, Ogawa T (1994) Angew Chem Int Ed Engl **33**: 1765
[4] Ennis SC, Fairbanks AJ, Tennant-Eyles RJ, Yeates HS (1999) Synlett 1387
[5] Ennis SC, Fairbanks AJ, Slinn CA, Tennant-Eyles RJ, Yeates HS (2001) Tetrahedron **57**: 4221
[6] Seward CMP, Cumpstey I, Aloui M, Ennis SC, Redgrave AJ, Fairbanks AJ (2000) J Chem Soc Chem Commun 1409
[7] Cumpstey I, Fairbanks AJ, Redgrave AJ (2001) Org Lett **3**: 2371
[8] Zhang YM, Mallet JM, Sinaÿ P (1992) Carbohydr Res **236**: 73
[9] Lichtentaler FW, Schneider-Adams T (1994) J Org Chem **59**: 6728
[10] Gordon DM, Danishefsky SJ (1990) Carbohydr Res **206**: 361
[11] Boons GJ, Isles S (1996) J Org Chem **61**: 4262
[12] Thiem J, Karl H, Schwenter J (1978) Synthesis 696
[13] Toshima K (2000) Carbohydr Res **327**: 15
[14] Matsumoto T, Maeta H, Suzuki K, Tsuchihashi G (1988) Tetrahedron Lett **29**: 3567
[15] Nicolaou KC, Chucholowski A, Dolle RE, Randall JL (1984) J Chem Soc Chem Commun 1155
[16] Mukaiyama T, Murai Y, Shoda S (1981) Chem Lett 431
[17] Ito Y, Ando H, Wada M, Kawai T, Ohnish Y, Nakahara Y (2001) Tetrahedron **57**: 4123
[18] It is not yet entirely clear exactly how **19** ($R = H$) is formed, since this formally requires loss of I^+ from the oxonium ion produced subsequently to intramolecular glycosylation

Received August 24, 2001. Accepted November 6, 2001

Zinc-Mediated Fragmentation of Methyl 6-Deoxy-6-iodo-hexopyranosides

Philip R. Skaanderup, Lene Hyldtoft, and **Robert Madsen***

Department of Chemistry, Technical University of Denmark, DK-2800 Lyngby, Denmark

Summary. An improved procedure was developed for the zinc-mediated fragmentation of protected and unprotected methyl 6-deoxy-6-iodo-hexopyranosides. The method employs sonication of the iodoglycoside with zinc dust in a *THF*/H_2O mixture.

Keywords. Fragmentation; Glycosides; Halogens; Sonication; Zinc.

Introduction

The zinc-mediated reductive fragmentation of methyl 6-deoxy-6-halo-hexopyranosides has been described for the first time by *Bernet* and *Vasella* in 1979 [1]. The 5,6-dideoxy-hex-5-enoses thus generated are useful chiral synthons and have found many applications in carbohydrate chemistry. Most notable is the synthesis of five-membered carbocycles by 1,3-dipolar cycloadditions [1,2] or radical cyclizations [3]. In addition, the fragmentation has also been used for the synthesis of azasugars [4] and complex natural products [5]. The corresponding methyl 5-deoxy-5-halo-pentofuranosides also undergo this fragmentation [6].

However, in spite of the many applications, the reductive fragmentation of ω-haloglycosides continues to cause significant experimental difficulties. The original procedure calls for reflux with acid-washed zinc powder in aqueous alcohols [1]. This is still the most widely used protocol, but the yields are often moderate due to the instability of 5,6-dideoxy-hex-5-enoses at reflux and the formation of several byproducts including methyl 6-deoxy-hexopyranosides (C-6 reduction), 2,5,6-trideoxy-hex-5-enoses (C-2 deoxygenation), and 5,6-dideoxy-hex-5-enose dialkyl acetals (acetal formation) [1,3]. Recently, the addition of vitamin B_{12} has been shown to give some improvement, allowing the fragmentation to be carried out at room temperature [7]. Using a better leaving group at the anomeric center, like the *p*-methoxyphenyl group, also gives higher yields [8], but this is not a convenient method for general use. Another procedure for the fragmentation employs the more reactive zinc-silver graphite in *THF* at room temperature [9]. Yields are here generally higher, and less byproducts are formed than when refluxing with zinc powder in aqueous alcohols.

---

* Corresponding author. E-mail: rm@kemi.dtu.dk

We have recently developed a domino reaction where methyl ω-haloglycosides are fragmented with zinc and the resulting aldehydes are subsequently alkylated in a *Barbier* reaction [10]. Zinc serves a dual function in this reaction. First, it mediates the fragmentation of the haloglycoside; secondly, it activates an alkyl halide for the *Barbier* alkylation. During this work we had to find an efficient and high-yielding protocol for the fragmentation of methyl ω-haloglycosides. Herein, we report a full account on these studies which resulted in the development of an improved fragmentation procedure.

Results and Discussion

We first studied the fragmentation with benzyl-protected methyl glucopyranosides **1** and **2** (Scheme 1). Refluxing bromide **1** with acid-washed zinc powder in aqueous 2-propanol gave enal **3** in about 65% yield in accordance with previous observations [1]. The major byproducts formed were 6-deoxy glucoside **4** together with a small amount of acetal **5**. In order to avoid these byproducts, more reactive forms of zinc were investigated. Unfortunately, bromide **1** and iodide **2** did not react with *Rieke* zinc, zinc graphite, or zinc-silver graphite in *THF* at room temperature or at reflux. Presumably, this lack of reactivity is due to the steric bulk of **1** and **2**, as sterically demanding protecting groups have previously been shown to inhibit the fragmentation [9].

Addition of an alcohol has been shown to enhance the rate of the fragmentation with these reactive forms of zinc [9b]. Indeed, treatment of iodide **2** with *Rieke* zinc in a *THF*/methanol mixture at reflux afforded complete fragmentation. Unfortunately, the product **3** had epimerized at C-2, maybe due to traces of potassium in the *Rieke* zinc. In order to avoid possible epimerizations, acetic acid was then added instead of methanol. However, this led to reduction of the iodide to afford deoxyglycoside **4**. Virtually complete reduction to **4** was also observed upon refluxing **2** with zinc powder and acetic acid in aqueous 2-propanol. Bromide **1** gave similar results; however, it is less reactive than iodide **2**, and the latter was therefore chosen for general use.

As it appeared important to have an alcohol and/or an acid present during the fragmentation, several experiments were then conducted with a *Lewis* acid. Treatment of **2** with *Rieke* zinc and *TMS*Cl in *THF*/2-propanol caused almost complete conversion to acetal **5**. Interestingly, no reduction of the iodide was observed under these conditions. To prevent acetal formation the reaction was repeated without 2-propanol. Gratifyingly, this now gave very clean formation of the desired enal **3**. The fragmentation was most conveniently carried out with zinc

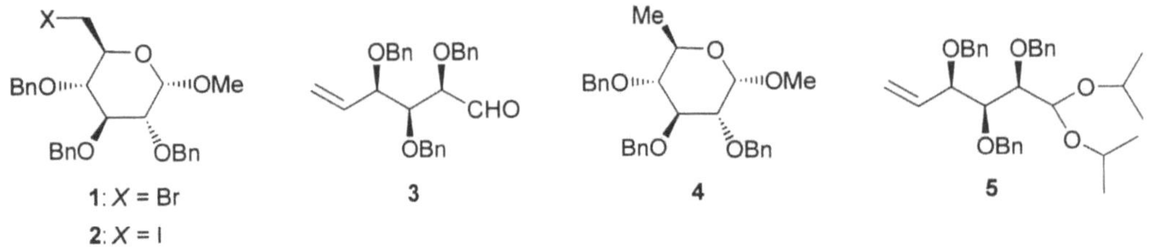

Scheme 1

Table 1. Zinc-mediated fragmentation of methyl 6-deoxy-6-iodo-hexopyranosides; conditions: A: Zn, *TMS*Cl, *THF*, sonication, 40°C, 4–6 h; B: Zn, *THF*:H_2O = 9:1, sonication, 40°C, 1 h; C: Zn, *THF*: H_2O = 4:1, sonication, 40°C, 1 h

Entry	Iodoglycoside	Conditions	Product	Yield/%
1	**2**	A	**3**	85
2	**2**	B	**3**	94
3	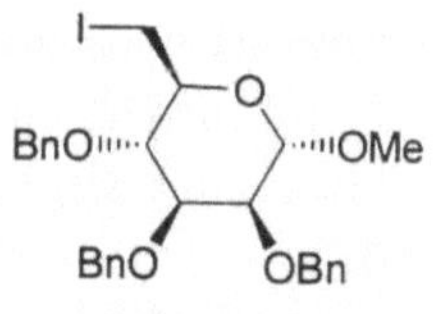 **6**	A	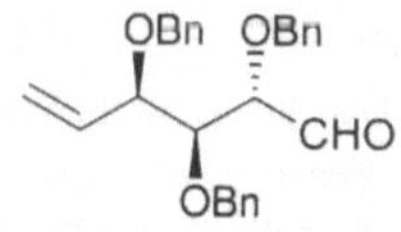 **11**	86
4		B		89
5	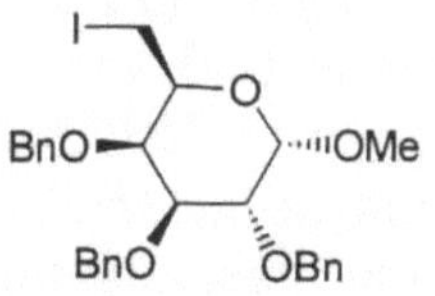 **7**	A	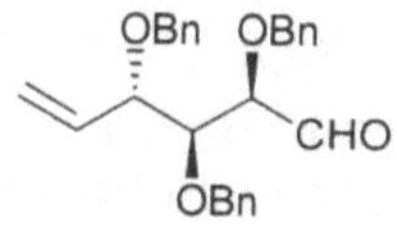 **12**	85
6		B		93
7	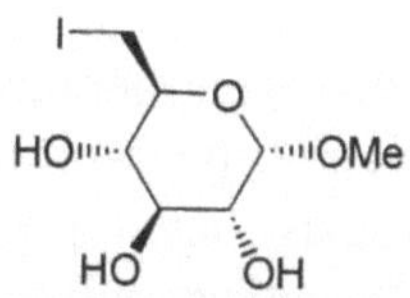 **8**	C	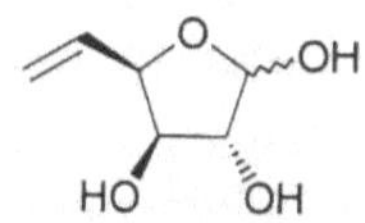 **13**	94
8	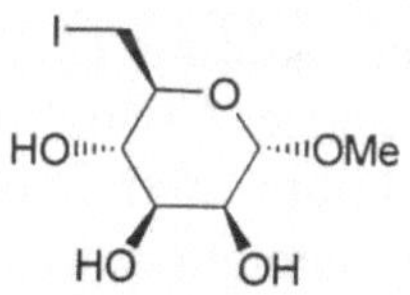 **9**	C	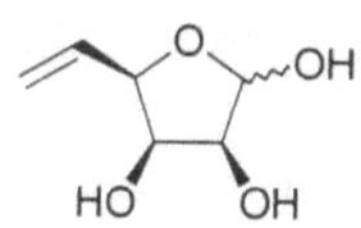 **14**	94
9	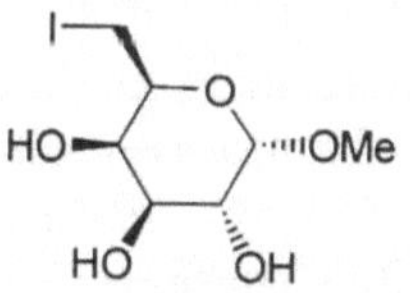 **10**	C	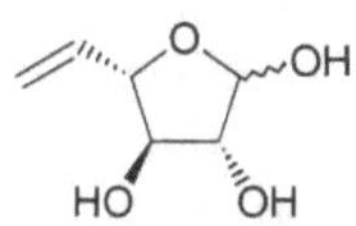 **15**	95

powder, as *Rieke* zinc did not provide better results and is more tedious to prepare. Refluxing **2** with zinc powder and *TMS*Cl worked well when starting with less than 1 g of **2**. On larger scale, however, the reaction proved less efficient, presumably due to precipitation of zinc salts onto the metallic zinc surface. Therefore, more reproducible conditions were sought and sonication was investigated. To the best of our knowledge this fragmentation has never been achieved under sonication conditions prior to our first report in 1999 [10a]. It turned out that sonication at 40°C caused clean and reproducible conversion into enal **3** (Table 1, entry 1). Even on a 10 g scale this procedure proved very reliable. A more finely grinded zinc (<10 μm) was used for these sonication experiments. Under reflux conditions this type of zinc was less efficient due to aggregation, and a more coarse powder was used for these experiments. The general applicability of the sonication protocol is further illustrated by the fragmentation of protected mannoside **6** and galactoside **7** (Table 1, entries 3 and 5).

Good yields were obtained under these conditions, but it still bothered us that some material was always lost. This was presumably due to slight decomposition of the enals **3**, **11**, and **12**, as no other products were isolated from these reactions. The problem seemed to be the acidic conditions and the 4–6 h of reaction time. *TMS*Cl was therefore replaced by H_2O hoping that a continuous wash of the zinc surface during the reaction would improve the reactivity. This turned out to work very well. The fragmentation was finished in less than 1 h giving slightly better yields of enals **3**, **11**, and **12** (Table 1, entries 2, 4, and 6).

The fragmentation in a *THF*/H_2O mixture also proved well suited for unprotected iodoglycosides **8–10** giving rise to furanoses **13–15** (Table 1, entries 7–9). The very high yields of all six fragmentation products **3** and **11–15** are noteworthy and constitute a testament to the efficiency of these reaction conditions. When unprotected iodoglucoside **8** was fragmented in anhydrous *THF* containing *TMS*Cl the product was not furanose **13**, but instead the corresponding methyl furanoside of **13** was formed. Apparently, a *Fischer* glycosylation with the liberated methanol takes place under these conditions.

Zinc appears to be the metal of choice for the reductive fragmentation of ω-haloglycosides. Magnesium has been shown to give *Wurtz*-type homocoupling, presumably by a radical mechanism [11]. In our previously developed domino reaction we were able to replace zinc with indium [10b]. However, indium is a less reactive metal. In fact, when we sonicated unprotected glucoside **8** with indium in a *THF*/H_2O mixture only a sluggish reaction was observed. Adding *TMS*Cl to the mixture gave full conversion of **8** in 12 h to give furanose **13** in 93% yield. Protected glucoside **2** could not be fragmented with indium, and only decomposition was observed on prolonged treatment. No fragmentation was also the result when **8** was sonicated with tin, antimony, or bismuth.

In conclusion, we have developed an improved procedure for the zinc-mediated fragmentation of methyl ω-iodoglycosides. The fragmentation is most efficiently carried out with zinc dust under sonication in a *THF*/H_2O mixture. For protected iodoglycosides a procedure under anhydrous conditions has also been developed employing *TMS*Cl. The latter method is useful when further synthetic transformations do not allow for the presence of H_2O [10b].

Experimental

Zinc was activated and dried immediately before use [10b]. Zinc powder (Fluka 96453) was used for fragmentations at reflux, whereas zinc dust (Aldrich 20.998-8) was used under sonication. All sonications were performed in a Branson 1210 ultrasonic bath. Thin-layer chromatography (TLC) was performed on aluminum plates precoated with silica gel (Merck 1.05554). Compounds were visualized by heating after dipping in a solution of 2.5 g $Ce(SO_4)_2$ and 6.25 g $(NH_4)_6Mo_7O_{24}$ in 250 cm^3 10% aqueous H_2SO_4. Flash chromatography was performed using silica gel 60. Reverse phase flash chromatography was carried out with silica gel 60-50 C_{18} (Macherey-Nagel). NMR spectra were recorded on a Varian Unity Inova 500 spectrometer. Microanalyses were conducted by the Department of Chemistry at the University of Copenhagen (data were in accordance with calculated values).

General procedure for the fragmentation of benzyl protected iodoglycosides (conditions A and B)

To a solution of 500 mg (0.87 mmol) methyl 2,3,4-tri-O-benzyl-6-deoxy-6-iodo-α-*D*-glycopyranoside in 20 cm^3 *THF*, 569 mg (8.7 mmol) pre-activated zinc and 0.11 cm^3 (0.87 mmol) *TMS*Cl (condition A) or 2.2 cm^3 H_2O (condition B) were added. The resulting suspension was sonicated at 40°C until TLC showed full conversion (A: 4–6 h, B: 1 h). Then, 30 cm^3 diethyl ether and 10 cm^3 H_2O were added. The resulting biphasic system was filtered, and the layers were separated. The organic phase was washed with 10 cm^3 H_2O, 10 cm^3 brine, dried (K_2CO_3), filtered, and concentrated to give a bright yellow syrup which was purified by flash chromatography (hexane:ethyl acetate = 5:1).

2,3,4-Tri-O-benzyl-5,6-dideoxy-D-xylo-hex-5-enose (**3**; $C_{27}H_{28}O_4$)

Colorless syrup; spectroscopic data in accordance with literature [7].

2,3,4-Tri-O-benzyl-5,6-dideoxy-D-lyxo-hex-5-enose (**11**; $C_{27}H_{28}O_4$)

Colorless syrup; spectroscopic data in accordance with literature [7].

2,3,4-Tri-O-benzyl-5,6-dideoxy-L-arabino-hex-5-enose (**12**; $C_{27}H_{28}O_4$)

Colorless syrup; $R_f = 0.45$ (hexane:ethyl acetate = 5:1); ^{1}H NMR ($CDCl_3$, 500 MHz): $\delta = 9.62$ (d, $J = 1.7$ Hz, 1 H), 7.40–7.20 (m, 15H), 5.90 (ddd, $J = 17.5$, 10.2, 7.7 Hz, 1 H), 5.44 (dd, $J = 17.5$, 1.7 Hz, 1 H), 5.42 (dd, $J = 10.2$, 1.7 Hz, 1 H), 4.65 (d, $J = 12.0$ Hz, 1 H), 4.61 (d, $J = 11.1$ Hz, 1 H), 4.56 (d, $J = 11.5$ Hz, 1 H), 4.55 (d, $J = 11.5$ Hz, 1 H), 4.50 (d, $J = 11.1$ Hz, 1 H), 4.20 (d, $J = 11.5$ Hz, 1 H), 4.11–4.08 (m, 2H), 3.89 (dd, $J = 7.7$, 3.8 Hz, 1 H) ppm; ^{13}C NMR ($CDCl_3$, 125 MHz): $\delta = 202.43$, 137.84, 137.50, 137.18, 135.47, 128.35, 128.23, 128.16, 128.07, 127.96, 127.73, 127.70, 127.52, 120.16, 83.92, 81.20, 79.23, 74.24, 73.39, 70.06 ppm.

General procedure for fragmentation of unprotected iodoglycosides (condition C)

To a solution of 200 mg (0.66 mmol) methyl 6-deoxy-6-iodo-α-*D*-glycopyranoside in 12 cm^3 *THF* and 3 cm^3 H_2O, 430 mg (6.6 mmol) pre-activated zinc were added, and the resulting suspension was sonicated at 40°C for 1 h. The reaction mixture was then filtered and concentrated to approximately 5 cm^3. The resulting solution was purified by reverse phase flash chromatography eluting with H_2O.

5,6-Dideoxy-D-xylo-hex-5-enofuranose (**13**; $C_6H_{10}O_4$)

Colorless syrup; $R_f = 0.21$ ($CHCl_3$:MeOH = 9:1); spectroscopic data in accordance with literature [12].

5,6-Dideoxy-D-lyxo-hex-5-enofuranose (**14**; $C_6H_{10}O_4$)

Colorless syrup; $R_f = 0.24$ ($CHCl_3$:MeOH = 9:1); 4:1 mixture of α:β anomers; ^{1}H NMR (D_2O, 500 MHz): α-anomer: $\delta = 5.86$ (ddd, $J = 17.5$, 10.5, 7.0 Hz, 1 H), 5.37–5.29 (m, 2H), 5.25 (d, $J = 4.5$ Hz, 1 H), 4.19–4.13 (m, 2H), 4.09 (t, $J = 4.5$ Hz, 1 H) ppm; β-anomer: $\delta = 5.23$ (d, $J = 5.0$ Hz, 1 H) ppm; ^{13}C NMR (D_2O, 125 MHz): α-anomer: $\delta = 133.32$, 120.18, 101.34, 82.03, 78.25, 73.50 ppm, β-anomer: $\delta = 134.18$, 119.90, 96.12, 82.21, 72.45, 72.13 ppm.

5,6-Dideoxy-L-arabino-hex-5-enofuranose (**15**: $C_6H_{10}O_4$)

Colorless syrup; $R_f = 0.18$ ($CHCl_3$:MeOH = 9:1); 1:1 mixture of α:β anomers; ^1H NMR (D_2O, 500 MHz, both anomers): $\delta = 5.91$–5.80 (m, 2H), 5.38–5.23 (m, 5H), 5.20 (d, $J = 3.0$ Hz, 1 H) 4.38–4.35 (m, 1 H), 4.07–3.91 (m, 4H), 3.84–3.82 (m, 1 H) ppm; ^{13}C NMR (D_2O, 125 MHz, both anomers): $\delta = 136.92$, 135.49, 120.18, 119.63, 101.44, 95.45, 84.23, 82.57, 82.18, 80.03, 78.20, 76.38 ppm.

Acknowledgments

We thank the Danish Natural Science Research Council for financial support.

References

[1] Bernet B, Vasella A (1979) Helv Chim Acta **62**: 1990
[2] a) Ferrier RJ, Furneaux RH, Prasit P, Tyler PC, Brown KL, Gainsford GJ, Diehl JW (1983) J Chem Soc Perkin Trans 1, 1621; b) Nakata M, Akazawa S, Kitamura S, Tatsuta K (1991) Tetrahedron Lett **32**: 5363; c) Dransfield PJ, Moutel S, Shipman M, Sik V (1999) J Chem Soc Perkin Trans 1, 3349
[3] Désiré J, Prandi J (2000) Eur J Org Chem 3075
[4] Bernotas RC, Papandreou G, Urbach J, Ganem B (1990) Tetrahedron Lett **31**: 3393
[5] a) Villalobos A, Danishefsky SJ (1990) J Org Chem **55**: 2776; b) Fürstner A, Baumgartner J (1993) Tetrahedron **49**: 8541
[6] Kobori Y, Myles DC, Whitesides GM (1992) J Org Chem **57**: 5899
[7] Kleban M, Kautz U, Greul J, Hilgers P, Kugler R, Dong H-Q, Jäger V (2000) Synthesis 1027
[8] Ovaa H, Codée JDC, Lastdrager B, Overkleeft HS, van der Marel GA, van Boom JH (1999) Tetrahedron Lett **40**: 5063
[9] a) Fürstner A, Weidmann H (1990) J Org Chem **55**: 1363; b) Fürstner A, Jumbam D, Teslic J, Weidmann H (1991) J Org Chem **56**: 2213; c) Fürstner A, Baumgartner J, Jumbam DN (1993) J Chem Soc Perkin Trans 1, 131
[10] a) Hyldtoft L, Poulsen CS, Madsen R (1999) Chem Commun 2101; b) Hyldtoft L, Madsen R (2000) J Am Chem Soc **122**: 8444
[11] a) Stout EI, Doane WM, Trinkus VC (1976) Carbohydr Res **50**: 282; b) Fürstner A, Weidmann H (1989) J Org Chem **54**: 2307
[12] de Raadt A, Ebner M, Ekhart CW, Fechter M, Lechner A, Strobl M, Stütz AE (1994) Catalysis Today **22**: 549

Received September 9, 2001. Accepted October 30, 2001

Reactivity Tournament of Isothiocyanato-Functionalized Saccharides with 1,6-Diamino-3,6-oxaoctane

Mark Walter and **Thisbe K. Lindhorst***

Christiana Albertina University of Kiel, Institute of Organic Chemistry, D-24098 Kiel, Germany

Summary. Thiourea-bridged glycoclusters and glycodendrimers are described in the literature as mimetics of (oligoantennary) oligosaccharides to serve as high-affinity lectin ligands. In extension of this concept, the assembly of different, structurally varied isothiocyanato-functionalized sugar derivatives on an oligoamine scaffold would lead to novel mixed glycoclusters. To control this approach, the relative reactivities of the isothiocyanates used in the thiourea-bridging reaction have to be known. Therefore, competition experiments with six different sugar isothiocyanates were carried out using 1,8-diamino-3,6-dioxaoctane as a symmetrical difunctionalized core molecule. Reactivities were ranked on the basis of integration ratios in the ^{1}H NMR spectra. A first mixed thiourea-bridged glycocluster was successfully prepared.

Keywords. Sugar isothiocyanates; Thiourea bridging; Carbohydrates; Bioorganic chemistry.

Introduction

The oligosaccharide portions found in glycoproteins and glycolipids have important functions in cell biology [1]. Due to their structural diversity they can serve to store biological information, *e.g.* in cell-cell recognition and adhesion processes, or to assist the control of biochemical events [2, 3]. Molecular recognition of carbohydrate ligands is mediated by specialized proteins called lectins and selectins [4]. Lectins may contain one or more carbohydrate recognition domains (CRDs) on a single peptide or protein chain, respectively, or they may occur as clusters, *e.g.* on cell surfaces [5]. Multivalent binding of the receptor protein CRDs to multiple copies of carbohydrate ligands as presented in oligoantennary glycoconjugates can lead to high avidities, a finding which has been named the cluster effect [6, 7]. Because multivalency has been recognized as an important characteristic of carbohydrate-protein interactions [8–10], numerous synthetic multivalent neoglycoconjugates of different architecture have been introduced [11–13], for example to prevent early adhesion of neutrophiles to endothelial surfaces [14] or to block microbial adhesion [15, 16].

* Corresponding author. E-mail: tklind@oc.uni-kiel.de

Representative classes of mimetics of oligosaccharides and oligoantennary glycoconjugates have been called glycoclusters and glycodendrimers [17–19]. A feasible approach for their synthesis includes a coupling reaction between an oligovalent core molecule and a sugar derivative. As glycosylation reactions suffer from the fact that they have to be optimized for every individual case, alternative approaches have been sought to link a carbohydrate derivative to a scaffold molecule, among which the thiourea-bridging strategy has become a very successful one [20–23]. This method combines an isothiocyanato-functionalized carbohydrate derivative [24] with a branched or dendritic oligoamine. Accordingly, thiourea-bridged glycodendrimers and glycoclusters have been prepared with glycosyl isothiocyanates and *p*-NCS-phenyl glycosides, and 6-deoxy-6-isothiocyanato-functionalized saccharides have been employed for the synthesis of thiourea-bridged glycoclusters [25, 26]. Usually, such reactions proceed with high yields and do not require tedious optimization of reaction conditions.

One of our projects involves the clustering of structurally different NCS-functionalized sugar derivatives to develop high-affinity ligands for various lectins. Using derivatives with different NCS-functionalized spacers would allow to vary the resulting glycoclusters with respect to their conformational flexibility. Furthermore, combination of different carbohydrate ligands in one thiourea-bridging reaction can eventually lead to highly diversified glycoclusters of a mixed type.

To allow for a rational synthesis of mixed glycoclusters of this kind in one pot, a grading of the reactivities of the reaction partners is necessary [27]. This paper describes a reactivity tournament of sugar isothiocyanates of different structural characteristics to obtain their relative reactivities in thiourea-bridging reactions.

Results and Discussion

Six different sugar isothiocyanates (**1–6**, Fig. 1) were chosen for comparison of their reactivities in thiourea-bridging: glycosyl isothiocyanates **1** and **2**, alkyl isothiocyanates **3** and **4**, and aryl isothiocyanates **5** and **6**. As we are especially interested in the inhibition of mannose-specific bacterial adhesion [28], mannose was chosen as starting material for all derivatives except for **1** which was included to obtain a hint about unexpected reactivity differences of analogous derivatives of different sugar series. Compounds **1** [29], **2** [29], **3** [30], **4** [25], and **6** [22] were synthesized according to standard procedures. A new route was followed for the preparation of **5**, involving direct glycosylation of 4-isothiocyanatobenzylalcohol instead of glycosylation of 4-nitrobenzylalcohol and subsequent derivatization of the *p*-nitro group. This method allowed to exclude the possibility of O $\rightarrow$ N-acetyl migration.

For the design of the competition experiments it has to be taken into account that in the thiourea-bridging reaction of sugar isothiocyanates with oligoamines the reactivity of the second amino group is influenced by reaction of the first one. Therefore, a diamine instead of a monoamine was chosen for the reactivity tournament to be closer to the real case scenario of the preparation of thiourea-bridged glycoclusters and glycodendrimers. The reaction between two different sugar isothiocyanates (**X** and **Y**, Fig. 2) and a diamine can lead to three products:

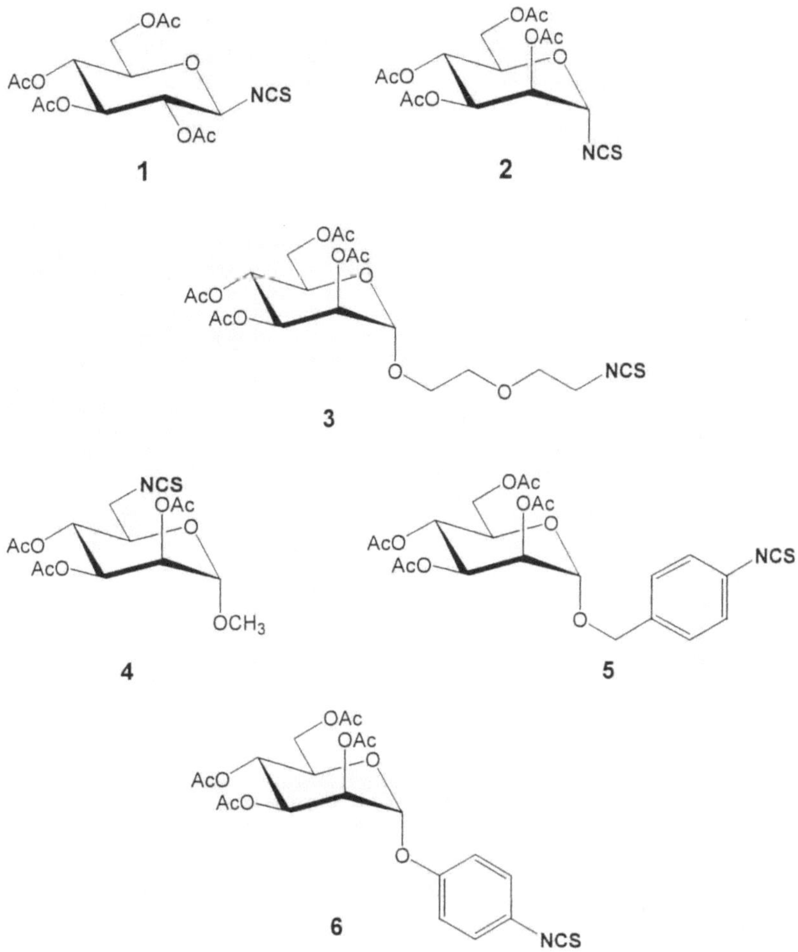

Fig. 1. Sugar isothiocyanates **1–6**

two homodimers and one heterodimer (Fig. 2). The diamine used in the reaction has to be soluble in an inert solvent. Because our first choice, 1,6-diaminohexane, had to be dissolved in protic solvents such as ethanol, 1,8-diamino-3,6-dioxaoctane was chosen instead. It is a symmetrical difunctionalized core molecule soluble in dichloromethane.

First, all homodimers (**1a–6a**) were prepared by thiourea-bridging with the above diamine (Fig. 3), mainly to be used as standards in NMR analysis (Table 1). Then, for each competition experiment a pair of sugar isothiocyanates chosen from compounds **1–6** (Fig. 1) was selected. Two equivalents of each relative to the diamine core molecule were dissolved in boiling dichloromethane, and the diamine was added dropwise as a solution in dichloromethane. The reaction mixture was stirred under reflux for three hours; then the solvent was removed under reduced pressure. Flash chromatography or HPLC did not allow the complete separation of the three different reaction products. Therefore, it was decided to use flash chromatography to separate the excess of educts from the product mixture. The two fractions thus obtained were analyzed by NMR spectroscopy, and the NMR spectra were used to estimate the relative reactivities of the isothiocyanates employed.

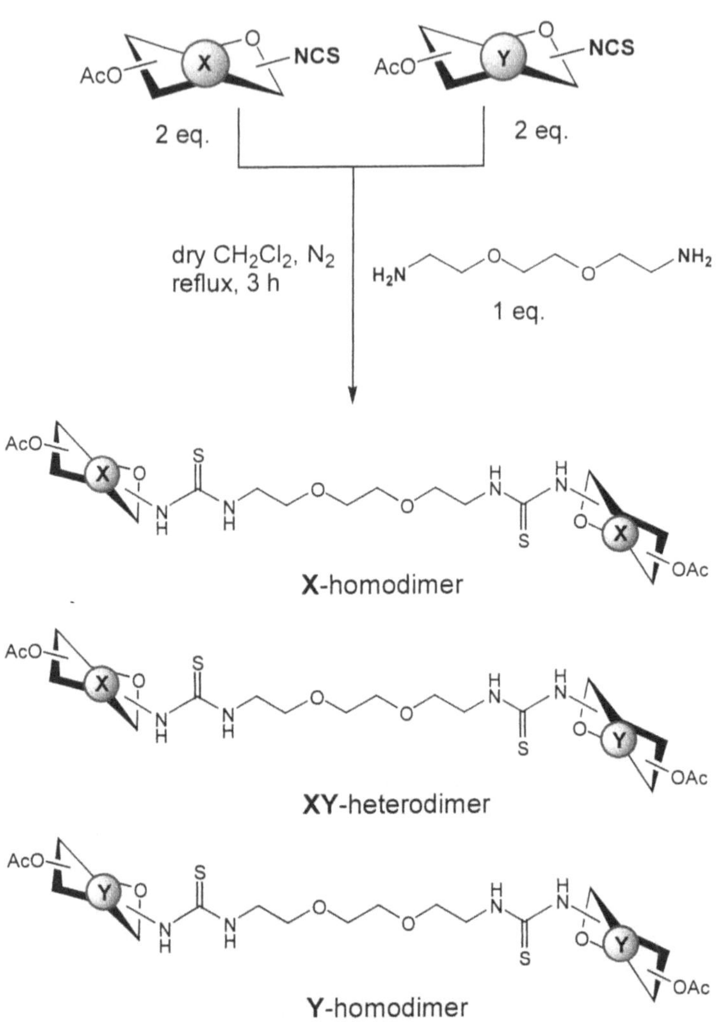

Fig. 2. The competition experiment: reaction of two different sugar isothiocyanates in the thiourea-bridging reaction

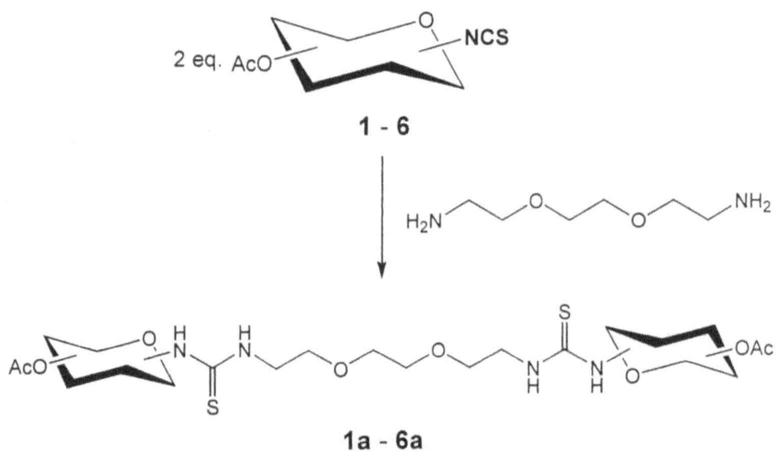

1a - 6a

Fig. 3. Thiourea-bridging reaction to obtain the homodimers **1a–6a**

Table 1. ^{1}H NMR shifts of H-1 signals of the employed sugar isothiocyanates and their respective dimers

	$\delta(^1H)$/ppm	B_0/MHz, solvent		$\delta(^1H)$/ppm	B_0/MHz, solvent
1	4.99	400, $CDCl_3$	**1a**	5.87	500, acetone-d_6
2	5.53	400, $CDCl_3$	**2a**	5.99	500, acetone-d_6
3	4.86	400, $CDCl_3$	**3a**	4.88	400, $CDCl_3$
4	4.72	400, $CDCl_3$	**4a**	4.65	400, $CDCl_3$
5	4.83	400, $CDCl_3$	**5a**	4.88	500, $CDCl_3$
6	5.47	400, $CDCl_3$	**6a**	5.52	500, $CDCl_3$

The ratio of unreacted starting materials (educts **X** and **Y**) as well as the ratio of glycosyl moieties **X** and **Y** present in the dimer mixtures could be obtained from the ^{1}H NMR spectra of the respective fractions. Two redundant informations were extracted from the spectra in each case: From the spectrum of the excess educts, the turnover of each educt isothiocyanate could be estimated, and from the spectrum of the product mixture it could be deduced which derivative had reacted with the amine to which extent. As in most cases the sugar H-1 signals were well separated from the other resonances, the integration of these peaks (Table 1) formed the basis for the determination of relative reactivities. The integration of the relevant signals of the NMR spectra of both fractions provided the same information for an individual competition experiment as exemplified for the competition reaction between **4** and **5** (C(**4**/**5**) in Table 2; Fig. 4).

The competition experiments clearly showed tremendous differences in the reactivity of different isothiocyanato-functionalized sugars when examined in the thiourea-bridging reaction (Table 2). The glycosyl isothiocyanates **1** and **2** were by far the most reactive derivatives due to the special reactivity of the anomeric

Table 2. Results of the reactivity tournament employing competition experiments between two different sugar isothiocyanates **X** and **Y**

C(**X**/**Y**)[1]	Educts		Products		Major product
	Turnover/ % **X**	Turnover/ % **Y**	% **X** in dimer mixture	% **Y** in dimer mixture	
C(**1**/**2**)	44	56	49	51	**2**
C(**1**/**5**)	83	17	80	20	**1**
C(**2**/**3**)	94	6	≥ 99	≤ 1	**2**
C(**2**/**4**)	97	3	99	1	**2**
C(**2**/**5**)	91	9	86	14	**2**
C(**2**/**6**)	88	12	–[2]	–[2]	**2**
C(**3**/**4**)	37	63	38	62	**4**
C(**3**/**5**)	14	86	–[2]	–[2]	**5**
C(**4**/**5**)	16	84	18	82	**5**
C(**5**/**6**)	54	46	56	44	**5**

[1] Competition reaction of sugar isothiocyanates **X** and **Y**; [2] Percentage in dimer mixture could not be determined because of overlapping signals in the ^{1}H NMR spectrum

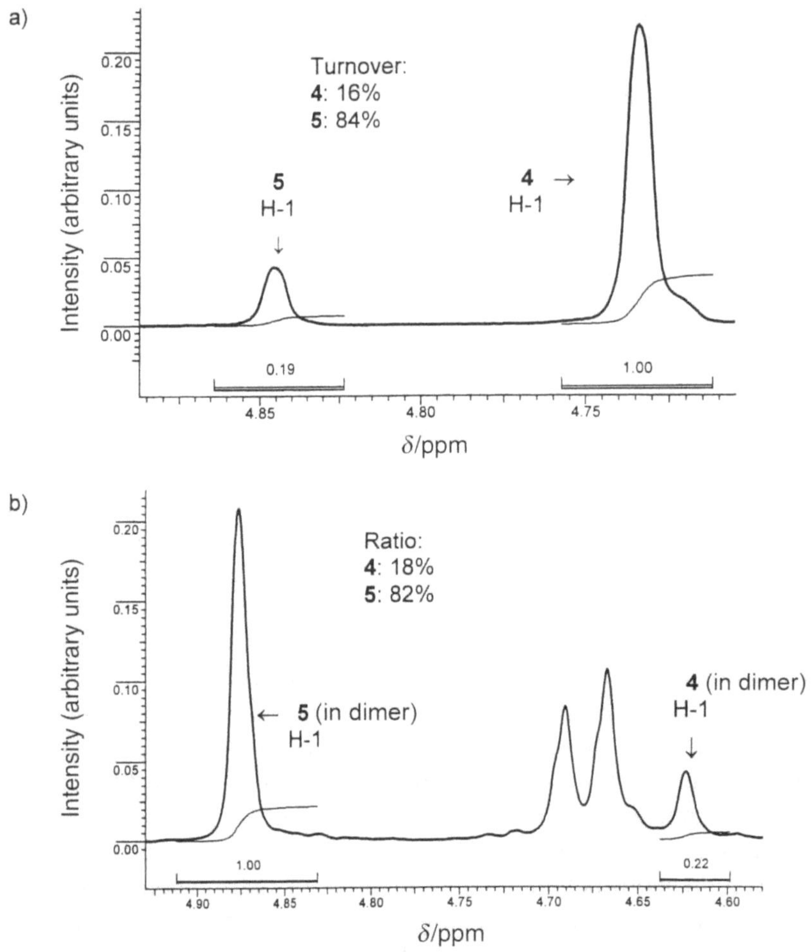

Fig. 4. (a) ^{1}H NMR signals of the H-1 protons of excess educts obtained in the competition experiment between **4** and **5**; (b) ^{1}H NMR signals of the H-1 protons of the product mixture obtained in the competition experiment between **4** and **5**; the signals were often broad, especially in the cases of **1a** and **2a** where acetone-d_6 was used as solvent for better resolved signals

position. Significant differences of the reactivities of the α-mannosyl (**2**) and β-glucosyl isothiocyanate (**1**) were not found. The derivatives with an aromatic aglycone formed a second group with similar reactivity. In this case, the mesomeric effect provided by the phenyl moiety may decrease the electrophilicity of the NCS-carbon. It could be shown that the *p*-NCS-benzyl glycoside **5** was slightly more reactive than the *p*-NCS-phenyl mannoside **6**. An explanation for this difference might be given on the basis of competing mesomeric effects. Among the alkyl isothiocyanates, the 6-NCS-modified derivative **4** is clearly more reactive than **3** which carries the NCS-group on the aliphatic spacer aglycone.

From integration of the broad H-1 signals in the measured ^{1}H NMR spectra the relative reactivities of the sugar isothiocyanates tested could be unequivocally deduced, leading to an order of reactivity of $\mathbf{2} \geq \mathbf{1} \gg \mathbf{5} > \mathbf{6} \gg \mathbf{4} > \mathbf{3}$.

Based on this ranking, a first rational synthesis of a mixed thiourea-bridged glycocluster was carried out. *Tris*-(2-aminoethyl)-amine was selected as branched

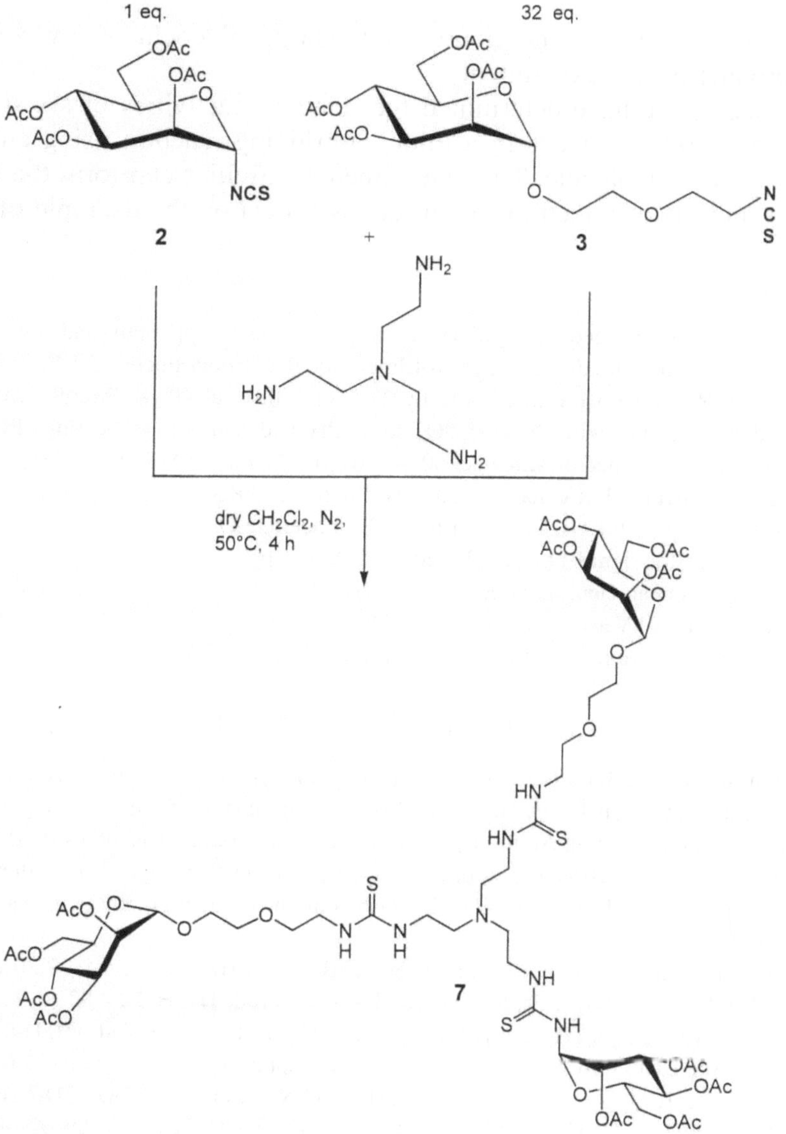

Fig. 5. Preparation of the defined mixed glycocluster $[(\mathbf{2})_1(\mathbf{3})_2]$ (**7**)

triamine to be reacted with the most reactive isothiocyanate **2** and the least reactive derivative **3** in one pot (Fig. 5). A molecule of the composition $[(\mathbf{2})_1(\mathbf{3})_2]$ (**7**) was selected as the target glycocluster. As **2** was shown to be 16 times more reactive than **3**, one equivalent of **2** and 32 equivalents of **3** were therefore reacted with one equivalent of the core molecule. The obtained product mixture was analyzed by MALDI-TOF mass spectrometry, revealing three different mass peaks at $m/z = 1401.44$ (corresponding to $[(\mathbf{2})_2(\mathbf{3})_1]$), $m/z = 1489.49$ (corresponding to the desired cluster $[(\mathbf{2})_1(\mathbf{3})_2]$), and $m/z = 1577.54$ (corresponding to $(\mathbf{3})_3$). The mass peak for the target cluster was clearly predominating over the others. Even though integration of MALDI-TOF peaks cannot necessarily form the basis of a quantitative analysis, there is not doubt that the target glycocluster **7** was formed as the main product in this experiment.

In conclusion, we have determined the relative reactivities of six structurally different sugar isothiocyanates in a thiourea-bridging reaction using an aliphatic diamine as the core molecule. The determined reactivities can form the basis of a controlled synthesis of mixed glycoclusters as shown on the example of **7**.

Experimental

Melting points were determined in capillary tubes in an Apotec apparatus and are uncorrected. Optical rotations were measured with a Perkin-Elmer model 241 polarimeter (20°C, 589 nm, length of cuvette: 1 dm). Reactions were monitored by TLC on silica gel GF_{254} (Merck) with detection under UV and by charring with 10% H_2SO_4 in EtOH and subsequent heating. Flash column chromatography was performed on silica gel 60 (40–63 μm, Merck). NMR spectra were recorded on Bruker AMX 400 or Bruker DRX 500 instruments. Chemical shifts are given relative to the solvent peaks of $CDCl_3$ (7.24 ppm for ^{1}H, 77.0 ppm for ^{13}C) or acetone-D_6 (2.04 ppm for ^{1}H, 29.3 ppm for ^{13}C). Where necessary, assignments were based on COSY experiments. IR spectra were taken with an ATI Mattson instrument, Genesis Series (KBr). MALDI-TOF mass spectra were measured with a Bruker Biflex III with 19 kV acceleration voltage. DHB ($c = 10\,\mu g/mm^3$ in 40% CH_3CN/H_2O) was used as the matrix. Ionization was effected with a nitrogen laser at 337 nm.

4-Isothiocyanatobenzyl 2,3,4,6-tetra-O-acetyl-mannopyranoside (**5**; $C_{22}H_{25}NO_{10}S$)

To a solution of 1-O-(2,3,4,6-tetra-O-acetyl-α-*D*-mannopyranosyl)-trichloroacetimidate (2.8 g, 5.85 mmol) and 4-isothiocyanatobenzylalcohol (1.0 g, 6.14 mmol) in 25 cm^3 dry CH_2Cl_2, two drops of *TMS-OTf* were added under an Ar atmosphere. The reaction mixture was then stirred for two days at room temperature. The reaction was stopped by addition of 1 cm^3 triethylamine. After evaporation of the solvent, the crude product was purified by flash column chromatography to give an amorphous yellowish solid.

Yield: 43%; IR (KBr): $\nu_{NCS} = 2123\,cm^{-1}$; ^{1}H NMR (400 MHz, $CDCl_3$): $\delta = 7.30$ (d, 2H, H_{ar}), 7.20 (d, 2H, H_{ar}), 5.33 (dd, 1H, H-3), 5.27 (t, 1H, H-4), 5.24 (dd, 1H, H-2), 4.83 (d, 1H, H-1), 4.67 (d, 1H, benzyl-C*H*H′), 4.51 (d, 1H, benzyl-CH*H*′), 4.25 (dd, 1H, H-6), 4.04 (dd, 1H, H-6′), 3.95 (ddd, 1H, H-5), 2.11, 2.08, 2.01, 1.96 (each s, each 3H, 4 OAc) ppm; $^3J_{1,2} = 1.5$, $^3J_{2,3} = 3.6$, $^3J_{3,4} = 9.6$, $^3J_{4,5} = 9.6$, $^3J_{5,6} = 5.1$, $^3J_{5,6'} = 2.5$, $^2J_{6,6'} = 12.2$ Hz; ^{13}C NMR (100.61 MHz, $CDCl_3$): $\delta = 170.54$, 169.97, 169.88, 169.65 (C=O), 135.78 (NCS), 135.47 (OCH_2C_{ar}), 131.09 (N–C_{ar}), 129.13 ($OCH_2C_{ar}(C_{ar})_2$), 125.87 (N–$C_{ar}(C_{ar})_2$), 96.73 (C-1), 69.42, 68.96, 68.77, 66.02 (C-2, C-3, C-4, C-5), 68.81 (CH_2), 62.36 (C-6), 20.81, 20.70, 20.62 (4 C(O)CH_3) ppm.

General procedure for the preparation of thiourea-bridged homodimers (**1a–6a**)

A solution of 2.2 equivalents of the sugar isothiocyanate in 10 cm^3 dry CH_2Cl_2 was heated to reflux temperature under a N_2 atmosphere. Then, a solution of 1,8-diamino-3,6-oxaoctane (1 equivalent in the 500-fold volume of dry CH_2Cl_2) was added dropwise. The course of the reaction was monitored by TLC (toluene/ethylacetate mixtures). After the reaction was complete (2–6 h) the solvent was removed *in vacuo* and the crude product was purified by flash column chromatography.

1a ($C_{36}H_{54}N_4O_{20}S_2$): ^{1}H NMR (500 MHz, acetone-d_6): $\delta = 7.45$ (s, 2H, sugar-NH), 7.33 (s, 2H, NH), 5.87 (s, 2H, H-1), 5.35 (t, 2H, H-3), 5.06–4.92 (m, 4H, H-2, H-4), 4.25 (d, 2H, H-6), 4.05 (dd, 2H, H-6′), 3.97 (bd, 2H, H-5), 3.83–3.66 (bd, 4H, HN–C*H*$_2$), 3.66–3.52 (bm, 8H, $HNCH_2CH_2OCH_2$, $HNCH_2CH_2$), 2.00, 1.99, 1.98, 1.94 (each s, each 6H, 8 OAc) ppm; $^3J_{2,3} = 9.8$, $^3J_{3,4} = 9.8$, $^3J_{4,5} = 9.8$, $^3J_{5,6'} = 2.2$, $^2J_{6,6'} = 12.3$ Hz; ^{13}C NMR (125.76 MHz, acetone-d_6): $\delta = 184.69$ (C=S), 170.18, 170.10, 169.62, 169.54 (C=O), 82.48 (C-1), 73.38, 73.29, 71.09, 68.70 (C-2, C-3, C-4, C-5),

70.30 ($HNCH_2CH_2O$), 69.16 ($HNCH_2CH_2OCH_2$), 62.09 (C-6), 44.75 ($HNCH_2$), 20.15, 20.10, 20.06, 19.99 (8 $C(O)CH_3$) ppm; MALDI-TOF-MS: $m/z = 927.35$ $[M+H]^+$ (calcd.: 927.29), 949.47 $[M+Na]^+$ (calcd.: 949.27), 965.47 $[M+K]^+$ (calcd.: 965.24).

2a ($C_{36}H_{54}N_4O_{20}S_2$): ^{1}H NMR (500 MHz, acetone-d$_6$): $\delta = 8.06$ (d, 2H, sugar-NH), 7.34 (s, 2H, NH), 5.99 (s, 2H, H-1), 5.36–5.30 (m, 4H, H-2, H-3), 5.24 (t, 2H, H-4), 4.27 (dd, 2H, H-6), 4.08 (dd, 2H, H-6′), 4.00 (bd, 2H, H-5), 3.84–3.69 (bs, 4H, $HNCH_2$), 3.69–3.56 (bm, 8H, $HNCH_2CH_2OCH_2$, $HNCH_2CH_2$), 2.12, 2.02, 2.01, 1.96 (s, each 6H, 8 OAc) ppm; $^3J_{2,3} = 3.2$, $^3J_{3,4} = 9.2$, $^3J_{4,5} = 9.2$, $^3J_{5,6} = 5.1$, $^3J_{5,6'} = 2.5$, $^2J_{6,6'} = 12.0$ Hz; ^{13}C NMR (125.76 MHz, acetone-d$_6$): $\delta = 184.36$ (C=S), 170.30, 169.81, 169.73, 169.55 (C=O), 80.45 (C-1), 70.36 ($HNCH_2CH_2O$), 70.11, 69.57, 69.21, 66.81, (C-2, C-3, C-4, C-5), 69.21 ($HNCH_2CH_2OCH_2$), 62.56 (C-6), 44.88 ($HNCH_2$), 20.20, 20.14, 20.05 (8 $C(O)CH_3$) ppm; MALDI-TOF-MS: $m/z = 927.29$ $[M+H]^+$ (calcd.: 927.29), 949.34 $[M+Na]^+$ (calcd.: 949.27), 965.31 $[M+K]^+$ (calcd.: 965.24).

3a ($C_{44}H_{70}N_4O_{20}S_2$): ^{1}H NMR (400 MHz, $CDCl_3$): $\delta = 6.91$ (bs, 4H, NH), 5.32–5.18 (m, 6H H-2, H-3, H-4), 4.88 (d, 2H, H-1), 4.24 (dd, 2H, H-6), 4.09 (dd, 2H, H-6′), 4.01 (ddd, 2H, H-5), 3.82–3.51 (m, 28H, spacer-CH_2), 2.13, 2.07, 2.02, 1.96 (s, each 6H, 8 OAc) ppm; $^3J_{1,2} = 1.5$, $^3J_{2,3} = 3.1$, $^3J_{3,4} = 10.2$, $^3J_{4,5} = 10.2$, $^3J_{5,6} = 5.1$, $^3J_{5,6'} = 2.6$, $^2J_{6,6'} = 12.2$ Hz; ^{13}C NMR (100.61 MHz, $CDCl_3$): $\delta = 170.73$, 170.39, 170.04, 169.73 (C=O), 97.49 (C-1), 70.00, 69.91, 69.82 (spacer-OCH_2), 69.66, 69.00, 68.43, 66.11 (C-2, C-3, C-4, C-5), 67.02 (sugar-OCH_2), 62.48 (C-6), 44.34 ($CH_2NHCSNHCH_2$), 20.93, 20.74, 20.69 (8 $C(O)CH_3$) ppm; MALDI-TOF-MS: $m/z = 1103.31$ $[M+H]^+$ (calcd.: 1103.39), 1125.35 $[M+Na]^+$ (calcd.: 1125.37), 1141.34 $[M+K]^+$ (calcd.: 1141.35).

4a ($C_{34}H_{54}N_4O_{18}S_2$): ^{1}H NMR (400 MHz, $CDCl_3$): $\delta = 5.30$ (dd, 2H, H-3), 5.19 (dd, 2H, H-2), 5.13 (t, 2H, H-4), 4.65 (d, 2H, H-1), 4.08 (bd, 2H, H-6), 4.00 (bd, 2H, H-6′), 3.92 (ddd, 2H, H-5), 3.66–3.61 (bq, 4H, $HNCH_2$), 3.61–3.47 (bm, 8H, $HNCH_2CH_2OCH_2$, $HNCH_2CH_2$), 3.36 (s, 6H, OCH_3), 2.12, 2.06, 1.96 (s, each 6H, 6 OAc) ppm; $^3J_{1,2} = 1.5$, $^3J_{2,3} = 3.6$, $^3J_{3,4} = 9.7$, $^3J_{4,5} = 9.7$, $^3J_{5,6} = 6.6$, $^3J_{5,6'} = 2.0$ Hz; ^{13}C NMR (100.61 MHz, $CDCl_3$): $\delta = 170.28$, 170.05, 169.83 (C=O), 98.37 (C-1), 69.95 ($HNCH_2CH_2O$), 69.87 ($HNCH_2CH_2OCH_2$), 69.59, 69.13, 68.79, 67.15 (C-2, C-3, C-4, C-5), 55.33 (OCH_3), 45.33 (C-6), 44.37 ($HNCH_2$), 20.88, 20.83, 20.66 (6 $C(O)CH_3$) ppm; MALDI-TOF-MS: $m/z = 871.10$ $[M+H]^+$ (calcd.: 871.30), 893.13 $[M+Na]^+$ (calcd.: 893.28), 909.10 $[M+K]^+$ (calcd.: 909.25).

5a ($C_{50}H_{66}N_4O_{22}S_2$): ^{1}H NMR (500 MHz, $CDCl_3$): $\delta = 7.33$ (d, 4H, H_{ar}), 7.14 (d, 4H, H_{ar}), 5.33 (dd, 2H, H-3), 5.29–5.24 (m, 4H, H-2, H-4), 4.88 (d, 2H, H-1), 4.69 (d, 2H, benzyl-CHH'), 4.51 (d, 2H, benzyl-CHH'), 4.26 (dd, 2H, H-6), 4.08 (dd, 2H, H-6′), 4.00 (ddd, 2H, H-5), 3.82–3.68 (bs, 4H, $HNCH_2$), 3.66–3.52 (m, 8H, $HNCH_2CH_2O$, $HNCH_2CH_2OCH_2$), 2.12, 2.09, 2.01, 1.96 (each s, each 6H, 8 OAc) ppm; $^3J_{1,2} = 1.5$, $^3J_{2,3} = 3.5$, $^3J_{3,4} = 10.1$, $^3J_{4,5} = 10.1$, $^3J_{5,6} = 5.1$, $^2J_{5,6'} = 2.2$ Hz; ^{13}C NMR (125.76 MHz, $CDCl_3$): $\delta = 170.59$, 170.04, 169.94, 169.64 (C=O), 129.39 ($C_{ar}N'H$), 128.98 ($C_{ar}CH_2O$), 128.17 ($C_{a,a'}C_{ar}CH_2O$), 124.41 ($C_{x,x'}C_{ar}NH$), 96.82 (C-1), 70.17 ($HNCH_2CH_2O$), 69.48 ($HNCH_2CH_2OCH_2$), 69.10 ($C_{ar}CH_2$), 69.48, 69.00, 68.73, 66.08 (C-2, C-3, C-4, C-5), 62.42 (C-6), 44.93 ($HNCH_2$), 20.84, 20.73, 20.65 (8 $C(O)CH_3$) ppm; MALDI-TOF-MS: $m/z = 1139.11$ $[M+H]^+$ (calcd.: 1139.37), 1161.18 $[M+Na]^+$ (calcd.: 1161.35), 1177.14 $[M+K]^+$ (calcd.: 1177.32).

6a ($C_{48}H_{62}N_4O_{22}S_2$): ^{1}H NMR (500 MHz, $CDCl_3$): $\delta = 8.12$ (s, 2H, ar-NH), 6.40 (s, 2H, NH), 7.12 (d, 4H, H_{ar}), 7.05 (d, 4H, H_{ar}), 5.52–5.46 (m, 4H, H-1, H-3), 5.38 (dd, 2H, H-2), 5.33 (t, 2H, H-4), 4.22 (dd, 2H, H-6), 4.07–3.99 (m, 4H, H-5, H-6′), 3.71 (bs, 4H, $HNCH_2$), 3.55 (bs, 4H, $HNCH_2CH_2OCH_2$), 3.50 (bs, 4H, $HNCH_2CH_2O$), 2.15, 2.01, 1.99, 1.98 (each s, each 6H, 8 OAc) ppm; $^3J_{3,4} = 9.8$, $^3J_{4,5} = 9.8$, $^3J_{5,6} = 5.4$, $^2J_{6,6'} = 12.6$ Hz; ^{13}C NMR (100.61 MHz, $CDCl_3$): $\delta = 180.81$ (C=S), 170.37, 169.90, 169.83, 169.52 (C=O), 153.95 (C_{ar}–O), 131.36 (C_{ar}–NH), 126.53 (HN–$C_{ar}C_{x,x'}$), 117.45 (O–$C_{ar}C_{a,a'}$), 95.79 (C-1), 70.02 ($HNCH_2CH_2O$), 69.20 ($HNCH_2CH_2OCH_2$), 69.12, 69.08, 68.63, 65.61 (C-2, C-3, C-4, C-5), 61.87 (C-6), 44.69 ($HNCH_2$), 20.73, 20.53 (8 $C(O)CH_3$) ppm; MALDI-TOF-MS: $m/z = 1111.25$ $[M+H]^+$ (calcd.: 1111.34), 1133.09 $[M+Na]^+$ (calcd.: 1133.32), 1149.01 $[M+K]^+$ (calcd.: 1149.29).

General procedure for the competition experiments

Under a N_2 atmosphere, 2 equivalents of each of the two competing sugar isothiocyanates were dissolved in 10 cm^3 dry CH_2Cl_2 at reflux temperature. Then, a solution of 1,8-diamino-3,6-oxaoctane (1 equivalent in a 500-fold volume of dry CH_2Cl_2) was added dropwise. Progress of the reaction was monitored by TLC (toluene/ethylacetate mixtures). After the reaction was complete (2–6 h) the solvent was removed *in vacuo* and the crude product was purified by flash column chromatography.

Mixed glycocluster (**7**; $C_{59}H_{91}N_7O_{31}S_3$)

Under a N_2 atmosphere, **2** (28 mg, 0.07 mmol, 0.33 equiv. per amino group) and **3** (1.06 g, 2.20 mmol, 16 equiv. per amino group) were dissolved in 15 cm^3 dry CH_2Cl_2. To this solution, *tris*-(2-aminoethyl)-amine (10 mg, 0.07 mmol, dissolved in 0.5 cm^3 of dry CH_2Cl_2) was added dropwise. This mixture was then heated under reflux for 3 h. After the removal of the solvent, the excess educts were separated from the products by flash column chromatography. The product mixture was then analyzed by MALDI-TOF-MS.

MALDI-TOF-MS: $m/z = 1402.75$ ([(**2**)$_2$(**3**)$_1$-cluster + H]$^+$, 30%, calcd.: 1402.45 for $C_{55}H_{83}N_7O_{29}S_3$ ($M = 1401.44$)), 1490.86 ([(**2**)$_1$(**3**)$_2$-cluster + H]$^+$, 57%, calcd.: 1490.50 for $C_{59}H_{91}N_7O_{31}S_3$ ($M = 1489.49$)), 1578.98 ([(**3**)$_3$-cluster + H]$^+$, 13%, calcd.: 1578.55 for $C_{53}H_{99}N_7O_{33}S_3$ ($M = 1577.54$)).

Acknowledgements

This work was kindly supported by the *Fonds der Chemischen Industrie* and the *Karl-Ziegler-Stiftung*.

References

[1] Varki A (1993) Glycobiology **3**: 97
[2] Dwek RA (1996) Chem Rev **96**: 683
[3] Kobata A (2000) Glycoconjugate J **17**: 443
[4] Lis H, Sharon N (1998) Chem Rev **98**: 637
[5] Weis WI, Drickamer K (1996) Annu Rev Biochem **65**: 441
[6] Lee YC (1978) Carbohydr Res **67**: 509
[7] Lee YC, Lee RT (1995) Acc Chem Res **28**: 321
[8] Kiessling LL, Pohl NL (1996) Chem Biol **3**: 71
[9] Mammen M, Choi SK, Whitesides GM (1998) Angew Chem **110**: 2908; Angew Chem Int Ed **37**: 2754
[10] Lee RT, Lee YC (2000) Glycoconjugate J **17**: 543
[11] Roy R (1996) Curr Opin Struct Biol **6**: 692
[12] Zanini D, Roy R (1998) In: Chapleur Y (ed) Carbohydrate Mimics, Concepts and Methods. Wiley-VCH, Weinheim, p 385
[13] Kiessling LL, Gestwicki JE, Strong LE (2000) Curr Opin Chem Biol **4**: 696
[14] Simanek EE, McGarvey GJ, Jablonski JA, Wong C-H (1998) Chem Rev **98**: 833
[15] Karlsson KA (1989) Ann Rev Biochem **58**: 309
[16] Lindhorst TK, Kieburg C, Krallmann-Wenzel U (1998) Glycoconjugate J **15**: 605
[17] Röckendorf N, Lindhorst TK (2001) Top Curr Chem **217**: 201
[18] Jayaraman N, Nepogodiev SA, Stoddart JF (1997) Chem Eur J **3**: 1193
[19] Roy R (1996) Polymer News **21**: 226
[20] Lindhorst TK, Kieburg C (1996) Angew Chem **108**: 2083; Angew Chem Int Ed Engl **35**: 1953

[21] Biessen EAL, Noorman F, van Teijlingen ME, Kuiper J, Barrett-Bergshoeff M, Bijsterbosch MK, Rijken DC, van Berkel TJC (1996) J Biol Chem **271**: 28024
[22] Kieburg C, Lindhorst TK (1997) Tetrahedron Lett **38**: 3885
[23] Roy R, Pagé D, Figueroa Perez S, Verez Bencomo V (1998) Glycoconjugate J **15**: 251
[24] Garcia Fernández JM, Ortiz Mellet C (2000) Adv Carbohydr Chem Biochem **55**: 36
[25] Kötter S, Krallmann-Wenzel U, Ehlers S, Lindhorst TK (1998) J Chem Soc Perkin Trans 1, 2193
[26] Baussanne I, Bemito JM, Ortiz Mellet C, Fernández García JM, Law H, Defaye J (2000) Chem Commun 1489
[27] Koeller KM, Wong C-H (2000) Chem Rev **100**: 4465
[28] Lindhorst TK (1999) In: Diederichsen U, Lindhorst TK, Westermann B, Wessjohann LA (eds) Bioorganic Chemistry. Wiley-VCH, Weinheim, p 133
[29] Lindhorst TK, Kieburg C (1995) Synthesis **10**: 1228
[30] Schönecker B, Ponsold K (1976) J prakt Chem **318**: 483

Received June 13, 2001. Accepted October 31, 2001

Photochemical Transformation of Levoglucosan Imides to Anhydrosugar-Annelated Azepanediones and Azocanediones

Christian Sund[1], **Swantje Thiering**[1], **Joachim Thiem**[1,*], **Jürgen Kopf**[2], and **Markus Stark**[3]

[1] Institute of Organic Chemistry, University of Hamburg, D-20146 Hamburg, Germany
[2] Institute of Inorganic Chemistry, University of Hamburg, D-20146 Hamburg, Germany
[3] Bell Flavors and Fragrances, Duft und Aroma GmbH, D-04205 Leipzig, Germany

Summary. By reaction of O-protected 3-amino-1,6-anhydro-*β*-*D*-glucopyranose with succinic, glutaric, and tartaric anhydride, respectively, the corresponding 3-substituted succinimido, glutarimido, and tatarimido derivatives were obtained. Irradiation of the succinimido and glutarimido derivatives at 254 nm gave 1,4-diradical intermediates by γ-hydrogen abstraction (*Norrish* Type II reaction) which subsequently recombined (*Yang* cyclization) to form azetidinols. These in turn fragmented and led to azepanedione and azocanedione derivatives. In contrast, irradiation of the tartarimido derivative resulted in elimination and led to both isomeric enolethers.

Keywords. 1,6-Anhydroglucopyranose; Photochemistry; Intramolecular alkylation; Anhydrosugar-annelated azepanediones and azocanediones.

Introduction

For access to complex heterocycles, C–C-bond formation by radical reactions leading to highly regio- and stereospecific products is increasingly gaining importance. A well established radical method for this purpose is comprised by the *Norrish* Type II reaction. This photochemical induced intramolecular alkylation of N-alkylated imides by 1,6- or 1,7-hydrogen abstraction leads to compounds containing the annelated ε-lactame structure, an integer part of a large number of important pharmaceuticals with multiple applications [1, 2]. Previously, *Kanaoka et al.* have described the synthesis of ε-lactames such as **5** by irradiation of N-alkylsuccinimides **1** [3] (Scheme 1).

Irradiation of imido derivatives **1** with UV light (254 nm) generates an excited carbonyl group. For steric reasons, hydrogen abstraction from the γ-position (1,6-abstraction) is preferred leading to a six-membered transition state **2**. The approach of the photochemically excited carbonyl function induces the intramolecular γ-hydrogen abstraction, thus establishing the diradical intermediate **3**. Its subsequent

* Corresponding author. E-mail: thiem@chemie.uni-hamburg.de

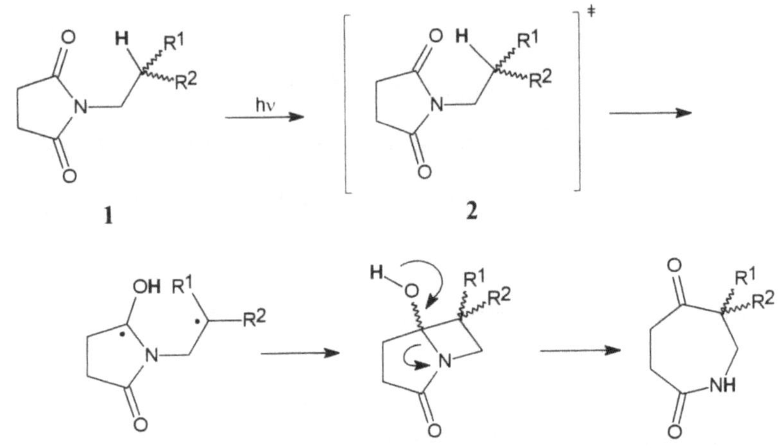

1 2 3 4 5

Scheme 1. Photocyclization of alkylimides

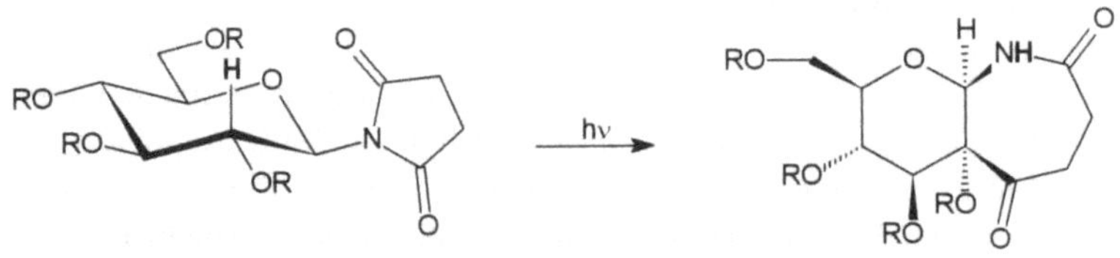

6 7

Scheme 2. Photocyclization of glycosylimides

Yang cyclization [4] results in formation of azetidinol **4** which in turn fragments in an irreversible retro-transannular ring opening and leads to the azepanedione **5**.

Previous studies of this photoreaction of 6-imidosaccharide derivatives [5] as well as N-glycosylimides [6–8] by *Thiem et al.* confirmed that corresponding carbohydrate compounds, *e.g.* **6**, can be transformed into sugar-derived azepanediones such as **7** (Scheme 2).

These investigations [5–8] have shown that the regiochemistry of the hydrogen abstraction is controlled by stereoelectronic and conformational factors. In order to elucidate the specific nature of these conformational effects, the bicyclic and thus conformationally fixed 1,6-anhydro-β-*D*-hexopyranoses were investigated.

Previous studies have focused on 2-imido-galactosan and 4-imido-mannosan derivatives [9]. In the following, we report about the synthesis of various 3-imido-glucosan derivatives and the results of their subsequent irradiation.

Results and Discussion

Preparation of saccharide imides

The synthesis of 3-imido-glucosan derivatives (Schemes 3 and 4) started from 1,6-anhydro-β-*D*-glucopyranose (levoglucosan, **8**) and followed the well-known pathway described by *Cerny et al.* [10, 11] leading to 1,6;3,4-dianhydro-β-*D*-allopyranose **9**. Regioselective opening of this epoxide by sodium azide gave the

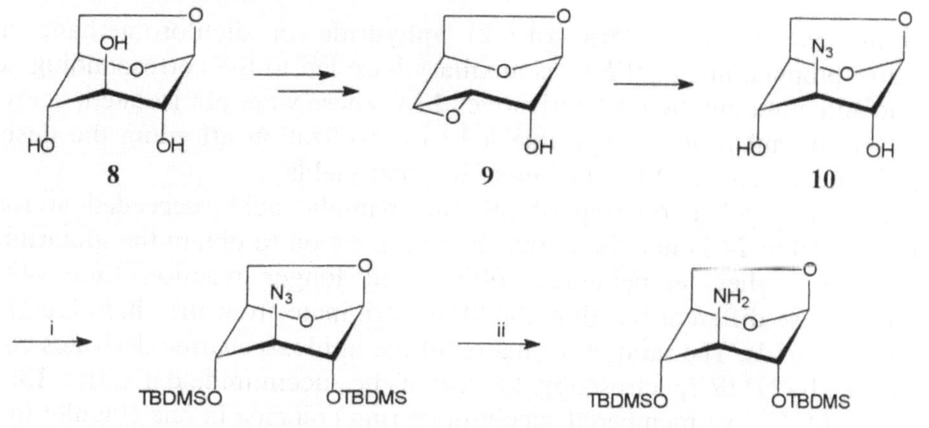

Scheme 3. *i*: *tert*-Butyldimethylsilyl triflate, pyridine, 0°C – room temperature, 20 h, 77%; *ii*: H_2, Pd/C, ethyl acetate/MeOH, room temperature, 24 h, 95%

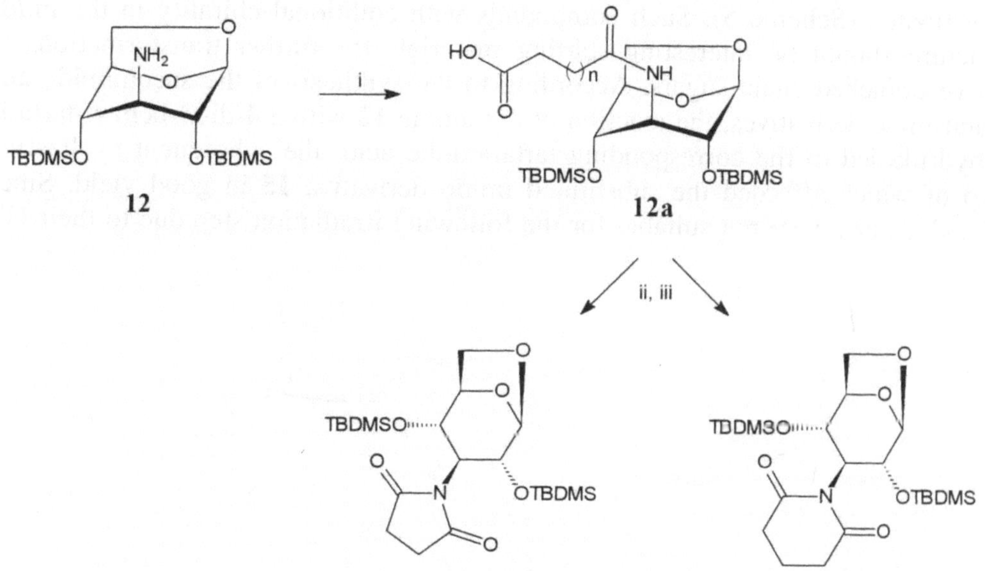

Scheme 4. *i*: Succinic anhydride ($n = 1$), glutaric anhydride ($n = 2$), *DIPEA*, room temperature, 24 h; *ii*: pyridine, Ac_2O, room temperature, 24 h, 65%; *iii*: pyridine, Ac_2O, 60°C, 48 h, 55%

gluco-configurated 3-azido derivative **10** [12]. Then the hydroxyl functions were protected as photochemically inert *tert*-butyldimethylsilyl (*TBDMS*) ethers using *TBDMS*-triflate in pyridine at 0°C. Further, reduction of the protected azide **11** by hydrogenolysis using palladium on activated charcoal led to the corresponding amine **12** in quantitive yield.

In a very versatile approach, imido-substituted derivatives are available by condensation of primary amines with different dicarbonic acid anhydrides [8, 13, 14]. Starting with the amine derivative **12**, reaction with 10 equivalents of succinic ($n = 1$) or glutaric ($n = 2$) anhydride in dichloromethane and N-ethyldiisopropylamine (*DIPEA*) as auxiliary base led to the corresponding succinamidic and glutaramidic acid derivatives **12a**. These were not isolated; acetylation with acetic anhydride in pyridine led to a cyclization affording the desired saccharide imides **13** and **14** in moderate to good yields.

Whereas the cyclization step of the succinamidic acid succeeded at room temperature within 24 hours, the corresponding reaction to obtain the glutarimide required both higher temperatures (60°C) and longer reaction time (48 h). Apparently, these different reaction conditions originate from the chain length of the dicarbonic acids. The ratio of formation of the imidosaccharide derivates could be monitored by NMR spectroscopy. In case of the succinimide derivative **13**, the four protons of the five-membered succinimide ring coincide in one singulet in the ^{1}H NMR spectrum due to the free rotation of the C3–N-bond and hence their chemical equivalence. In contrast, the six protons of the glutarimide ring in derivative **14** appear in two characteristic signals, one triplet for the four *meta*-positioned protons and one quintet for the two *para*-positionend protons.

In further attempts, another approach to functionalized imido derivatives was investigated (Scheme 5). Such compounds with additional chirality in the imide structure should be interesting starting materials for further transformations to derive branched-chain sugars. According to the synthesis of the succinimide and glutarimide derivatives, the reaction of the amine **12** with 3,4-di-O-benzyl-tartaric anhydride led to the corresponding tartaramidic acid, the subsequent cyclization step of which afforded the substituted imido derivative **15** in good yield. Since benzyl groups were not suitable for the following irradiation step due to their UV

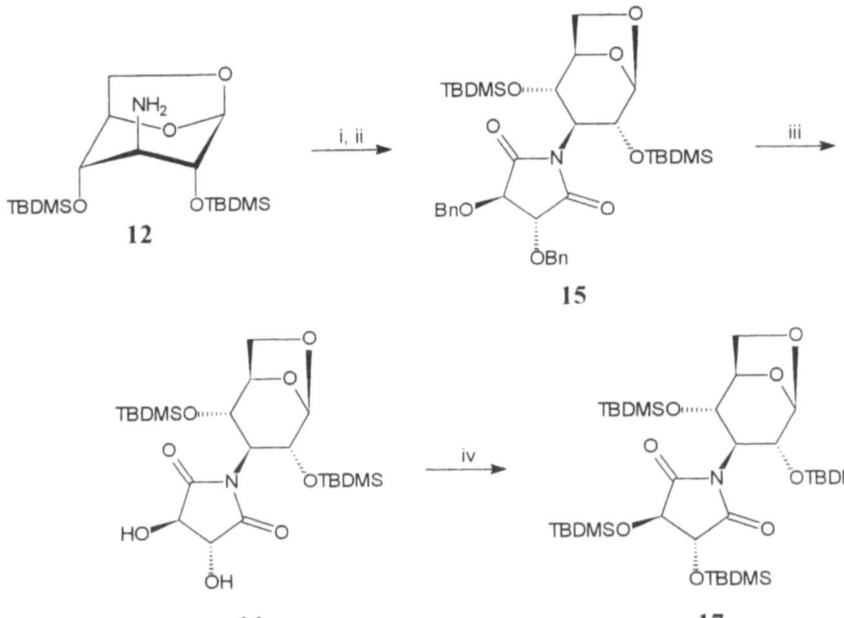

Scheme 5. *i*: 3,4-Di-O-benzyl-tartaric anhydride, *DIPEA*, room temperature, 20 h; *ii*: Ac_2O, pyridine, room temperature, 24 h, 76%; *iii*: H_2, Pd/C, MeOH, room temperature, 48 h, 86%; *iv*: *tert*-butyldimethylsilyl triflate, 2,6-dimethylpyridine, −40°C – room temperature, 48 h, 55%

absorption, a change of the protecting groups had to take place. By hydrogenolysis using palladium on activated charcoal the deprotected tartarimido derivative **16** was obtained, the free hydroxyl groups of which were subsequently protected as *TBDMS* ethers with *tert*-butyldimethylsilyltrifluoromethanesulfonate in 2,6-dimethylpyridine at −40°C to give compound **17**.

Irradiation of saccharide imides

All irradiations were performed with a low pressure mercury lamp at 254 nm under an argon atmosphere on solutions in anhydrous acetonitrile. To avoid heating of the reaction system by radiationless energy transfer, water cooling was required.

An abstraction in terms of a *Norrish* Type II reaction can be obtained only from a *cis*-positioned hydrogen [5–9]. Considering the geometry of the glucosan derivatives **13** and **14**, two suitable *cis*-hydrogen atoms (H-2 and H-4) are present for abstraction. Irradiation of **13** was completed after 6 hours as indicated by TLC. In accordance with its geometry, both isomers **18** and **19** were obtained in almost equal yields in addition to 11% of starting material (Scheme 6).

The ^{1}H NMR spectrum of the heterocyclic sugar derivatives showed characteristic signals suitable for the assignment of the above structures. The new NH protons were located as doublets at 6.31 ppm in **18** and 5.69 ppm in **19**, respectively. The coupling constants of these protons with H-3 of the sugar system (highlighted in Scheme 6) were determined to 6.0 (**18**) and 6.5 (**19**) Hz. The original singulet of the succinimide CH_2-groups was split into two multiplets due to formation of the novel seven-membered ring. The formation of a new quarternary center C-9 was evident from the disappearance of proton H-9 and verified by the corresponding signals in the ^{13}C NMR spectrum at 77.65 (**18**) and 80.17 (**19**) ppm. Furthermore, the ^{13}C NMR spectrum displayed two typical signals of the amide groups at 175.67 (**18**) and 175.82 (**19**) ppm; the keto functions led to resonances at 205.68 (**18**) and 209.13 (**19**) ppm in addition to two new signals of secondary carbons instead of the former identical secondary carbon atoms of the succinimide ring. The coupling constants of the sugar systems in these novel tricyclic structures did not show significant changes compared to the starting material, thus indicating a chair conformation of the pyranose ring.

Both photoproducts crystallized and allowed X-ray structural studies; the corresponding data are summarized in Table 1. Figure 1 shows an ORTEP drawing of compound **19** to support the structure as deduced from the NMR data.

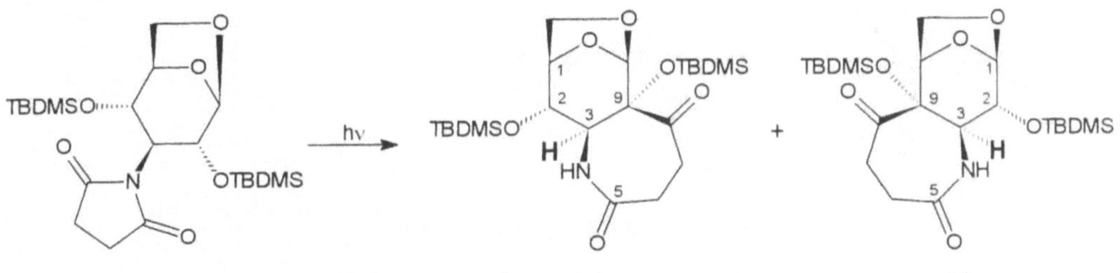

13 **18** **19**

Scheme 6. *i*: 254 nm, CH_3CN, 18°C, 6 h; **18**: 26%, **19**: 28%

Table 1. Crystal data, data collection parameters, and refinement results for **19** (Enraf-Nonius CAD4 diffractometer)

Formula	$C_{22}H_{41}NO_6Si_2$
Formula weight	471.74
Crystal dimension (mm)	$0.6 \times 0.5 \times 0.1$
Crystal system	orthorhombic
Space group	$P2_12_12_1$ (No. 19)
a (pm)	835.4 (1)
b (pm)	1229.2 (1)
c (pm)	2664.7 (2)
V ($10^6\,pm^3$)	2736.3 (4)
Z	4
ρ_{calcd} ($g \cdot cm^{-1}$)	1.145
$F(000)$	1024
Radiation	CuK_α
λ (pm)	154.178
μ (mm^{-1})	1.45
T (°C)	−100
$2\theta_{max}$ (°)	152.74
Unique reflexions	3268
Observed reflexions	3014
R_{all}	0.0558
Flack parameter	−0.06 (5)
$S = GooF$	0.857

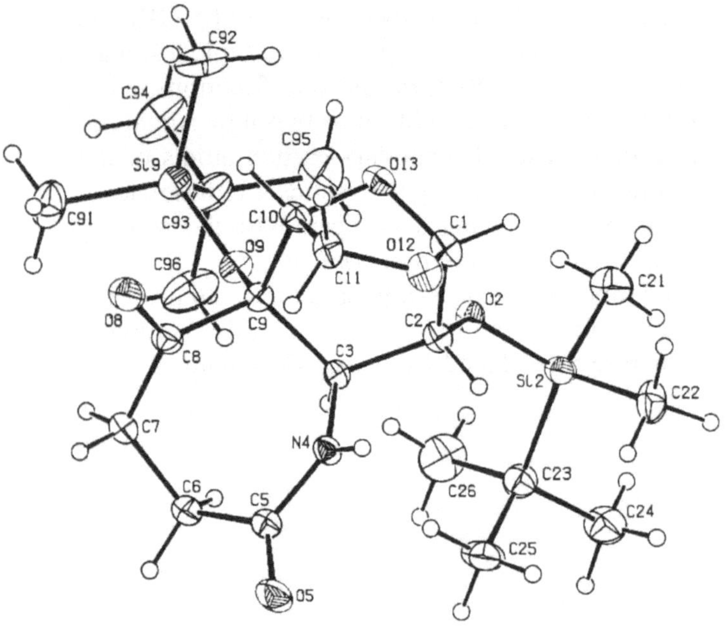

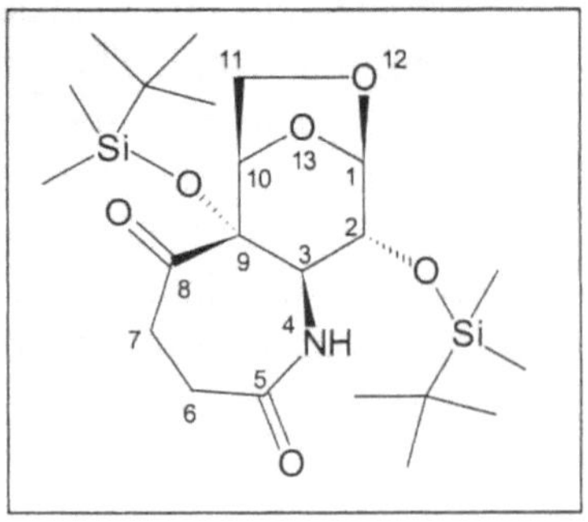

Fig. 1. ORTEP drawing of **19**

Even though the R value in case of compound **18** turned out to be less convincing compared to the corresponding value of derivative **19**, its structure is also shown here (Fig. 2). Apparently, the single crystal of **18** was of lower quality;

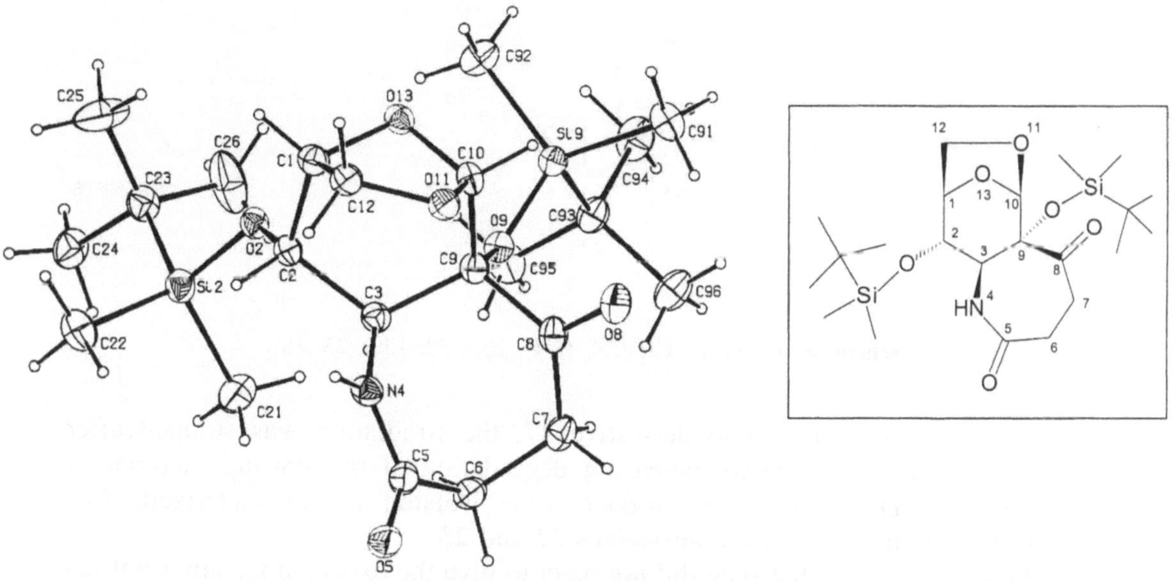

Fig. 2. ORTEP drawing of **18**

however, the data set obtained from NMR spectroscopy was rather similar to the corresponding data set of compound **19**.

The reaction of the glutarimido derivative **14** was complete after 4 hours of irradiation (Scheme 7). According to geometric reasons, hydrogen abstraction from position 2 and position 4 proceeded as expected and gave the heterocyclic sugars **20** and **21**.

Both structures were elucidated by NMR spectroscopy showing the characteristic signals of the amide group (NH proton as a doublet at 5.65 (**20**) and 5.97 (**21**) ppm; lactame carbon at 172.52 (**20**) and 177.23 (**21**) ppm) and the keto function at 207.89 ppm for derivative **20** and 210.70 ppm for compound **21**. The new quarternary centres at C-10 (78.20 (**20**) and 81.69 (**21**) ppm) were in accordance with the absence of the signals of H-10 in the ^{1}H NMR spectra. The former glutarimide system appeared in case of isomer **20** at 23.40 (C-7), 33.83 (C-6), and 38.44 (C-8) ppm, now fixed in the annelated eight membered ring. Similar data were found for compound **21**: 24.53 (C-7), 34.33 (C-6), and 38.20 (C-8) ppm. Again, no changes concerning the coupling constants of the sugar system were noted which proves that the conformation of the pyranose ring did not change.

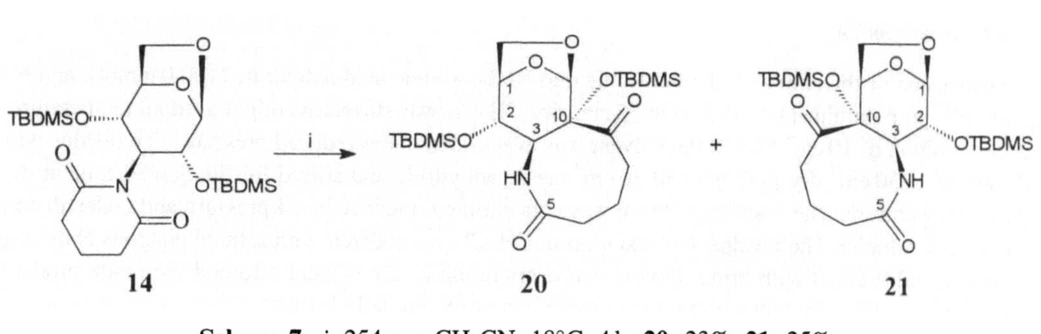

14 **20** **21**

Scheme 7. *i*: 254 nm, CH_3CN, 18°C, 4 h; **20**: 23%, **21**: 25%

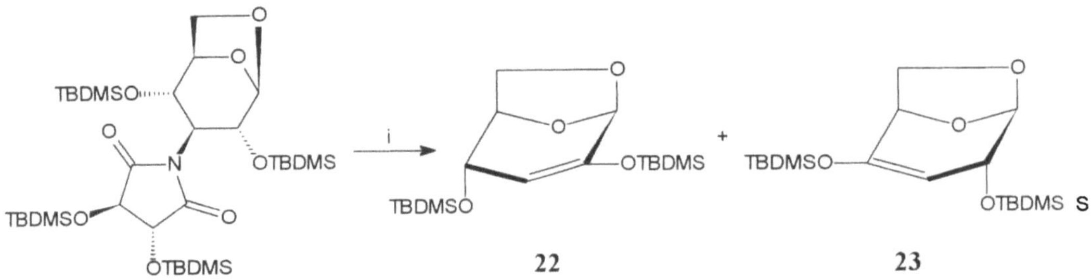

17 22 23

Scheme 8. *i*: 254 nm, CH_3CN, 18°C, 20 h; **22**: 12%, **23**: 18%

In case of the tartarimido derivative **17**, the irradiation was stopped after 20 hours (Scheme 8) due to increasing degradation of the starting material as monitored by TLC. Two main products were isolated and characterized; they prooved to be the unsaturated derivatives **22** and **23**.

Obviously, the excited state did not react to give the expected lactames but led after elimination of the protected imide to the enolethers **22** and **23**. It may be tentatively assumed that the *TBDMS* protecting groups could stabilize the five-membered ring and assist an elimination in contrast to the photoreactions of the succinimide and glutarimide derivatives **13** and **14**.

Experimental

All reactions were monitored by TLC analysis with silica gel 60-coated sheets (Merck). Compounds were visualized by spraying with 20% H_2SO_4 in EtOH followed by heating and/or UV irradiation. Column chromatography was performed on silica gel 60 (230–240 mesh, grain size 0.040–0.063 mm, Merck). All irradiations were performed in dry acetonitrile (Fluka) deoxygenated by degassing (30 min) with Ar in an ultrasonic bath employing a 60 W low-pressure Hg vapour lamp ($\lambda = 254$ nm) from Heraeus. The photoreactor was made of quartz glass and measured 35 cm in length and 4.5 cm in diameter. To maintain the temperature at 18°C, H_2O cooling was used. Melting points were measured on a ST-apotec apparatus and are reported uncorrected. Optical rotations were determined at room temperature with a Perkin Elmer Polarimeter 243. Elemental analyses were performed by the microanalytical laboratory of the Institute of Organic Chemistry of the University of Hamburg; their results agreed favourably with the calculated values. IR spectra were recorded on an ATI Mattson FTIR instrument (Genesis Series). NMR spectra were recorded on a Bruker AMX-400 spectrometer (^1H: 400 MHz, ^{13}C: 100 MHz); chemical shifts are referred to the solvent used, *J*-values are given in Hz.

Imide formation (A)

A suspension of the amine (10.0 mmol), the quoted dicarbonic acid anhydride (100.0 mmol), and N-ethyldiisopropylamine (10.0 mmol) in 75 cm^3 dry CH_2Cl_2 was stirred overnight at room temperature. After addition of 10 cm^3 MeOH the solvent was evaporated under reduced pressure. The residue was dissolved in 30 cm^3 dry pyridine and 30 cm^3 acetic anhydride and stirred for the quoted time at the given temperature. The reaction mixture was concentrated under reduced pressure and codestillated twice with toluene. The residue was taken up in CH_2Cl_2, neutralized with satured aqueous $NaHCO_3$ solution, and washed with brine. Drying and evaporation of the solvent afforded the crude product which was purified by column chromatography using the quoted eluent.

Irradiation (B)

The imido-substituted sugar (0.5–2.0 mmol) was dissolved in 200 cm^3 dry, deoxygenated acetonitrile and irradiated at 18°C under an Ar atmosphere for the quoted time. Evaporation of the solvent and purification by column chromatography using the quoted eluent led to the product.

1,6-Anhydro-3-azido-2,4-bis-O-(tert-butyldimethylsilyl)-3-deoxy-β-D-glucopyranose
(**11**: $C_{18}H_{37}N_3O_4Si_2$)

A solution of **10** (700 mg, 3.74 mmol) in 20 cm^3 pyridine was cooled to 0°C and treated with *tert*-butyldimethylsilyltrifluoromethanesulfonate (8.23 mmol, 2.2 equiv.). After stirring at room temperature for 20 h the solvent was evaporated under reduced pressure. The residue was codestilled with toluene, taken up in CH_2Cl_2, and washed with saturated aqueous $NaHCO_3$ and brine. Drying and evaporation of the solvent left the crude product. Purification by column chromatography using toluene:petroleum ether = 15:1 as eluent afforded the product **11** (1.2 g, 2.89 mmol, 77%) as light yellow crystals.

M.p.: 54°C; $[\alpha]_D^{20} = -24.7$ ($c = 1.0$, $CHCl_3$); IR (KBr): $\nu = 2112$ (N_3) cm^{-1}; ^{1}H NMR ($CDCl_3$): $\delta = 0.07$, 0.09, 0.13, 0.14 (4s, 12H, 2 Me–Si), 0.93 (s, 18H, 2 *tert*-Bu–Si), 3.31–3.32 (m, 2H, $J_{2,3} = 4.0$ Hz, $J_{2,4} = 7.5$ Hz, H-3, H-4), 3.37 (dd, 1H, H-2), 3.60–3.67 (m, 2H, $J_{5,6a} = 5.0$ Hz, $J_{6a,6b} = 7.0$ Hz, H-6a, H-6b), 4.33 (d, 1H, H-5), 5.20 (s, 1H, H-1) ppm; ^{13}C NMR ($CDCl_3$): $\delta = -4.80$, -3.54, -2.72, -2.62 (4C, 2 Me–Si), 16.89, 16.94 (2C, q, 2 *tert*-Bu–Si), 24.66, 24.68 (6C, 2 *tert*-Bu–Si), 66.63 (1C, –CH_2, C-6), 67.36 (1C, C-4), 73.87 (1C, C-2), 74.07 (1C, C-3), 77.99 (1C, C-5), 103.39 (1C, C-1) ppm.

3-Amino-1,6-anhydro-2,4-bis-O-(tert-butyldimethylsilyl)-3-deoxy-β-D-glucopyranose
(**12**; $C_{18}H_{39}NO_4Si_2$)

Compound **11** (2.23 g, 5.36 mmol) was dissolved in a mixture of 25 cm^3 dry ethyl acetate and 25 cm^3 dry MeOH and hydrogenated using 0.22 g (10% of weight) Pd/C (10%) by stirring at room temperature under an atmosphere of H_2 for 24 h. Removal of the catalyst by filtration and evaporation of the solvents yielded the amine which was used in the next step without further purification. **12** (1.98 g, 5.08 mmol, 95%) was obtained as a colourless oil.

$[\alpha]_D^{20} = -26.2°$ ($c = 0.5$, $CHCl_3$); ^{1}H NMR ($CDCl_3$): $\delta = 0.08$, 0.10, 0.11, 0.12 (4s, 12H, 2 Me–Si), 0.91, 0.92 (2s, 18H, 2 *tert*-Bu–Si), 2.81 (t, 1H, $J_{2,4} = 7.0$ Hz, H-4), 3.25 (d, 1H, $J_{2,3} = 1.0$ Hz, H-2), 3.33 (dd, 1H, $J_{3,4} = 7.0$ Hz, H-3), 3.61 (dd, 1H, $J_{5,6a} = 4.5$ Hz, $J_{6a,6b} = 7.0$ Hz, H-6a), 3.69 (m, 1H, H-6b), 4.29 (d, 1H, H-5), 5.18 (s, 1H, H-1) ppm; ^{13}C NMR ($CDCl_3$): $\delta = -4.63$, -4.54, -4.51, -4.43 (4C, 2 Me–Si), 17.92, 17.98 (2C, q, 2 *tert*-Bu–Si), 25.75, 25.77, 25.83, 25.86 (6C, 2 *tert*-Bu–Si), 56.17 (1C, C-3), 66.42 (1C, –CH_2, C-6), 72.00, 73.08 (2C, C-2, C-4), 78.00 (1C, C-5), 103.37 (1C, C-1) ppm.

1,6-Anhydro-2,4-bis-O-(tert-butyldimethylsilyl)-3-deoxy-3-N-succinimido-β-D-gluco-pyranose (**13**; $C_{22}H_{41}NO_6Si_2$)

The protected 3-amino derivative **12** (150 mg, 0.38 mmol) was reacted with succinic anhydride (0.38 g, 3.8 mmol) and N-ethyldiisopropylamine (0.1 cm^3, 0.38 mmol) according to procedure A. The cyclization was performed by stirring overnight at room temperature. The crude product was purified by column chromatography using toluene:acetone = 50:1 as eluent to give **13** (117 mg, 0.25 mmol, 65%) as white needles.

M.p.: 112°C; $[\alpha]_D^{20} = -27.2°$ ($c = 1.0$, $CHCl_3$); IR (KBr): $\nu = 1779$, 1698 (C=O) cm^{-1}; ^{1}H NMR ($CDCl_3$): $\delta = -0.12$, -0.11, 0.04, 0.07 (4s, 12H, 2 Me–Si), 0.84 (s, 18H, 2 *tert*-Bu–Si), 2.65 (s, 4H,

–CH_2–CH_2–, imide), 3.72 (dd, 1H, $J_{5,6a}=5.0$ Hz, $J_{6a,6b}=7.5$ Hz, H-6a), 3.80 (dd, 1H, $J_{5,6b}=1.0$ Hz, H-6b), 4.01 (d, 1H, $J_{2,4}=8.5$ Hz, H-2), 4.10–4.19 (m, 2H, $J_{3,4}=8.5$ Hz, H-3, H-4), 4.35 (d, 1H, H-5), 5.22 (s, 1H, H-1) ppm; ^{13}C NMR ($CDCl_3$): $\delta=-4.61$, -3.77, -2.87, -2.68 (4C, 2 Me–Si), 16.58, 16.64 (2C, q, 2 *tert*-Bu–Si), 24.44 (6C, 2 *tert*-Bu–Si), 26.88 (2C, –CH_2–CH_2–, imide), 56.27 (1C, C-4), 66.51 (1C, –CH_2, C-6), 67.81, 70.16 (2C, C-2, C-3), 78.32 (1C, C-5), 104.21 (1C, C-1) ppm.

1,6-Anhydro-2,4-bis-O-(tert-butyldimethylsilyl)-3-deoxy-3-N-glutarimido-β-D-glucopyranose (**14**; $C_{23}H_{43}NO_6Si_2$)

The 3-amino-derivative **12** (1.14 g, 2.93 mmol) was reacted with glutaric anhydride (3.34 g, 29.28 mmol) and N-ethyldiisopropylamine (0.5 cm^3, 2.93 mmol) according to procedure A. The cyclization was performed by stirring for 48 h at 60°C. The crude product was purified by column chromatography using petroleum ether:ethyl acetate = 7:1 as eluent to give **14** (250 mg, 0.25 mmol, 65%) as yellow oil.

$[\alpha]_D^{20}=-57.6°$ ($c=1.0$, $CHCl_3$); ^{1}H NMR ($CDCl_3$): $\delta=-0.1$, -0.09, -0.03, 0.02 (4s, 12H, 2 Me–Si), 0.82, 0.83 (2s, 18H, 2 *tert*-Bu–Si), 1.89–1.96 (m, 2H, –CH_2–CH_2–CH_2–, imide), 2.63 (t, 4H, –CH_2–CH_2–CH_2–, imide), 3.71 (dd, 1H, $J_{5,6a}=6.0$ Hz, $J_{6a,6b}=7.5$ Hz, H-6a), 3.80 (dd, 1H, $J_{5,6b}=1.5$ Hz, H-6b), 3.99 (d, 1H, $J_{2,3}=8.0$ Hz, H-2), 4.08 (d, 1H, $J_{3,4}=8.0$ Hz, H-4), 4.33 (dd, 1H, H-5), 4.84 (t, 1H, H-3), 5.20 (s, 1H, H-1) ppm; ^{13}C NMR ($CDCl_3$): $\delta=-5.09$, -4.95, -4.59, -4.47 (4C, 2 Me–Si); 18.06 (2C, q, 2 *tert*-Bu–Si); 25.46 (6C, 2 *tert*-Bu–Si), 33.0, 33.66 (2C, –CH_2–CH_2–CH_2–, imide), 57.56 (1C, C-3), 67.01 (1C, –CH_2, C-6), 71.70 (2C, C-2, C-4), 79.33 (1C, C-5), 105.31 (1C, C-1) ppm.

1,6-Anhydro-2,4-bis-O-(tert-butyldimethylsilyl)-3-deoxy-3-N-((3′R,4′R)-3′,4′-bis-O-benzyl)-tatarimido-β-D-glucopyranose (**15**; $C_{36}H_{53}NO_8Si_2$)

Compound **12** (700 mg, 1.8 mmol) was reacted with 3,4-di-O-benzyltataric anhydride (1.12 g, 3.59 mmol) and N-ethyldiisopropylamine (0.31 cm^3, 1.8 mmol) according to procedure A. The cyclization step was carried out by stirring overnight at room temperature. The crude product was purified by column chromatography using petroleum ether:acetone = 30:1 as eluent to give compound **15** (939 mg, 76%) as colourless oil.

$[\alpha]_D^{20}=-100.5°$ ($c=0.008$, $CHCl_3$); IR (KBr): $\nu=1721$, 1679 (C=O) cm^{-1}; ^{1}H NMR ($CDCl_3$): $\delta=-0.12$, -0.11, 0.00, 0.03 (4s, 12H, 2 Me–Si), 0.82 (s, 18H, 2 *tert*-Bu–Si), 3.73 (dd, 1H, $J_{5,6a}=5.0$ Hz, $J_{6a,6b}=7.0$ Hz, H-6a), 3.79 (dd, 1H, $J_{5,6b}=4.5$ Hz, H-6b), 3.94 (dd, 1H, $J_{2,4}=7.0$ Hz, $J_{3,4}=2.0$ Hz, H-4), 4.09–4.16 (m, 2H, H-2, H-3), 4.26 (s, 2H, –CH–CH–, imide), 4.35 (d, 1H, H-5), 4.74 (d, 2H, –CH_2, –OBn), 4.99 (d, 2H, –CH_2, –O*Bn*), 5.23 (s, 1H, H-1), 7.31–7.36 (m, 10H, –CH, –O*Bn*) ppm; ^{13}C NMR ($CDCl_3$): $\delta=-4.67$, -4.53, -4.24, -3.98 (4C, 2 Me–Si), 17.96, 18.04 (2C, q, 2 *tert*-Bu–Si), 25.87, 25.93 (6C, 2 *tert*-Bu–Si), 57.76 (1C, C-4), 68.04 (1C, –CH_2, C-6), 71.41, 79.64 (3C, C-2, C-3, C-5), 73.66 (2C, 2 -CH_2, –O*Bn*), 78.76 (2C, –CH–CH–, imide), 105.50 (1C, C-1), 127.35 (10C, –CH, –O*Bn*), 135.67 (2C, q, –O*Bn*), 172.00 (2C, 2 C=O) ppm.

1,6-Anhydro-2,4-bis-O-(tert-butyldimethylsilyl)-3-deoxy-3-N-((3′R,4′R)-3′,4′-bis-O-hydroxy)-tatarimido-β-D-glucopyranose (**16**; $C_{22}H_{41}NO_8Si_2$)

A solution of **15** (900 mg, 1.32 mmol) in 30 cm^3 dry MeOH was hydrogenated using 90 mg (10% of weight) Pd/C (10%) by stirring at room temperature under an atmosphere of H_2 for 24 h. Removal of the catalyst by filtration and evaporation of the solvents yielded compound **16** (574 mg, 86%), which was used in the next step without further purification.

M.p.: 85°C; $[\alpha]_D^{20}=-83.3°$ ($c=1.0$, $CHCl_3$); IR (KBr): $\nu=1754$ (C=O) cm^{-1}; ^{1}H NMR ($CDCl_3$): $\delta=-0.09$, -0.01, 0.05 (3s, 12H, 2 Me–Si), 0.84 (s, 18H, 2 *tert*-Bu–Si), 3.10 (s, 2H, 2 OH), 3.71 (dd, 1H, $J_{5,6a}=5.0$ Hz, $J_{6a,6b}=7.5$ Hz, H-6a), 3.80 (dd, 1H, $J_{5,6b}=4.5$ Hz, H-6b), 3.87–3.93

(m, 1H, $J_{2,4}=6.5$ Hz, $J_{3,4}=2.0$ Hz, H-4), 4.11–4.14 (m, 2H, H-2, H-3), 4.35 (d, 1H, H-5), 4.49 (s, 2H, –CH–CH–, imide), 5.23 (s, 1H, H-1) ppm; ^{13}C NMR ($CDCl_3$): $\delta=-4.50$, −4.53, −4.17, −3.97 (4C, 2 Me–Si), 17.94, 18.04 (2C, q, 2 *tert*-Bu–Si), 25.83, 25.88 (6C, 2 *tert*-Bu–Si), 58.30 (1C, C-4), 68.21 (1C, $-CH_2$, C-6), 71.31, 71.55 (2C, C-2, C-3), 74.95 (2C, –CH–CH–, imide), 79.65 (1C, C-5), 105.37 (1C, C-1), 175.00 (1C, 2 C=O) ppm.

1,6-Anhydro-2,4-bis-O-(tert-butyldimethylsilyl)-3-deoxy-3-N-((3′R,4′R)-3′,4′-bis-O-(tert-butyldimethylsilyl))-tatarimido-β-D-glucopyranose (**17**; $C_{34}H_{69}NO_8Si_4$)

Compound **16** (574 mg, 1.14 mmol) was dissolved in 10 cm^3 dry CH_2Cl_2 cooled to −40°C and treated with *tert*-butyldimethylsilyltrifluoromethanesulfonate (0.65 cm^3, 2.85 mmol) and 2,6-dimethylpyridine (0.33 cm^3, 2.85 mmol). After stirring under warming to room temperature for 48 h the mixture was washed with saturated aqueous NH_4Cl, saturated aqueous $NaHCO_3$, and brine. Drying and evaporation of the solvent afforded the crude product which was purified by column chromatography with petroleum ether:ethyl acetae = 100:1 to give **17** (460 mg, 55%) as a white solid.

M.p.: 101°C; $[\alpha]_D^{20} = -103.3°$ ($c=1.0$, $CHCl_3$); IR (KBr): $\nu=1725$ (C=O) cm^{-1}; ^{1}H NMR ($CDCl_3$): $\delta=-0.10$, −0.09, 0.00, 0.03, 0.14, 0.20 (6s, 24H, 4 Me–Si), 0.83, 0.83, 0.93 (3s, 36H, 4 *tert*-Bu–Si), 3.68 (dd, 1H, $J_{5,6a}=5.0$ Hz, $J_{6a,6b}=7.0$ Hz, H-6a), 3.80 (dd, 1H, H-6b), 3.91 (d, 1H, $J_{2,4}=8.5$ Hz, H-2), 4.08–4.14 (m, 2H, H-3, H-4), 4.32 (d, 1H, H-5), 4.34 (s, 2H, –CH–CH–, imide), 5.20 (s, 1H, H-1) ppm; ^{13}C NMR ($CDCl_3$): $\delta=-5.53$, −4.11 (8C, 2 Me–Si), 16.56–17.84 (4C, q, 2 *tert*-Bu–Si), 25.83, 25.93, 26.03 (12C, 4 *tert*-Bu–Si), 57.37 (1C, C-4), 68.14 (1C, $-CH_2$, C-6), 71.27, 71.49 (2C, C-2, C-3), 77.17 (2C, –CH–CH–, imide), 79.71 (1C, C-5), 105.54 (1C, C-1), 176.20 (2C, 2 C=O) ppm.

(1R,2R,8S,9S,10R)-4-Aza-2,9-bis-tert-butyldimethylsiloxy-11,13-dioxa-tricyclo [8.2.1.0^{3,9}]tridecane-5,8-dione (**18**; $C_{22}H_{41}NO_6Si_2$)
and *(1R,2S,8R,9R,10R)-4-Aza-2,9-bis-tert-butyldimethylsiloxy-12,13-dioxa-tricyclo [8.2.1.0^{3,9}]tridecane-5,8-dione* (**19**; $C_{22}H_{41}NO_6Si_2$)

A solution of succinimide **13** (500 mg, 1.06 mmol) was irradiated for 6 h according to procedure B. Purification by column chromatography with toluene:acetone = 50:1 afforded **19** (146 mg, 0.31 mmol, 28%) and **18** (134 mg, 0.28 mmol, 26%), both as white crystals.

18: M.p.: 153°C; $[\alpha]_D^{20} = +16.0°$ ($c=0.5$, $CHCl_3$); IR (KBr): $\nu=3241$ (NH), 1711, 1679 (C=O) cm^{-1}; ^{1}H NMR ($CDCl_3$): $\delta=-0.07$, 0.00, 0.11 (3s, 12H, 2 Me–Si), 0.79, 0.81 (2s, 18H, 2 *tert*-Bu–Si), 2.30 (dd, 1H, H-4a), 2.45–2.55 (m, 1H, H-5a), 2.88–2.98 (m, 2H, H-4b, H-5b), 3.57 (s, 1H, H-9), 3.61 (dd, 1H, $J_{10,11a}=5.0$ Hz, $J_{11a,11b}=8.5$ Hz, H-11a), 3.72–3.76 (m, 2H, $J_{3,NH}=6.0$ Hz, H-3, H-11b), 4.34 (dd, 1H, H-10), 5.62 (s, 1H, H-1), 6.31 (d, 1H, NH) ppm; ^{13}C NMR ($CDCl_3$): $\delta=-2.18$, −1.03, −0.85 (4C, 2 Me–Si), 22.69 (2C, q, 2 *tert*-Bu–Si), 28.33, 28.49 (6C, 2 *tert*-Bu–Si), 31.73, 38.05 (2C, 2-CH_2, C-4, C-5), 65.87 (1C, C-8), 68.08 (1C, $-CH_2$, C-11), 74.45 (1C, C-9), 77.65 (1C, q, C-2), 80.80 (1C, C-10), 102.87 (1C, C-1), 175.67 (1C, q, NHC=O), 205.68 (1C, q, C=O) ppm.

19: M.p.: 148°C; $[\alpha]_D^{20} = -44.4°$ ($c=0.5$, $CHCl_3$); IR (KBr): $\nu=3369$ (NH), 1704, 1684 (C=O) ppm; ^{1}H NMR ($CDCl_3$): $\delta=0.04$, 0.11, 0.12, 0.24 (4s, 12H, 2 Me–Si), 0.92 (s, 18H, 2 *tert*-Bu–Si), 2.47–2.61 (m, 2H, H-6a, H-7a), 3.02–3.12 (m, 1H, H-7b), 3.19–3.27 (m, 1H, H-6b), 3.52 (s, 1H, H-2), 3.79 (dd, 1H, $J_{10,11a}=5.5$ Hz, $J_{11a,11b}=8.5$ Hz, H-11a), 3.83 (d, 1H, $J_{3,NH}=6.5$ Hz, H-3), 3.90 (dd, 1H, $J_{10,11b}=1.0$ Hz, H-11b), 4.73 (dd, 1H, H-10), 5.31 (s, 1H, H-1), 5.69 (d, 1H, NH) ppm; ^{13}C NMR ($CDCl_3$): $\delta=-2.00$, −1.93, −0.89, 3.85 (4C, 2 Me–Si), 21.24 (2C, q, 2 *tert*-Bu–Si), 28.58, 28.78 (6C, 2 *tert*-Bu–Si), 32.05, 36.69 (2C, $-CH_2-$, C-6, C-7), 63.52 (1C, C-3), 68.20 (1C, $-CH_2-$, C-11), 74.04 (1C, C-10), 77.70 (1C, C-2), 80.17 (1C, q, C-9), 104.98 (1C, C-1), 175 (1C, q, CO, NHC=O), 209.13 (1C, q, C=O) ppm.

(1R,2R,9S,10S,11R)-4-Aza-2,10-bis-tert-butyldimethylsiloxy-12,14-dioxa-tricyclo [9.2.1.0^{3,10}]tetradecane-5,7-dione (**20**; $C_{23}H_{43}NO_6Si_2$) and *(1R,2S,9R,10R,11R)-4-Aza-2,10-bis-tert-butyldimethylsiloxy-13,14-dioxa-tricyclo [9.2.1.0$^{3.10}$]tetradecane-5,9-dione* (**21**; $C_{23}H_{43}NO_6Si_2$)

A solution of the 3-glutarimide **14** (250 mg, 0.51 mmol) was irradiated for 4 h according to procedure B. Purification by column chromatography with toluene:acetone = 30:1 afforded **20** (58.29 mg, 0.12 mmol, 23%) and **21** (61.93 mg, 0.13 mmol, 25%), both as colourless oils.

20: $[\alpha]_D^{20} = -123.5°$ ($c = 0.5$, $CHCl_3$); ^{1}H NMR ($CDCl_3$): $\delta = 0.02$, 0.11, 0.13, 0.23 (4s, 12H, 2 Me–Si), 0.92, 0.95 (2s, 18H, 2 *tert*-Bu–Si), 1.95–1.98 (m, 1H, H-7a), 2.08–2.13 (m, 1H, H-7b), 2.22–2.26 (m, 1H, H-8a), 2.43–2.54 (m, 2H, H-6a, H-6b), 3.31–3.36 (dt, 1H, H-8b), 3.62 (s, 1H, H-2), 3.79–3.83 (m, 2H, $J_{3,NH} = 10.0$ Hz, $J_{11,12a} = 5.0$ Hz, $J_{12a,12b} = 8.5$ Hz, H-3, H-12a), 4.26 (d, 1H, H-12b), 4.55 (d, 1H, H-11), 5.34 (s, 1H, H-1), 5.98 (d, 1H, NH) ppm; ^{13}C NMR ($CDCl_3$): $\delta = -4.96$, -4.71, -3.37, -2.68 (4C, 2 Me–Si), 18.53 (2C, q, 2 *tert*-Bu–Si), 24.53 (1C, $-CH_2$, C-7), 25.96, 26.00 (6C, 2 *tert*-Bu–Si), 34.33 (1C, $-CH_2$, C-6), 38.20 (1C, $-CH_2$, C-8), 58.30 (1C, C-3), 65.32 (1C, C-12), 72.20 (1C, C-2), 75.37 (1C, C-11), 79.01 (1C, C-10), 102.42 (1C, C-1), 174.55 (1C, q, NH–C=O), 210.70 (1C, q, C=O) ppm.

21: $[\alpha]_D^{20} = +34.7°$ ($c = 1.0$, $CHCl_3$); ^{1}H NMR ($CDCl_3$): $\delta = 0.02$, 0.12, 0.13, 0.20 (4s, 12H, 2 Me–Si), 0.93 (2s, 18H, 2 *tert*-Bu–Si), 2.00 (m, 1H, $-CH_2$, H-7a), 2.25 (m, 2H, $-CH_2$, H-8a, H-7b), 2.45 (m, 1H, $-CH_2$, H-6a), 2.58 (m, 1H, $-CH_2$, H-6b), 3.20 (m, 1H, $-CH_2$, H-8b), 3.70 (s, 1H, H-2), 3.76 (dd, 1H, $J_{1,13a} = 5.0$ Hz, $J_{13a,13b} = 9.0$ Hz, H-13a), 3.90 (d, 1H, $J_{3,NH} = 11.50$ Hz), 3.95 (d, 1H, H-13b), 4.42 (d, 1H, H-1), 5.62 (s, 1H, H-11), 5.64 (d, 1H, NH) ppm; ^{13}C NMR ($CDCl_3$): $\delta = -4.87$, -4.73, -3.26, -2.37 (4C, 2 Me–Si), 17.85 (2C, q, 2 *tert*-Bu–Si), 23.40 (1C, $-CH_2$, C-7), 25.88, 25.90 (6C, 2 *tert*-Bu–Si), 33.83 (1C, $-CH_2$, C-6), 38.44 1C, $-CH_2$, C-8), 60.08 (1C, C-3), 65.88 (1C, $-CH_2$, C-13), 73.04 (1C, C-2), 78.20 (2C, C-1, C-10), 100.30 (1C, C-11), 172.52 (1C, q, NH–C=O), 207.89 (1C, q, C=O) ppm.

1,6-Anhydro-2,4-bis-O-(tert-butyldimethylsilyl)-3-deoxy-β-D-erythro-hex-2-enopyranose (**22**; $C_{18}H_{36}O_4Si_2$) and *1,6-Anhydro-2,4-bis-O-(tert-butyldimethylsilyl)-3-deoxy-β-D-erythro-hex-3-enopyranose* (**23**; $C_{18}H_{36}O_4Si_2$)

Compound **17** (410 mg, 0.56 mmol) was irradiated for 20 h according to procedure B. Subsequent column chromatography with petroleum ether:ethyl acetate = 100:1 yielded compounds **22** (38 mg, 18%) and **23** (25 mg, 12%).

22: ^{1}H NMR ($CDCl_3$): $\delta = 0.1$, 0.17, 0.22, (3s, 12H, 2 Me–Si), 0.90, 0.92 (2s, 18H, 2 *tert*-Bu–Si), 3.60–3.72 (m, 2H, $J_{5,6a} = 4.0$ Hz, $J_{6a,6b} = 6.0$ Hz, H-6a, H-6b), 3.90 (dd, 1H, $J_{3,4} = 4.0$ Hz, $J_{4,5} = 1.0$ Hz, H-4), 4.41 (d, 1H, H-5), 4.59 (d, 1H, H-3), 5.34 (s, 1H, H-1) ppm; ^{13}C NMR ($CDCl_3$): $\delta = -4.83$, -3.64, -2.35, -1.98 (4C, 2 Me–Si), 16.97, 17.30 (2C, q, 2 *tert*-Bu–Si), 24.51, 24.90 (6C, 2 *tert*-Bu–Si), 67.22 (1C, C-4), 67.94 (1C, $-CH_2$, C-6), 72.74 (1C, C-5), 98.20 (1C, C-3), 101.94 (1C, C-1), 152.51 (1C, q, C-2) ppm.

23: ^{1}H NMR ($CDCl_3$): $\delta = 0.11$, 0.17, 0.18 (3s, 12H, 2 Me–Si), 0.91, 0.92 (2s, 18H, 2 *tert*-Bu–Si), 3.37 (dd, 1H, $J_{5,6a} = 1.5$ Hz, $J_{6a,6b} = 7.5$ Hz, H-6a), 3.84 (dd, $J_{5,6b} = 6.5$ Hz, H-6b), 3.87 (dd, 1H, $J_{1,2} = 1.5$ Hz, $J_{2,3} = 5.0$ Hz, H-2), 4.49–4.52 (m, 1H, H-5), 4.61–4.64 (m, 1H, H-3), 5.20 (d, 1H, H-1) ppm; ^{13}C NMR ($CDCl_3$): $\delta = -4.64$, -3.37, -2.81, -1.99 (4C, 2 Me–Si), 17.03, 17.29 (2C, q, 2 *tert*-Bu–Si), 24.51, 24.69 (6C, 2 *tert*-Bu–Si), 62.43 (1C, $-CH_2$, C-6), 67.59 (1C, C-2), 76.31 (1C, C-5), 97.95 (1C, C-1), 98.09 (1C, C-3), 152.70 (1C, q, C-4) ppm.

Complete crystallographic data have been deposited with the Cambridge Crystallographic Data Centre, CCDC no. 167716 (**19**). Copies may be obtained free of charge from The Director, CCDC, 12 Union Road, Cambridge CB2 1EZ, UK (fax: +44-1223-336033, e-mail: deposit@ccdc.cam.ac.uk or http://www.ccdc.cam.ac.uk).

Acknowledgements

Support of this work by the *Fonds der Chemischen Industrie* and the *Deutsche Forschungsgemeinschaft* (GRK 464) is gratefully acknowledged.

References

[1] Triggle DJ, Langs DA, Janis RA (1989) Med Res Rev **9**: 123
[2] Merluzzi VJ, Hargrave KD, Labadia M, Grozinger K, Skoog M, Wu JC, Shih KC, Eckner K, Hattox S, Adams J, Rosenthal AS, Faanes R, Eckner RJ, Koup KR, Sullivan JL (1990) Science **250**: 1411
[3] Kanaoka Y (1978) Acc Chem Res **11**: 407
[4] Yang NC, Elliot SP, Kim B (1969) J Am Chem Soc **91**: 7551
[5] Sowa CE, Stark M, Heidelberg T, Thiem J (1996) Synlett **3**: 227
[6] Sowa CE, Thiem J (1994) Angew Chem **106**: 2041; (1994) Angew Chem Int Ed Engl **33**: 1979
[7] Sowa CE, Stark M, Heidelberg T, Thiem J (1996) Synlett **3**: 227
[8] Thiering S, Sowa CE, Thiem J (2001) J Chem Soc Perkin Trans 1, 801
[9] Thiering S, Sund C, Thiem J, Giesler A, Kopf J (2001) Carbohydr Res **336**: 271
[10] Cerny M, Stanek J (1972) Synthesis Commun 698
[11] Cerny M, Stanek J (1977) Adv Carbohydr Chem Biochem **34**: 23
[12] Paulsen H, Koebernick H, Stenzel W, Köll P (1975) Tetrahedron Lett **18**: 1493
[13] Lemieux RU, Abbas SZ, Burzynska MH, Ratcliffe RM (1982) Can J Chem **60**: 63
[14] Schmidt RR, Grundler G (1985) Carbohydr Res **135**: 201

Received September 3, 2001. Accepted October 31, 2001

Synthesis and Biological Activity of a Series of Methylene-Expanded Oxetanocin Nucleoside Analogues

Michael E. Jung[1,*], **Akemi Toyota**[1], **Erik de Clercq**[2], and **Jan Balzarini**[2]

[1] Department of Chemistry and Biochemistry, University of California, Los Angeles, CA 90095-1569, USA

[2] Rega Institute, Katholieke Universiteit Leuven, B-3000 Leuven, Belgium

Summary. A series of methylene-expanded oxetanocin nucleoside analogues, *e.g.* analogues of **2** and the known antiviral nucleosides *AZT*, *FLT*, and *ddC* (**3**) were prepared by a very direct route beginning with the readily available (*S*)-glycidol **4** and proceeding *via* the dihydrofuran-3-methanols **9a**,**b**. Biological testing of these modified nucleosides indicates that they are non-cytotoxic compounds with generally weak antiviral activity. However, the guanosine analogue **2G** showed pronounced activity *vs.* herpes simplex virus type 1 (HSV-1) in cell culture and was HSV-1-encoded thymidine kinase dependent. This compound is therefore an interesting new lead structure for the development of new anti-HSV agents.

Keywords. Nucleoside synthesis; Anhydro nucleosides; Cyclic stereocontrol; Alkoxyepoxide chemistry; Antiviral evaluation; Anti-HSV-1 activity.

Introduction

In the search for ever more potent and selective agents for antiviral therapy, scientists have turned more and more towards modified nucleosides since many such compounds have quite good antiviral activity [1]. In the last several years, nucleosides with abnormal structures, *e.g.* *L*-nucleosides, isonucleosides, α-nucleosides, oxetanocins, cyclopropyl nucleosides, acyclic nucleosides, *etc.* have been shown to be very useful antiviral agents [2]. For quite some time now we have been interested in the synthesis of potentially antiviral modified nucleosides of several different structural classes and have developed some new efficient synthetic methods for their preparation [3]. For example, we have published a general synthesis of several methylene-expanded oxetanocin isonucleosides (**1**) in which a methylene group was inserted between the ring oxygen and the carbon bearing the base. One of these – the thymidine analogue **1a** – showed moderate anti-HIV activity in the anti-HIV drug testing system of the National Cancer Institute ($IC_{50} > 2 \times 10^{-4}\,M$, $EC_{50} = 8 \times 10^{-7}\,M$, $TI_{50} > 250$) [3g]. Recently, we have reported on a very efficient synthesis of a different class of methylene-expanded oxetanocin analogues

* Corresponding author. E-mail: jung@chem.ucla.edu

(**2**) in which the methylene group is inserted between the oxygen and the carbon bearing the hydroxymethyl group [4]. We have also described the synthesis of several modified nucleosides in this class containing azido, fluoro, or hydrido groups in place of the secondary hydroxyl (**3**) [4]. We now report the full details of this synthesis and the results of the testing of these new modified nucleosides, which identified the guanosine analogue **2G** as an interesting new lead structure for an anti-herpes agent.

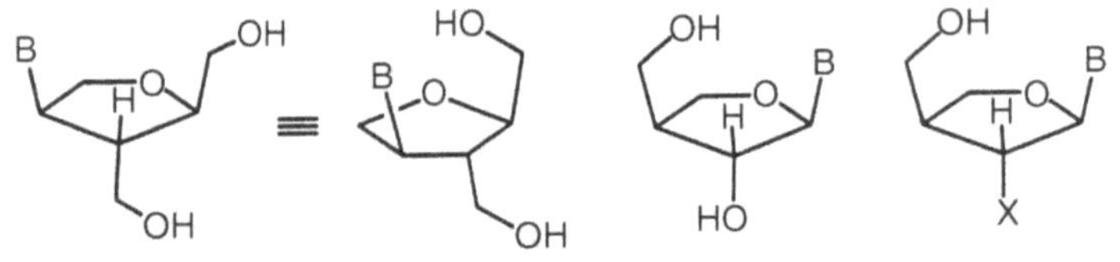

1 (**1a**: *B* = *Th*) **2** **3**

Results and Discussion

Synthesis

Our *de novo* synthesis of these important modified nucleosides uses a novel approach in which all of the asymmetry required is derived from the inexpensive precursor (*S*)-glycidol (**4**). *Sharpless* asymmetric epoxidation of allyl alcohol using *D-(+)-DIPT* and cumyl hydroperoxide afforded (*S*)-glycidol **4** in 43% yield and >90% *ee* [5] (Scheme 1). This compound has been prepared in very large scale by an industrial application of this process [6]. Treatment of the anion of **4** with 2-chloroethenyl phenyl sulfone (**5**; prepared in three steps and 82% overall yield from 1,1,2-trichloroethane [7]) afforded the addition-elimination product **6** in 78% yield [7]. Treatment of **6** with *LHMDS* gave in 60% yield the cyclized product **7**, the alcohol function of which was protected as either the *tert*-butyldiphenylsilyl (*TBDPS*) or monomethoxytrityl (*MMTr*) ether, **8a,b**, in quantitative yield. Sodium amalgam reduction gave the desired enol ethers **9a,b** also quantitatively. Epoxidation with dimethyl dioxirane gave the epoxides **10a,b** as the major isomers in an 8:1 ratio with the opposite diastereomers **11a,b** in quantitative yield. These two key

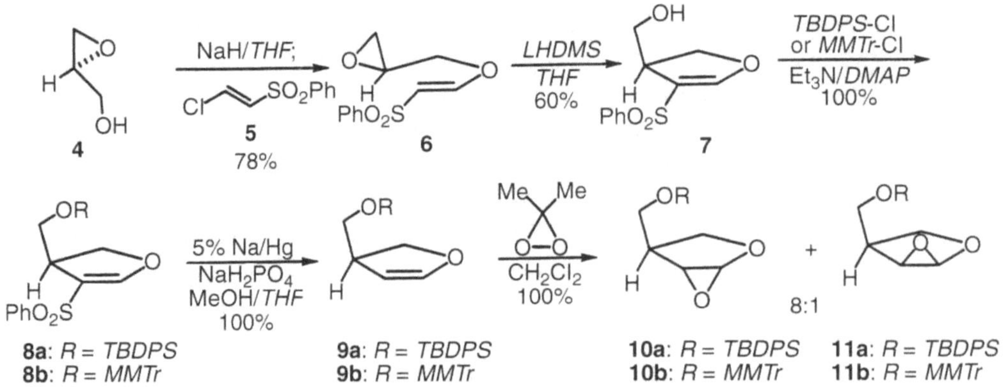

Scheme 1

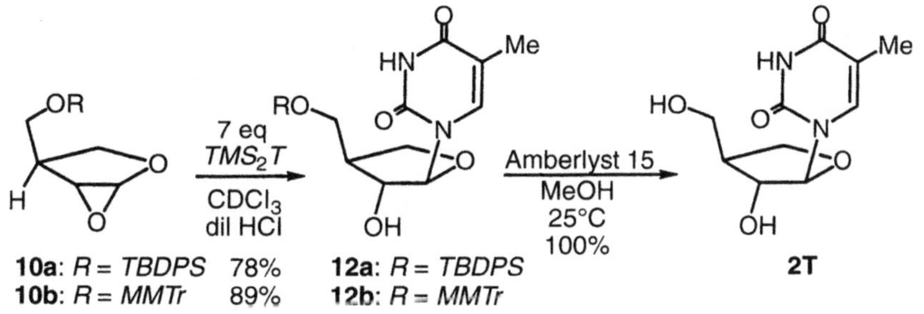

Scheme 2

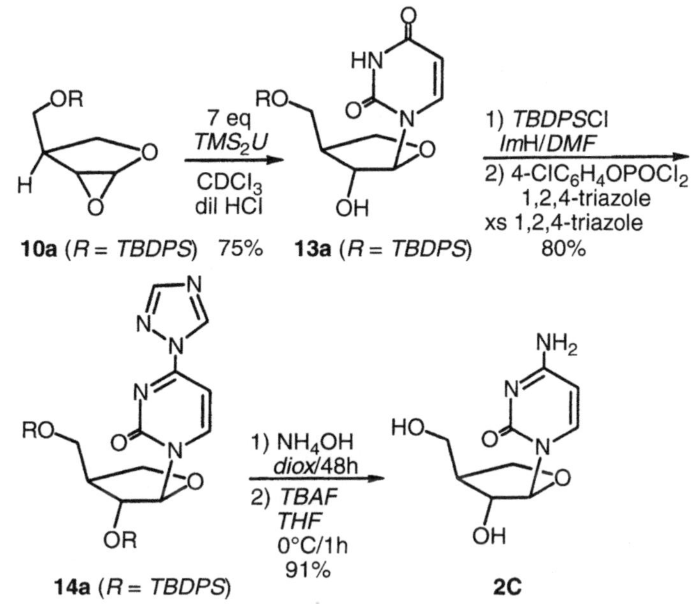

Scheme 3

compounds were then converted into the desired modified nucleosides as shown in Schemes 2–5. Both of the ethers **10a,b** could be opened with excess *bis*-silylated thymine to give in good yields the protected nucleoside analogues **12a,b**, the latter of which could be hydrolyzed to give **2T** in quantitative yield (Scheme 2). The cytidine analogue **2C** could also be prepared by treatment of the epoxide **10a** with excess *bis*-silylated uracil to give the protected nucleoside analogue **13a** in 75% yield (Scheme 3). Protection of the secondary alcohol as the *TBDPS* ether and conversion of the amide into the triazole afforded the *bis*-silyl ether triazole **14a** in 80% yield. Final conversion to the cytosine ring and deprotection of both silyl ethers gave the cytidine analogue **2C** in 91% yield. The purine analogues were prepared by a different route, again starting with the enol ether **9b** (Scheme 4). Dihydroxylation using catalytic amounts of osmium tetroxide and *NMO* followed by acetylation gave a quantitative yield of a 20:1 ratio of the two isomeric diacetates **15**, as a 1.5:1 mixture of α and β anomers, and **16** as only the α-anomer. Reaction of this mixture with the N^6-benzoyl *bis*-silylated adenine in the presence of trimethylsilyl triflate (*TMSOTf*) afforded in 93% yield the desired β anomer **17** which was debenzoylated and then deprotected to give the adenosine analogue **2A**.

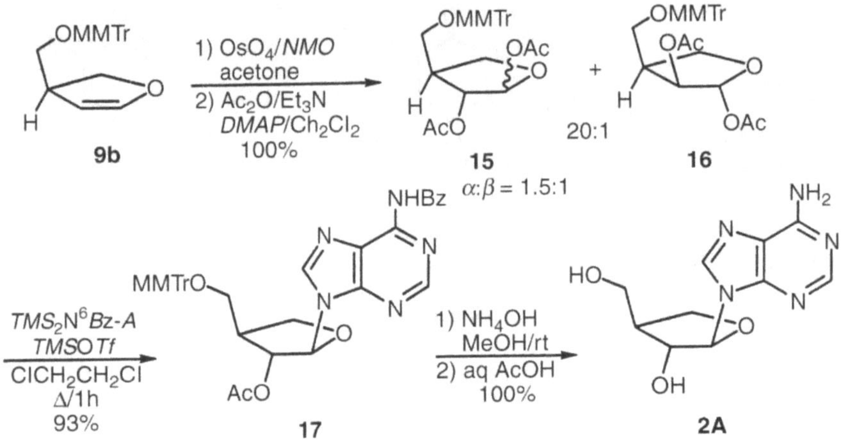

Scheme 4

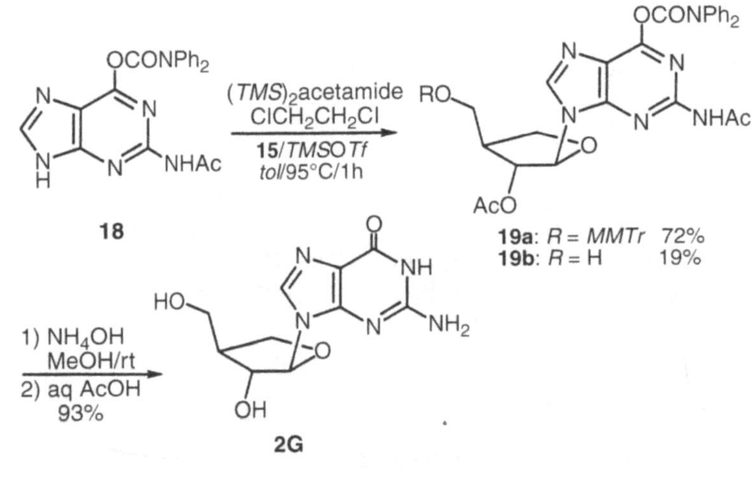

Scheme 5

Finally, a similar route afforded the guanosine analogue (Scheme 5). Thus, the known carbamate **18** (prepared in two steps from guanine by acetylation followed by O-acylation with diphenylcarbamoyl chloride [8]) was *bis*-silylated using *bis*-(trimethylsilyl)-acetamide and then treated with the diacetate **15** and *TMSOTf* to give a mixture of the *MMTr* ether and the alcohol **19a,b** in yields of 72% and 19%, respectively. These compounds were both treated with ammonium hydroxide; then, the protecting groups were removed to give the guanosine analogue **2G**.

In addition to the simple analogues of the natural nucleosides, we also wished to prepare several analogues which resembled more closely well-known antiviral agents such as *AZT, FLT, ddC, etc*. Therefore, we decided to prepare compounds **3a–c** in which azido, fluoro, and hydrido groups replace the secondary hydroxyl group of the appropriate nucleosides. The syntheses of these compounds are shown in Schemes 6–8. Thus, an internal *Mitsunobu* reaction on the two protected thymidine analogues **12a,b** using diisopropyl azodicarboxylate (*DIAD*) afforded the 1′,2′-anhydronucleosides **20a,b** in excellent yield. Opening of the silyl ether

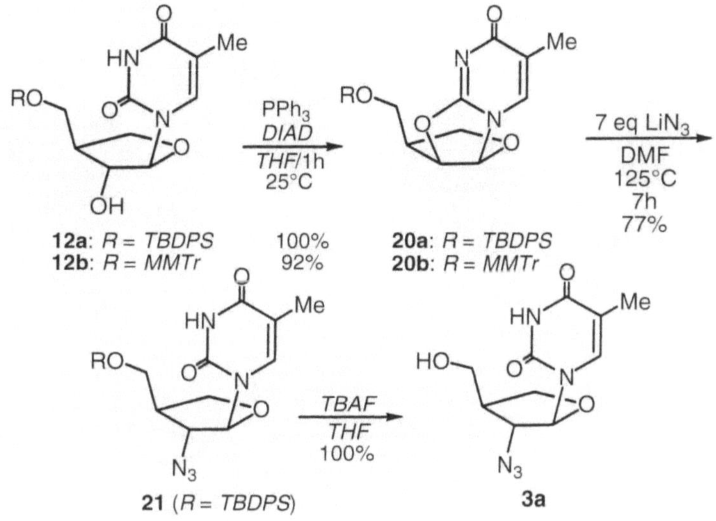

Scheme 6

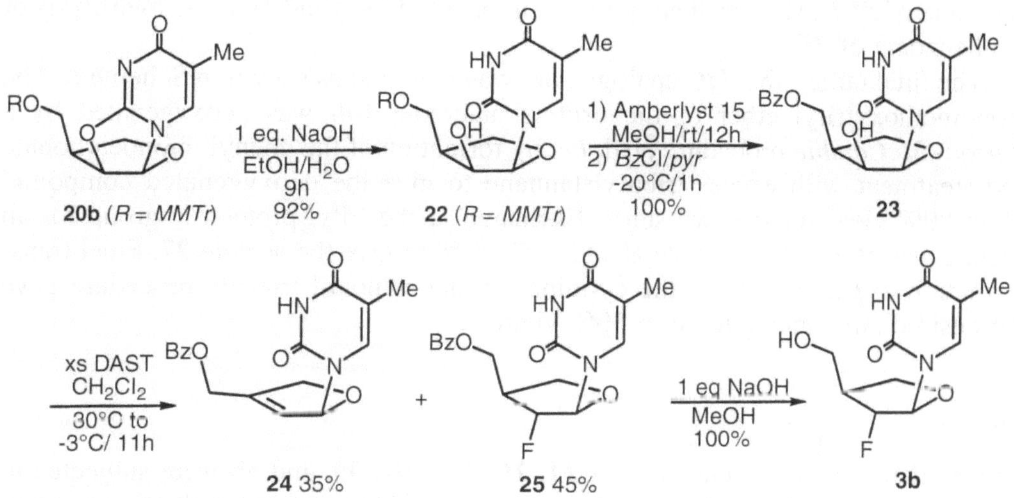

Scheme 7

20a with excess lithium azide in *DMF* at 125°C furnished the azidothymidine **21** in 77% yield along with 17% recovered starting material. Cleavage the *TBDPS* protecting group afforded the desired *AZT* analogue **3a** quantitatively. The preparation of the fluoro analogue was somewhat more difficult since we were unable to achieve a clean opening of the 1′,2′-anhydronucleoside with fluoride ion (*e.g.* by using the method of *Green* and *Blum* [9]). Therefore, we had to use a longer but more certain route (Scheme 7). Basic hydrolysis of the anhydro nucleoside **20b** afforded the secondary alcohol of retained stereochemistry **22** in 92% yield. The protecting group on the primary alcohol was exchanged for a benzoate in two steps to give the alcohol **23**. Treatment of **23** with excess *DAST* gave 45% of the inverted fluoride **25** along with 33% of the elimination product **24**. The desired nucleoside

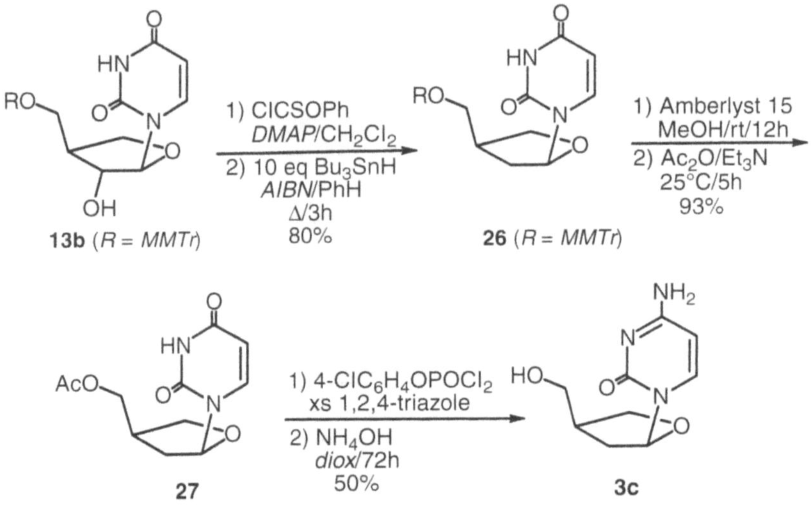

Scheme 8

analogue of *FLT* (**3b**) was then prepared in quantitative yield by basic hydrolysis of the benzoate of **25**.

The final target, the *ddC* analogue **3c**, was prepared as shown in Scheme 8. The monomethoxytrityl ether of the uridine analogue **13b** was deoxygenated by a *Barton–McCombie* procedure [10], *i.e.* by formation of the phenyl thionocarbonate and treatment with excess tributylstannane to give the deoxygenated compound **26** in 80% yield for the two steps. Exchange of the trityl protecting group for an acetate was effected in two steps and 93% yield to give the acetate **27**. Final transformation of the uridine to the cytidine *via* the standard triazole procedure gave the desired *ddC* analogue **3c** in 50% yield.

Testing

The synthetic modified nucleosides **2T**, **2C**, **2A**, **2G**, **3a**, and **3b** were subjected to extensive cytotoxicity and antiviral screening. In addition, some well known standard antiviral agents were also used as controls. The results of the testing is shown in Tables 1–3.

Table 1 gives the cytotoxicity and antiviral activity of the compounds in E_6SM cell cultures. Five viruses were included in the study: Herpes Simplex Virus type 1 (HSV-1), Herpes Simplex Virus type 2 (HSV-2), Vaccinia Virus, Vesicular Stomatitis Virus, and Thymidine Kinase-deficient Herpes Simplex Virus type 1 (HSV-1/TK^- KOS-ACV^R). As Table 1 shows, only two of the synthetic compounds showed activity in these screens, the thymidine and guanosine analogues **2T** and **2G**, with the guanosine analogue being preferably active *vs.* HSV-1 and fivefold less so *vs.* HSV-2. Compound **2G** is inactive against the thymidine kinase deficient HSV-1 strain, implying that it must be phosphorylated by the virus-encoded thymidine kinase before showing antiviral activity. Even though this compound has significantly lower activity than BVDU, acyclovir (ACV), or

Table 1. Cytotoxicity and antiviral activity in E_6SM cell cultures

Compound	Minimum cytotoxic concentration[a] ($\mu g/cm^3$)	Minimum inhibitory concentration[b] ($\mu g/cm^3$)				
		Herpes simplex virus-1 (KOS)	Herpes simplex virus 2 (G)	Vaccinia virus	Vesicular stomatitis virus	Herpes simplex virus-1 TK^-KOS ACV^R
2T	> 400	9.6	> 400	> 400	> 400	240
3a	≥ 400	> 80	> 80	> 80	> 80	240
3b	≥ 400	> 80	> 80	240	> 80	240
2C	> 400	240	> 400	240	> 400	240
2A	≥ 400	> 80	> 80	240	> 80	240
2G	≥ 400	1.9	9.6	240	> 400	240
BVDU	400	0.001	> 80	0.026	> 80	> 80
Ribavirin	> 400	240	240	48	48	240
ACV	400	0.077	0.128	> 80	> 80	48
GCV	> 100	0.001	0.032	> 100	> 100	0.48

[a] Required to cause a microscopically detectable alteration of normal cell morphology; [b] required to reduce virus-induced cytopathogenicity by 50%

Table 2. Cytotoxicity and antiviral activity in Vero cell cultures

Compound	Minimum cytotoxic concentration[a] ($\mu g/cm^3$)	Minimum inhibitory concentration[b] ($\mu g/cm^3$)				
		Parainfluenza-3 virus	Reovirus-1	Sindbis virus	Coxsackie virus B4	Punta Toro virus
2T	400	> 80	> 80	> 80	> 80	> 80
3a	400	> 80	> 80	> 80	> 80	> 80
3b	400	> 80	> 80	> 80	> 80	> 80
2C	400	> 80	> 80	> 80	> 80	> 80
2A	400	> 80	> 80	> 80	> 80	> 80
2G	≥ 80	> 16	> 16	> 16	> 16	> 16
BVDU	> 400	> 400	> 400	> 400	> 400	> 400
(*S*)-*DHPA*	≥ 400	400	48	> 80	> 80	> 80
Ribavirin	> 400	80	80	400	80	16

[a] Required to cause a microscopically detectable alteration of normal cell morphology; [b] required to reduce virus-induced cytopathogenicity by 50%

ganciclovir (GCV), its activity *vs.* HSV-1 is potent and selective enough to postulate that certain derivatives might have increased activity and perhaps be useful anti-herpes agents.

Table 2 lists the cytotoxicity and antiviral activity of the compounds in Vero cell cultures. Five RNA viruses were used: Parainfluenza-3 virus, Reovirus-1, Sindbis virus, Coxsackie virus B4, and Punta Toro virus. As Table 2 shows, all of the synthetic compounds were inactive in these evaluations at subtoxic

Table 3. Cytotoxicity and antiviral activity in HeLa cell cultures

Compound	Minimum cytotoxic concentration[a] ($\mu g/cm^3$)	Minimum inhibitory concentration[b] ($\mu g/cm^3$)		
		Vesicular stomatitis virus	Coxsackie virus B4	Respiratory syncytial virus
2T	400	>80	>80	>80
3a	400	>80	>80	>80
3b	400	>80	>80	>80
2C	400	>80	>80	>80
2A	400	>80	>80	>80
2G	400	>80	>80	>80
BVDU	≥400	>80	>80	>80
(*S*)-*DHPA*	≥400	>240	>400	>400
Ribavirin	≥400	9.6	240	16

[a] Required to cause a microscopically detectable alteration of normal cell morphology; [b] required to reduce a virus-induced cytopathogenicity by 50%

concentrations (80 $\mu g/cm^3$ for **2G** and ~400 $\mu g/cm^3$ for the other compounds). Also, the known standard antiviral agents (*E*)-5-(2-bromovinyl)-2′-deoxyuridine (BVDU), the (*S*)-enantiomer of 9-(2,3-dihydroxypropyl)-adenine (*DHPA*), and ribavirin were not markedly inhibitory.

Table 3 gives the cytotoxicity and antiviral activity of the compounds in HeLa cell cultures. As expected, these compounds show no activity *vs.* the *RNA* viruses Vesicular Stomatitis Virus, Coxsackie Virus B4, and Respiratory Syncytial Virus, as is the case also for the reference compounds *BVDU* and (*S*)-*DHPA*. Ribavirin showed modest antiviral activity.

Thus, the testing results indicate that four of the synthetic nucleoside analogues – **2C**, **2A**, **3a**, and **3b** – are inactive in all antiviral test systems. The thymidine analogue **2T** has weak antiviral activity *vs.* HSV-1. However, the major finding of the testing is that the guanosine analogue **2G** is reasonably active ($MIC = 1.9$–$9.6\,\mu g/cm^3$) *vs.* HSV-1 and HSV-2 with some selectivity *vs.* HSV-1 in E_6SM cell cultures. None of the compounds proved inhibitory against human immunodeficiency virus type 1 (HIV-1) (strain III_B) and type 2 (HIV-2) (strain ROD) in CEM cell cultures (data not shown). These data indicate that the compounds, particularly the fluoro- and azido-substituted **2T** derivatives, were not efficiently recognized by cellular nucleos(t)ide kinases and/or are not inhibitory against HIV-reverse transcriptase, the antiviral target for the triphosphates of AZT and FLT.

Conclusions

Efficient *de novo* syntheses of a series of methylene-expanded oxetanocin nucleoside analogues – **2C**, **2A**, **2T**, **2G**, **3a**, **3b**, and **3c** – beginning with (*S*)-glycidol **4** are reported. Evaluation of these compounds in cell cultures indicates that the guanosine analogue **2G** is reasonably inhibitory against herpes simplex virus. This compound is an interesting new lead structure for the development of new anti-HSV agents.

Experimental

(R)-2-((2-Phenylsulfonyl)-ethenyloxymethyl)-oxirane (**6**; $C_{11}H_{12}O_4S$)

To a stirred suspension of sodium hydride (60% in oil dispersion, 465 mg, 10.7 mmol) in *THF* (5.5 cm^3), a solution of (*S*)-glycidol (**4** [5]; 660 mg, 8.91 mmol) in *THF* (5.5 cm^3) was added at −23°C, and the mixture was stirred for 30 min at the same temperature. To the mixture was added a solution of (*E*)-chlorovinyl phenyl sulfone (**5** [7]; 2.16 g, 10.7 mmol) in *THF* (11 cm^3), and the stirred mixture was gradually warmed to 0°C over 1 h. After the reaction was quenched with sat. aq. NH_4Cl (11 cm^3), the mixture was extracted with diethyl ether (3 × 15 cm^3) and dried over $MgSO_4$. After removal of the solvent, the residue was purified by flash chromatography (hexane:AcOEt = 1:2) to give the epoxy vinyl ether **6** (1.68 g, 78%).

$[\alpha]_D^{23} = -12.0°$ ($c = 0.6$, $CHCl_3$); IR (neat): $\nu = 3200$, 1636, 1615, 1450, 1306, 1217, 1144, 1086 cm^{-1}; ^{1}H NMR (400 MHz, $CDCl_3$): $\delta = 2.64$ (1H, dd, $J = 4.8$ and 2.6 Hz), 2.84 (1H, dd, $J = 4.8$ and 4.3 Hz), 3.22 (1H, m), 3.72 (1H, dd, $J = 11.6$ and 6.1 Hz), 4.17 (1H, dd, $J = 11.6$ and 2.5 Hz), 5.77 (1H, d, $J = 12.2$ Hz), 7.50 (2H, m), 7.57 (1H, d, $J = 12.2$ Hz), 7.57 (1H, m), 7.85 (2H, m) ppm; ^{13}C NMR (100 MHz, $CDCl_3$): $\delta = 44.16$ (CH of epoxy), 49.25 (CH_2 of epoxy), 72.32 (CH_2O-vinyl), 107.80 (CHS), 126.93 (2C of Ph), 129.21 (2C of Ph), 132.94 (1C of Ph), 142.24 (C-S), 160.36 (vinyl CH–O) ppm.

(3S)-4-Phenylsulfonyl-2,3-dihydrofuran-3-methanol (**7**; $C_{11}H_{12}O_4S$)

To a solution of **6** (1.17 g, 4.87 mmol) in *THF* (28 cm^3), *LHMDS* (1 *M* in hexane, 5.75 cm^3) was added at −78°C, and the stirred mixture was gradually warmed to 0°C over 30 min. After the reaction was quenched with sat. aq. $(NH_4)_2SO_4$ (23 cm^3), the mixture was extracted with AcOEt (3 × 15 cm^3) and dried over $MgSO_4$. After removal of the solvent, the residue was purified by flash chromatography (hexane:AcOEt = 1:2) to give the dihydrofuran **7** (697 mg, 60%).

$[\alpha]_D^{20} = +48.0°$ ($c = 1.6$, $CDCl_3$); IR (neat): $\nu = 3522$, 1605, 1447, 1304, 1125, 1073, 1038, 727, 688 cm^{-1}; (400 MHz, ^{1}H NMR, $CDCl_3$): $\delta = 2.64$ (1H, t, $J = 7$ Hz, OH), 3.16 (1H, m), 3.68 (1H, ddd, $J = 11.6$, 7 and 5.1 Hz), 3.77 (1H, ddd, $J = 11.7$, 7 and 4.2 Hz), 4.49 (1H, dd, $J = 9.6$ and 7.3 Hz), 4.66 (1H, dd, $J = 10.7$ and 9.6 Hz), 7.34 (1H, d, $J = 1.4$ Hz), 7.56 (2H, m), 7.64 (1H, m), 7.92 (2H, m) ppm; ^{13}C NMR (100 MHz, $CDCl_3$): $\delta = 44.02$ (C4), 62.94 (CH_2OH), 77.42 (C5), 117.93 (C3), 127.21 (C3′ and C5′ of Ph), 129.39 (C2′ and C6′ of Ph), 133.41 (C4′ of Ph), 140.55 (C1′ of Ph), 159.53 (C2) ppm.

(3R)-3-(((4-Methoxyphenyl)-diphenylmethoxy)-methyl)-4-phenylsulfonyl-2,3-dihydrofuran (**8b**; $C_{31}H_{28}O_5S$)

To a solution of **7** (256 mg, 1.07 mmol) in CH_2Cl_2 (2.5 cm^3), triethylamine (0.19 cm^3, 1.4 mmol), *DMAP* (11 mg, 0.09 mmol), and 4-monomethoxy-trityl chloride (416 mg, 1.30 mmol) were added at room temperature, and the mixture was stirred at the same temperature for 2 h. After the reaction was quenched with H_2O, the mixture was extracted with CH_2Cl_2 (3 × 15 cm^3) and dried over $MgSO_4$. After removal of the solvent, the residue was purified by flash chromatography (hexane:AcOEt = 3:1) to give the monomethoxytrityl ether **8b** (548 mg, 100%).

$[\alpha]_D^{23} = +57.6°$ ($c = 0.5$, $CHCl_3$); IR (neat): $\nu = 1607$, 1510, 1447, 1306, 1252, 1144, 1073, 1034, 727 cm^{-1}; ^{1}H NMR (400 MHz, $CDCl_3$): $\delta = 2.95$ (1H, dd, $J = 9.4$ and 9.4 Hz), 3.23 (1H, m), 3.37 (1H, dd, $J = 9.4$ and 3.8 Hz), 3.81 (3H, s), 4.53 (1H, dd, $J = 9.6$ and 6 Hz), 4.65 (1H, dd, $J = 9.6$ and 9.6 Hz), 6.82 (2H, ddd, $J = 9.0$, 3.1 and 2.1 Hz), 7.18 (2H, ddd, $J = 9.0$, 3.1 and 2.1 Hz), 7.25–7.35 (10H, m), 7.32 (1H, d, $J = 1.4$ Hz), 7.42 (2H, m), 7.57 (1H, m), 7.71 (2H, m) ppm; ^{13}C NMR (100 MHz, $CDCl_3$): $\delta = 42.07$ (C4), 55.26 (OCH_3), 64.21 (CH_2OH), 78.68 (C5), 86.64 (CAr_3), 113.13 (*ortho*-C × 2 of MeOAr), 118.00 (C3), 126.98 (Tr), 127.02 (Tr), 127.27 (C3′ and C5′ of PhS),

127.85 (2C of *Tr*), 127.87 (2C of *Tr*), 128.26 (2C of *Tr*), 128.32 (2C of *Tr*), 129.10 (C2′ and C6′ of PhS), 130.31 (2C of Tr), 132.95 (C4′ of PhS), 135.29 (*para*-C of MeOAr), 141.32 (C1′ of Ph), 144.12 (*meta*-C of MeOAr), 144.23 (*meta*-C of MeOAr), 158.63 (C–OMe), 158.79 (C2) ppm.

(3R)-3-(((4-Methoxyphenyl)-diphenylmethoxy)-methyl)-2,3-dihydrofuran (**9b**; $C_{25}H_{24}O_3$)

To a solution of **8b** (264 mg, 0.515 mmol) in MeOH (3.6 cm^3) *THF* (0.72 cm^3), $NaH_2PO_4 \cdot H_2O$ (808 mg, 5.86 mmol) and 5% Na–Hg (1.43 g, 3.11 mmol) were added at −25°C, and the stirred mixture was gradually warmed to +5°C over 45 min. After the mixture was extracted with hexane (3 × 10 cm^3) and hexane:ether = 5:1 (10 cm^3), the combined extracts were dried over $MgSO_4$. After removal of the solvent, the residue was purified by flash chromatography (hexane:ether = 5:1) to give the dihydrofuran **9b** (192 mg, 100%).

$[\alpha]_D^{23} = -61.0°$ (c = 0.6, $CHCl_3$); IR (neat): ν = 2950, 1609, 1510, 1491, 1447, 1300, 1252, 1181, 1138, 1067, 1034, 831, 708 cm^{-1}; ^1H NMR (400 MHz, $CDCl_3$): δ = 3.00 (1H, dd, J = 8.5 and 8.5 Hz), 3.16 (1H, dd, J = 8.5 and 5.6 Hz), 3.30 (1H, m), 3.81 (3H, s), 4.16 (1H, dd, J = 9.5 and 6.3 Hz), 4.40 (1H, dd, J = 9.5 and 9.5 Hz), 4.94 (1H, dd, J = 2.5 and 2.5 Hz), 6.36 (1H, dd, J = 2.5 and 2.2 Hz), 6.86 (2H, ddd, J = 8.9, 3.1 and 2.1 Hz), 7.18 (2H, ddd, J = 9.0, 3.1 and 2.1 Hz), 7.2–7.4 (8H, m), 7.47 (4H, m) ppm; ^{13}C NMR (100 MHz, $CDCl_3$): δ = 43.17 (C4), 55.24 (OCH_3), 66.53 (CH_2OH), 73.13 (C5), 86.08 (CAr_3), 101.45 (C3), 113.09 (*ortho*-C × 2 of MeOAr), 126.88 (2C of Ph), 127.82 (4C of Ph), 128.46 (2C of Ph), 128.47 (2C of Ph), 130.40 (2C of Ph), 135.82 (*para*-C of MeOAr), 144.60 (*meta*-C of MeOAr), 144.70 (*meta*-C of MeOAr), 146.86 (C2), 158.55 (C–OMe) ppm.

5-Methyl-1-((1R,2R,3S)-tetrahydro-2-hydroxy-3-(((4-methoxyphenyl)-diphenylmethoxy)-methyl)-1-furanyl)-2,4(1H,3H)-pyrimidinedione (**12b**; $C_{30}H_{30}N_2O_6$)

To a solution of **9b** (104 mg, 0.28 mmol) in CH_2Cl_2 (4 cm^3), dimethyl dioxirane (*ca.* 0.1 *M* solution in acetone, 3.5 cm^3) was added at −78°C, and the stirred mixture was gradually warmed to 0°C over 1 h. After removal of the solvent, the residue (the epoxide **10b**) was coevaporated with C_6H_6 and dissolved in $CDCl_3$ (1.5 cm^3). To the solution of the epoxide **10b** was added a solution of *bis*-silylated thymine (457 mg, 1.69 mmol) in $CDCl_3$ (0.2 cm^3), and the mixture was stirred at room temperature for 36 h. After removal of the solvent, the residue was purified by flash chromatography (hexane:AcOEt = 2:1, 1:1, 1:2) to give the silyl ether (122 mg, 71%) and the alcohol **12b** (17 mg, 12%).

Silyl ether of **12b**: $[\alpha]_D^{20} = -2.4°$ (c = 1.0, $CDCl_3$); IR (neat): ν = 2957, 1690, 1510, 1464, 1449, 1254, 1181, 1111, 1074, 872, 843, 706 cm^{-1}; ^1H NMR (200 MHz, $CDCl_3$): δ = 0.03 (9H, s), 1.77 (3H, s), 2.60 (1H, m), 3.10 (2H, d, J = 6.4 Hz), 3.80 (3H, s), 4.06 (1H, dd, J = 9 and 5.7 Hz), 4.35 (1H, m), 4.35 (1H, m), 5.62 (1H, d, J = 3.8 Hz), 6.82 (2H, m), 6.99 (1H, d, J = 1 Hz), 7.2–7.45 (12H, m), 8.26 (1H, br) ppm; ^{13}C NMR (50 MHz, $CDCl_3$): δ = 0.04 (*TMS*), 12.49 (5-Me), 47.32 (C3′), 55.24 (OCH_3), 62.51 (C4′), 71.25 (C2′), 77.47 (C5′), 86.55 (CAr_3), 93.02 (C1′), 110.69 (C5), 113.22 (*ortho*-C × 2 of MeOAr), 127.15 (2C of Ph), 127.94 (4C of Ph), 128.32 (4C of Ph), 130.25 (2C of Ph), 135.15 (C4), 135.46 (*para*-C of MeOAr), 144.03 (*meta*-C of MeOAr), 144.11 (*meta*-C of MeOAr), 150.63 (C6), 158.70 (C–OMe), 164.33 (C2) ppm.

To a solution of the above silyl ether (122 mg, 0.20 mmol) in MeOH (1.4 cm^3)/*THF* (0.6 cm^3), 1 *N* aq. HCl (0.002 cm^3) in MeOH (1 cm^3) was added at 0°C, and the stirred mixture was gradually warmed to room temperature over 30 min. After the reaction was neutralized with 1 *N* aq. NaOH (0.002 cm^3), the mixture was evaporated *in vacuo*. The residue was purified by flash chromatography (hexane:AcOEt = 1:2) to give the alcohol **12b** (103 mg, 100%).

$[\alpha]_D^{20} = +11.8°$ (c = 1.0, $CDCl_3$); IR (neat): ν = 3401, 1694, 1510, 1468, 1252, 1181, 1089, 1034, 911, 831, 729 cm^{-1}; ^1H NMR (200 MHz, $CDCl_3$): δ = 1.84 (3H, s), 2.72 (1H, m), 3.11 (1H, dd, J = 9.1 and 7.3 Hz), 3.24 (1H, dd, J = 9.1 Hz and 5.8 Hz), 3.78 (3H, s), 3.97 (1H, dd, J = 8.7 and

7.7 Hz), 4.15 (1H, m), 4.36 (1H, dd, $J = 8.7$ and 7.7 Hz), 5.57 (1H, d, $J = 3.5$ Hz), 6.81 (2H, m), 7.16 (1H, d, $J = 1.1$ Hz), 7.2–7.45 (12H, m), 9.31 (1H, br) ppm; ^{13}C NMR (50 MHz, $CDCl_3$): $\delta = 12.49$ (5-Me), 46.49 (C3′), 55.23 (OCH_3), 62.02 (C4′), 71.71 (C2′), 79.12 (C5′), 86.44 (CAr_3), 94.22 (C1′), 110.62 (C5), 113.17 (*ortho*-C × 2 of MeOAr), 127.05 (2C of Ph), 127.90 (4C of Ph), 128.29 (4C of Ph), 130.24 (2C of Ph), 134.55 (*para*-C of MeOAr), 135.25 (C4), 144.10 (*meta*-C of MeOAr), 144.20 (*meta*-C of MeOAr), 151.47 (C6), 158.62 (C–OMe), 164.25 (C2) ppm.

5-Methyl-1-((1R,2R,3S)-tetrahydro-2-hydroxy-3-hydroxymethyl-1-furanyl)-2,4 (1H,3H)-pyrimidinedione (**2T**; $C_{10}H_{14}N_2O_4$)

To a solution of **12b** (39 mg, 0.076 mmol) in MeOH (0.8 cm^3), Amberlyst 15-H (18 mg) was added at room temperature, and the mixture was stirred for 14 h. After filtration, the solvent was evaporated *in vacuo*. The residue was purified by flash chromatography (AcOEt:MeOH = 10:1) to give the thymidine analogue **2T** (19 mg, 100%).

$[\alpha]_D^{20} = -9.5°$ ($c = 0.4$, MeOH); IR (neat): $\nu = 3386$, 1696, 1480, 1267, 1100, 1070 cm^{-1}; ^{1}H NMR (200 MHz, $CDCl_3$:CD_3OD = 20:1): $\delta = 1.86$ (3H, d, $J = 0.9$ Hz), 2.48 (1H, m), 3.63 (2H, d, $J = 5.7$ Hz), 3.96 (1H, dd, $J = 8.7$ Hz and 8.3 Hz), 4.17 (1H, dd, $J = 6.6$ and 4.4 Hz), 4.23 (1H, dd, $J = 8.4$ and 8.3 Hz), 5.56 (1H, d, $J = 4.4$ Hz), 7.24 (1H, d, $J = 1.1$ Hz) ppm; ^{13}C NMR (50 MHz, CD_3OD): $\delta = 10.92$ (5-Me), 48.05 (C3′), 60.22 (C4′), 70.02 (C2′), 75.45 (C5′), 92.04 (C1′), 110.20 (C5), 136.91 (C4), 151.33 (C6), 165.01 (C2).

(3R)-3-((((1,1-Dimethylethyl)-diphenyl)-silyloxy)-methyl)-4-phenylsulfonyl-2,3-dihydrofuran (**8a**; $C_{27}H_{30}O_4SSi$)

To a solution of **7** (198 mg, 0.82 mmol) in CH_2Cl_2 (1.9 cm^3), triethylamine (0.14 cm^3, 1.0 mmol), *DMAP* (5 mg, 0.04 mmol), and *tert*-butyldiphenyl-silyl chloride (*TBDPS*Cl; 0.25 cm^3, 0.96 mmol) were added at room temperature, and the mixture was stirred at the same temperature for 20 h. After the reaction was quenched with H_2O, the mixture was extracted with CH_2Cl_2 (3 × 12 cm^3) and dried over $MgSO_4$. After removal of the solvent, the residue was purified by flash chromatography (hexane:AcOEt = 10:1) to give the *TBDPS* ether (376 mg, 96%).

$[\alpha]_D^{22} = +38.9°$ ($c = 0.55$, $CHCl_3$); IR (neat): $\nu = 2934$, 2860, 1605, 1477, 1447, 1429, 1316, 1306, 1144, 1113, 1034, 727, 704, 606 cm^{-1}; ^{1}H NMR (400 MHz, $CDCl_3$): $\delta = 1.02$ (9H, s), 3.18 (1H, m), 3.57 (1H, dd, $J = 10.3$ and 8.3 Hz), 3.76 (1H, dd, $J = 10.3$ and 3.8 Hz), 4.59 (1H, dd, $J = 9.7$ and 9.7 Hz), 4.69 (1H, dd, $J = 9.6$ and 5.5 Hz), 6.82 (2H, ddd, $J = 9.0$, 3.1 and 2.1 Hz), 7.31 (1H, d, $J = 1.3$ Hz), 7.35–7.5 (8H, m), 7.5–7.6 (5H, m), 7.7–7.75 (2H, m) ppm; ^{13}C NMR (100 MHz, $CDCl_3$): $\delta = 19.22$ (*t*-Bu), 26.84 (*t*-Bu), 43.91 (C4), 64.08 (CH_2OSi), 78.11 (C5), 117.92 (C3), 127.16 (C3′ and C5′ of PhS), 127.76 (Ph), 127.80 (Ph), 129.12 (C2′ and C6′ of PhS), 129.82 (Ph), 129.86 (Ph), 132.93 (Ph), 133.07 (C4′ of PhS), 133.26 (Ph), 135.57 (Ph), 141.45 (C1′ of Ph), 158.96 (C2) ppm.

5-Methyl-1-((1R,2R,3R)-tetrahydro-2-hydroxy-3-(1,1-dimethylethyldiphenyl)-silyloxymethyl-1-furanyl)-2,4(1H,3H)-pyrimidinedione (**12a**; $C_{26}H_{32}N_2O_5Si$)

To a solution of **8a** (94.0 mg, 0.195 mmol) in MeOH (1.4 cm^3)/*THF* (0.28 cm^3), $NaH_2PO_4 \cdot H_2O$ (298 mg, 2.16 mmol) and 5% Na–Hg (528 mg, 1.15 mmol) were added at −30°C, and the stirred mixture was gradually warmed to +5°C over 1 h. After the mixture was extracted with hexane (3 × 5 cm^3), the combined extracts were dried over $MgSO_4$. After removal of the solvent, the residue was purified by flash chromatography (hexane:ether = 10:1) to give the dihydrofuran **9a** (66.0 mg, 100%).

To a solution of **9a** (66.0 mg, 0.195 mmol) in CH_2Cl_2 (2.8 cm^3), dimethyldioxirane (*ca.* 0.1 *M* solution in acetone, 2.3 cm^3) was added at −78°C, and the stirred mixture was gradually warmed to 0°C over 35 min. After removal of the solvent, the residue (the epoxide **10a**) was coevaporated with C_6H_6 and dissolved in $CHCl_3$ (0.8 cm^3). To the solution of **10a**, *bis*-silylated thymine (281 mg, 1.04 mmol) in $CDCl_3$ (0.2 cm^3) was added, and the mixture was stirred at room temperature for 36 h. After removal of the solvent, the residue was purified by flash chromatography (hexane:AcOEt = 5:1, 5:2, 1:1) to give the trimethylsilyl ether of **12a** (76 mg, 71%) and the alcohol **12a** (7 mg, 7%).

Trimethylsilyl ether of **12a**: $[\alpha]_D^{20} = +14.0°$ ($c = 1.0$, $CDCl_3$); IR (neat): $\nu = 2959$, 1690, 1472, 1429, 1391, 1265, 1254, 1113, 845, 702 cm^{-1}; ^1H NMR (400 MHz, $CDCl_3$): $\delta = 0.04$ (9H, s), 1.08 (9H, s), 1.86 (3H, s), 2.49 (1H, ddd, $J = 6$, 6 and 5.2 Hz), 3.61 (1H, dd, $J = 10.5$ and 5.2 Hz), 3.68 (1H, dd, $J = 10.5$ Hz and 6 Hz), 4.08 (1H, dd, $J = 8.8$ and 7 Hz), 4.20 (1H, dd, $J = 8.8$ and 8.5 Hz), 4.41 (1H, dd, $J = 6$ and 4.6 Hz), 5.72 (1H, d, $J = 4.6$ Hz), 7.07 (1H, s), 7.35–7.5 (6H, m), 7.55–7.7 (4 H, m), 8.06 (1H, br) ppm; ^{13}C NMR (100 MHz, $CDCl_3$): $\delta = -0.01$ (*TMS*), 12.53 (5-Me), 19.28 (*t*-Bu), 26.90 (*t*-Bu), 48.74 (C3′), 61.56 (C4′), 70.07 (C2′), 76.12 (C5′), 92.16 (C1′), 111.02 (C5), 127.85, 127.90, 129.98, 130.04, 132.82, 132.95, 135.47, 135.49 (Ph), 135.52 (C4), 150.55 (C6), 163.94 (C2) ppm.

To a solution of the above trimethylsilyl ether (127 mg, 0.23 mmol) in MeOH (3 cm^3)/*THF* (0.6 cm^3), 1 *N* aq. HCl (0.003 cm^3) in MeOH (1 cm^3) was added at 0°C, and the stirred mixture was gradually warmed to room temperature over 30 min. After the reaction was neutralized with 1 *N* aq. NaOH (0.003 cm^3), the mixture was evaporated *in vacuo*. The residue was purified by flash chromatography (hexane:AcOEt = 1:1) to give the alcohol **12a** (111 mg, 100%).

$[\alpha]_D^{20} = +9.7°$ ($c = 0.5$, $CDCl_3$); IR (neat): $\nu = 3407$, 2932, 1694, 1472, 1429, 1265, 1113, 739, 702 cm^{-1}; ^1H NMR (400 MHz, $CDCl_3$): $\delta = 1.03$ (9H, s), 1.91 (3H, d, $J = 1.2$ Hz), 2.62 (1H, m), 3.75 (1H, dd, $J = 10.5$ and 5.5 Hz), 3.78 (1H, dd, $J = 10.5$ Hz and 4.6 Hz), 4.03 (1H, dd, $J = 8.5$ and 8 Hz), 4.18 (1H, dd, $J = 7$ and 4.1 Hz), 4.28 (1H, dd, $J = 8.5$ and 8 Hz), 5.54 (1H, d, $J = 4.1$ Hz), 7.26 (1H, s), 7.35–7.5 (6H, m), 7.55–7.7 (4 H, m), 8.66 (1H, br) ppm; ^{13}C NMR (100 MHz, $CDCl_3$): $\delta = 12.57$ (5-Me), 19.29 (*t*-Bu), 26.84 (*t*-Bu), 48.05 (C3′), 61.46 (C4′), 70.62 (C2′), 77.98 (C5′), 93.62 (C1′), 110.69 (C5), 127.84 (Ph), 129.90 (Ph), 129.92 (Ph), 133.04 (Ph), 134.71 (C4), 135.50 (Ph), 135.53 (Ph), 151.65 (C6), 164.24 (C2) ppm.

2,2′-Anhydro-5-methyl-1-((1R,2S,3R)-tetrahydro-2-hydroxy-3-(1,1-dimethylethyldiphenyl)-silyloxymethyl-1-furanyl)-2,4(1H,3H)-pyrimidinedione (**20a**; $C_{26}H_{30}N_2O_4Si$)

To a solution of **12a** (104 mg, 0.216 mmol) and triphenyl phosphine (90 mg, 0.34 mmol) in *THF* (1.5 cm^3), *DIAD* (0.07 cm^3, 0.34 mmol) was added at room temperature, and the mixture was stirred for 1 h. After removal of the solvent, the residue was purified by flash chromatography (hexane: AcOEt = 2:1, 1:1, 1:2; AcOEt) to give the anhydrothymidine **20a** (97 mg, 97%).

^1H NMR (400 MHz, $CDCl_3$): $\delta = 1.05$ (9H, s), 1.99 (3H, d, $J = 1.3$ Hz), 2.68 (1H, m), 3.44 (1H, dd, $J = 11.6$ and 9.3 Hz), 3.79 (1H, dd, $J = 10.5$ Hz and 7.2 Hz), 4.03 (1H, dd, $J = 10.5$ and 7.4 Hz), 4.17 (1H, dd, $J = 9.3$ and 7.2 Hz), 5.33 (1H, dd, $J = 5.2$ and 5.2 Hz), 6.07 (1H, d, $J = 5.2$ Hz), 7.19 (1H, d, $J = 1.3$ Hz), 7.35–7.45 (6H, m), 7.6–7.7 (4 H, m) ppm.

5-Methyl-1-((1R,2R,3R)-tetrahydro-2-azido-3-(1,1-dimethylethyldiphenyl)-silyloxymethyl-1-furanyl)-2,4(1H,3H)-pyrimidinedione (**21**; $C_{26}H_{31}N_5O_3Si$)

To a solution of **20a** (95 mg, 0.20 mmol) in *DMF* (1.6 cm^3), LiN_3 (70 mg, 1.4 mmol) was added at room temperature, and the mixture was stirred at 125–130°C for 7 h. After removal of the solvent, the residue was purified by flash chromatography (hexane:AcOEt = 2:1, AcOEt, AcOEt:MeOH = 10:1) to give the azide **21** (75 mg, 72%), the desilylated compound (3 mg, 5%), and starting material (16 mg, 17%).

$[\alpha]_D^{20} = +2.3°$ ($c = 1.2$, $CDCl_3$); IR (neat): $\nu = 2932$, 2108, 1694, 1472, 1429, 1265, 1113, 741, 702 cm^{-1}; ^1H NMR (200 MHz, $CDCl_3$): $\delta = 1.07$ (9H, s), 1.89 (3H, d, $J = 0.9$ Hz), 2.54 (1H, m), 3.68 (1H, dd, $J = 10.6$ and 5.1 Hz), 3.77 (1H, dd, $J = 10.6$ Hz and 5.1 Hz), 4.08 (1H, dd, $J = 9$ and 7.6 Hz), 4.17 (1H, dd, $J = 9$ and 7.7 Hz), 4.17 (1H, dd, $J = 4$ and 4 Hz), 5.77 (1H, d, $J = 4.9$ Hz), 7.07 (1H, d, $J = 1.2$ Hz), 7.35–7.5 (6H, m), 7.55–7.7 (4 H, m), 8.79 (1H, br) ppm; ^{13}C NMR (50 MHz, $CDCl_3$): $\delta = 12.63$ (5-Me), 19.30 (*t*-Bu), 26.87 (*t*-Bu), 47.02 (C3′), 61.27 (C4′), 66.51 (C2′), 70.04 (C5′), 91.04 (C1′), 111.48 (C5), 127.00 (Ph), 130.17 (Ph), 132.82 (Ph), 132.66 (Ph), 134.63 (C4), 135.50 (Ph), 151 (C6), 163 (C2) ppm.

5-Methyl-1-((1R,2R,3S)-tetrahydro-2-azido-3-hydroxymethyl-1-furanyl)-2,4(1H,3H)-pyrimidinedione (**3a**; $C_{10}H_{13}N_5O_3$)

To a solution of **21** (77 mg, 0.15 mmol) in *THF* (0.8 cm^3), *TBAF* (1 *M* in *THF*, 0.15 cm^3) was added at 0°C, and the mixture was stirred at the same temperature for 1 h. After removal of the solvent, the residue was purified by flash chromatography (hexane:AcOEt = 2:1, AcOEt) to give the *AZT* analogue **3a** (42 mg, 100%).

$[\alpha]_D^{20} = -11.5°$ ($c = 2$, $CDCl_3$); IR (neat): $\nu = 3450$, 2957, 2108, 1694, 1470, 1267, 1115, 1067 cm^{-1}; ^1H NMR (200 MHz, $CDCl_3$): $\delta = 1.92$ (3H, d, $J = 0.9$ Hz), 2.56 (1H, m), 3.75 (1H, dd, $J = 11$ and 5.6 Hz), 3.85 (1H, dd, $J = 11$ Hz and 5.1 Hz), 4.13 (1H, dd, $J = 8.5$ and 7.2 Hz), 4.23 (1H, dd, $J = 8.5$ and 8.5 Hz), 4.44 (1H, dd, $J = 5.8$ and 4.3 Hz), 5.61 (1H, d, $J = 4.3$ Hz), 7.18 (1H, d, $J = 1.2$ Hz) ppm; ^{13}C NMR (50 MHz, $CDCl_3$): $\delta = 12.49$ (5-Me), 47.30 (C3′), 60.53 (C4′), 66.86 (C2′), 70.94 (C5′), 93.32 (C1′), 110.67 (C5), 136.47 (C4), 150.82 (C6), 164.42 (C2) ppm.

2,2′-Anhydro-5-Methyl-1-((1R,2S,3R)-tetrahydro-2-hydroxy-3-(((4-methoxyphenyl)-diphenylmethoxy)-methyl)-1-furanyl)-2,4(1H,3H)-pyrimidinedione (**20b**; $C_{30}H_{28}N_2O_5$)

To a solution of **12b** (101 mg, 0.197 mmol) and triphenyl phosphine (78 mg, 0.29 mmol) in *THF* (1.5 cm^3), *DIAD* (0.06 cm^3, 0.30 mmol) was added at room temperature, and the mixture was stirred for 1.5 h. After removal of the solvent, the residue was purified by flash chromatography (hexane: AcOEt = 2:1, 1:1, 1:2, AcOEt, AcOEt:MeOH = 10:1) to give the anhydro thymidine **12b** (90 mg, 92%).

IR (neat): $\nu = 1649$, 1561, 1509, 1483, 1449, 1288, 1252, 1229, 1181, 1132, 1080, 1034, 980, 833, 733, 700 cm^{-1}; ^1H NMR (200 MHz, $CDCl_3$): $\delta = 1.97$ (3H, d, $J = 0.9$ Hz), 2.63 (1H, m), 3.29 (1H, dd, $J = 9.5$ and 7.6 Hz), 3.41 (1H, dd, $J = 11.45$ Hz and 9.2 Hz), 3.54 (1H, dd, $J = 9.5$ and 6.7 Hz), 3.80 (3H, s), 4.24 (1H, dd, $J = 9.2$ and 7.1 Hz), 5.33 (1H, dd, $J = 5.2$ and 5.2 Hz), 6.07 (1H, d, $J = 5.1$ Hz), 6.84 (2H, m), 7.18 (1H, d, $J = 1.3$ Hz), 7.2–7.45 (12H, m) ppm.

5-Methyl-1-((1R,2S,3S)-tetrahydro-2-hydroxy-3-(((4-methoxyphenyl)-diphenylmethoxy)-methyl)-1-furanyl)-2,4(1H,3H)-pyrimidinedione (**22**; $C_{30}H_{30}N_2O_6$)

To a solution of **20b** (89 mg, 0.18 mmol) in EtOH (4.3 cm^3)/H_2O (2.1 cm^3), 1 *N* aq. NaOH (0.43 cm^3) was added at room temperature, and the mixture was stirred at the same temperature for 9 h. After the reaction was neutralized with 1 *N* aq. HCl (0.43 cm^3), the mixture was extracted with AcOEt and dried over $MgSO_4$. After removal of the solvent, the residue was purified by flash chromatography (hexane:AcOEt = 1:2) to give the alcohol **22** (85 mg, 92%).

IR (neat): $\nu = 3384$, 2928, 1698, 1667, 1510, 1478, 1279, 1252, 1181, 1057, 700 cm^{-1}; ^1H NMR (200 MHz, $CDCl_3$): $\delta = 1.83$ (3H, s), 2.76 (1H, m), 3.22 (1H, dd, $J = 9$ and 6.4 Hz), 3.52 (1H, dd, $J = 9$ Hz and 7.6 Hz), 3.78 (3H, s), 3.88 (1H, dd, $J = 11.1$ and 7.7 Hz), 4.09 (1H, m), 4.78 (1H, m), 6.06 (1H, d, $J = 2.3$ Hz), 6.82 (2H, m), 7.2–7.5 (13H, m), 10.39 (1H, br) ppm.

5-Methyl-1-((1R,2S,3S)-tetrahydro-2-hydroxy-3-hydroxymethyl-1-furanyl)-2,4(1H,3H)-pyrimidinedione ($C_{10}H_{14}N_2O_5$)

To a solution of **22** (93 mg, 0.18 mmol) in MeOH (1.8 cm^3), Amberlyst 15-H (28 mg) was added at room temperature, and the mixture was stirred for 12 h. After filtration, the solvent was evaporated *in vacuo*. The residue was purified by flash chromatography (AcOEt:MeOH = 10:1) to give the diol (44 mg, 100%).

^{1}H NMR (200 MHz, $CDCl_3$:CD_3OD = 20:1): δ = 1.89 (3H, d, J = 0.8 Hz), 2.53 (1H, m), 3.67 (2H, d, J = 5.8 Hz), 3.97 (1H, dd, J = 8.7 Hz and 8.5 Hz), 4.18 (1H, dd, J = 6.7 and 4.4 Hz), 4.26 (1H, dd, J = 8.5 and 8.3 Hz), 5.55 (1H, d, J = 4.4 Hz), 7.26 (1H, s) ppm.

5-Methyl-1-((1R,2S,3S)-tetrahydro-2-hydroxy-3-benzoyloxymethyl-1-furanyl)-2,4(1H,3H)-pyrimidinedione (**23**; $C_{17}H_{18}N_2O_6$)

To a solution of the above diol (44 mg, 0.18 mmol) in CH_2Cl_2 (2.1 cm^3)/pyridine (0.8 cm^3), benzoyl chloride (0.07 cm^3, 0.60 mmol) was added at −40°C, and the stirred mixture was gradually warmed to −5°C over 2 h. After the reaction was quenched with MeOH (0.3 cm^3, 10 mmol) the solvent was evaporated *in vacuo*. The residue was purified by flash chromatography (hexane:AcOEt = 1:2) to give the benzoate **23** (59 mg, 95%).

$[\alpha]_D^{20}$ = +61.7° (c = 0.3, $CDCl_3$); IR (neat): ν = 3362, 2924, 1694, 1665, 1480, 1275, 1144, 1113 cm^{-1}; ^{1}H NMR (200 MHz, $CDCl_3$): δ = 1.79 (3H, s), 2.89 (1H, m), 4.09 (1H, dd, J = 10.9 and 7.8 Hz), 4.22 (1H, dd, J = 7.9 Hz and 7.8 Hz), 4.47 (1H, dd, J = 11.1 and 6.8 Hz), 4.69 (1H, dd, J = 11.1 and 7.6 Hz), 4.86 (1H, m), 6.04 (1H, d, J = 2.5 Hz), 7.37 (1H, s), 7.4–7.65 (3H, m), 8–8.1 (2H, m), 10.64 (1H, br) ppm; ^{13}C NMR (50 MHz, $CDCl_3$): δ = 12.27 (5-Me), 44 (C3′), 61.59 (C4′), 71 (C2′), 89.08 (C5′), 100 (C1′), 108.25 (C5), 128.41 (2C of Ph), 129.59 (2C of Ph), 129.75 (Ph), 133.22 (Ph), 138.52 (C4), 152 (C6), 166 (COPh), 166.74 (C2) ppm.

5-Methyl-1-((1R,2R,3S)-tetrahydro-3-benzoyloxymethyl-2-fluoro-1-furanyl)-2,4(1H,3H)-pyrimidinedione (**25**; $C_{17}H_{17}FN_2O_5$)

To a solution of **23** (58 mg, 0.17 mmol) in CH_2Cl_2 (2.5 cm^3), DAST (0.60 cm^3, 4.5 mmol) was added at −30°C, and the stirred mixture was gradually warmed to −10°C over 2 h and stirred at −10°C ~ 3°C for 17 h. After the reaction was quenched with 2 N aq. NaOH (5 cm^3)/sat. aq. $NaHCO_3$, the mixture was extracted with CH_2Cl_2 and dried over $MgSO_4$. After removal of the solvent, the residue was purified by flash chromatography (hexane:AcOEt = 2:1, 1:1, 1:2) to give the fluoride **25** (25 mg, 45%), the dehydrated product **24** (19 mg, 35%) and starting material **23** (7 mg, 12%).

25: $[\alpha]_D^{20}$ = +22° (c = 0.4, $CDCl_3$); IR (neat): ν = 1694, 1468, 1453, 1275, 1115, 1110, 714 cm^{-1}; ^{1}H NMR (400 MHz, $CDCl_3$): δ = 1.89 (3H, d, J = 1.2 Hz), 3.09 (1H, dm, J = 27 Hz), 4.20 (1H, dd, J = 9 and 7.1 Hz), 4.39 (1H, dd, J = 9 Hz and 8.2 Hz), 4.49 (1H, dd, J = 11.5 and 6.7 Hz), 4.52 (1H, dd, J = 11.5 and 7.0 Hz), 5.50 (1H, ddd, J = 54, 4 and 3 Hz), 5.70 (1H, dd, J = 18 and 3 Hz), 7.10 (1H, d, J = 1.2 Hz), 7.44 (2H, m), 7.58 (1H, m), 8.01 (2H, m), 8.50 (1H, br) ppm; ^{13}C NMR (100 MHz, $CDCl_3$): δ = 12.41 (5-Me), 45.24 (d, J = 86 Hz, C3′), 62.52 (d, J = 17 Hz, C4′), 71.49 (d, J = 13.5 Hz, C5′), 94.40 (d, J = 147 Hz, C1′), 96.85 (d, J = 738 Hz, C2′), 111.35 (C5), 128.53 (2C of Ph), 129.43 (Ph), 129.65 (2C of Ph), 133.43 (Ph), 137.16 (C4), 150.51(C6), 164.03 (COPh), 166.23 (C2) ppm.

24: IR (neat): ν = 1694, 1470, 1453, 1275, 1256, 1111, 1061, 714 cm^{-1}; ^{1}H NMR (200 MHz, $CDCl_3$): δ = 1.92 (3H, d, J = 0.9 Hz), 4.75 (1H, dm, J = 14 Hz), 4.91 (1H, dm, J = 14 Hz), 5.08 (2H, s), 5.82 (1H, d, J = 1.5 Hz), 6.93 (1H, d, J = 1.4 Hz), 7.04 (1H, m), 7.44 (2H, m), 7.58 (1H, m), 8.06 (2H, m), 8.35 (1H, br) ppm.

5-Methyl-1-((1R,2R,3S)-tetrahydro-3-hydroxymethyl-2-fluoro-1-furanyl)-2,4(1H,3H)-pyrimidinedione (**3b**; $C_{10}H_{13}FN_2O_4$)

To a solution of **25** (20.5 mg, 0.0588 mmol) in MeOH (0.4 cm^3), 1 *N* aq. NaOH (0.07 cm^3) was added at room temperature, and the mixture was stirred at the same temperature for 30 min. After removal of the solvent, the residue was purified by flash chromatography (AcOEt) to give the fluoro nucleoside **3b** (14 mg, 100%).

$[\alpha]_D^{20} = -16°$ ($c = 0.4$, $CDCl_3$); IR (neat): $\nu = 3389$, 2928, 1692, 1470, 1269, 1113 cm^{-1}; ^1H NMR (400 MHz, $CDCl_3$): $\delta = 1.91$ (3H, d, $J = 1.1$ Hz), 2.66 (1H, br), 2.79 (1H, dm, $J = 28.5$ Hz), 3.81 (1H, dd, $J = 11$ and 5.5 Hz), 3.88 (1H, dd, $J = 11$ Hz and 5 Hz), 4.17 (1H, dd, $J = 9$ and 6 Hz), 4.32 (1H, dd, $J = 9$ and 8.2 Hz), 5.50 (1H, ddd, $J = 54$, 3.2 and 2.6 Hz), 5.67 (1H, dd, $J = 19$ and 2.6 Hz), 7.17 (1H, d, $J = 1.2$ Hz) ppm; ^{13}C NMR (100 MHz, $CDCl_3$): $\delta = 12.42$ (5-Me), 47.48 (d, $J = 82$ Hz, C3′), 60.57 (d, $J = 25$ Hz, C4′), 71.57 (d, $J = 10$ Hz, C5′), 94.43 (d, $J = 151$ Hz, C1′), 97.40 (d, $J = 722$ Hz, C2′), 110.60 (C5), 137.51 (C4), 150.93 (C6), 164.45 (C2) ppm.

1-((1R,2R,3R)-Tetrahydro-2-hydroxy-3-(1,1-dimethylethyldiphenyl)-silyloxymethyl-1-furanyl)-2,4(1H,3H)-pyrimidinedione (**13a**; $C_{25}H_{30}N_2O_5Si$)

To a solution of **10a** (138 mg, 0.39 mmol) in $CDCl_3$ (1.6 cm^3), *bis*-silylated uracil (705 mg, 2.75 mmol) in $CDCl_3$ (0.4 cm^3) was added, and the mixture was stirred at room temperature for 36 h. After removal of the solvent, the residue was purified by flash chromatography (hexane:AcOEt = 3:1, 2:1, 1:1) to give the trimethylsilyl ether of **13a** (113 mg, 54%) and the alcohol **13a** (12 mg, 7%).

Trimethylsilyl ether of **13a**: $[\alpha]_D^{20} = +20.0°$ ($c = 0.70$, $CHCl_3$); IR (neat): $\nu = 3054$, 2957, 1686, 1460, 1429, 1383, 1265, 1254, 1113, 843, 702 cm^{-1}; ^1H NMR (200 MHz, $CDCl_3$): $\delta = 0.07$ (9H, s), 1.07 (9H, s), 2.52 (1H, m), 3.60 (2H, d, $J = 6.1$ Hz), 4.07 (1H, dd, $J = 8.7$ and 6.2 Hz), 4.24 (1H, dd, $J = 8.7$ and 8 Hz), 4.40 (1H, dd, $J = 4.7$ and 4 Hz), 5.60 (1H, dd, $J = 8.1$ and 2.3 Hz), 5.67 (1H, d, $J = 4$ Hz), 7.21 (1H, $J = 8.1$ Hz), 7.3–7.5 (6H, m), 7.55–7.65 (4 H, m), 8.14 (1H, br) ppm; ^{13}C NMR (50 MHz, $CDCl_3$): $\delta = 0.02$ (*TMS*), 19.25 (*t*-Bu), 26.89 (*t*-Bu), 49.09 (C3′), 61.93 (C4′), 70.67 (C2′), 76.74 (C5′), 93.09 (C1′), 102.34 (C5), 127.88, 127.93, 130.02, 130.07, 132.78, 132.91, 135.51, 135.56 (Ph), 139.73 (C4), 150.50 (C6), 163.72 (C2) ppm.

To a solution of the above silyl ether (108 mg, 0.200 mmol) in MeOH (2 cm^3)/*THF* (0.5 cm^3), 1 *N* aq. HCl (0.002 cm^3) in MeOH (0.6 cm^3) was added at 0°C, and the stirred mixture was gradually warmed to room temperature over 30 min. After the reaction was neutralized with 1 *N* aq. NaOH (0.002 cm^3), the mixture was evaporated *in vacuo*. The residue was purified by flash chromatography (hexane:AcOEt = 1:2) to give the alcohol **13a** (93 mg, 100%).

$[\alpha]_D^{20} = +24.9°$ ($c = 0.43$, $CHCl_3$); IR (neat): $\nu = 1696$, 1472, 1429, 1267, 1113, 741, 702 cm^{-1}; ^1H NMR (200 MHz, $CDCl_3$): $\delta = 1.03$ (9H, s), 2.62 (1H, m), 3.67 (1H, dd, $J = 10.5$ and 6.4 Hz), 3.77 (1H, dd, $J = 10.5$ Hz and 4.9 Hz), 3.91 (1H, br s), 4.04 (1H, dd, $J = 8.7$ and 8.5 Hz), 4.18 (1H, ddd, $J = 6.2$, 3.6 and 3 Hz), 4.28 (1H, dd, $J = 8.5$ and 8 Hz), 5.57 (1H, d, $J = 3.6$ Hz), 5.67 (1H, d, $J = 8.2$ Hz), 7.3–7.5 (7H, m), 7.55–7.7 (4 H, m), 9.63 (1H, br) ppm; ^{13}C NMR (50 MHz, $CDCl_3$): $\delta = 19.29$ (*t*-Bu), 26.87 (*t*-Bu), 48.27 (C3′), 61.65 (C4′), 71.00 (C2′), 78.27 (C5′), 94.21 (C1′), 102.20 (C5), 127.81 (Ph), 129.96 (Ph), 129.98 (Ph), 133.04 (Ph), 135.52 (Ph), 135.56 (Ph), 138.92 (C4), 151.65 (C6), 163.99 (C2) ppm.

1-((1R,2R,3R)-Tetrahydro-2-(1,1-dimethylethyldiphenyl)-silyloxy-3-(1,1-dimethylethyldiphenyl)-silyloxymethyl-1-furanyl)-2,4(1H,3H)-pyrimidinedione ($C_{41}H_{48}N_2O_5Si_2$)

To a solution of **13a** (74 mg, 0.16 mmol) in *DMF* (0.8 cm^3), imidazole (56 mg, 0.82 mmol) and *TBDPS*Cl (0.11 cm^3, 0.42 mmol) were added at room temperature, and the mixture was stirred at the same temperature for 23 h. After the reaction was quenched with H_2O, the mixture was extracted with Et_2O ($3 \times 6\,cm^3$) and dried over $MgSO_4$. After removal of the solvent, the residue was purified by flash chromatography (hexane:AcOEt = 3:1) to give the *bis*-silyl ether (71.5 mg, 81%).

$[\alpha]_D^{22} = +57.6°$ ($c = 0.25$, $CHCl_3$); IR (neat): $\nu = 2920$, 2861, 1696, 1472, 1429, 1250, 1113, 824, 741, 702, 613 cm^{-1}; ^1H NMR (400 MHz, $CDCl_3$): $\delta = 0.97$ (9H, s), 1.01 (9H, s), 2.64 (1H, m), 3.45 (1H, dd, $J = 10.4$ and 7.1 Hz), 3.53 (1H, dd, $J = 10.4$ and 4.2 Hz), 4.02 (1H, dd, $J = 9$ and 6.3 Hz), 4.18 (1H, dd, $J = 9$ and 8.0 Hz), 4.30 (1H, dd, $J = 5.5$ and 5 Hz), 5.26 (1H, dd, $J = 8.1$ and 2.4 Hz), 5.85 (1H, d, $J = 5$ Hz), 6.74 (1H, d, $J = 8$ Hz), 7.25–7.6 (20H, m), 8.1 (1H, br) ppm; ^{13}C NMR (100 MHz, $CDCl_3$): $\delta = 19.04$ and 19.21 (*t*-Bu), 26.79 and 26.85 (*t*-Bu), 49.09 (C3′), 62.61 (C4′), 70.16 (C2′), 76.78 (C5′), 92.40 (C1′), 102.51 (C5), 127.80, 127.83, 127.99, 129.89, 129.95, 130.05, 130.20, 132.14, 132.92, 133.03, 133.05, 135.51, 135.70, 135.83 (Ph), 140.19 (C4), 150.13 (C6), 163.20 (C2) ppm.

1-((1R,2R,3R)-Tetrahydro-2-(1,1-dimethylethyldiphenyl)-silyloxy-3-(1,1-dimethylethyldiphenyl)-silyloxymethyl-1-furanyl)-4-(1,2,4-triazol-1-yl)-2(1H)-pyrimidinone (**14a**; $C_{43}H_{49}N_5O_4Si_2$)

To a solution of the above *bis*-silyl ether (19 mg, 0.027 mmol) in pyridine ($0.5\,cm^3$), 4-chlorophenyl dichlorophosphate ($0.020\,cm^3$, 0.12 mmol) and 1,2,4-triazole (37 mg, 0.54 mmol) were added at 0°C, and the mixture was stirred at room temperature for 28 h. After the reaction was quenched with H_2O, the mixture was extracted with CH_2Cl_2 ($3 \times 5\,cm^3$) and dried over $MgSO_4$. After removal of the solvent, the residue was purified by flash chromatography (hexane:AcOEt = 1:1) to give the triazolyl nucleoside **14a** (19 mg, 93%).

^1H NMR (200 MHz, $CDCl_3$): $\delta = 0.94$ (9H, s), 1.08 (9H, s), 2.57 (1H, m), 3.10 (1H, dd, $J = 8.7$ and 8.7 Hz), 3.20 (1H, m), 4.23 (1H, m), 4.4 (2H, m), 5.93 (1H, d, $J = 2.6$ Hz), 6.53 (1H, d, $J = 7.2$ Hz), 7.25–7.6 (21H, m), 8.12 (1H, s), 9.21 (1H, s) ppm.

4-Amino-1-((1R,2R,3R)-tetrahydro-2-(1,1-dimethylethyldiphenyl)-silyloxy-3-(1,1-dimethylethyldiphenyl)-silyloxymethyl-1-furanyl)-2(1H)-pyrimidinone ($C_{41}H_{49}N_3O_4Si_2$)

To a solution of **14a** (58 mg, 0.077 mmol) in dioxane ($1.7\,cm^3$), sat. aq. NH_3 ($0.85\,cm^3$) was added, and the mixture was stirred at room temperature for 3 days. After removal of the solvent, the residue was purified by flash chromatography ($CHCl_3$:MeOH = 20:1) to give the protected cytidine (50 mg, 93%).

$[\alpha]_D^{23} = +72.7°$ ($c = 0.22$, $CHCl_3$); IR (neat): $\nu = 3300$, 2932, 2859, 1634, 1487, 1474, 1428, 1250, 1113, 741, 702, 613 cm^{-1}; ^1H NMR (400 MHz, $CDCl_3$): $\delta = 0.94$ (9H, s), 1.02 (9H, s), 2.56 (1H, m), 3.23 (1H, dd, $J = 10.2$ and 8.7 Hz), 3.34 (1H, dd, $J = 10.2$ and 4.9 Hz), 4.10 (1H, dd, $J = 8.9$ and 5.6 Hz), 4.24 (1H, dd, $J = 8.9$ and 7.7 Hz), 4.29 (1H, dd, $J = 4.5$ and 4 Hz), 5.27 (1H, d, $J = 7.4$ Hz), 5.90 (1H, d, $J = 4$ Hz), 6.87 (1H, d, $J = 7.4$ Hz), 7.2–7.6 (20H, m) ppm; ^{13}C NMR (100 MHz, $CDCl_3$): $\delta = 19.10$ and 19.15 (*t*-Bu), 26.81 and 26.85 (*t*-Bu), 49.65 (C3′), 62.89 (C4′), 70.77 (C2′), 77.57 (C5′), 94.27 (C1′), 94.88 (C5), 127.74, 127.83, 129.74, 129.79, 129.87, 132.47, 133.17, 133.19, 133.39, 135.50, 135.70, 135.85 (Ph), 141.26 (C4), 155.82 (C6), 165.62 (C2) ppm.

4-Amino-1-((1R,2R,3S)-tetrahydro-2-hydroxy-3-hydroxymethyl-1-furanyl)-2(1H)-pyrimidinone (**2C**; $C_9H_{13}N_3O_4$)

To a solution of the above protected cytidine (56 mg, 0.079 mmol) in C_6H_6 ($0.8\,cm^3$), Amberlite A26 F^- (226 mg) was added, and the mixture was stirred under reflux for 6 h. After the resin was filtered off, washed with C_6H_6, Et_2O, and AcOEt, and extracted with MeOH, the MeOH extract was evaporated *in vacuo*. The residue was purified by flash chromatography ($CHCl_3$:MeOH = 10:1) to give the cytidine analogue **2C** (17 mg, 94%).

$[\alpha]_D^{20} = +65.6°$ ($c = 0.13$, $CHCl_3$:$CH_3OH = 5{:}1$); IR (neat): $\nu = 3347$, 2928, 2859, 1653, 1526, 1491, 1289, 1111, 785 cm^{-1}; ^1H NMR (400 MHz, $CDCl_3$:$CD_3OD = 5{:}1$): $\delta = 2.45$ (1H, m), 3.48 (1H, dd, $J = 11.2$ and 7.0 Hz), 3.60 (1H, dd, $J = 11.2$ and 5.1 Hz), 3.94 (1H, dd, $J = 8.6$ and 8.2 Hz), 4.03 (1H, dd, $J = 5.8$ and 3.8 Hz), 4.25 (1H, dd, $J = 8.6$ and 7.9 Hz), 5.50 (1H, d, $J = 3.5$ Hz), 5.77 (1H, d, $J = 7$ Hz), 7.51 (1H, d, $J = 7$ Hz) ppm; ^{13}C NMR (100 MHz, CD_3OD): $\delta = 48.65$ (C3′), 60.41 (C4′), 70.64 (C2′), 77.09 (C5′), 94.21 (C1′), 94.51 (C5), 141.17 (C4), 157.39 (C6), 166.32 (C2) ppm.

1-((1R,2R,3S)-Tetrahydro-2-hydroxy-3-(((4-methoxyphenyl)-diphenylmethoxy)-methyl)-1-furanyl)-2,4(1H,3H)-pyrimidinedione (**13b**; $C_{29}H_{28}N_2O_6$)

To a solution of **10b** (89 mg, 0.23 mmol) in $CDCl_3$ (0.95 cm^3), a solution of *bis*-silylated uracil (450 mg, 1.75 mmol) in $CDCl_3$ (0.24 cm^3) was added, and the mixture was stirred at room temperature for 36 h. After removal of the solvent, the residue was purified by flash chromatography (hexane:AcOEt = 2:1, 1:1, 1:2) to give the silyl ether of **13b** (84 mg, 64%) and the alcohol **13b** (13 mg, 11%).

Trimethylsilyl ether of **13b**: $[\alpha]_D^{20} = +7.2°$ ($c = 1.0$, $CHCl_3$); IR (neat): $\nu = 2930$, 1684, 1609, 1510, 1449, 1254, 1181, 1107, 1071, 1034, 843, 702 cm^{-1}; ^1H NMR (200 MHz, $CDCl_3$): $\delta = 0.07$ (9H, s), 2.61 (1H, m), 3.06 (2H, m), 3.81 (3H, s), 4.04 (1H, dd, $J = 9$ and 5.1 Hz), 4.32 (1H, dd, $J = 3.2$ and 3.2 Hz), 4.36 (1H, dd, $J = 9$ and 9 Hz), 5.48 (1H, dd, $J = 8.2$ and 2.2 Hz), 5.90 (1H, d, $J = 3.2$ Hz), 6.83 (2H, m), 7.09 (1H, d, $J = 8.2$ Hz), 7.2–7.45 (12H, m), 8.51 (1H, br) ppm; ^{13}C NMR (50 MHz, $CDCl_3$): $\delta = 0.01$ (*TMS*), 47.45 (C3′), 55.28 (OCH_3), 62.33 (C4′), 71.62 (C2′), 77.81 (C5′), 86.59 (CAr_3), 93.66 (C1′), 102.08 (C5), 113.24 (*ortho*-C × 2 of MeOAr), 127.14 (2C of Ph), 127.96 (4C of Ph), 128.27 (4C of Ph), 130.23 (2C of Ph), 135.05 (C4), 135.58 (*para*-C of MeOAr), 143.91 (*meta*-C of MeOAr), 144.04 (*meta*-C of MeOAr), 150.29 (C6), 158.71 (C–OMe), 163.55 (C2) ppm.

To a solution of the above silyl ether (86 mg, 0.15 mmol) in MeOH (1.3 cm^3)/*THF* (0.5 cm^3), 1 *N* aq. HCl (0.001 cm^3) in MeOH (0.7 cm^3) was added at 0°C, and the stirred mixture was gradually warmed to room temperature over 30 min. After neutralization with 1 *N* aq. NaOH (0.001 cm^3), the mixture was evaporated *in vacuo*. The residue was purified by flash chromatography (hexane: AcOEt = 1:4) to give the alcohol **13b** (75 mg, 100%).

$[\alpha]_D^{20} = +26.5°$ ($c = 0.4$, $CHCl_3$); IR (neat): $\nu = 3393$, 1694, 1609, 1510, 1464, 1449, 1265, 1254, 1181, 1088, 1034, 833, 737, 708 cm^{-1}; ^1H NMR (200 MHz, $CDCl_3$): $\delta = 2.72$ (1H, m), 3.09 (1H, dd, $J = 9.2$ and 7.5 Hz), 3.27 (1H, dd, $J = 9.2$ Hz and 5.7 Hz), 3.64 (1H, br s), 3.79 (3H, s), 3.95 (1H, dd, $J = 8.7$ and 8.2 Hz), 4.10 (1H, m), 4.35 (1H, dd, $J = 8.7$ and 7.5 Hz), 5.52 (1H, d, $J = 3.5$ Hz), 5.64 (1H, d, $J = 8.0$ Hz), 6.82 (2H, m), 7.2–7.4 (13H, m), 8.87 (1H, br) ppm; ^{13}C NMR (50 MHz, $CDCl_3$): $\delta = 46.49$ (C3′), 55.26 (OCH_3), 61.83 (C4′), 71.96 (C2′), 79.17 (C5′), 86.48 (CAr_3), 94.62 (C1′), 102.10 (C5), 113.21 (*ortho*-C × 2 of MeOAr), 127.07 (2C of Ph), 127.95 (4C of Ph), 128.26 (4C of Ph), 130.28 (2C of Ph), 135.17 (*para*-C of MeOAr), 138.71 (C4), 144.02 (*meta*-C of MeOAr), 144.21 (*meta*-C of MeOAr), 151.46 (C6), 158.66 (C–OMe), 163.84 (C2) ppm.

1-((1R,3S)-Tetrahydro-3-(((4-methoxyphenyl)-diphenylmethoxy)-methyl)-1-furanyl)-2,4(1H,3H)-pyrimidinedione (**26**; $C_{29}H_{28}N_2O_5$)

To a solution of **13b** (55 mg, 0.11 mmol) in CH_2Cl_2 (3.2 cm^3) *DMAP* (50 mg, 0.41 mmol) and phenyl chlorothionoformate (0.03 cm^3, 0.22 mmol) were added at room temperature, and the mixture was stirred at the same temperature for 3 h. After the reaction was quenched with H_2O, the mixture was extracted with CH_2Cl_2 and dried over $MgSO_4$. After removal of the solvent, the residue was purified by flash chromatography (hexane:AcOEt = 1:1) to give the thionoformate (63 mg, 90%).

^1H NMR (200 MHz, $CDCl_3$): $\delta = 2.94$ (1H, m), 3.30 (1H, dd, $J = 9.4$ and 6.5 Hz), 3.41 (1H, dd, $J = 9.4$ Hz and 6.1 Hz), 3.79 (3H, s), 4.17 (1H, dd, $J = 9$ and 5.6 Hz), 4.29 (1H, dd, $J = 9$ and 7.8 Hz), 5.52 (1H, dd, $J = 8.1$ and 2.2 Hz), 5.84 (1H, dd, $J = 4$ and 4 Hz), 5.97 (1H, d, $J = 4$ Hz), 6.84 (2H, m), 7.05–7.5 (18H, m), 8.53 (1H, br) ppm.

Argon gas was bubbled through a solution of the above thionoformate (58 mg, 0.091 mmol) in C_6H_6 (1.5 cm^3) for 20 min. *AIBN* (16 mg, 0.095 mmol) and Bu_3SnH (0.24 cm^3, 0.87 mmol) were added to the mixture, and the mixture was stirred under reflux for 3 h. After removal of the solvent, the residue was purified by flash chromatography (hexane:AcOEt = 1:1) to give the deoxygenated product **26** (43 mg, 97%).

$[\alpha]_D^{20} = +29.5°$ ($c = 0.6$, $CHCl_3$); IR (neat): $\nu = 2924$, 1686, 1609, 1510, 1462, 1449, 1265, 1252, 1181, 1074, 1034, 710, 702 cm^{-1}; ^1H NMR (200 MHz, $CDCl_3$): $\delta = 1.70$ (1H, m), 2.61 (1H, m), 2.74 (1H, m), 3.05 (1H, dd, $J = 9$ and 6.8 Hz), 3.18 (1H, dd, $J = 9$ Hz and 5.9 Hz), 3.81 (3H, s), 3.90 (1H, dd, $J = 8.5$ and 6.6 Hz), 4.13 (1H, dd, $J = 8.5$ and 7.2 Hz), 5.58 (1H, dd, $J = 8.2$ and 2.2 Hz), 5.97 (1H, dd, $J = 6$ and 6 Hz), 6.82 (2H, m), 7.2–7.45 (13H, m), 8.17 (1H, br) ppm; ^{13}C NMR (50 MHz, $CDCl_3$): $\delta = 35.80$ (C3′), 38.92 (C2′), 55.26 (OCH_3), 63.75 (C4′), 71.96 (C5′), 86.44 (CAr_3), 86.96 (C1′), 102.14 (C5), 113.19 (*ortho*-C × 2 of MeOAr), 127.08 (2C of Ph), 127.93 (4C of Ph), 128.27 (4C of Ph), 130.29 (2C of Ph), 135.22 (*para*-C of MeOAr), 139.03 (C4), 144.12 (*meta*-C of MeOAr), 144.21 (*meta*-C of MeOAr), 150.24 (C6), 158.67 (C–OMe), 163.19 (C2) ppm.

4-Amino-1-((1R,2R,3S)-tetrahydro-3-hydroxymethyl-1-furanyl)-2(1H)-pyrimidinone
(**3c**; $C_9H_{13}N_3O_3$)

To a solution of **26** (32.5 mg, 0.67 mmol) in MeOH (1.0 cm^3), Amberlyst 15-H (11 mg) was added at room temperature, and the mixture was stirred for 4 h. After filtration, the solvent was evaporated *in vacuo*. The residue was purified by flash chromatography (AcOEt:MeOH = 10:1) to give the alcohol (14 mg, 100%).

^1H NMR (200 MHz, $CDCl_3$:CD_3OD = 10:1): $\delta = 1.77$ (1H, m), 2.56 (2H, m), 3.56 (1H, dd, $J = 10.9$ and 6.1 Hz), 3.62 (1H, dd, $J = 10.9$ Hz and 5.3 Hz), 3.91 (1H, dd, $J = 8.6$ and 6.8 Hz), 4.09 (1H, dd, $J = 8.6$ and 8 Hz), 5.69 (1H, d, $J = 8.2$ Hz), 5.95 (1H, dd, $J = 6.2$ and 6.2 Hz), 7.49 (1H, d, $J = 8.2$ Hz) ppm.

To a solution of the above alcohol (13 mg, 0.061 mmol) in CH_2Cl_2 (0.77 cm^3), Ac_2O (0.008 cm^3, 0.09 mmol), NEt_3 (0.013 cm^3, 0.093 mmol), and *DMAP* (1 mg, 0.008 mmol) were added at room temperature, and the mixture was stirred at the same temperature for 5 h. After the reaction was quenched with H_2O, the mixture was extracted with CH_2Cl_2 and dried over $MgSO_4$. After removal of the solvent, the residue was purified by flash chromatography (AcOEt) to give the acetate **27** (14.5 mg, 93%).

IR (neat): $\nu = 2926$, 1690, 1464, 1379, 1273, 1242, 1038, 814 cm^{-1}; ^1H NMR (200 MHz, $CDCl_3$): $\delta = 1.72$ (1H, m), 2.07 (3H, s), 2.73 (2H, m), 3.87 (1H, dd, $J = 8.8$ and 7.2 Hz), 4.12 (3H, m), 5.77 (1H, dd, $J = 8.2$ and 2 Hz), 5.98 (1H, dd, $J = 6.4$ and 6.4 Hz), 7.42 (1H, d, $J = 8.2$ Hz), 8.79 (1H, br) ppm.

To a solution of **27** (14.5 mg, 0.057 mmol) in pyridine (0.6 cm^3) 4-chlorophenyl dichlorophosphate (0.03 cm^3, 0.18 mmol) and 1,2,4-triazole (44 mg, 0.64 mmol) were added at 0°C, and the mixture was stirred at room temperature for 25 h. After the reaction was quenched with H_2O, the mixture was extracted with CH_2Cl_2 (3 × 5 cm^3) and dried over $MgSO_4$. After removal of the solvent, the residue was purified by flash chromatography ($CHCl_3$:MeOH = 20:1) to give the triazolyl nucleoside (11 mg, 63%).

^1H NMR (200 MHz, $CDCl_3$): $\delta = 1.80$ (1H, ddd, $J = 13.5$, 7.2 and 6 Hz), 2.04 (3H, s), 2.82 (1H, m), 3.01 (1H, ddd, $J = 13.5$, 8.2 and 6 Hz), 3.95 (1H, dd, $J = 8.8$ and 7.0 Hz), 4.03 (2H, m), 4.26 (1H, dd, $J = 8.8$ and 7.4 Hz), 6.01 (1H, dd, $J = 6.1$ and 5.7 Hz), 6.09 (1H, d, $J = 7.2$ Hz), 8.12 (1H, s), 8.19 (1H, d, $J = 7.2$ Hz), 9.27 (1H, s) ppm.

To a solution of the above triazolyl nucleoside (11 mg, 0.036 mmol) in dioxane (0.8 cm^3), sat. aq. NH_3 (0.4 cm^3) was added, and the mixture was stirred at room temperature for 3 days. After removal of the solvent, the residue was purified by flash chromatography ($CHCl_3$:MeOH = 20:1) to give the *ddC* analogue **3c** (6 mg, 80%).

$[\alpha]_D^{23} = +95°$ ($c = 0.35$, $CHCl_3$); IR (neat): $\nu = 3343$, 2924, 1653, 1609, 1528, 1489, 1287, 1115, 1053, 789 cm^{-1}; ^1H NMR (400 MHz, $CDCl_3$:CD_3OD = 10:1): $\delta = 1.64$ (1H, m), 2.57 (2H, m), 3.45

(1H, dd, $J = 10.8$ and 6.5 Hz), 3.52 (1H, dd, $J = 10.8$ Hz and 5.2 Hz), 3.87 (1H, dd, $J = 8.5$ and 7.1 Hz), 4.09 (1H, dd, $J = 8.5$ and 7.5 Hz), 5.72 (1H, d, $J = 7.5$ Hz), 5.88 (1H, dd, $J = 6.3$ and 5.8 Hz), 7.52 (1H, d, $J = 7.5$ Hz) ppm; ^{13}C NMR (100 MHz, $CDCl_3$:CD_3OD = 10:1): $\delta = 35.82$ (C3′), 40.84 (C2′), 62.62 (C4′), 71.55 (C5′), 88.18 (C1′), 94.30 (C5), 140.32 (C4), 156.4 (C6), 165.73 (C2) ppm.

(1RS,2R,3S)-Tetrahydro-1,2-bis-acetoxy-3-(((4-methoxyphenyl)-diphenylmethoxy)-methyl)-furan (**15**; $C_{29}H_{30}O_7$)

To a solution of the dihydrofuran **9b** (150 mg, 0.40 mmol) in acetone (2 cm^3), OsO_4/*t*-BuOH solution (0.16 cm^3, prepared from OsO_4 (60 mg, 0.24 mmol), *t*-BuOH (6 cm^3), and 30% H_2O_2 (1 drop)) and 4-methylmorpholine N-oxide (*NMO*, 50% in H_2O, 0.20 cm^3, 1.0 mmol) were added at 0°C, and the mixture was stirred at the same temperature for 1.5 h. After the reaction was quenched with sat. aq. NH_4Cl, the mixture was extracted with CH_2Cl_2 and dried over $MgSO_4$. After removal of the solvent, the residue was purified by flash chromatography (hexane:AcOEt = 1:2) to give a mixture of the 1,2-diols (165 mg, 100%). To a solution of this diol mixture (165 mg, 0.406 mmol) in CH_2Cl_2 (1.6 cm^3), NEt_3 (0.41 cm^3, 2.9 mmol), Ac_2O (0.17 cm^3, 1.4 mmol), and *DMAP* (3 mg, 0.025 mmol) were added at room temperature, and the mixture was stirred at the same temperature for 22 h. After the reaction was quenched with H_2O, the mixture was extracted with CH_2Cl_2 and dried over $MgSO_4$. After removal of the solvent, the residue was purified by flash chromatography (hexane:AcOEt = 2:1) to give the diacetates **15** and **16** (199 mg, 100%) as a 12:8:1 mixture of (1*R*,2*R*,3*S*)-, (1*S*,2*R*,3*S*)-, and (1*R*,2*S*,3*S*)-isomers.

IR (neat): $\nu = 2932$, 1748, 1609, 1510, 1449, 1372, 1300, 1250, 1221, 1181, 1078, 1032, 961, 833, 700 cm^{-1}; ^{1}H NMR (400 MHz, $CDCl_3$): (1*R*,2*R*,3*S*)-**15**: $\delta = 2.04$ (3H, s), 2.08 (3H, s), 2.78 (1H, m), 3.18 (1H, dd, $J = 9.2$ and 6.4 Hz), 3.34 (1H, dd, $J = 9.2$ and 6.9 Hz), 3.80 (3H, s), 3.8 (1H, m), 4.23 (1H, dd, $J = 8.7$ and 6.7 Hz), 5.04 (1H, dd, $J = 8.8$ and 4.2 Hz), 6.33 (1H, d, $J = 4.2$ Hz), 6.85 (2H, m), 7.2–7.45 (12H, m) ppm; (1*R*,2*R*,3*S*)-**15**: $\delta = 1.91$ (3H, s), 2.09 (3H, s), 2.61 (1H, m), 3.24 (2H, m), 3.79 (3H, s), 3.8 (1H, m), 4.28 (1H, dd, $J = 8.6$ and 7.9 Hz), 5.15 (1H, d, $J = 2.5$ Hz), 6.09 (1H, s), 6.85 (2H, m), 7.2–7.45 (12H, m) ppm; (1*R*,2*S*,3*S*)-**15**: $\delta = 1.87$ (3H, s), 2.09 (3H, s), 2.78 (1H, m), 3.24 (2H, m), 3.79 (3H, s), 3.8 (1H, m), 4.28 (1H, dd, $J = 8.6$ and 7.9 Hz), 5.35 (1H, d, $J = 4.8$ Hz), 6.07 (1H, s), 6.85 (2H, m), 7.2–7.45 (12H, m) ppm.

6-N-Benzoyl-9-((1R,2R,3S)-tetrahydro-2-acetoxy-3-(((4-methoxyphenyl)-diphenylmethoxy)-methyl)-1-furanyl)-9H-adenine (**17**; $C_{39}H_{35}N_5O_6$)

To a suspension of 6-benzoyl-adenine (25 mg, 0.10 mmol) in 1,2-dichloroethane (1 cm^3), N,O-*bis*-(trimethylsilyl)-acetamide (0.080 cm^3, 0.32 mmol) was added, and the mixture was stirred under reflux for 15 min. To the above solution, a mixture of **15** and **16** (21 mg, 0.034 mmol) in 1,2-dichloroethane (0.2 cm^3) and trimethylsilyl triflate (*TMSOTf*, 0.065 cm^3) were added, and the mixture was stirred under reflux for 1 h. After the reaction was neutralized with aq. $KHCO_3$, the mixture was extracted with CH_2Cl_2 and dried over $MgSO_4$. After removal of the solvent, the residue was purified by flash chromatography (AcOEt) to give the acetate **17** (21 mg, 93%) and the detritylated product (1.5 mg, 4%).

$[\alpha]_D^{23} = -2°$ ($c = 0.1$, $CHCl_3$); IR (neat): $\nu = 1744$, 1701, 1611, 1582, 1510, 1456, 1250, 1181, 1074, 1034, 812, 708 cm^{-1}; ^{1}H NMR (400 MHz, $CDCl_3$): $\delta = 2.09$ (3H, s), 2.87 (1H, m), 3.42 (1H, dd, $J = 9$ and 7.5 Hz), 3.46 (1H, dd, $J = 9$ and 6.3 Hz), 3.79 (3H, s), 4.26 (1H, dd, $J = 8.8$ and 8.4 Hz), 4.36 (1H, dd, $J = 8.8$ and 7.8 Hz), 5.84 (1H, dd, $J = 5.8$ and 3.6 Hz), 6.01 (1H, d, $J = 3.6$ Hz), 6.83 (2H, m), 7.2–7.35 (8H, m), 7.42 (4H, m), 7.53 (2H, t, $J = 7.5$ Hz), 7.61 (1H, d, $J = 7.5$ Hz), 7.98 (1H, s), 8.02 (1H, d, 7.5 Hz), 8.64 (1H, s), 8.95 (1H, br s) ppm; ^{13}C NMR (100 MHz, $CDCl_3$): $\delta = 20.74$ (CH_3CO), 45.82 (C3′), 55.27 (OCH_3), 61.54 (C4′), 71.69 (C2′), 78.97 (C5′), 86.67 (CAr_3), 90.28 (C1′), 113.20 (*ortho*-C $\times$ 2 of MeOAr), 124(C5), 127.09 (2C of Ph), 127.87 (2C of *Bz*), 127.92 (4C of Ph), 128.32 and 128.35 (4C of Ph), 128.85 (2C of *Bz*), 130.26 (2C of Ph), 132.74 (1C of *Bz*), 134

(1C of *Bz*), 135.28 (*para*-C of MeOAr), 141.86 (C8), 144.12 (2 × *meta* C of MeOAr), 149.5 (C4), 151.3 (C6), 152.68 (C2), 158.67 (C–OMe), 164.66 (PhCO), 170.27 (COMe) ppm.

9-((1R,2R,3S)-Tetrahydro-2-hydroxy-3-(((4-methoxyphenyl)-diphenylmethoxy)-methyl)-1-furanyl)-9H-adenine ($C_{30}H_{29}N_5O_4$)

To a solution of **17** (40.0 mg, 0.0597 mmol) in MeOH (3.2 cm^3), sat. aq. NH_3 (2.2 cm^3) was added, and the mixture was stirred at room temperature for 24 h. After removal of the solvent, the residue was purified by flash chromatography (AcOEt:MeOH = 20:1) to give the monomethoxytrityl ether (31.5 mg, 100%).

$[\alpha]_D^{23.5} = +8.8°$ ($c = 0.16$, $CHCl_3$); IR (neat): $\nu = 3335$, 3187, 2930, 1653, 1605, 1576, 1510, 1447, 1418, 1300, 1252, 1181, 1088, 1034, 831, 797, 729, 700 cm^{-1}; ^1H NMR (200 MHz, $CDCl_3$): $\delta = 2.88$ (1H, m), 3.20 (1H, dd, $J = 9$ and 7.6 Hz), 3.46 (1H, dd, $J = 9$ and 4.8 Hz), 3.79 (3H, s), 4.08 (1H, dd, $J = 9.2$ and 9.1 Hz), 4.45 (1H, dd, $J = 8.3$ and 6.5 Hz), 4.45 (1H, m), 5.51 (1H, br), 5.68 (2H, br s), 5.75 (1H, d, $J = 5.6$ Hz), 6.82 (2H, m), 7.2–7.45 (12H, m), 7.94 (1H, s), 8.27 (1H, s) ppm; ^{13}C NMR (50 MHz, $CDCl_3$): $\delta = 46.08$ (C3′), 55.23 (OCH_3), 62.38 (C4′), 71.84 (C2′), 77.99 (C5′), 86.48 (CAr_3), 92.68 (C1′), 113.17 (*ortho*-C × 2 of MeOAr), 120 (C5), 127.01 (2C of Ph), 127.89 (4C of Ph), 128.31 (4C of Ph), 130.29 (2C of Ph), 135.32 (*para*-C of MeOAr), 138.34 (C8), 144.15 and 144.24 (2 × *meta*-C of MeOAr), 149.12 (C4), 152.51 (C6), 155.60 (C2), 158.61 (C–OMe) ppm.

9-((1R,2R,3S)-Tetrahydro-2-hydroxy-3-(hydroxymethyl)-1-furanyl)-9H-adenine (**2A**; $C_{10}H_{13}N_5O_3$)

To the above monomethoxytrityl ether (12.0 mg, 0.023 mmol), AcOH (0.64 cm^3) and H_2O (0.16 cm^3) were added, and the mixture was stirred at room temperature for 5 h. After removal of the solvent, the residue was coevaporated with toluene and purified by flash chromatography ($CHCl_3$:MeOH = 5:1) to give the adenosine analogue **2A** (6 mg, 100%).

$[\alpha]_D^{20} = -21.6°$ ($c = 0.25$, $CHCl_3$); IR (neat): $\nu = 3335$, 3187, 2924, 1651, 1605, 1576, 1478, 1420, 1061 cm^{-1}; ^1H NMR (200 MHz, $CDCl_3$:CD_3OD = 10:1): $\delta = 2.57$ (1H, m), 3.65 (2H, d, $J = 5.8$ Hz), 4.11 (1H, dd, $J = 9$ and 8 Hz), 4.30 (1H, dd, $J = 9$ and 8 Hz), 4.54 (1H, dd, $J = 6$ and 4.4 Hz), 5.77 (1H, d, $J = 4.4$ Hz), 7.96 (1H, s), 8.18 (1H, s) ppm; ^{13}C NMR (50 MHz, $CDCl_3$:CD_3OD = 10:1): $\delta = 52.55$ (C3′), 64.48 (C4′), 74.71 (C2′), 80.93 (C5′), 96.10 (C1′), 123 (C5), 143.09 (C8), 153 (C4), 156.42 (C6), 159.59 (C2) ppm.

2-N-Acetyl-6-O-diphenylcarbamoyl-9-((1R,2R,3S)-tetrahydro-2-acetoxy-3-(((4-methoxyphenyl)-diphenylmethoxy)-methyl)-1-furanyl)-9H-guanine (**19a**; $C_{47}H_{42}N_6O_7$)

To a suspension of 2-N-acetyl-6-O-diphenylcarbamoylguanine **18** (40 mg, 0.10 mmol) in 1,2-dichloroethane (1 cm^3), N,O-*bis*-(trimethylsilyl)-acetamide (0.080 cm^3, 0.32 mmol) was added, and the mixture was stirred under reflux for 15 min. After removal of the solvent, the residue was dissolved in toluene (0.5 cm^3). To the solution, a mixture of diacetates **15** and **16** (19 mg, 0.039 mmol) in toluene (0.5 cm^3) and trimethylsilyl triflate (*TMSOTf*, 0.065 cm^3) were added, and the mixture was stirred at 90°C (bath temperature). After neutralization with aq. $KHCO_3$, the mixture was extracted with AcOEt and dried over $MgSO_4$. After removal of the solvent, the residue was purified by flash chromatography (hexane:AcOEt = 1:2, AcOEt:MeOH = 30:1) to give the acetate **19a** (23 mg, 72%) and the detritylated product (4 mg, 19%).

$[\alpha]_D^{23} = +6°$ ($c = 0.13$, $CHCl_3$); IR (neat): $\nu = 3310$, 2900, 1748, 1622, 1590, 1510, 1493, 1449, 1412, 1372, 1298, 1231, 1183, 1063, 1034, 984, 758, 737, 700 cm^{-1}; ^1H NMR (400 MHz, $CDCl_3$): $\delta = 2.07$ (3H, s), 2.44 (3H, s), 2.79 (1H, m), 3.29 (1H, dd, $J = 9.7$ and 4 Hz), 3.32 (1H, dd, $J = 9.7$ and 3.5 Hz), 3.77 (3H, s), 4.17 (1H, dd, $J = 8.8$ and 7.9 Hz), 4.31 (1H, dd, $J = 8.8$ and 7.9 Hz), 5.75

(1H, dd, $J = 5.5$ and 3.6 Hz), 5.93 (1H, d, $J = 3.6$ Hz), 6.81 (2H, m), 7.2–7.55 (22H, m), 7.78 (1H, br s), 7.87 (1H, s) ppm; ^{13}C NMR (100 MHz, $CDCl_3$): $\delta = 20.74$ (CH_3COO), 25.09 (CH_3CON), 45.71 (C3′), 55.24 (OCH_3), 61.67 (C4′), 71.23 (C2′), 78.61 (C5′), 86.74 (CAr_3), 89.93 (C1′), 113.25 (*ortho*-C × 2 of MeOAr), 121.27 (C5), 127.15 (2C of *Tr*), 127.94 (4C of *Tr*), 128.23 and 128.25 (4C of Ph), 129.24 (Ph), 130.20 (2C of *Tr*), 135.14 (*para*-C of MeOAr), 142.16 (C8), 143.98 (Ph), 144.04 (Ph), 150.28 (C4), 152.02 (C2), 154.45 (C6), 156.23 (OCON), 158.70 (C–OMe), 169.97 ($NCOCH_3$) ppm.

9-((1R,2R,3S)-Tetrahydro-2-hydroxy-3-(((4-methoxyphenyl)-diphenylmethoxy)-methyl)-1-furanyl)-9H-guanine ($C_{30}H_{29}N_5O_5$)

To a solution of the above acetate (54 mg, 0.066 mmol) in MeOH (5.9 cm^3), sat. aq. NH_3 (2 cm^3) was added, and the mixture was stirred at room temperature for 66 h. After removal of the solvent, the residue was purified by flash chromatography (AcOEt:MeOH = 10:1) to give the monomethoxytrityl ether (33 mg, 93%).

$[\alpha]_D^{22} = +72°$ ($c = 0.15$, $CHCl_3$); IR (neat): $\nu = 3337$, 3179, 2922, 1694, 1636, 1607, 1534, 1509, 1489, 1447, 1368, 1250, 1177, 1034, 779, 698 cm^{-1}; ^{1}H NMR (400 MHz, $CDCl_3$:CD_3OD = 10:1): $\delta = 2.70$ (1H, m), 3.11 (1H, dd, $J = 9$ and 8 Hz), 3.29 (1H, dd, $J = 8$ and 6 Hz), 3.73 (3H, s), 3.99 (1H, dd, $J = 8.8$ and 8.8 Hz), 4.31 (1H, dd, $J = 8.8$ and 8.8 Hz), 4.35 (1H, dd, $J = 7$ and 4.5 Hz), 5.57 (1H, d, $J = 4.5$ Hz), 6.77 (2H, m), 7.1–7.55 (13H, m) ppm; ^{13}C NMR (100 MHz, $CDCl_3$:CD_3OD = 10:1): $\delta = 50.43$ (C3′), 59.08 (OCH_3), 66.34 (C4′), 75.28 (C2′), 81.26 (C5′), 90.35 (CAr_3), 95.33 (C1′), 117.04 (*ortho*-C × 2 of MeOAr), 120.8 (C5), 130.90 (2C of *Tr*), 131.75 (4C of *Tr*), 132.19 (4C of Ph), 134.15 (2C of *Tr*), 139.26 (*para*-C of MeOAr), 139.5 (C8), 148.04 and 148.12 (2 × *meta*-C of MeOAr), 154.9 (C4), 157.32 (C2), 162.2 (C6), 162.45 (C–OMe) ppm.

9-((1R,2R,3S)-Tetrahydro-2-hydroxy-3-hydroxymethyl-1-furanyl)-9H-guanine (**2G**; $C_{10}H_{13}N_5O_4$)

To the above monomethoxytrityl ether (18.0 mg, 0.033 mmol), AcOH (0.92 cm^3) and water (0.23 cm^3) were added, and the mixture was stirred at room temperature for 6 h. After removal of the solvent, the residue was coevaporated with toluene and washed with Et_2O and AcOEt to give the guanosine analogue **2G** (9 mg, 100%).

$[\alpha]_D^{22} = -23°$ ($c = 0.06$, MeOH); IR (neat): $\nu = 3341$, 3127, 1696, 1603, 1534, 1485, 1364, 1250, 1180, 1044, 781 cm^{-1}; ^{1}H NMR (400 MHz, *DMSO*-d_6): $\delta = 2.33$ (1H, m), 3.51 (1H, ddd, $J = 11$, 7.6 and 5.5 Hz), 3.61 (1H, ddd, $J = 11$, 5 and 4.8 Hz), 3.92 (1H, dd, $J = 8.8$ and 8.2 Hz), 4.02 (1H, dd, $J = 8.2$ and 8.2 Hz), 4.39 (1H, ddd, $J = 7.5$, 5.5 and 5.5 Hz, C2′–H), 4.72 (1H, dd, $J = 5.5$ and 5 Hz, C5′–OH), 5.537 (1H, d, $J = 5.5$ Hz), 5.542 (1H, d, $J = 5.5$ Hz), 6.41 (2H, br s), 7.83 (1H, s), 10.57 (1H, br s) ppm.

Antiviral assays

The antiviral assays, other than HIV, were based on inhibition of virus-induced cytopathicity in either E_6SM (herpes simplex virus type 1 (HSV-1) (KOS), HSV-2 (G), vaccinia virus, vesicular stomatitis virus and thymidine kinase-deficient HSV-1 (KOS-ACV^R)), Vero (parainfluenza-3, reovirus-1, Sindbis, Coxsackie B4, Punta Toro virus), or HeLa (Vesicular stomatitis virus, Coxsackie virus B4, respiratory syncytial virus) cell cultures. Confluent cell cultures in microtiter 96-well plates were inoculated with 100 $CCID_{50}$ of virus, 1 $CCID_{50}$ being the virus dose to infect 50% of the cell cultures. After a 1 h virus adsorption period, residual virus was removed, and the cell cultures were incubated in the presence of varying concentrations (400, 200, 100, ... μg/cm^3) of the test compounds. Viral cytopathicity was recorded as soon as it reached completion in the control virus-infected cell cultures that were not treated with the test compounds. The methodology of the anti-HIV assays was as follows: human CEM ($\sim 3 \times 10^5$ cells · cm^{-3}) cells were infected with $CCID_{50}$

(HIV) (III_B) or HIV-2 (ROD)/cm^3 and seeded in 200 cm^3 wells of a microtiter plate containing appropriate dilutions of the test compounds. After 4 days of incubation at 37°C, HIV-induced CEM giant cell formation was examined microscopically.

Acknowledgements

We thank the National Institutes of Health (GM47228), the Universitywide AIDS Research Program (R97-LA-139), the Belgische Federatie tegen Kanker, and the Geconcerteerde Onderzoeksacties Vlaanderen (GOA 00/12) for generous financial support.

References

[1] a) Ichikawa E, Kato K (2001) Curr Med Chem **8**: 385; b) Balzarini J (2000) Pharmacol Ther **87**: 175; c) Shuter J (1999) Cancer Invest **17**: 145; d) Koszalka GW, Daluge SM, Boyd FL (1998) Annu Rep Med Chem **33**: 163; e) Darby G (1995) Antiviral Chem Chemother **6**: 54; f) De Clercq E (1994) Nucleosides Nucleotides **13**: 1271; g) Perigaud C, Gosselin G, Imbach JL (1992) Nucleosides Nucleotides **11**: 903; h) Huryn DM, Okabe M (1992) Chem Rev **92**: 1745

[2] a) Van Calenbergh S, Herdewijn P (2000) Expert Opin Ther Pat **10**: 289; b) Zemlicka J (2000) Pharmacol Ther **85**: 251; c) Graciet J-CG, Schinazi RF (1999) Adv Antiviral Drug Des **3**: 1; d) Walczak K (1999) Pol J Chem **73**: 1613; e) Bera S, Mickle T, Nair V (1999) Nucleosides Nucleotides **18**: 2379; f) Zheng XP, Nair V (1999) Nucleosides Nucleotides **18**: 1961; g) Cavalcanti SCH, Xiang XJ, Newton MG, Schinazi RF, Cheng YC, Chu CK (1999) Nucleosides Nucleotides **18**: 2233; h) Wang PY, Gullen B, Newton MG, Cheng YC, Schinazi RF, Chu CK (1999) J Med Chem **42**: 3390; i) Lee K, Choi Y, Hong JH, Schinazi RF, Chu CK (1999) Nucleosides Nucleotides **18**: 537; j) Soike KF, Huang JL, Russell JW, Whiterock VJ, Sundeen JE, Stratton LW, Clark JM (1994) Antiviral Res **23**: 219; k) Tino JA, Clark JM, Field AK, Jacobs GA, Lis KA, Michalik TL, McGeeverrubin B, Slusarchyk WA, Spergel SH, Sundeen JE, Tuomari AV, Weaver ER, Young MG, Zahler R (1993) J Med Chem **36**: 1221; l) Pankiewicz KW, Krzeminski J, Watanabe KA (1992) J Org Chem **57**: 7315

[3] a) Jung ME, Toyota A (2001) J Org Chem **66**: 2624; b) Rhee H, Yoon DO, Jung ME (2000) Nucleosides Nucleotides **19**: 619; c) Castro C, Chen C, Jung ME (1999) Nucleosides Nucleotides **18**: 2415; d) Jung ME, Nichols CJ, Kretschik O, Xu Y (1999) Nucleosides Nucleotides **18**: 541; e) Jung ME, Kiankarimi M (1998) J Org Chem **63**: 8133; f) Jung ME, Castro C, S. I. Khan SI (1998) Nucleosides Nucleotides **17**: 2383; g) Jung ME, Nichols CJ (1998) J Org Chem **63**: 347; h) Jung ME, Xu Y (1998) Heterocycles **47**: 349; i) Jung ME, Kretschik O (1998) J Org Chem **63**: 2975; j) Jung ME, Nichols CJ (1998) Tetrahedron Lett **39**: 4615

[4] Jung ME, Toyota A (2000) Tetrahedron Lett **41**: 3577

[5] a) Burgos CE, Ayer DE, Johnson RA (1987) J Org Chem **52**: 4973; b) Katsuki T, Sharpless KB (1980) J Am Chem Soc **102**: 5976

[6] Sharpless KB (private communication)

[7] a) McCombie SW, Shankar BB, Ganguly AK (1985) Tetrahedron Lett **26**: 6301; b) Pelter A, Ward RS, Little GM (1990) J Chem Soc Perkin Trans 1, 2775

[8] Zou R, Robins MJ (1987) Can J Chem **65**: 1436

[9] Green K, Blum DM (1991) Tetrahedron Lett **32**: 2091

[10] a) Barton DHR, McCombie SW (1975) J Chem Soc Perkin Trans 1, 1574; b) For the use of phenyl thionocarbonates, see: Robins MJ, Wilson JS, Hansske F (1983) J Am Chem Soc **105**: 4059; For more recent procedures, see: c) Quiclet-Sire B, Zard SZ (1998) Tetrahedron Lett **39**: 9435; d) Lopez RM, Hays DS, Fu GC (1997) J Am Chem Soc **119**: 6949

Received September 3, 2001. Accepted September 17, 2001

Thiosugars VI: A Simple Stereoselective Approach to (1→3)-3-S-Thiodisaccharides from Levoglucosenone

Zbigniew J. Witczak[1,*], **Peter Kaplon**[2], and **Mark Kolodziej**[2]

[1] Department of Pharmaceutical Sciences, Nesbitt School of Pharmacy, Wilkes University, Wilkes-Barre, PA 18766, USA

[2] Department of Pharmaceutical Sciences, School of Pharmacy, University of Connecticut, Storrs, CT 06269-2092, USA

Summary. A new stereoselective synthesis of (1→3)-3-S-thiodisaccharides is described. Levoglucosenone-derived chiral building blocks produced by the selenium-mediated iodination at C-3 afforded 3-iodoketones in moderate yield. Iodine displacement with sulfur nucleophiles (1-thiols) resulted in 3-thiodisaccharides. Following reduction of the C-2 keto function constituted a new two-step general approach to these biologically important thiosugars.

Keywords. Thiosugars; Levoglucosenone; S_N2 Displacement; (1 → 3)-3-S-Thiodisaccharides.

Introduction

In continuation of our program on C-glycosyl compounds [1] and C-disaccharides [2] as non-hydrolyzable epitopes, we focused our attention on the development of new strategies for the synthesis of (1→4), (1→2), and (1→3)-S-thiodisaccharides [3, 4]. Thiodisaccharides [5] containing sulfur in the glycosidic linkage have been synthesized previously by a variety of methods [6–14], including S_N2-type reactions involving the action of a thiolate anion and a glycosyl halide, *i.e.* the displacement of a leaving group by 1-thio-glucopyranose.

Taking into account that existing synthetic methods of thiodisaccharide synthesis are multistep [6–17] and overall low-yield approaches, an urgent need for shorter and more convergent syntheses exists. Moreover, the biological relevance and therapeutic potential of thiosugars [5], including antitumor activity of (1→2)-S-linked thiodisaccharides on selected cell culture systems [9], make them attractive new biological targets and constitutes the rationale for their synthesis.

In our efforts towards preparing thiodisaccharides [3, 4], we employed the new chiral building blocks **1–6** derived from levoglucosenone [19–21] (1,6-anhydro-3,4-dideoxy-*β*-*D*-glycero-hex-3-enopyranos-2-ulose). These new 'synthons' are

* Corresponding author. E-mail: witczak@wilkes.edu

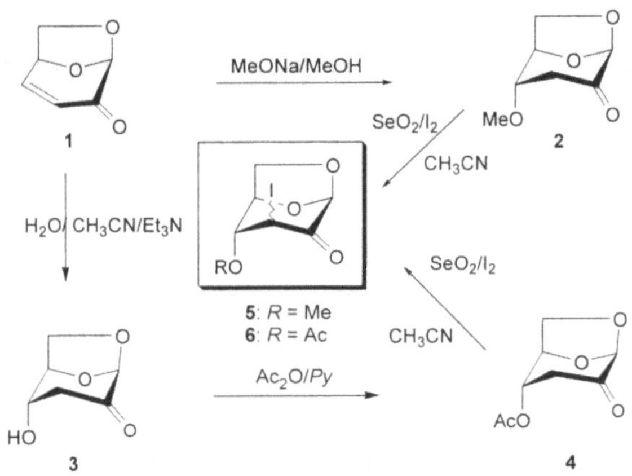

Scheme 1

extremely attractive and versatile building blocks because of their high functionality and conformational rigidity (Scheme 1).

The 1,6-anhydro bridge in **1**–**6** eliminates the need for protecting groups at the anomeric carbon and the C-6-OH. Moreover, the bridge fixes the conformation of the system and sterically hinders the *β*-*D*-face of the molecules. Therefore, these 'synthons' have already been used in the synthesis of various natural products and their intermediates [22–24].

Results and Discussion

The new approach presented here features ongoing studies on the utility of the new, convenient chiral 'synthons' **5** and **6** produced from levoglucosenone [19–21]. *Michael* addition of methanol to levoglucosenone under basic conditions proceeds easily with the direct formation of adduct **2** [25]. Analogous addition of water afforded adduct **3** [19] which by conventional *in situ* acetylation furnished the acetyl derivative **4**. Direct *α*-iodination of **2** and **4** using iodine/selenium dioxide [26] led to the isolation of a diastereoisomeric mixture (1:1) of the 3-iodoketones **5** and **6** in 72% yield.

Taking into account that 3-iodoketone **5** will require additional deprotection of the C-4 methyl group after coupling reaction with thiols, the 3-iodoketone **6** was chosen as a more versatile and reactive substrate in the stereoselective displacement reaction with glycosyl thiols **7** [27] and **8** [28–30].

The reaction of **6** with thiols **7** [27] and **8** [28–30] proceeded smoothly with the formation of *β*-(1→3)-3-S-thiodisaccharides **9** and **10** in 60–72% yield. Proton–proton couplings in the ^{1}H NMR spectra of **9** and **10** confirmed that only the 3-equatorial products was obtained as the single product in both cases. This stereospecificity has been observed previously in levoglucosenone conjugate addition [3, 4] and proceeds by the attack of incoming nucleophile at the iodine face opposite the 1,6-anhydro ring.

The advantage of the stereoselective iodide displacement is the exclusive formation of an S-linkage from the less hindered face of the molecule with inversion of the configuration at C-3. The shielding effect of the 1,6-anhydro bridge in **6** effectively prevents the formation of the 3-axial (*a*) product, thus yielding only the 3-equatorial (*e*) product. This equatorial attack of the sulfur nucleophiles was expected [12–14], but a new chiral center at C-3 surprisingly stabilizes the molecule, as no epimerization [11, 13] or β-elimination was observed during the coupling reaction. Also, the 4-acetyl protecting group of **6** is sufficiently stable under the above reaction conditions, and no deacetylation or epoxide formation has been observed. This important observation additionally indicates the preferential stereochemistry at C-3 of **6**. The sterically hindered 1,6-anhydro bridge is therefore assumed to effectively prevent an attack from the upper side of the molecule, which could give an access to the alternative formation of the 3-axial product.

Indeed, the proton–proton couplings in the ^{1}H NMR spectra of **9** and **10** confirmed the D-allo stereochemistry with coupling constants of $J_{3a,4} = 7.6$ Hz, indicating an equatorial orientation of the substituent at C-3. The ^{1}H NMR spectra of the products did not show signals corresponding to the D-gluco isomer, clearly demonstrating that the stereochemistry of the reaction of thiols **7** and **8** with 3-ketone **6** is completely controlled by the steric bulk of the 1,6-anhydro bridge. The ^{13}C NMR spectra of both **9** and **10** showed no $-CH_2-$ group signal, and the C-3 signal appeared at *ca.* 23.0–23.4 ppm.

The reduction of the C-2 keto function of ketones **9** and **10** with L-Selectride®, followed by conventional *in situ* acetylation, proceeded stereoselectively under formation of the D-altro isomers **12** and **13** in 79% yield. Only a trace amount of the corresponding D-altro isomers **11a** and **12a** were detected by ^{1}H NMR spectroscopy. Additionally, the absence of a coupling between H-4 and H-5 and coupling constants of $J_{1,2} = 8.2$ and $J_{2,3} = 5.2$ Hz for **11** and **12** indicate an equatorial orientation of the new substituent at C-2.

Unlike levoglucosenone, where the C-2 keto group reduction was predictably controlled [31] by the 1,6-anhydro bridge, the analogous reduction of **9** and **10** was expected to be dominated by the relative steric contribution of the bulky equatorial substituent at C-3 as well as by the 1,6-anhydro bridge. This is in full agreement with earlier observations [14, 15] of high stereoselectivity and another classical example of the preferential attack of the reducing agent from the bottom face on this bicyclic molecule. Moreover, the ^{1}H and ^{13}C NMR spectra of **11** and **12** firmly support their assignments to the D-allo configuration. Particularly, the coupling constants between the equatorially disposed H-2 and the axially oriented H-3 are of great diagnostic value ($J_{2,3} = 5$ Hz).

The cleavage of the 1,6-anhydro ring in **11** and **12** was examined under various reaction conditions. Acetolysis using trifluoroacetic acid and acetic anhydride

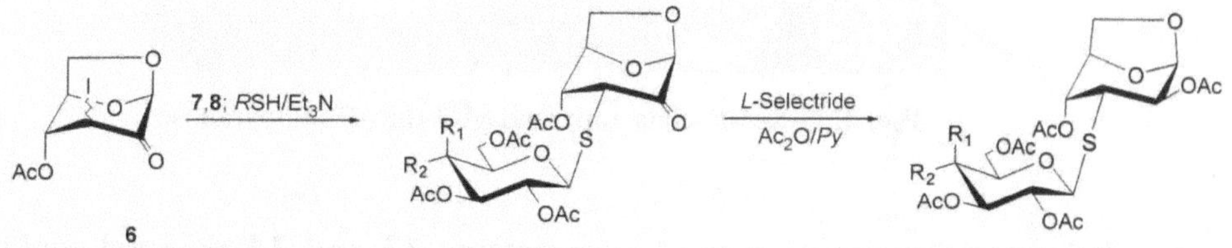

7: *R* = 2,3,4,6-tetra-O-acetyl-*D*-glucopyranosyl
8: *R* = 2,3,4,6-tetra-O-acetyl-*D*-galactopyranosyl

9: R^1 = H, R^2 = OAc
10: R^1 = OAc, R^2 = H

11: R^1 = H, R^2 = OAc
12: R^1 = OAc, R^2 = H

Scheme 2

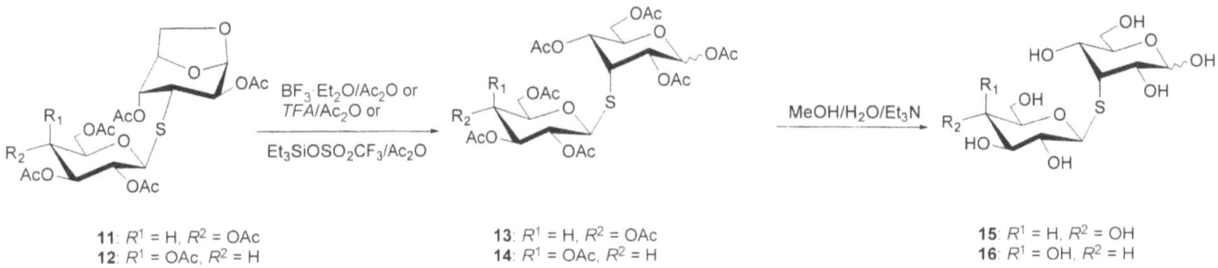

Scheme 3

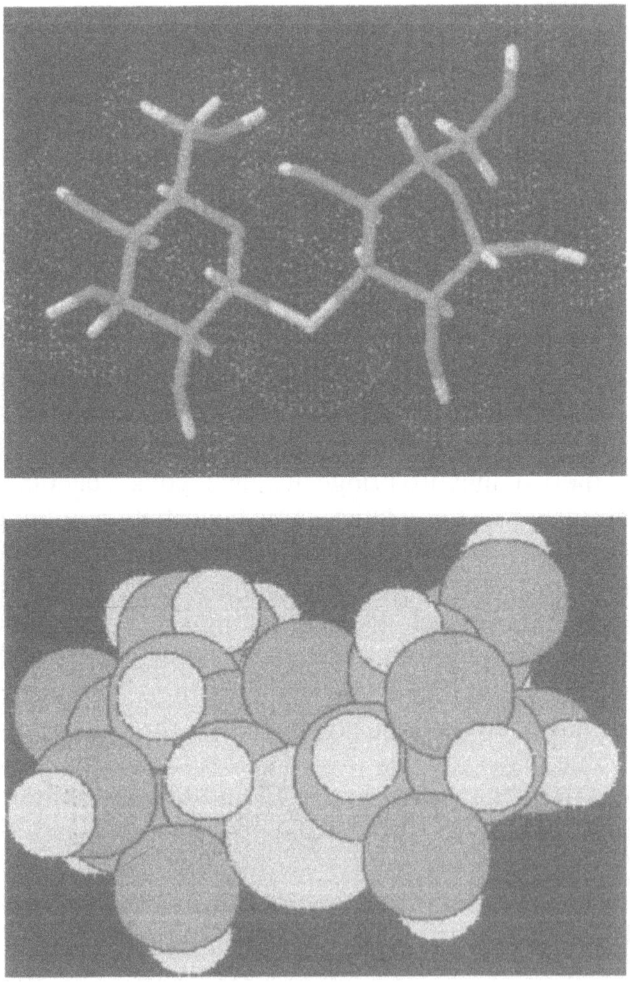

Fig. 1. 3-S-*β*-*D*-Galactopyranosyl-3-thio-*D*-allopyranose

afforded an anomeric mixture of octaacetates **13** and **14** in good yield (62%). Employing acetic anhydride and a catalytic amount of borontrifluoride etherate also gave acceptable results (68%). However, chromatographic purification of the crude material was required.

The method of choice was acetolysis using acetic anhydride solution and a catalytic amount of trifluoromethanesulfonate and proceeding according to the convenient protocol of *Fraser-Reid* [32]. This resulted in the formation of an anomeric mixture of octaacetates **13** and **14** (α:β = 1:5) in 89% yield. The presence of the sulfur bridge at C-3 clearly indicates the strong influence of the 1,6-anhydro ring on the cleavage; a prolonged reaction time (up to the 12 h) as compared with the literature data reported [32] is required. Separation of the anomers proved impossible because of their almost identical R_f values. However, the ^{1}H and ^{13}C NMR spectra and mass spectroscopic data for the mixture firmly established their identity.

Final deprotection of octaacetates **13** and **14** was performed with an aqueous methanolic solution of triethylamine ($MeOH:Et_3N:H_2O = 4:1:5$) at room temperature for 8 h, resulting in the new thiodisacharides 3-S-(β-D-glucopyranosyl)-3-thio-D-allopyranose (**15**) and 3-S-(β-D-galactopyranosyl)-3-thio-*D*-allopyranose **16** in 89% yield.

A molecular model of **16** (Fig. 1; generated by ACD Chemsketch 4.0, 3D software program) clearly confirms the structure as determined by NMR spectroscopy. The length of the C-1–S bond amounts to 1.8132 Å, whereas the C-3–S distance is 1.8086 Å; the dihedral and torsional angles for the C–S–C bridge are 107.887 and 61.305°, respectively. Calculations for the sulfur bridge of other thiodisaccharides give similar results and differ substantially from those of their oxygen counterparts as previously postulated in the literature [33, 34].

Conclusions

The S_N2 nucleophilic displacement reaction of 3-iodoketone **6** with protected 1-thiosugars, followed by C-2 keto group reduction and deprotection, constitutes a new short and stereoselective approach to (1→3)-3-S-thiodisacharides. All new functionalized thiodisaccharides are stable glycomimetics of potential biological interest.

Experimental

General

Unless otherwise noted, starting materials were obtained from commercial suppliers and used without purification. All melting points are uncorrected and were measured in open capillary tubes. Optical rotations were determined on a Jasco Model DIP-370 polarimeter in $CHCl_3$. Thin-layer chromatography (TLC) was performed on precoated Silica gel $60F_{254}$ plates from E. Merck; spots were visualized by spraying with 10% ethanolic H_2SO_4 and subsequent heating. Column chromatography was performed on Silica Gel 60 (70–230 mesh, Merck No. 34). NMR samples were prepared in $CDCl_3$ (99.8 at.% D), filtered, freeze-thawed, and sealed in a 5 mm tube. *TMS* was used as an internal chemical shift reference. High-resolution NMR spectra were obtained on a Bruker DMX-500 spectrometer. Mass spectra were obtained either in the EI mode at 70 eV or using CI (NH_3). Levoglucosenone was produced according to the published methodology [19–21]. 1,6-Anhydro-4-O-methyl-β-*D*-erythro-hexopyranos-2-ulose (**2**) [25] and 1,6-anhydro-4-hydroxy-β-*D*-erythro-hexopyranos-2-ulose (**3**) [14] were prepared according to literature methods. 2,3,4,6-Tetra-O-acetyl-1-thio-β-*D*-glucopyranose (**7**) [27] and 2,3,4,6-tetra-O-acetyl-1-thio-β-*D*-galactopyranose (**8**) [28, 29] were prepared by reacting the corresponding α- or β-glycosyl halides with thiourea and the reduction of the isothiouronium salts with potassium pyrosulfite according to the published procedure [30].

4-O-Acetyl-1,6-anhydro-2,3-dideoxy-β-D-erythro-hexopyranos-2-ulose (**4**; $C_8H_{10}O_5$)

Compound **3** (0.1 mmol) was conventionally acetylated at room temperature for 24 h with acetic anhydride (5 cm^3) in pyridine (4 cm^3). The reaction mixture was poured into ice water and extracted with ethyl acetate. The extract was washed and dried. Removal of the solvent on a rotary evaporator (coevaporation with toluene:EtOH = 1:1; $5 \times 30\,cm^3$) afforded an oily residue that was subjected to column chromatography on silica gel. The fraction eluted with 20% ethyl acetate in hexane (v/v) gave a syrupy product which crystallized from ether/hexane.

Yield: 205 mg (83%); m.p.: 108–110.2°C; $R_f = 0.45$ (EtOAc); $[\alpha]_D = +231°$ ($c = 1.0$, $CHCl_3$); for ^1H and ^{13}C NMR data, see Tables 1 and 2.

General method for the preparation of 3-iodoketones **5** *and* **6**

Compounds **5** and **6** were produced by the *Bekaert* [26] methodology as described below. To a solution of **4** (0.1 mol) in 50 cm^3 of CH_3CN, 6.1 g powdered selenium dioxide (55 mmol) and 13.97 g iodine (55 mmol) were added. The mixture was heated at 80°C for 8 h. The red selenium precipitate was removed by filtration through a short silica gel column. The dark-red solution was diluted with 200 cm^3 of H_2O and extracted with CH_2Cl_2. The organic layer was extracted with 100 cm^3 of a 10% solution of $Na_2S_2O_3$ and washed with H_2O. After drying with anhydrous magnesium sulfate, the solvent was removed on a rotary evaporator, and the residue was purified by column chromatography on silica gel. The fraction eluted with 20% ethyl acetate in hexane (v/v) gave pure syrupy products which crystallized from ether/hexane.

1,6-Anhydro-2,3-dideoxy-3-iodo-4-methoxy-β-D-erythro-hexopyranos-2-ulose
(**5**; $C_7H_9IO_4$)

Yield: 205 mg (73%); m.p.: 80–82.5°C; $R_f = 0.51$ (EtOAc); $[\alpha]_D = +301°$ ($c = 1.0$, $CHCl_3$); for ^1H and ^{13}C NMR data, see Tables 1 and 2.

1,6-Anhydro-2,3-dideoxy-3-iodo-4-O-acetyl-β-D-erythro-hexopyranos-2-ulose
(**6**; $C_8H_9IO_5$)

Yield: 235 mg (73%); m.p.: 60–61.5°C; $R_f = 0.59$ (EtOAc); $[\alpha]_D = +318°$ ($c = 1.0$, $CHCl_3$); for ^1H and ^{13}C NMR data, see Tables 1 and 2.

General method for the preparation of (1→3)-3-S-thiodisacharides **9** *and* **10**

To a solution of 126 mg 3-iodoketone **6** (0.1 mmol) in 10 cm^3 CH_3CN, a solution of 125 mg **7** [26] or **8** [27, 28] (0.34 mmol) in 5 cm^3 of CH_3CN and 2 cm^3 of thiethylamine was added dropwise. The reaction mixture was stirred at room temperature for 24 h; then, pyridine (4 cm^3) and acetic anhydride (5 cm^3) were added, and the mixture was stirred at room temperature overnight, poured into ice water, and extracted with ethyl acetate. The extract was washed and dried. Removal of the solvent *in vacuo* after coevaporation with $5 \times 30\,cm^3$ toluene:EtOH = 1:1 afforded an oily residue that was subjected to column chromatography on silica gel. The fraction eluted with 20% ethyl acetate in hexane (v/v) gave syrupy products **9** and **10** which crystallized from ether/hexane.

Table 1. ^{1}H NMR chemical shifts (ppm) and coupling constants (Hz) for compounds **5**, **6**, and **9–16** (500 MHz, $CDCl_3$, *TMS*)

	H-1′ $J_{1,2}$	H-2′ $J_{2,3}$	H-3′ $J_{2,3'}$	H-4′ $J_{3,4}$	H-5′ $J_{5,6'}$	H-6′ $J_{5,6''}$	H-6′ $J_{5,6''}$	H-1 $J_{1,2}$	H-2 $J_{2,3}$	H-3a $J_{3a,4}$	H-4 J	H-5 $J_{5,6exo}$	H-6*endo* $J_{5,6endo}$	H-6*exo* $J_{6exo,6endo}$	$COCH_3$
5								5.3 d		3.12 dd	3.56 d	4.50 dd	3.93 dd	4.04 dd	3.82
								–		7.6	8.0	4.9	1.1	8.2	
6								5.3 d		3.12 dd	3.56 d	4.50 dd	3.93 dd	4.04 dd	2.02
								–		7.6	8.0	4.9	1.1	8.2	
9	4.66 d	5.19 dd	5.23 t	5.04 d	4.82 d	4.26 dd	4.26 dd	5.3 d		3.12 dd	3.56 d	4.90 dd	3.93 dd	4.04 dd	2.02–1.98
	8.6	9.6	9.9	9.5	5.0	–	–	–		7.6	8.0	4.9	1.1	8.2	5s, 5 × OAc
10	4.66 d	5.19 dd	5.23 t	5.04 d	4.82 d	4.26 dd	4.26 dd	5.3 d		3.12 dd	3.56 d	4.90 dd	3.93 dd	4.04 dd	2.02–2.1,
	8.6	9.6	9.9	9.5	5.0	–	–			7.6	8.0	4.9	1.1	8.2	5s, 5 × OAc
11	4.66 d	5.19 dd	5.23 t	5.04 d	4.82 d	4.26 dd	4.26 dd	5.3 d	3.66 d	3.12 dd	3.56 d	4.90 dd	3.93 dd	4.04 dd	2.02–2.1,
	8.6	9.6	9.9	9.5	5.0	–	–	8.2	5.0	7.6	8.0	4.9	1.1	8.2	6s, 6 × OAc
12	4.66 d	5.19 dd	5.23 t	4.65 d	4.82 d	4.26 dd	4.26 dd	5.4 d	3.66 d	3.12 dd	4.35 d	4.48 dd	3.93 dd	4.04 dd	2.02–2.1,
	8.6	9.6	9.9	9.5	5.0	–		8.2	5.0	7.6	8.0	4.9	1.1	8.2	6s, 6 × OAc
13	4.66 d	5.19 dd	5.23 t	5.04 d	4.82 d	4.26 dd	4.26 dd	4.8 d	3.66 d	3.12 dd	3.56 d	4.90 dd	3.93 dd	4.04 dd	2.01–2.1
	8.6	9.6	9.9	9.5	5.0	–	–	8.2	5.0	7.6	8.0	4.9	1.1	8.2	8s, 8 × OAc
14	4.6 d	5.19 dd	5.23 t	5.04 d	4.82 d	4.26 dd	4.26 dd	4.8 d	3.66 d	3.12 dd	3.56 d	4.90 dd	3.93 dd	4.04 dd	2.01–2.1
	8.6	9.6	9.9	9.5	5.0	–	–	8.2	5.0	7.6	8.0	4.9	1.1	8.2	8s, 8 × OAc
15[a]	4.86 d	5.21 dd	5.26 t	5.08 d	4.84 d	4.28 dd	4.28 dd	4.8 d	3.58 d	3.19 dd	3.59 d	4.84 dd	3.96 dd	4.06 dd	
	8.2	9.8	10.0	9.7	5.2	–	–	8.2	5.2	7.4	8.1	4.8	1.2	8.4	
16[a]	4.86 d	5.21 dd	5.26 t	5.08 d	4.84 d	4 28 dd	4.28 dd	4.8 d	3.58 d	3.19 dd	3.59 d	4.84 dd	3.96 dd	4.06 dd	
	8.2	9.8	10.0	9.7	5.2	–	–	8.2	5.2	7.4	8.1	4.8	1.2	8.4	

[a] D_2O, *TMSPA*-Na

Table 2. ^{13}C NMR chemical shifts (ppm) for compounds **5**, **6**, and **10–16** (125 MHz, $CDCl_3$, *TMS*)

	C-1	C-2	C-3	C-4	C-5	C-6	C-1′	C-2′	C-3′	C-4′	C-5′	C-6′	–OCO	CH_3
5	102.1	207.2	23.0	78.3	85.0	65.7								50.8
6	102.6	207.1	23.1	76.2	85.1	65.7							171.2	17.6
9	104.8	207.7	63.4	71.9	78.1	68.1	83.7	67.1	48.4	70.6	76.2	68.8	5 × 171.1	17.6, 17.5, 17.4
10	104.6	207.7	63.1	71.8	78.1	68.6	83.7	67.1	48.1	70.6	76.2	68.8	5 × 171.1	17.6, 17.5, 17.5, 17.4
11	104.7	73.7	63.1	71.9	78.0	68.1	83.7	65.7	48.7	72.1	75.0	68.8	6 × 176.0	6 × 17.6
12	104.8	73.7	63.1	71.9	78.0	68.6	83.9	65.7	48.7	72.1	72.0	68.8	6 × 176.0	6 × 17.6
13	98.8	73.7	63.1	65.9	78.0	63.1	84.9	70.7	48.7	69.9	72.0	68.8	8 × 176.0	8 × 17.6
14	98.8	73.7	63.1	65.9	75.0	68.1	84.8	70.7	48.7	69.9	75.0	68.0	8 × 176.0	8 × 17.6
15[a]	98.5	74.4	63.7	68.4	75.7	64.8	84.9	75.6	48.2	68.5	75.4	64.8		
16[a]	98.6	74.4	63.7	68.4	75.7	64.8	84.8	77.0	48.1	68.8	72.4	64.8		

[a] D_2O, *TMSPA*-Na

1,6-Anhydro-3-S-(2,3,4,6-tetra-O-acetyl-β-glucopyranosyl)-β-D-glycero-hexopyranos-4-ulose (**9**; $C_{20}H_{26}O_{12}S$)

Yield: 436 mg (89%); m.p.: 157–158.5°C; $R_f = 0.59$ (hexane:EtOAc = 1:4); $[\alpha]^{30} = -120.21°$ (c = 0.84, $CHCl_3$); HRMS: $m/z = 490.11$ (calcd.: 490.47); for ^1H and ^{13}C NMR data: see Tables 1 and 2.

1,6-Anhydro-3-S-(2,3,4,6-tetra-O-acetyl-β-galactopyranosyl)-β-D-glycero-hexopyranos-4-ulose (**10**; $C_{20}H_{26}O_{12}S$)

Yield: 385 mg (78.5%); m.p.: 151–152.5°C; $R_f = 0.41$ (hexane:EtOAc = 1:4); $[\alpha]^{30} = -122.2°$ ($c = 0.8$, $CHCl_3$); HRMS: $m/z = 490.11$ (calcd.: 490.47); for ^1H and ^{13}C NMR data, see Tables 1 and 2.

General method for the reduction of (1→3)-3-S-thiodisacharides **9** *and* **10**

To a cooled and stirred solution of 210 mg thiodisaccharides **9** or **10** (0.428 mmol) in *THF*, 1.0 cm^3 *L*-Selectride® (1 *M* in *THF*) was added at −78°C under an Ar atmosphere. The reaction mixture was stirred for 3 h; then, pyridine (4 cm^3) and acetic anhydride (5 cm^3) were added, and stirring at room temperature was continued overnight. The reaction mixture was poured into ice water and extracted with ethyl acetate. The extract was washed and dried. Removal of the solvent *in vacuo* after coevaporation with 5 × 30 cm^3 toluene:EtOH = 1:1 afforded an oily residue that was subjected to column chromatography on silica gel. The fraction eluted with 20% ethyl acetate in hexane (v/v) gave syrupy products **11** and **12**.

4-O-Acetyl-1,6-anhydro-3-S-(2,3,4,6-tetra-O-acetyl-3-thio-β-glucopyranosyl)-β-D-allo-hexo-pyranose (**11**; $C_{24}H_{32}O_{15}S$)

Yield: 190 mg (79%); $R_f = 0.42$ (hexane:EtOAc = 1:4); $[\alpha]^{30} = -34.2°$ ($c = 0.84$, $CHCl_3$); HRMS: $m/z = 592.14$ (calcd.: 592.57); for ^1H and ^{13}C NMR data, see Tables 1 and 2.

4-O-Acetyl-1,6-anhydro-3-S-(2,3,4,6-tetra-O-acetyl-3-thio-β-galactopyranosyl)-β-D-allo-hexo-pyranose (**12**, $C_{24}H_{33}O_{15}S$)

Yield: 170 mg (81.5%); $R_f = 0.33$ (hexane:EtOAc = 1:4); $[\alpha]^{30} = -30.4°$ ($c = 0.68$, $CHCl_3$); HRMS: $m/z = 592.17$ (calcd.: 592.57); for ^1H and ^{13}C NMR data, see Tables 1 and 2.

General method for the hydrolysis of (1→3)-S-thiodisaccharides **11** *and* **12**

a) 0.2 g Thiodisaccharide **11** or **12** (0.375 mmol) were dissolved in Ac_2O (8.5 cm^3) and CF_3COOH (6.2 cm^3), and the mixture was stirred at room temperature for 10 h. After the solvent was evaporated, the resulting brown syrup was chromatographed (hexane:Et_2OAc = 2:1) to afford an anomeric mixture of 160 mg (62%) heptaacetates **13** and **14** as colorless syrup (α:β-ratio: 1:5).
b) 0.4 g Thiodisaccharide **11** or **12** (0.75 mmol) were dissolved in Ac_2O (8.5 cm^3). Boron trifluoride etherate ($BF_3 \cdot Et_2O$, 0.1 cm^3) was added, and the mixture was stirred at room temperature for 10 h. After the solvent was evaporated, the resulting brown syrup was chromatographed (hexane:Et_2OAc = 2:1) to afford 0.30 g (68%) of an anomeric mixture of heptaacetates **13** and **14** as a colorless syrup (α:β-ratio = 1:5).
c) To a cooled solution (0°C) of 0.5 g of **11** or **12** (0.93 mmol) in acetic anhydride (10 cm^3) stirred under argon, two drops (3 mm^3) of trimethylsilyl trifluoromethanesulfonate were added. TLC

(EtOAc:hexane = 1:1) indicated the completion of the reaction after 12 h. A solution of saturated sodium bicarbonate was added, and the mixture was stirred for 30 min and was extracted with $3 \times 20\,cm^3$ ethyl acetate. The combined extracts were washed with $20\,cm^3$ saturated $NaHCO_3$ and $20\,cm^3$ brine and dried over Na_2SO_4. Evaporation of the solvent under reduced pressure yielded 0.35 g (89%) of an inseparable anomeric mixture (α:β-ratio $\approx$ 1:6).

1,2,4,6-Tetra-O-acetyl-3-S-(2,3,4,6-tetra-O-acetyl-β-glucopyranosyl)-3-thio-α,β-D-allo-pyranose (**13**; $C_{28}H_{38}O_{18}S$)

Yield: 0.648 g (91%); colorless syrup; $R_f = 0.34$, 0.31 (hexane:EtOAc = 1:4); $[\alpha]^{30} = -12.4°$ (c = 0.82, $CHCl_3$); HRMS: $m/z = 636.17$ (calcd.: 636.62); for ^1H and ^{13}C NMR data (mainly the β-anomer), see Tables 1 and 2.

1,2,6-Tri-O-acetyl-3-S-(2,3,4,6-tetra-O-acetyl-β-galactopyranosyl)-3-thio-α,β-D-allopyranose (**14**; $C_{26}H_{38}O_{18}S$)

Yield: 0.641 g (86%); colorless syrup; $R_f = 0.42$, 0.41 (hexane:EtOAc = 1:4); $[\alpha]^{30} = -12.4°$ (c = 0.82, $CHCl_3$); HRMS: $m/z = 636.17$ (calcd.: 636.62); for ^1H and ^{13}C NMR data (mainly the β-anomer), see Tables 1 and 2.

General method for the deprotection of (1→3)-3-S-thiodisacharides **13** *and* **14**

0.250 mg of thiodisaccharides **13** or **14** (0.15 mmol) were dissolved in $15\,cm^3$ of a MeOH:Et_3N:H_2O = 4:1:5 mixture and stirred at room temperature. TLC indicated the completion of the reaction after 6 h. Evaporation of the solvent produced an inseparable anomeric mixture (α:β-ratio = 1:6).

3-S-(β-D-Glucopyranosyl)-3-thio-D-allopyranose (**15**; $C_{12}H_{22}O_{10}S$)

Yield: 119 mg (89%); colorless syrup; $[\alpha]^{30} = -10.2° \rightarrow -16.2°$ ($c = 0.82$, H_2O); HRMS: $m/z = 342.09$ (calcd.: 342.36); for ^1H and ^{13}C NMR data (mainly the β-anomer), see Tables 1 and 2.

3-S-(β-D-Galactopyranosyl)-3-thio-D-allopyranose (**16**; $C_{12}H_{22}O_{10}S$)

Yield: 88 mg (66%); colorless syrup; $[\alpha]^{30} = -12.6° \rightarrow -14°$ ($c = 0.8$, H_2O); HRMS: $m/z = 342.12$ (calcd.: 342.36); for ^1H and ^{13}C NMR data (mainly the β-anomer), see Tables 1 and 2.

Acknowledgments

Financial support from the American Cancer Society (Institutional Grant to University of Connecticut Health Center, ACSIN 152L-132 grant program) is gratefully acknowledged.

References

[1] Witczak ZJ (1994) Pure Appl Chem **66**: 2189
[2] Witczak ZJ, Chhabra R, Chojnacki J (1997) Tetrahedron Lett **38**: 2215
[3] Witczak ZJ, Sun J, Mielguj R (1995) Bioorg Med Chem Lett **5**: 2169
[4] Witczak ZJ, Chhabra R, Chen H, Xie X-Q Carbohydr Res **301**: 167; Witczak ZJ, Chen H, Kaplon P (2000) Tetrahedron Asymm **11**: 519; Witczak ZJ, Chhabra R, Boryczewski D (2000) J Carbohydr Chem **19**: 543

[5] Witczak ZJ (1999) Curr Med Chem **6**: 165
[6] For reviews see a) Defaye J, Gelas J (1991) In: Atta-ur-Rahman (ed) Studies in Natural Products Chemistry, vol 8E. Elsevier, Amsterdam, p 315; b) Driguez H (1997) In: Driguez H, Thiem J (eds) Topics in Current Chemistry, vol 187. Springer Berlin Heidelberg New York, p 85
[7] Blanc-Muesser M, Defaye J, Driguez H (1982) J Chem Soc Perkin Trans 1, 15
[8] Defaye J, Guillot J-M (1994) Carbohydr Res **253**: 185
[9] Comber RN, Friedrich JD, Dunshee DA, Petty SL, Secrist III JA (1994) Carbohydr Res **262**: 245
[10] Hashimoto H, Shimada K, Horito S (1993) Tetrahedron Lett **34**: 4953
[11] Hashimoto H, Shimada K, Horito S (1994) Tetrahedron Asymm **12**: 2351
[12] Calvo-Asin JA, Calvo-Flores FG, Exposito-Lopez JM, Hernandez-Mateo F, Garcia-Lopez JJ, Isac-Garcia J, Santoyo-Gonzalez F, Vargas-Berenguel A (1997) J Chem Soc Perkin Trans 1, 1079
[13] Defaye J, Guillot J-M, Biely P (1992) Carbohydr Res **228**: 47
[14] Petrusova M, Lattova E, Matulova M, Petrus L, BeMiller JN (1996) Carbohydr Res **283**: 73
[15] Michael K, Kessler H (1996) Tetrahedron Lett **37**: 3453
[16] Becker B, Thimm J, Thiem J (1996) J Carbohydr Chem **15**: 1179
[17] Johnston BD, Pinto BM (2000) J Org Chem **65**: 4607
[18] Knapp S, Malolanarsimhan K (2000) Org Lett **1**: 611
[19] Ward DD, Shafizadeh F (1981) Carbohydr Res **95**: 155
[20] Shafizadeh F, Furneaux RH, Stevenson TT (1979) Carbohydr Res **71**: 161
[21] Witczak ZJ (1994) In: Witczak ZJ (ed) Levoglucosenone and Levoglucosans. Chemistry and Applications. ATL Press, Mount Prospect, p 7
[22] Matsumoto K, Ebata T, Koseki K, Okano K, Kawakami H, Matsushita H (1992) Heterocycles **34**: 1935
[23] Mori K, Chuman T, Kato K (1984) Carbohydr Res **129**: 73
[24] Bamba M, Nishikawa T, Isobe M (1996) Tetrahedron Lett **37**: 8199
[25] Forsyth AC, Gould RO, Paton RM, Sadler JH, Watt I (1993) J Chem Soc Perkin Trans 1, 2737
[26] Beakert A, Barberan O, Gervais M, Brion J-D (2000) Tetrahedron Lett **41**: 2903
[27] Horton D (1963) Methods Carbohydr Chem **2**: 433
[28] Černý M (1963) Monatshefte Chem **94**: 290
[29] Fragala J, Černý M, Stanek J (1975) Collect Czech Chem Commun **40**: 1411
[30] Joseph B, Rollin P (1993) J Carbohydr Chem **12**: 719
[31] Okano K, Ebata T, Koseki K, Kawakami H, Matsumoto K, Matsushita H (1993) Chem Pharm Bull **41**: 861
[32] Zotola M, Venkateswara B, Fraser-Reid B (1991) J Chem Soc Chem Commun 969
[33] Geyer A, Hummel G, Eisele T, Reinehardt S, Schmidt RR (1996) Chem Eur J **2**: 981
[34] Espinosa JF, Cañada FJ, Asensio JL, Dietrich H, Martin-Lomas M, Schmidt RR, Jiménez-Barbero J (1996) Angew Chem Int Ed Engl **35**: 303

Received November 11, 2001. Accepted November 22, 2001

Synthesis of Sulfur-Containing Analogues of *α*Gal*NAc (Tn-Antigen) and *β*Gal*1,3*α*Gal*NAc (T-Antigen)

Irmgard Wenzl[1], **Norbert Neuwirth**[1], **Alexander G. Hedenetz**[1], **Christian Fiedler**[1], **Hansjörg Streicher**[2], **Frank M. Unger**[3], and **Walther Schmid**[1,*]

[1] Institut für Organische Chemie, Universität Wien, A-1090 Wien, Austria

[2] Fakultät für Chemie, Universität Konstanz, D-78457 Konstanz, Germany

[3] Zentrum für Ultrastrukturforschung, Universität für Bodenkultur, A-1180 Wien, Austria

Summary. A method for the preparation of thio-analogues of the T- and Tn-antigen was developed. Thus, starting from a known N-acetamido-glucoside derivative, the epidithio analogue of the Tn-antigen was accessible in a four-step reaction sequence. The corresponding epidithio analogue and the thioanhydro derivative of the T-antigen were synthesized starting from a disaccharide derivative. For the preparation of the epidithio analogue the sulfur atoms were introduced *via* thiocyanates in a stepwise fashion, using mesylate as the leaving group at C-6 and triflate as the leaving group at C-4 in the reducing carbohydrate moiety. The synthesis of the thioanhydro analogue was achieved by introducing a thiocyanate group at C-6 into the glucose moiety, followed by subsequent displacement of a mesylate group at C-4 under inversion of configuration utilizing sodium methoxide.

Keywords. Carbohydrates; T-Antigen; Tn-Antigen; Thio analogues; Glycosides.

Introduction

The development of synthetic strategies towards sulfur-containing carbohydrates remains a challenging task in organic synthesis [1]. Thiosugar derivatives are produced by the introduction of sulfur in the place of oxygen atoms in defined positions of carbohydrate molecules. Due to specific interactions with enzymes and other proteins they are of interest as enzyme inhibitors or pharmaceutical lead structures [1–3]. Here we report on the synthesis of thio analogues of the Tn-(*α*Gal*NAc) and T-antigen (*β*Gal*1,3*α*Gal*NAc) prepared as the allyl glycosides. The T-antigen, first described by *Thomsen* [4] and *Friedenreich* [5], has been identified to be responsible for erythrocyte agglutination with serum antibodies [6]. Subsequently, this antigen has been discovered on the surface of various carcinoma cells [7, 8]. Therefore, the synthesis of thio analogues of the T- and Tn-antigen is of interest for medicinal chemistry as well as immunological studies. These

* Corresponding author. E-mail: walther.schmid@univie.ac.at

compounds may potentially act as photooxidants [9] and immunomodulators [10]. Furthermore, they should be suitable for studies of various lectin binding sites [11, 12]. Indeed, thiosugar derivatives of the T-antigen have been shown to specifically bind to legume lectins [13].

Results and Discussions

Recently we have reported on the synthesis of the allyl N-acetyl-4,6-epidithio-galactosamine analogue **4** [14], which was prepared following essentially a procedure developed by *Hill et al.* [15]. Here we first present an improved synthesis in terms of overall yield of the target compound **4**. We started the reaction sequence with the well known 2-acetamido-3-O-benzoyl-2-deoxy-α-D-glucopyranoside **1** [16], which upon treatment with 1.1 equivalents of toluenesulfonyl chloride gave the monotosylate **2** (Scheme 1). Esterification of the hydroxyl group on C-4 afforded its corresponding triflate. This was followed by double displacement of the leaving groups by thiocyanate to yield the galacto derivative **3**. Conversion of the dithiocyanate derivative **3** to the 4,6-epidithio analogue **4** was achieved by applying the *Zemplén* saponification procedure, providing **4** in 27% overall yield from **1**.

The synthesis of the epidithio analogue **12** of the T-antigen was performed starting from the known disaccharide **7** [17, 18], which was easily prepared by *Helferich* glycosylation [19] of the saccharide **6** with acetobromogalactose (**5**) [20] (Scheme 2). The sulfur atoms of the target **10** were introduced in a stepwise fashion as thiocyanate groups, a strategy significantly superior over the one-step substitution procedure described above for the Tn-analogue. Thus, mesylating the primary hydroxyl group of the disaccharide **7** yielded the monomesylate **8**. The leaving group at the primary position was subjected to a thiocyanate displacement in dimethoxy ethane (*DME*) at elevated temperature to yield the derivative **9**. Alcohol **9** was transformed to the corresponding triflate **10**, which in a second displacement step using potassium thiocyanate was converted into the dithiocyanato disaccharide derivative **11**. For confirmation of the change in configuration at

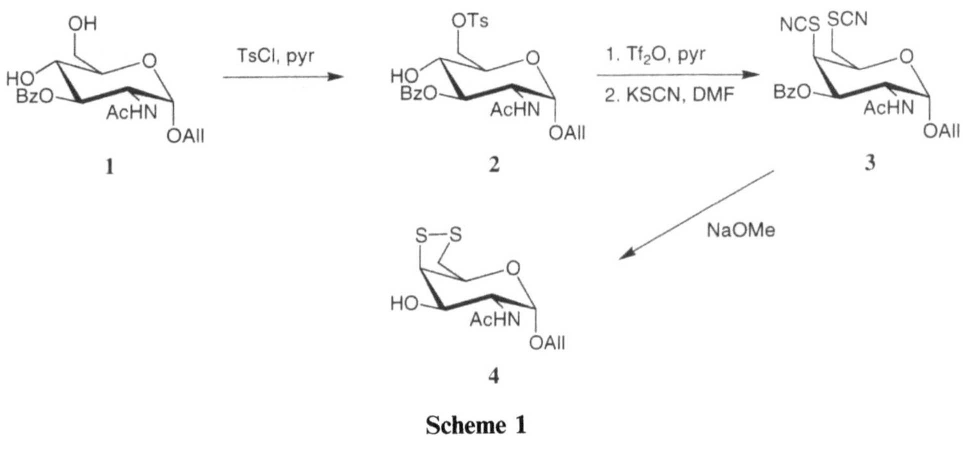

Scheme 1

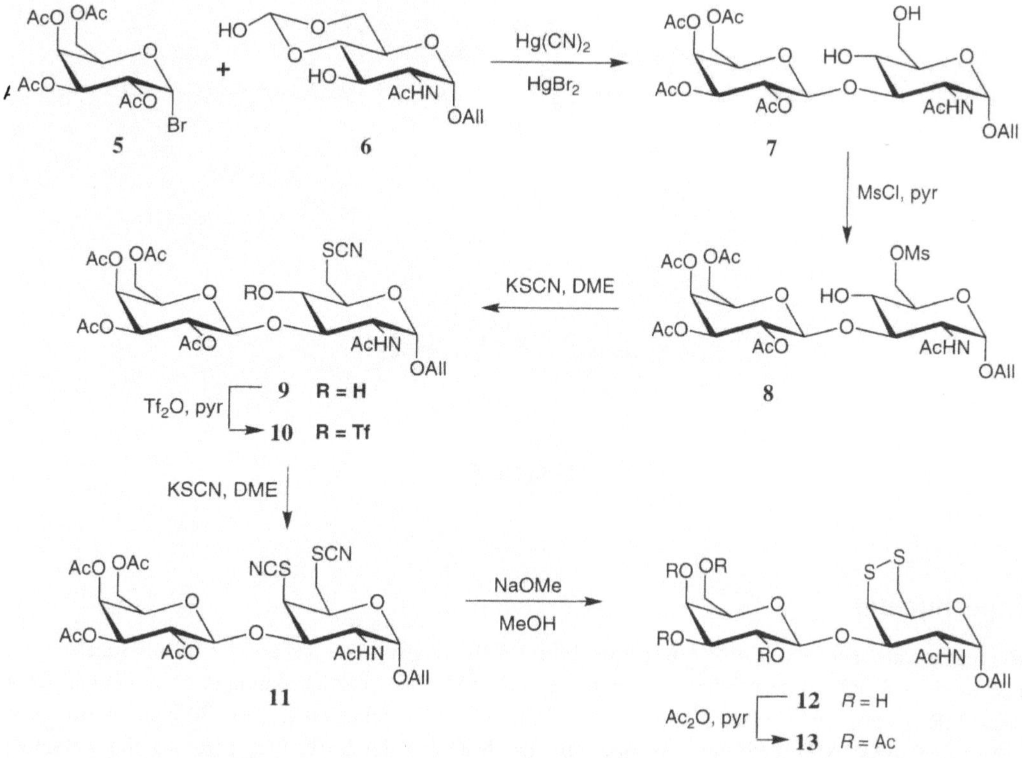

Scheme 2

C-4 of derivative **9**, various two dimensional NMR techniques were employed [21, 22]. Thus, all ^{1}H and ^{13}C signals were assigned, and small coupling constants $J_{3,4}$ and $J_{4,5}$ (4.0 and 1.1 Hz, respectively) indicate the galacto-configuration of the dithiocyanato moiety of **11**. Conversion into the epidithio analogue **12** was achieved by treatment of **11** with sodium methoxide in dry methanol. The presence of the intramolecular disulfide bridge in the T-antigen thio analogue **12** was assigned unambiguously on the basis of NMR and electrospray mass spectroscopy. To facilitate structural assignment, compound **12** was also transformed into its peracetylated derivative **13**.

The synthesis of the 4,6-thioanhydro derivative **16** was achieved starting from disaccharide **7** and applying a slightly modified strategy (Scheme 3). Thus, mesylation of both hydroxyl groups of derivative **7** followed by monosubstitution of the primary mesylate leaving group using potassium thiocyanate gave compound **15**. Treatment of **15** with sodium methoxide in methanol at elevated temperature followed by reacetylation of the crude thioanhydro analogue for simplification of structural assignments afforded compound **16** in moderate yields.

In conclusion, various sulfur-containing analogues of the T- and Tn-antigen were synthesized in a straightforward way. These compounds show selective binding to plant lectins [13] and are interesting candidates for further bioorganic studies. Investigations aiming at the exploration of carbohydrate binding sites of various biomolecules are under way.

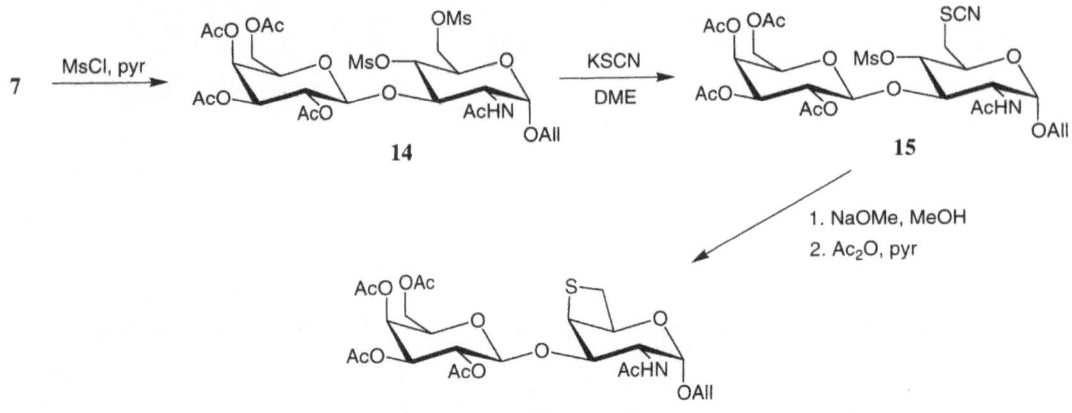

Scheme 3

Experimental

IR spectra were recorded on a Perkin Elmer 1600 FT-IR spectrometer. ^1H and ^{13}C NMR spectra were measured in $CDCl_3$ unless otherwise given using $CHCl_3$ as internal standard on a Bruker DRX 400 NMR spectrometer at 400.13 and 100.61 MHz, or on a Bruker DPX 250 NMR spectrometer at 250.13 and 62.90 MHz. Primed assignments for disaccharide derivatives refer to the galactose moiety, double primed ones to the allyl aglycon, and all others to the *Glc*NAc residues, respectively. Mass spectra were recorded on a Finnigan MAT 900S spectrometer. Electrospray interface mass spectra were measured on a LC-MS/MS PE Sciex API 365 using MeOH (10% NH_4OAc) as solvent. Melting points were determined on a Kofler apparatus and are uncorrected. Optical rotations were measured on a Perkin Elmer 361 polarimeter in a 1 dm cell. TLC was performed on precoated Merck plates (silica F_{254}). Detection was performed using UV light, I_2, and by dipping the plates into a solution of 2% $Ce(SO_4)_2 \cdot 4H_2O$ in 5% H_2SO_4 with subsequent heating. Flash chromatography was performed with Merck silica gel 60 (230–400 mesh). Elemental analyses agreed favourably with the calculated data. All solvents were of reagent grade. Reagents were obtained from commercial suppliers. Compound **1** was prepared according to Ref. [16], compound **5** according to Ref. [20], and compound **6** according to Ref. [23].

Allyl 2-acetamido-3-O-benzoyl-2-deoxy-6-p-toluenesulfonyl-α-D-glucopyranoside
(**2**; $C_{25}H_{29}NO_9S$)

To a stirred solution of 624 mg of **1** (1.71 mmol) in 7 cm^3 dry pyridine and 10 mg of dimethylamino pyridine (*DMAP*), a solution of 358 mg *p*-toluenesulfonic acid (1.88 mmol) in 2 cm^3 dry pyridine was added at 0°C. The reaction mixture was kept at 0°C for 5 days. The solvent was removed by coevaporation with toluene, and the crude product was purified by flash chromatography (gradient: hexanes:ethyl acetate = 1:1 to 1:2) to yield 80% of **2** (12% of starting material were recovered).

TLC (ethyl acetate): $R_f = 0.74$; ^1H NMR (400 MHz, δ, $CDCl_3$): 7.98–7.32 (m, $C_6H_4CH_3$, C_6H_5), 5.84 (m, $H_{2'}$), 5.78 (d, $J_{NH,2} = 8.86$, NHAc), 5.24 (m, H_3), 4.80 (d, $J_{1,2} = 3.45$, H_1), 4.34 (m, $J_{5,6a} = 1.97$, $J_{6a,6b} = 11.32$, H2, H_{6a}, H_{6b}), 4.13 (m, $H_{1'}$), 3.93 (m, $H_{1'}$), 3.88 (ddd, $J_{4,5} = 9.85$, $J_{5,6b} = 4.92$, H_5), 3.78 (dt, $J_{4,OH} = 4.43$, H_4), 3.11 (br s, OH), 2.42 (s, $C_6H_4CH_3$), 1.81 (s, $NHCOCH_3$) ppm; ^{13}C NMR (100 MHz, δ, $CDCl_3$): 169.95, 167.92, 144.97, 133.59, 133.09, 132.82, 129.95, 129.82, 129.03, 128.51, 127.97, 118.32, 96.43, 74.81, 70.69, 70.08, 68.66, 68.57, 68.54, 51.36, 23.10, 21.62 ppm.

Allyl 2-acetamido-3-O-benzoyl-4,6-di-thiocyanato-2,4,6-trideoxy-α-D-galactopyranoside (**3**; $C_{20}H_{21}N_3O_5S_2$)

To a solution of 710 mg (1.37 mmol) of derivative **2** in 18 cm^3 dry pyridine, 0.35 cm^3 (2.08 mmol) of triflic anhydride were added dropwise at 30°C. After 10 min of stirring the reaction flask was transferred to an ice bath, and stirring was continued for 4 h. The reaction was quenched by the addition of 14 cm^3 saturated aq. NaCl solution and 25 cm^3 ethyl acetate. The phases were separated, and the aqueous phase was extracted twice with 25 cm^3 of ethyl acetate. The combined organic phases were dried over $MgSO_4$, filtered, and evaporated to dryness. The crude material obtained was dissolved in 10 cm^3 *DMF*, and 1.33 g of KSCN (13.7 mmol) were added. The reaction mixture was stirred for 16 h at room temperature (r.t.), followed by heating to 110°C for 72 h. The mixture was cooled to r.t., diluted with 10 cm^3 H_2O, and extracted with three portions of 20 cm^3 ethyl acetate. The combined organic phases were dried over $MgSO_4$, filtered, and the solvent was removed under reduced pressure. After flash chromatography (hexanes:ethyl acetate = 1:1), 360 mg of dithiocyanate **3** (59%) were obtained. Additionally, 76 mg of **2** (12%) were recovered.

TLC (ethyl acetate:hexanes = 4:1): $R_f = 0.77$; $[\alpha]_D^{20} = +33.78°$ ($c = 1.19$, $CHCl_3$); IR: $\nu = 3306$, 2941, 2158, 1724, 1666, 1537 cm^{-1}; ^{1}H NMR (400 MHz, δ, $CDCl_3$): 8.08–7.45 (m, C_6H_5), 5.89 (m, $H_{2'}$), 5.77 (d, $J_{NH,2} = 9.35$, NHAc); 5.56 (dd, $J_{2,3} = 11.32$, $J_{3,4} = 3.94$, H_3), 5.33 (m, $H_{3'}$), 4.97 (d, $J_{1,2} = 3.94$, H_1), 4.71 (ddd, H_2), 4.60 (ddd, $J_{4,5} = 1.48$, $J_{5,6b} = 3.94$, $J_{5,6a} = 8.86$, H_5), 4.33 (m, $H_{1'}$), 4.25 (dd, H_4), 4.09 (m, $H_{1'}$), 3.28 (ddd, $J_{6a,6b} = 13.78$, H_{6a}, H_{6b}), 1.89 (s, NHCOCH_3) ppm; ^{13}C NMR (100 MHz, δ, $CDCl_3$): 170.01, 166.31, 134.01, 132.32, 130.20, 128.66, 128.21, 119.56, 110.92, 110.57, 96.65, 69.69, 69.30, 68.05, 55.58, 48.06, 35.72, 23.16 ppm; MS (FI, 7 kV, 3 mA, 140°C): $m/z = 448.3$ (M^+H), 395.2.

Allyl 2-acetamido-4,6-epidithio-2,4,6-trideoxy-α-D-galactopyranoside (**4**; $C_{11}H_{17}NO_4S_2$)

To a stirred solution of 557 mg dithiocyanate **3** (1.24 mmol) in 16 cm^3 dry MeOH, 9 cm^3 of a 0.1 *M* solution of MeONa in dry MeOH were added. The reaction mixture was heated for 10 min to reflux, followed by stirring at r.t. for 1 h. The mixture was neutralized by the addition of Dowex 50 W (H^+). The resin was filtered off, and the filtrate was evaporated to dryness. The crude product obtained was purified by flash chromatography (ethyl acetate:hexanes = 4:1; $R_f = 0.32$) to yield 200 mg (55%) **4** as yellowish crystals.

M.p.: 195°C (decomposition); $[\alpha]_D^{20} = +202.58°$ ($c = 1.24$, CH_3OH); IR: $\nu = 3294.3$, 1648.2, 1549.4 cm^{-1}; ^{1}H NMR (250 MHz, δ, D_2O/CD_3OD): 5.94 (m, $H_{2'}$), 5.26 (m, $H_{3'}$), 4.88 (d, $J_{1,2} = 3.34$, H_1), 4.85 (m, H_5), 4.45 (dd, $J_{2,3} = 10.96$, $J_{3,4} = 4.79$, H_3), 4.35 (dd, H_2), 4.22 (m, $H_{1'}$), 4.10 (dd, $J_{4,5} = 2.74$, H_4), 4.02 (m, $H_{1'}$), 3.43 (dd, $J_{6a,5} = 4.57$, $J_{6a,6e} = 11.88$, H_{6a}), 3.24 (dd, $J_{6b,5} = 1.37$, H_{6b}), 1.99 (s, NHCOCH_3) ppm; ^{13}C NMR (62.9 MHz, δ, D_2O/CD_3OD): 135.37, 117.73, 99.08, 75.82, 69.70, 66.80, 66.76, 51.57, 44.45, 22.61 ppm; MS (FI, 7 kV, 3 mV, 120°C): m/z (rel. intensity) = 291.1 (100, M^+-H), 267.2 (18).

Allyl tetra-O-acetyl-β-D-galactopyranosyl-(1-3)-2-acetamido-2-deoxy-α-D-glucopyranoside (**7**; $C_{25}H_{37}NO_{15}$)

To 80 cm^3 of a mixture of toluene and nitromethane (1:1), 800 mg glucopyranoside **6** (2.29 mmol) were added in one portion and dissolved by heating to reflux. The solution was transferred under Ar into a flask containing 580 mg $Hg(CN)_2$ (2.30 mmol), 83 mg $HgBr_2$ (0.23 mmol), and 2.7 g molecular sieves (3 Å). After stirring for 1 h at 60°C, 940 mg galactosylbromide **5** (2.29 mmol) dissolved in 4 cm^3 nitromethane were added dropwise over a period of 2 h. After stirring for 16 h at 60°C, further additions were made of 580 mg of $Hg(CN)_2$ (2.30 mmol), 83 mg $HgBr_2$ (0.23 mmol), and 940 mg bromide **5** dissolved in 4 cm^3 nitromethane, and heating was continued until TLC (toluene:acetone = 3:1) showed that all starting material **6** had been consumed (2 h). The suspension obtained was filtered over Celite®, and the solvent was removed by evaporation. The residue was redissolved in CH_2Cl_2 (50 cm^3) and washed with a 5% solution of $NaHCO_3$. The organic phase was dried over $MgSO_4$, filtered, and the solvent was removed under vacuum. After flash chromatography (toluene:acetone = 3:1), 1.37 g of benzylidene protected disaccharide were obtained. The product obtained was dissolved in 120 cm^3 of 80% acetic acid and heated to 90°C for 45 min. The solvent was removed by coevaporation with toluene (3 times); the crude product obtained (1.16 g; 86%) was pure enough for further conversions. The physical and spectroscopic data were in full agreement with those reported in Refs. [17] and [18].

Allyl tetra-O-acetyl-β-D-galactopyranosyl-(1-3)-2-acetamido-2-deoxy-6-methanesulfonyl-α-D-glucopyranoside (**8**; $C_{26}H_{39}NO_{17}S$)

To a solution of 460 mg (0.78 mmol) of compound **7** in 6 cm^3 dry pyridine, 0.06 cm^3 (0.78 mmol) methanesulfonyl chloride were added dropwise at 0°C. After 2 h of stirring the solvent was removed by coevaporation with toluene, and the crude product obtained was purified by flash chromatography (ethyl acetate:hexanes = 9:1) to yield 461 mg of **8** (88%).

M.p.: 92°C; $[\alpha]_D^{20} = +56.2°$ ($c = 0.87, CH_2Cl_2$); IR: ν = 3468.1, 2938.2, 1751.9, 1654.7 cm^{-1}; ^{1}H NMR (400 MHz, δ, $CDCl_3$): 5.87 (m, $H_{2''}$), 5.53 (d, $J_{NH,2}$ = 9.85, NHAc), 5.35 (d, $J_{3',4'}$ = 2.95, $H_{4'}$), 5.27 (m, $H_{3''}$), 5.20 (dd, $J_{1',2'}$ = 7.88, $J_{2',3'}$ = 10.34, $H_{2'}$), 4.97 (dd, $J_{3',4'}$ = 3.44, $H_{3'}$), 4.80 (d, J = 3.94, H_1), 4.52 (d, $H_{1'}$), 4.50 (dd, J = 1.48), 4.38 (dd, J = 5.42, J = 10.83), 4.28 (ddd, $J_{2,3}$ = 9.85, H_2), 4.13 (m, $H_{1''}$), 3.97 (m, $H_{1''}$), 3.80 (m, 1H), 3.71 (s, 1H, OH), 3.56 (m, $H_{6'}$), 3.03 (s, SO_2CH_3), 2.14, 2.06, 2.04, 1.99, 1.96 (5s, $COCH_3$) ppm; ^{13}C NMR (100 MHz, δ, $CDCl_3$): 170.40, 170.10, 169.40, 169.34, 133.02, 118.78, 101.58, 96.62, 82.72, 71.08, 70.66, 69.81, 68.62, 68.58, 68.51, 68.32, 66.79, 61.23, 51.17, 37.56, 23.40, 20.63, 20.59, 20.56, 20.48 ppm; MS (FI, 7 kV, 3 mA, 230°C): m/z (rel. intensity) = 670.2 (82), 612.0 (18), 347.0 (20), 331.0 (36), 166.9 (36), 95.8 (63), 59.9 (100).

Allyl tetra-O-acetyl-β-D-galactopyranosyl-(1-3)-2-acetamido-2,6-dideoxy-6-thiocyanato-α-D-glucopyranoside (**9**; $C_{26}H_{36}N_2O_{14}S$)

A suspension of 445 mg (0.66 mmol) of monomesylate **8** and 276 mg (2.84 mmol) of KSCN in 10 cm^3 1,2-dimethoxyethane (*DME*) was heated at 70°C for 12 h. Additional 280 mg (2.88 mmol) of KSCN and 10 cm^3 *DME* were added, and heating at reflux was continued for 20 h. The solvent was removed under vacuum, and the residue was dissolved in CH_2Cl_2 (35 cm^3) and washed 3 times with 20 cm^3 portions of H_2O. The organic layer was dried over $MgSO_4$, filtered, and the solvent was evaporated. The crude product obtained was purified by flash chromatography (toluene:acetone = 3:1) to yield 381 mg (91%) of **9** as colorless crystals.

M.p.: 177°C; $[\alpha]_D^{20} = +76.6°$ ($c = 1.23, CH_2Cl_2$); IR: ν = 3468.0, 2933.9, 2156.4, 1752.6, 1658.6 cm^{-1}; ^{1}H NMR (400 MHz, δ, $CDCl_3$): 5.89 (m, $H_{2''}$), 5.53 (d, $J_{NH,2}$ = 9.53, NHAc), 5.35 (dd, $J_{4',5'}$ = 1.00, $J_{3',4'}$ = 3.51, $H_{4'}$), 5.28 (m, $H_{3''}$), 5.20 (dd, $J_{1',2'}$ = 8.03, $J_{2',3'}$ = 10.54, $H_{2'}$), 4.97 (dd, $H_{3'}$), 4.79 (d, $J_{1,2}$ = 4.02, H_1), 4.51 (d, $H_{1'}$), 4.30 (ddd, $J_{2,3}$ = 10.04, H_2), 4.25 (m, $H_{1''}$), 4.12 (m, $H^*_{6'}$), 4.00 (m, $H^*_{5'}$, $H_{1''}$), 3.84 (ddd, $J_{5,6a}$ = 2.51, H_5), 3.73 (s, OH), 3.56 (dd, $J_{3,4}$ = 8.53, H^*_3), 3.51 (dd, $J_{6a,6b}$ = 14.06, H_{6a}), 3.43 (t, J = 9.03, H_4), 3.01 (dd, $J_{5,6b}$ = 8.53, H_{6b}), 2.13, 2.06, 2.05, 2.00, 1.95 (5s, $COCH_3$) ppm; ^{13}C NMR (100 MHz, δ, $CDCl_3$): 170.41, 170.09, 169.42, 169.30, 132.87, 119.10, 112.17, 101.61, 96.41, 82.60, 71.21, 71.16, 70.65, 70.45, 68.61, 68.45, 66.82, 61.23, 51.18, 35.95, 23.42, 20.62, 20.59, 20.47 ppm; assignments marked with an asterisk may be interchanged.

Allyl tetra-O-acetyl-β-D-galactopyranosyl-(1-3)-2-acetamido-2,6-dideoxy-6-thiocyanato-4-trifluoromethanesulfonyl-α-D-glucopyranoside (**10**; $C_{27}H_{35}F_3N_2O_{16}S_2$)

To a solution of 237 mg (0.37 mmol) of **9** in 6 cm^3 dry pyridine, 0.11 cm^3 (0.65 mmol) trifluoromethanesulfonic anhydride were added dropwise at −15°C. After 15 min the temperature was raised to 0°C, and stirring was continued for 3 h. The reaction mixture was diluted with 30 cm^3 CH_2Cl_2 and washed twice with saturated aqueous $NaHCO_3$. The organic layer was dried over $MgSO_4$, filtered, and evaporated to dryness. The crude product obtained could be purified by flash chromatography (hexanes:ethyl acetate = 1:2) to yield 178 mg (63%) of **10** as colorless crystals.

M.p.: 139–140°C; $[\alpha]_D^{20} = +45.67°$ ($c = 0.9, CH_2Cl_2$); IR: ν = 3369.2, 2938.0, 2158.8, 1750.1, 1683.2 cm^{-1}; ^{1}H NMR (400 MHz, δ, $CDCl_3$): 5.91 (m, $H_{2''}$), 5.59 (d, $J_{NH,2}$ = 10.04, NHAc), 5.35 (m, $H_{4'}$, $H_{3''}$), 5.10 (dd, $J_{1',2'}$ = 8.03, $J_{2',3'}$ = 10.54, $H_{2'}$), 4.92 (dd, $J_{3',4'}$ = 3.51, $H_{3'}$), 4.80 (d, $J_{1,2}$ = 4.02, H_1), 4.64 (d, $H_{1'}$), 4.57 (t, J – 9.53, H_5), 4.47 (ddd, $J_{2,3}$ = 10.54, H_2), 4.30 (m, $H_{1''}$), 4.20 (dd, J = 6.03, J = 11.55), 4.15 (dt, $J_{5a,6}$ = 2.51, H_5), 4.05 (m, 3 H), 3.85 (t, J = 7.03), 3.40 (dd, $J_{6a,6b}$ = 14.06, H_{6a}), 3.07 (dd, $J_{5,6b}$ = 9.04, H_{6b}), 2.10, 2.05, 2.03, 1.95 (4s, $COCH_3$) ppm; ^{13}C NMR (100 MHz, δ, $CDCl_3$): 170.12, 169.61, 169.27, 132.24, 120.00, 111.31, 100.41, 96.03, 83.16, 72.82, 70.82, 70.45, 69.29, 68.35, 68.27, 66.52, 60.69, 52.40, 35.12, 23.36, 20.65, 20.52, 20.34 ppm.

Allyl tetra-O-acetyl-β-D-galactopyranosyl-(1-3)-2-acetamido-4,6-di-thiocyanato-2,4,6-trideoxy-α-D-galactopyranoside (**11**; $C_{27}H_{35}N_3O_{13}S_2$)

A suspension of 73 mg (0.095 mmol) of **10** and 46 mg (0.473 mmol) of KSCN in 10 cm^3 *DME* was heated to reflux for 12 h. Then the solvent was removed under vacuum. The crude material obtained was redissolved in 30 cm^3 of CH_2Cl_2 and washed 3 times with 20 cm^3 portions of H_2O. The organic layer was dried over $MgSO_4$, filtered, and evaporated. After purification by flash chromatography (toluene:acetone = 3:1), 45 mg (83%) of **11** were obtained as colorless crystals.

M.p.: 142°C; $[\alpha]_D^{20} = +86.30°$ ($c = 0.52, CH_2Cl_2$); IR: ν = 3306.4, 2936.2, 2157.6, 1750.3, 1659.9 cm^{-1}; ^{1}H NMR (600 MHz, δ, $CDCl_3$): 5.91 (m, $H_{2''}$), 5.58 (d, $J_{NH,2}$ = 9.5, NHAc), 5.37 (dd, $J_{4',5'}$ = 0.8, $J_{3',4'}$ = 3.3, $H_{4'}$), 5.34 (m, $H_{3''}$), 5.32 (dd, $J_{1',2'}$ = 8.4, $J_{2',3'}$ = 10.4, $H_{2'}$), 5.01 (dd, $H_{3'}$), 4.90 (d, $J_{1,2}$ = 3.8, H_1), 4.74 (d, $H_{1'}$), 4.41 (m, $J_{4,5}$ = 1.1, $J_{5,6a}$ = 3.9, $J_{5,6b}$ = 8.8, H_5), 4.38 (ddd, $J_{2,3}$ = 9.7, H_2), 4.30 (m, $J_{6a',6b'}$ = 11.5, $H_{6a'}$, $H_{1''}$), 4.18 (dd, $J_{4,5}$ = 1.1, $J_{3,4}$ = 4.0, H_4), 4.14 (dd, $J_{5',6b'}$ = 5.6, $H_{6b'}$), 4.06 (m, H_3, $H_{1''}$), 3.89 (m, $H_{5'}$), 3.29 (m, $J_{6a,6b}$ = 14.4, H_{6a}, H_{6b}), 2.15 (s, $COCH_3$), 2.02 (s, 2 × $COCH_3$), 2.00, 1.98 (2s, $COCH_3$) ppm; ^{13}C NMR (150 MHz, δ, $CDCl_3$): 170.63, 170.58, 170.31, 169.37, 169.33, 132.55 (C2″), 119.72 (C3″), 111.29 (–SCN), 110.94 (–SCN), 101.91 (C1′), 96.70 (C1), 74.96 (C3), 71.80 (C5′), 70.65 (C3′), 69.29 (C1″), 68.90 (C2′), 68.03 (C5), 67.17 (C4′), 61.86 (C6′), 57.75 (C4), 48.55 (C2), 35.87 (C6), 23.37, 20.93, 20.74, 20.67, 20.64 ppm; MS (FI, 7 kV, 3 mA, 190°C): m/z (rel. intensity) = 674.5 (78), 621.6 (100), 589.6 (19).

Allyl β-D-galactopyranosyl-(1-3)-2-acetamido-4,6-epidithio-2,4,6-trideoxy-α-D-galactopyranoside (**12**; $C_{17}H_{27}NO_9S_2$)

To a solution of 101 mg (0.15 mmol) of dithiocyanate derivative **11** in 50 cm^3 dry MeOH, 0.45 cm^3 of a 1 *M* methanolic MeONa solution were added. After 4 h of stirring the mixture was neutralized by addition of solid CO_2. The solvent was concentrated under vacuum to a volume of 2 cm^3, filtered over silica gel (2 g), and the silica gel layer was eluted with 10 cm^3 of MeOH. The product was obtained by removing the solvent under vacuum as a colorless glass.

Yield: 54 mg (79%); IR: ν = 3319.5, 2948.3, 2162.7, 1629.0, 1538.1 cm^{-1}; ^{1}H NMR (400 MHz, δ, D_2O): 6.03 (m, $H_{2''}$), 5.37 (m, $H_{3''}$), 5.09 (s, H_5), 5.01 (d, $J_{1,2}$ = 3.01, H_1), 4.59 (m, $J_{2,3}$ = 10.54, H_2, H_3, H_4), 4.48 (d, $J_{1',2'}$ = 8.0, $H_{1'}$), 4.31 (m, $H_{1''}$), 4.13 (m, $H_{1''}$), 3.95 (d, $J_{3',4'}$ = 3.51, $H_{4'}$), 3.81 (m, $H_{6'}$), 3.71 (dd, $J_{5,6a}$ = 4.51, $J_{5,6b}$ = 7.53, $H_{5'}$), 3.65 (dd, $J_{2',3'}$ = 10.03, $H_{3'}$), 3.52 (m, $H_{2'}$, H_{6a}), 3.42 (dd, $J_{5,6b}$ = 1.01, $J_{6a,6b}$ = 12.55, H_{6b}), 2.06 (s, $NHCOCH_3$) ppm; ^{13}C NMR (100 MHz, δ, D_2O): 175.13, 133.91 (C2″), 118.58 (C3″), 105.43 (C1′), 97.80 (C1), 75.54 (C5′), 75.29 (C3′), 75.11 (C5), 72.95, 70.93 (C2′), 69.24 (C1″), 69.01 (C4′), 63.06 (C4), 61.37 (C6′), 48.81 (C2), 43.70 (C6), 22.38 ppm; ES-MS (negative mode): m/z = 452.0 (calcd.: 452.5; M-H$^-$).

Allyl tetra-O-acetyl-β-D-galactopyranosyl-(1-3)-2-acetamido-4,6-epidithio-2,4,6-trideoxy-α-D-galactopyranoside (**13**; $C_{25}H_{35}NO_{13}S_2$)

To a solution of 50 mg (0.11 mmol) of **12** in 4 cm^3 dry pyridine, 0.4 cm^3 acetic anhydride were added, and the reaction mixture was stirred at r.t. for 16 h. The solvents were removed under vacuum, and the crude product was purified by flash chromatography (ethyl acetate) to yield 65 mg (96%) of **13**.

M.p.: 223–224°C; $[\alpha]_D^{20} = +87.65°$ ($c = 0.85$, $CHCl_3$); IR: $\nu = 3306.8$, 2931.0, 1751.1, 1654.2 cm^{-1}; ^1H NMR (400 MHz, δ, $CDCl_3$): 5.84 (m, $H_{1''}$), 5.51 (d, $J_{NH,2} = 9.03$, NHAc), 5.31 (d, $H_{4'}$), 5.22 (m, $H_{3''}$), 5.08 (dd, $J_{1',2'} = 8.03$, $J_{2',3'} = 10.54$, $H_{2'}$), 4.91 (dd, $J_{3',4'} = 3.51$, $H_{3'}$), 4.87 (d, $J_{1,2} = 3.52$, H_1), 4.72 (m, H_5^*), 4.58 (m, H_2, $H_{1'}$), 4.27 (dd, $J = 5.02$, $J = 10.5$, H_3^*), 4.14 (m, $H_{6a'}^*$, $H_{1''}$), 4.05 (m, H_4^*, $H_{6b'}^*$), 3.96 (m, $H_{1''}$), 3.83 (dt, $J = 6.53$, $H_{5'}^*$), 3.24 (m, H_{6a}, H_{6b}), 2.11, 2.01, 1.98, 1.93, 1.90 (5s, $COCH_3$) ppm; ^{13}C NMR (100 MHz, δ, $CDCl_3$): 170.35, 170.24, 170.15, 169.51, 169.36, 133.33, 118.34, 101.45, 98.04, 74.68, 73.34, 70.88, 70.73, 68.74, 68.50, 66.86, 62.88, 61.18, 48.23, 43.51, 23.40, 20.66, 20.64, 20.61, 20.50 ppm; MS (FI, 7 kV, 3 mA, 130°C): m/z (rel. intensity) = 621.36 (100; M^+), 530.63 (18); assignments marked with an asterisk may be interchanged.

Allyl tetra-O-acetyl-β-D-galactopyranosyl-(1-3)-2-acetamido-2-deoxy-4,6-dimethanesulfonyl-α-D-glucopyranoside (**14**; $C_{27}H_{41}NO_{19}S_2$)

To a solution of 15 mg (0.144 mmol) of disaccharide **7** in 3 cm^3 dry pyridine, 0.06 cm^3 (0.775 mmol) methanesulfonyl chloride and 10 mg *DMAP* were added at 0°C. After stirring for 20 min the reaction mixture was allowed to warm to r.t. Stirring was continued for 6 h; then the reaction was quenched by addition of 3 cm^3 H_2O. The solvents were removed by coevaporation with toluene. The residue was redissolved in 10 cm^3 of H_2O and extracted 3 times with 10 cm^3 portions of ethyl acetate. The combined organic phases were dried over $MgSO_4$, filtered, and evaporated to dryness. The dimesylated derivative **14** was obtained after flash chromatography (ethyl acetate:hexanes = 5:1).

Yield: 90 mg (84%); colorless glass; $[\alpha]_D^{20} = +45.45°$ ($c = 0.77$, $CHCl_3$); IR: $\nu = 3386.9$, 2940.0, 1752.6, 1680.0 cm^{-1}; ^1H NMR (400 MHz, δ, $CDCl_3$): 5.88 (m, $H_{2''}$), 5.60 (d, $J_{NH,2} = 10.04$, NHAc), 5.36 (d, $H_{4'}$), 5.30 (m, $H_{3''}$), 5.10 (dd, $J_{1',2'} = 8.03$, $J_{2',3'} = 10.54$, $H_{2'}$), 4.93 (dd, $J_{3',4'} = 3.51$, $H_{3'}$), 4.75 (d, $J_{1,2} = 3.51$, H_1), 4.64 (d, $H_{1'}$), 4.46 (dd, $J = 2.51$, $J = 12.04$), 4.37 (dt, $J_{2,3} = 10.54$, H_2), 4.32 (m, 2 H), 4.20 (dd, $J = 6.03$, $J = 11.55$), 4.14 (m, H1″), 3.98 (m, 4 H), 3.85 (t, $J = 7.02$), 3.22, 3.04 (2s, SO_2CH_3), 2.10, 2.09, 2.04, 2.02, 1.95 (5s, $COCH_3$) ppm; ^{13}C NMR (100 MHz, δ, $CDCl_3$): 170.37, 170.05, 170.01, 169.56, 169.29, 132.65, 119.46, 100.86, 96.36, 76.54, 74.63, 70.53, 70.40, 68.96, 68.13, 67.92, 67.63, 66.88, 60.61, 52.15, 38.69, 37.58, 23.38, 20.62, 20.55, 20.47 ppm; MS (FI, 7 kV, 3 mA, 175°C): m/z (rel. intensity) = 748.4 (79; M^+), 712.7 (64), 687.4 (100), 651.4 (30).

Allyl tetra-O-acetyl-β-D-galactopyranosyl-(1-3)-2-acetamido-2,6-dideoxy-4-methanesulfonyl-6-thiocyanato-α-D-glucopyranoside (**15**; $C_{27}H_{38}N_2O_{16}S_2$)

A suspension of 64 mg (0.086 mmol) of **14** and 82 mg (0.84 mmol) of KSCN in 10 cm^3 *DME* was heated to 100°C for 15 h. Subsequently, the solvent was removed under vacuum. The residue was redissolved in 10 cm^3 H_2O and extracted 3 times with 15 cm^3 portions of ethyl acetate. The combined organic phases were dried over $MgSO_4$, filtered, and evaporated to dryness. The crude product obtained was purified by flash chromatography (ethyl acetate:hexanes = 5:1) to yield 50 mg (82%) of **15** as a colorless glass.

$[\alpha]_D^{20} = +68.67°$ ($c = 1.20$, $CHCl_3$); IR: $\nu = 3350.7$, 3019.8, 2938.1, 2157.6, 1753.8, 1681.8 cm^{-1}; ^1H NMR (400 MHz, δ, $CDCl_3$): 5.90 (m, $H_{2''}$), 5.60 (d, $J_{NH,2} = 10.04$, NHAc), 5.33 (m, $H_{4'}$, $H_{3''}$), 5.01 (dd, $J_{1',2'} = 8.04$, $J_{2',3'} = 10.54$, $H_{2'}$), 4.93 (dd, $J_{3',4'} = 3.51$, $H_{3'}$), 4.79 (d, $J_{1,2} = 4.02$, H_1), 4.62 (d, $H_{1'}$), 4.44 (dt, $J_{2,3} = 10.54$, H_2), 4.27 (m, 3 H), 4.05 (m, 3 H), 4.00 (dd, $J_{3,4} = 9.53$, H_3), 3.89 (t, 1H), 3.47 (dd, $J_{5,6a} = 2.51$, $J_{6a,6b} = 14.06$, H_{6a}), 3.22 (s, SO_2CH_3), 3.10 (dd,

$J_{5,6b} = 8.03$, H_{6b}), 2.10, 2.09, 2.05, 2.02, 1.95 (5s, $COCH_3$) ppm; ^{13}C NMR (62.9 MHz, δ, $CDCl_3$): 170.04, 169.53, 169.29, 132.43 ($C_{2''}$), 119.75 ($C_{3''}$), 112.03 (SCN), 100.91 ($C_{1'}$), 96.20 (C_1), 78.62, 74.54, 70.47, 70.39, 69.02 ($C_{1''}$), 68.90, 67.87, 66.86, 60.59 ($C_{6'}$), 52.17 (C_2), 38.56 (SO_2CH_3), 35.23 (C_6), 23.38, 20.64, 20.55, 20.47 ppm; MS (FI, 7 kV, 3 mA): m/z (rel. intensity) = 711.3 (48, M^+), 650.3 (100).

Allyl tetra-O-acetyl-β-D-galactopyranosyl-(1-3)-2-acetamido-4,6-thioanhydro-2,4,6-trideoxy-α-D-galactopyranoside (**16**; $C_{25}H_{35}NO_{13}S$)

To a solution of 27 mg (0.038 mmol) of **15** in 3 cm^3 dry MeOH, 0.08 cm^3 of a 1 *M* methanolic MeONa solution were added at r.t. The reaction mixture was heated at reflux for 2 h and subsequently neutralized by addition of Dowex 50 W (H^+). The resin was filtered off, and the solvent was removed under vacuum. The crude product obtained was redissolved in 2 cm^3 dry pyridine, and 0.5 cm^3 acetic anhydride as well as 5 mg *DMAP* were added. The reaction mixture was stirred for 20 h at r.t. The solvents were removed by coevaporation with toluene, and the product obtained was purified by flash chromatography (ethyl acetate:hexanes = 6:1).

Yield: 8 mg (36%); IR: ν = 3308.0, 2938.0, 1750.1, 1654.2, 1544.2 cm^{-1}; ^{1}H NMR (400 MHz, δ, $CDCl_3$): 5.84 (m, $H_{2''}$), 5.52 (d, $J_{NH,2} = 8.95$, NHAc), 5.34 (dd, $J_{3',4'} = 3.53$, $J_{4',5'} = 0.95$, $H_{4'}$), 5.23 (m, $H_{3''}$), 5.06 (dd, $J_{1',2'} = 8.01$, $J_{2',3'} = 10.36$, $H_{2'}$), 4.99 (d, $J_{1,2} = 2.83$, H_1), 4.94 (dd, $H_{3'}$), 4.72 (ddd, $J_{2,3} = 10.13$, H_2), 4.59 (d, $H_{1'}$), 4.56 (t, $J_{5,6a} = 4.71$, H_5), 4.40 (t, H_4^*), 4.14 (m, $H_{6'}^*$, $H_{1''}$), 3.97 (m, $H_{1''}$), 3.91 (dd, $J = 6.12$, H_3^*), 3.85 (dt, $H_{5'}$), 3.42 (dd, H_{6a}), 2.59 (d, $J_{6a,6b} = 10.12$, H_{6b}), 2.19, 2.10, 2.08, 2.03, 1.99 (5s, $COCH_3$) ppm; ^{13}C NMR (100 MHz, δ, $CDCl_3$): 170.37, 170.27, 170.18, 169.54, 133.54 ($C_{2''}$), 117.94 ($C_{3''}$), 100.95 ($C_{1'}$), 97.21 (C_1), 72.96, 72.04, 70.86, 70.84, 68.63 ($C_{1''}$), 68.55, 66.92, 61.27 ($C_{6'}$), 49.54, 43.45, 28.48 (C_6), 23.45, 20.68, 20.60, 20.55 ppm; MS (FI, 7 kV, 3 mA, 120–170°C): m/z (rel. intensity) = 590.2 (100, M^+), 488.3 (10), 142.1 (20); assignments marked with an asterisk may be interchanged.

Acknowledgements

This work was supported by the *Jubiläumsfonds der Oesterreichischen Nationalbank* (Project Nos. 6011 to *F. M. Unger* and 4663 to *W. Schmid*).

References

[1] Witzak ZJ (1999) Current Medicinal Chemistry **6**: 163
[2] Defaye J, Gelas J (1991) In: Atta-ur-Rahman (ed) Studies in Natural Product Chemistry, vol 8E. Elsevier, Amsterdam, p 315
[3] Driguez H (1997) Top Curr Chem **187**: 85
[4] Thomsen O (1927) Z Immun Forsch **52**: 85
[5] Friedenreich V (1930) The Thomsen Hemagglutination Phenomenon. Levin & Munksgaard, Copenhagen
[6] Vaith P, Uhlenbruck G (1978) Z Immun Forsch **154**: 1
[7] Springer GF (1984) Science **224**: 1198
[8] Springer GF, Taylor CR, Howard DR, Tegtmeyer H, Desai PR, Murthy SM, Felder B, Scanlon EF (1985) Cancer **55**: 561
[9] Prompers JJ, Hilbers CW, Pepermans HAM (1999) FEBS Lett **456**: 409
[10] Unger FM, Liehl E (1985) German Patent Appl DE 3438585, A1, May 2
[11] Sharma V, Vigayon M, Surolia A (1996) J Biol Chem **271**: 21209
[12] Surolia A, Sharon N, Schwarz FP (1996) J Biol Chem **271**: 17697
[13] Streicher H, Schmid W, Wenzl I, Fiedler C, Kählig H, Unger FM (2000) Bioorg Med Chem Lett **10**: 1369

[14] Wenzl I, Kählig H, Unger FM, Schmid W (1999) Monatsh Chem **130**: 1137
[15] Hill J, Hough L, Richardson AC (1968) Carbohydr Res **8**: 19
[16] Jeanloz RW (1954) J Am Chem Soc **76**: 555
[17] Lee RT, Lee YC (1994) Carbohydr Res **251**: 69
[18] Gambert U, Lio RG, Farkas E, Thiem J, Bencomo VV, Liptak A (1997) Bioorg Med Chem **5**: 1285
[19] Helferich B, Zirner J (1962) Chem Ber **34**: 957
[20] Jaenloz RW, Stoffyn PJ (1962) Methods Carbohydr Chem **1**: 224
[21] Kessler H, Gehrke M, Griesinger C (1988) Angew Chem Int Ed Engl **27**: 490
[22] Rance M, Sörensen OW, Bodenhausen G, Wagner G, Ernst RR, Wüthrich K (1983) Biochem Biophys Res Commun 479
[23] Warren CD, Jeanloz RW (1977) Carbohydr Res **53**: 67

Received October 16, 2001. Accepted November 7, 2001

The Chameleon of Retaining Glycoside Hydrolases and Retaining Glycosyl Transferases: The Catalytic Nucleophile

Robert V. Stick* and **Andrew G. Watts**

Department of Chemistry, The University of Western Australia, Crawley WA, 6009 Australia

Summary. This paper addresses the existence and role of a catalytic nucleophile in retaining glycoside hydrolases and retaining glycosyl transferases. Although the former has now been established beyond doubt, such is not the case with the latter. We report reliable procedures for the synthesis of various 2-deoxy-2-fluoro glycosyl nucleoside diphosphates, useful donor analogues for the study of the mechanism of action of retaining glycosyl transferases.

Keywords. Glycoside hydrolases; Glycosyl transferases; 2-Deoxy-2-fluoro hexopyranoses.

Introduction

The burgeoning field of glycobiology arose from the recognition that carbohydrates play a pivotal role in an overwhelming number of biological processes [1–6]. Carbohydrates are now known to be of key importance in cell adhesion, infection, differentiation, and regulation as well as in many intercellular communication and signal transduction events [7]. In order to gain a deeper understanding of these events, it was natural for chemists, biochemists, and molecular biologists to probe the mechanism of action of the enzymes that process carbohydrates, mainly the glycoside hydrolases and glycosyl transferases.

This article, therefore, will first address some recent efforts with glycoside hydrolases, and then turn to the more vexed question regarding the mechanism of action of retaining glycosyl transferases.

Results and Discussion

Glycoside hydrolases

The mode of action of glycoside hydrolases, enzymes that catalyze the hydrolytic cleavage of glycosidic bonds in both simple and complex carbohydrate structures, is thoroughly understood thanks to the early and seminal efforts of *Koshland* [8] and *Legler* [9], and more recently by *Sinnott* [10, 11], *Vasella* [12], *Davies* [13, 14],

* Corresponding author. E-mail: rvs@chem.uwa.edu.au

and *Withers* [15–20]. The glycoside hydrolases have been classified into families based on amino acid sequence homology [13, 21, 22], and most of these families more broadly into clans [23]. The members of a family all exhibit the same mechanism of action, whether they be *exo*- or *endo*-hydrolases and whether they act on similar or unrelated substrates. More recently, glycoside hydrolases have also been classified according to the direction from which the glycosidic oxygen is protonatated during hydrolysis (*syn*- or *anti*-protonators) [24].

Glycoside hydrolases catalyze the cleavage of the glycosidic linkage with either inversion or retention of configuration at the anomeric centre; the well established steps in the two separate processes are shown below (for a β-D-glycoside) [8].

Inversion:

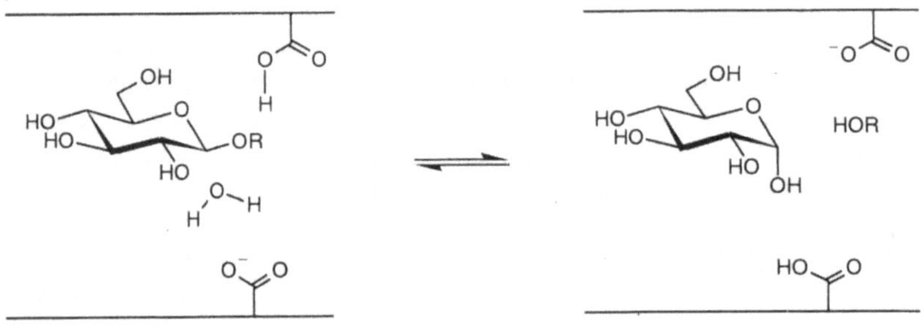

Retention:

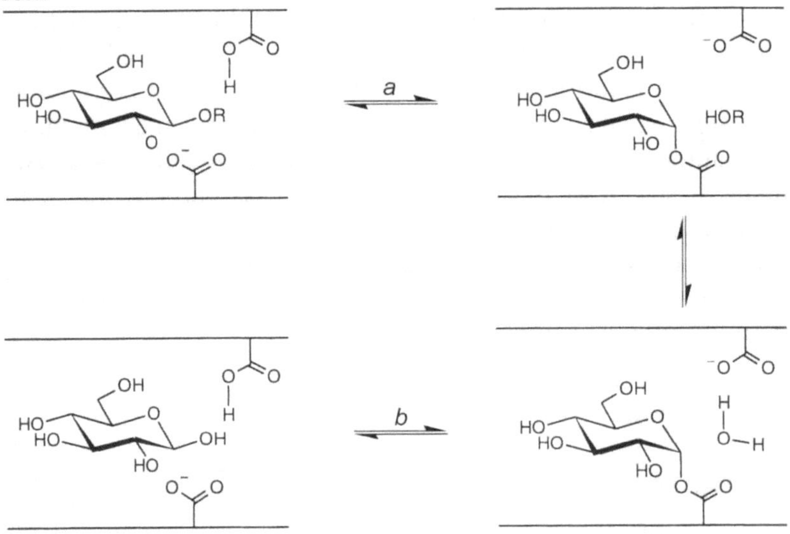

For an inverting glycoside hydrolase, general acid-base catalysis prevails, and there is a need for pK_a modulation of the two relevant carboxylic acid residues. The active site, in order to accommodate the substrate and a molecule of water, is some 10 Å across [13]. The salient points of a retaining glycoside hydrolase are the smaller active site (6 Å), the formation of a true glycosyl-enzyme intermediate through the agency of a catalytic nucleophile (aspartate or glutamate), and the stabilization of the transition states associated with steps *a* and *b* by the hydroxyl group present at C2 [13, 19].

The main (and obvious) difference between an inverting and a retaining hydrolase is the presence of a carboxylic acid residue capable of acting as a catalytic nucleophile. This nucleophile was first detected and, in some instances, identified by the use of affinity labels such as epoxyalkyl glycosides [9] – these molecules somewhat resemble the natural substrate for the enzyme and, additionally, are capable of (sometimes indiscriminate) attachment to a carboxylic residue within the enzyme [25, 26].

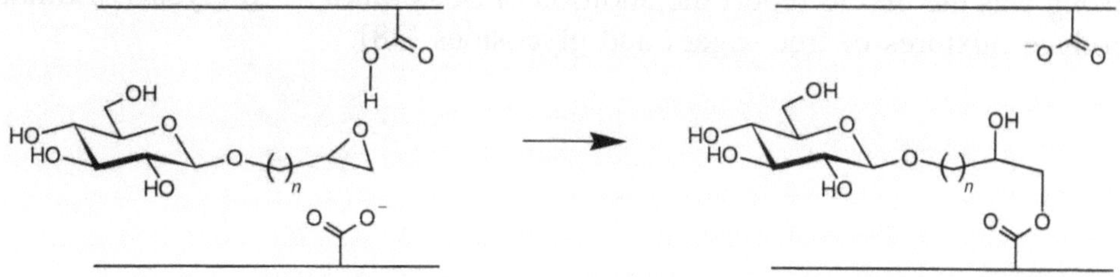

Much more impressive as labels of the catalytic nucleophile are the deoxy fluoro sugars recently developed by *Withers*. A 2-deoxy-2-fluoro sugar, suitably activated at the anomeric carbon with an excellent leaving group, is able to label the catalytic nucleophile specifically [27–29].

The glycosyl-enzyme intermediate is sufficiently long-lived (owing to the stabilizing effect of the fluorine atom) as to be detected and observed by nuclear magnetic resonance, mass spectrometric, or X-ray crystallographic techniques. Nowhere has this approach been put to better use than in the recent labelling of the catalytic nucleophile of hen egg-white lysozyme, long accepted as the paradigm for retaining glycoside hydrolases and now shown to operate in common with other members of the family [30].

For somewhat related reasons, 5-fluoro sugars, also activated at the anomeric carbon, were found to be effective reagents for the labelling of the catalytic nucleophile [18, 31, 32]. The establishment of a mechanism for retaining glycoside hydrolases has led to effective procedures for glycoside synthesis through the action of transglycosylation and the invention of glycosynthases [33, 34].

We were attracted to the synthesis of 2-deoxy-2-fluoro sugars some years ago, but sorely missed the boat in terms of publication. However, we still feel that some comment is warranted here. Although there are various methods available for the synthesis of 2-deoxy-2-fluoro sugars, none are general, and some involve the use of obnoxious reagents such as acetyl hypofluorite or elemental fluorine [35–37]. The whole situation changed with the advent of Selectfluor™, a fluoroammonium salt:

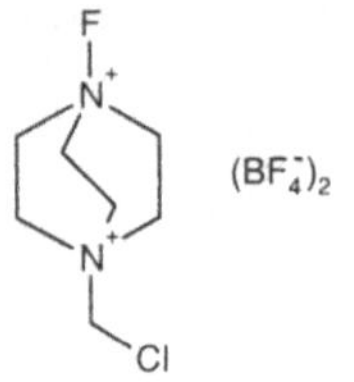

Wong was the first to report the addition of Selectfluor™ to glycals, ultimately to produce mixtures of free sugars and glycosides [38].

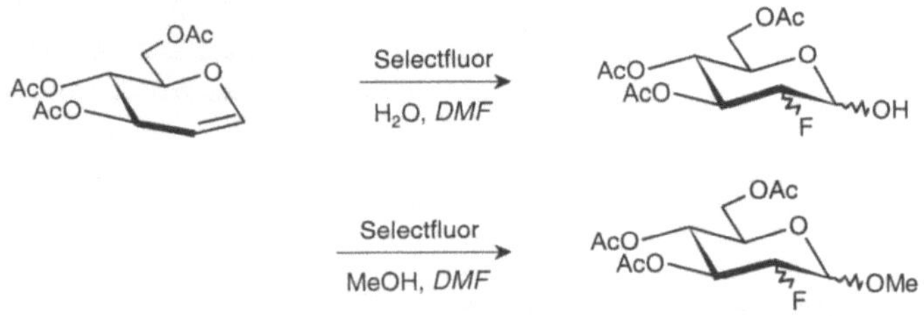

The stereoselectivity of the process can be improved by using bulky protecting groups on the glycal or through the use of a favourable substrate [39, 40].

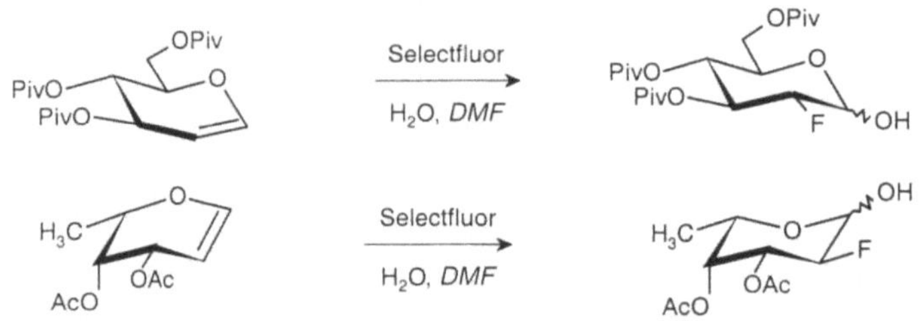

Most of the mechanistic studies involving the addition of Selectfluor™ to glycals have been conducted by *Dax* [39, 41] and *Wong* [40] and, together with these results, we propose a rationalization for the process. The addition, by necessity, is a concerted process, and with a substrate such as tri-O-acetyl-D-glucal, two glycosylammonium salts result [39].

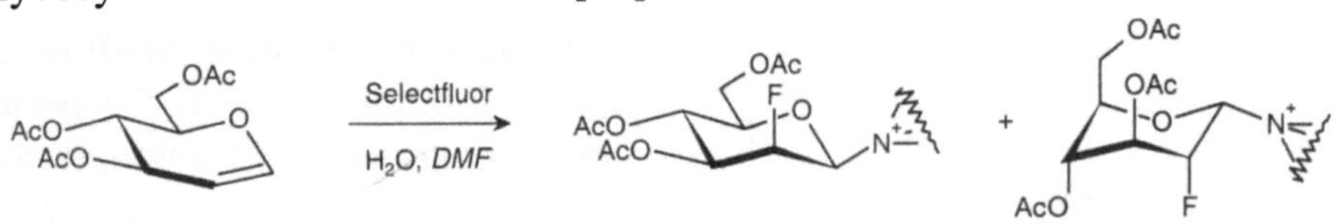

The 2-deoxy-2-fluoro-D-mannopyranosyl salt would be present in the $^4C_1$ conformation, and there is some evidence that the corresponding D-glucopyranosyl salt is also dominated by a reverse anomeric effect and found in the $^1C_4$ conformation [40]. A subsequent hydrolysis, for example, yields the observed mixtures of free sugars. With glycals such as tri-O-pivaloyl-D-glucal and tri-O-acetyl-D-galactal, the addition of Selectfluor™ occurs from the less hindered face; di-O-acetyl-L-fucal is an ideal substrate in several ways.

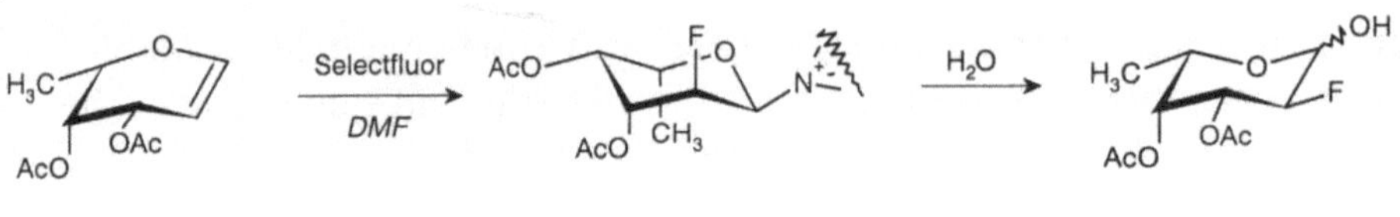

It is also worth mentioning that some glycal derivatives bearing substituents at C2 are found to be unreactive towards Selectfluor™. Tri-O-acetyl-1,2-dideoxy-2-fluoro-*D*-arabino-hex-1-enopyranose, when treated with Selectfluor™ for several days, failed to produce tri-O-acetyl-2-deoxy-2,2-difluoro-*D*-glucopyranose. A similar lack of reactivity has been observed with tri-O-acetyl-1,2-dideoxy-2-(N,N-diacetylamino)-*D*-arabino-hex-1-enopyranose [42].

Glycosyl transferases

The discussion here will be restricted to those enzymes that catalyze the transfer of a sugar residue from a nucleoside diphosphate to an acceptor molecule, usually another carbohydrate.

Investigations into the mechanism of action of such glycosyl transferases have sorely lagged behind those of the glycoside hydrolases. The reasons for this lag are numerous, including difficulties in access and supply and the fact that many of the glycosyl transferases are membrane associated – crystallization of the membrane-free enzyme has been, until recently, a difficult affair.

However, in spite of the above problems, glycosyl transferases have been classified into families (along lines similar to those for the glycoside hydrolases), even though the amino acid sequence homology for transferases can be virtually non-existent [43, 44]. The transferases are also classed as inverting or retaining, and this characteristic of the enzyme is easily assessed (in contrast to the glycoside hydrolases) by a simple inspection of the reaction products.

Perhaps one of the follies of carbohydrate chemists and biochemists alike in the late twentieth century will be the belief that the mechanistic principles established for glycoside hydrolases could be applied directly, almost without question, to glycosyl transferases.

Inversion:

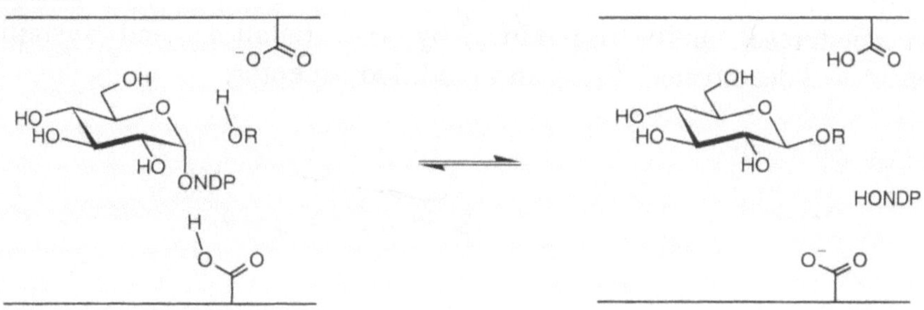

Retention:

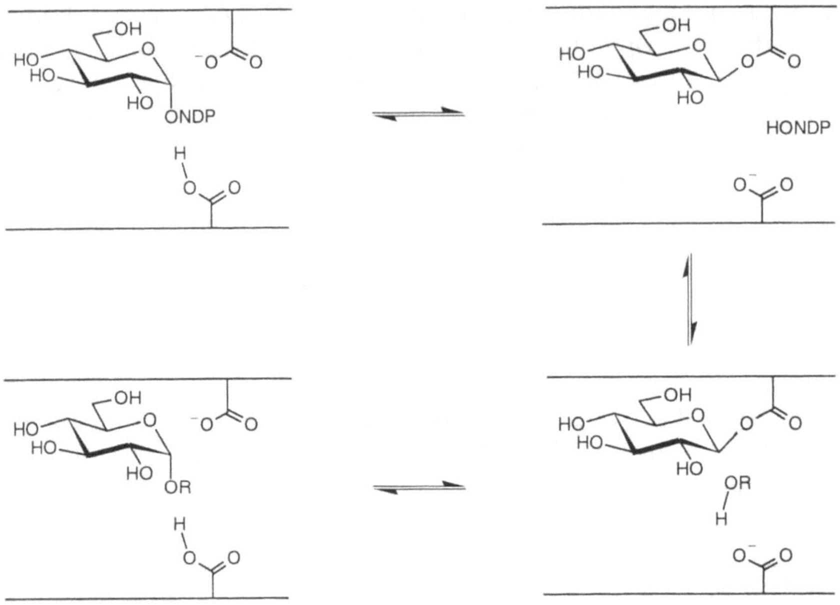

The application of such mechanistic principles presents few problems for an inverting glycosyl transferase, with some (limited) experimental evidence to support the case [45–47]. However, a retaining enzyme again requires the involvement of a catalytic nucleophile, generally suggested to be a carboxylic acid (aspartate or glutamate) residue. Although the presence of a carboxylic catalytic nucleophile is suggested on the basis of results from non-specific labelling reagents (carbodiimides) and amino acid sequence alignments (for glycogen synthase) [48], there has never been reported the actual labelling and identification of such a residue for a glycosyl transferase.

Until recently, no glycosyl transferase had been crystallized, so as to allow mechanistic investigations along the lines of the beautiful piece of work on glycoside hydrolases by *Davies* and *Withers* [49]. However, in the last decade, six glycosyl transferases, all inverting in their action, were crystallized and investigated by X-ray crystallography [50–55]. Several of these enzymes required the presence of a nucleoside diphosphate donor for a successful outcome [51–53]. Unfortunately, no great insights into the enzymes were revealed, particularly when the actual identity of some of the natural substrates remained unknown [52].

The seventh glycosyl transferase to be crystallized, however, was a retaining enzyme [56]. The successful experiment was conducted with a bacterial 1,4-α-galactosyl transferase (family 8), modified by point mutations, and crystallized in the presence of a deactivated donor and modified acceptor:

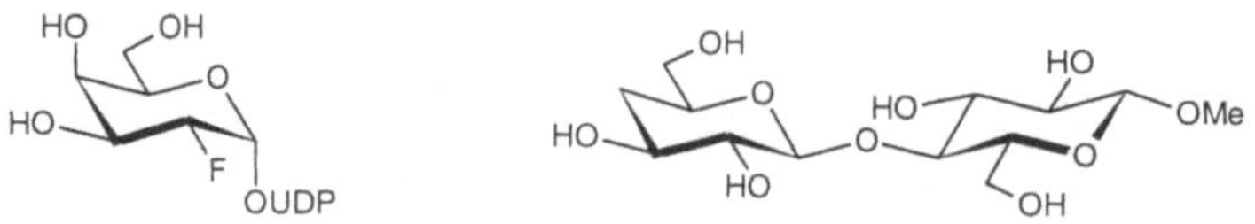

Surprisingly, the authors could find no evidence for the presence of any catalytic nucleophile. In the absence of a catalytic carboxylic acid (*Asp* and *Glu*) residue, several other contenders were considered, including a glutamine residue and the 6-OH of the acceptor molecule. In some desperation, a chemically unusual S_Ni mechanism was finally proposed [56].

The most striking points for us in this very important paper were the absence of an obvious catalytic nucleophile and the necessity for a metal ion (Mn^{2+}) for successful catalysis. We have been working for some time now on another family 8 glycosyl transferase, glycogenin, a self-glucosylating enzyme that initiates the synthesis of macro-glycogen [57, 58]:

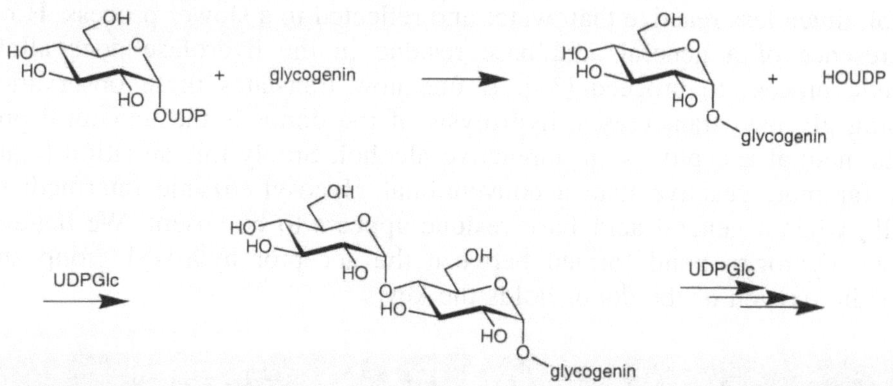

Glycogenin is an α-retaining enzyme, and we have sought to label the catalytic nucleophile by incubation of the enzyme with *UDP*-2-deoxy-2-fluoro-α-D-glucose – the results have been, at best, equivocal. However, a notable fact is that glycogenin is also metal ion (Mn^{2+}) dependant [59]. If it were to eventuate that some retaining glycosyl transferases do not employ a nucleophile in their catalytic event, what then is their mechanism of action? For family 8 retaining transferases, one of the few conserved residues (*DXD*) is doubtlessly involved in the binding of the metal ion (Mn^{2+}). This leads us to suggest a general process for the catalytic action of (family 8) retaining glycosyl transferases:

(1) The initial event involves binding of the glycosyl donor into the active site, partly stabilized by an interaction between the pyrophosphate residue and the metal ion (Mn^{2+}).
(2) Next follows the binding of the acceptor, which triggers a change in the shape of the enzyme at the active site, ensuring an anhydrous environment [52, 53].
(3) This change in conformation also allows a hydrogen bond to form between the hydroxyl group of the acceptor and the glycosidic oxygen atom, causing cleavage of the glycosidic linkage and generating an oxacarbenium ion that may be stabilized through non-bonding interactions with adjacent amino acid residues.
(4) The oxacarbenium ion, through the agency of the aforementioned conformational change, is now adjacent to the acceptor and forms the new glycosidic linkage from the appropriate face.

(5) The enzyme undergoes another conformational change, first releasing the new glycoside, followed by the nucleoside diphosphate, thus presenting the enzyme in its native form ready to repeat the cycle.

In the case of the bacterial 1,4-α-galactosyl transferase mentioned previously, the oxacarbenium ion predicted in (3) is likely to be stabilized by a highly conserved glutamine residue (Fig. 1).

We feel that some other points are worth mentioning in the context of the above proposal. With retaining glycoside hydrolases, water is the natural acceptor molecule for the glycosyl-enzyme intermediate; when such hydrolases are used in an unnatural process, *i.e.* glycoside synthesis, the acceptor molecule is generally an alcohol, much less reactive than water and reflected in a slower process. However, the presence of a general acid/base residue in the hydrolase does allow the synthetic process to proceed [34]. If one now translates these observations to retaining glycosyl transferases, hydrolysis of the donor is the unnatural process, and the natural acceptor is an unreactive alcohol. Surely this situation requires a donor far more reactive than a conventional glycosyl-enzyme intermediate, especially when a general acid/base residue appears to be absent. We believe that the new hydrogen bond formed between the acceptor hydroxyl group and the glycosidic oxygen of the donor holds the key.

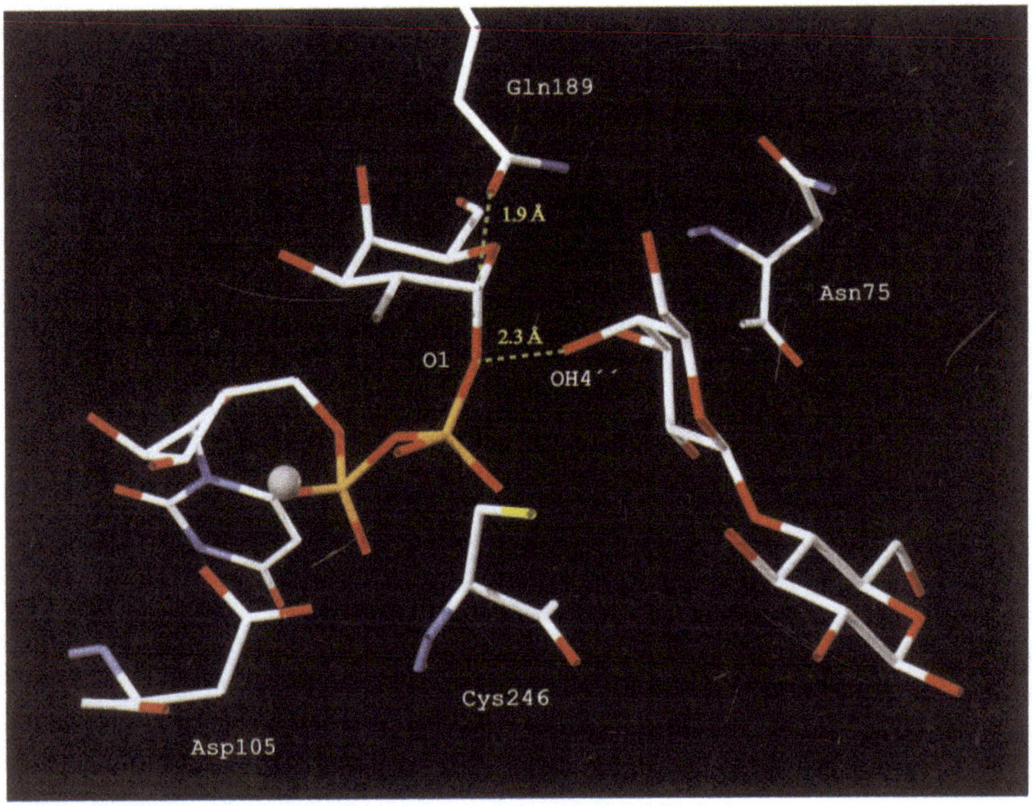

Fig. 1. Depiction of a modified active site arrangement for bacterial 1,4-α-galactosyl transferase (PDB: 1GA8) showing bond distances between C1 and Gln189 (rotated) as well as O1 and OH4$''$ (inserted)

Syntheses

Our final words in this paper are concerned with the synthesis of the glycosyl donors (and related molecules) used in the above investigations: 2-deoxy-2-fluoro glycosyl nucleoside diphosphates.

For the synthesis of *NDP*-2-deoxy-2-fluoro-α-D-glucose derivatives, tri-O-acetyl-D-glucal (Scheme 1) was treated with Selectfluor™ to give, after selective acetylation and purification, tetra-O-acetyl-2-deoxy-2-fluoro-α-D-glucopyranose. Inversion of configuration at the anomeric centre *via* the agency of the bromide gave the β-D-anomer, ready for a successful *MacDonald* phosphorylation to provide the α-D-glucosyl phosphate isolated as the di(cyclohexylammonium) salt [60]. This salt was then converted into the desired nucleoside derivative (*UDP* or *ADP*), employing a modified morpholidate coupling procedure described by *Wittmann* and *Wong* [61]. It should be noted that this transformation could be conveniently monitored by HPLC, and was found to proceed well using the tri-*n*-butylammonium salt of the α-D-glucopyranosyl phosphate.

Interestingly, the major product from the Selectfluor™ reaction on tri-O-acetyl-D-glucal, tetra-O-acetyl-2-deoxy-2-fluoro-α-D-mannopyranose (Scheme 1), is inert to the MacDonald phosphorylation procedure – a result of the double deactivation at the anomeric carbon by an axial acetyl group and an adjacent fluorine atom [42]. Instead, tetra-O-acetyl-2-deoxy-2-fluoro-α-D-mannopyranose was converted *via* the bromide exclusively to the α-D-anomer of the hemiacetal (Scheme 2). Treatment of the hemiacetal with diphenyl chlorophosphate then gave the diphenyl phosphate that was convertible, after catalytic hydrogenolysis, into the desired *GDP*-2-deoxy-2-fluoro-α-D-mannopyranose.

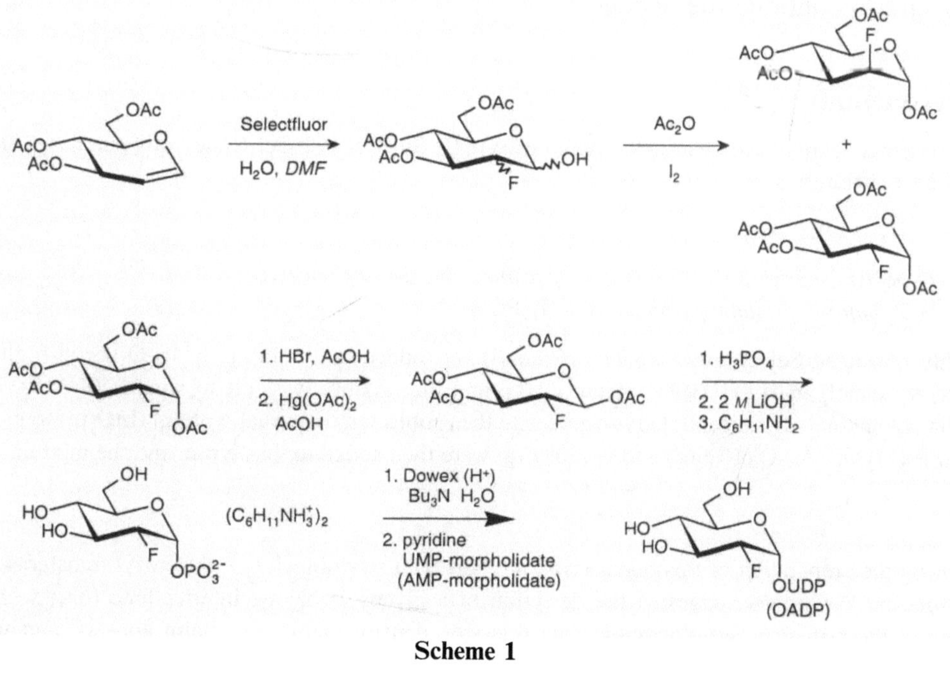

Scheme 1

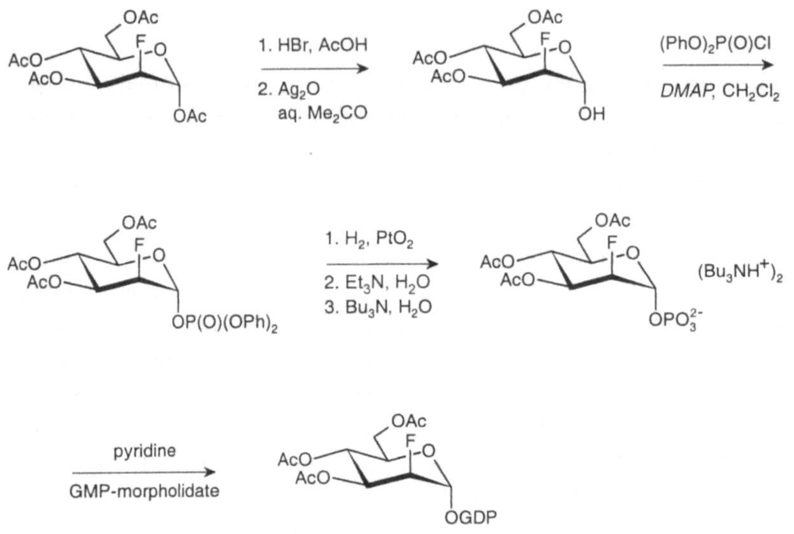

Scheme 2

Conclusions

There is no doubt that two great papers have appeared in the first year of the new century – the first, on the mechanism of action of hen egg-white lysozyme, all but closes the chapter on retaining glycoside hydrolases; the second, on the search for a catalytic nucleophile in a retaining glycosyl transferase, answers some questions and asks of others.

Much work remains to be done, and an upcoming report on the X-ray structure determination of glycogenin [62], another retaining glycosyl transferase of family 8, is sure to continue the debate[1].

Experimental

Experimental details have been given previously [64]. Bio-Gel P-2 Gel (Fine) was purchased from Bio-Rad Laboratories.

Tetra-O-acetyl-2-deoxy-2-fluoro-α-D-glucopyranose and *Tetra-O-acetyl-2-deoxy-2-fluoro-α-D-mannopyranose* ($C_{14}H_{19}FO_9$)

Freshly powdered Selectfluor™ (5.7 g, 16 mmol) was added to a solution of tri-O-acetyl-D-glucal (2.2 g, 8.0 mmol) in H_2O/*DMF* (100 cm^3, 1:1), and the mixture was left to stir at 40° (8 h). The resultant solution was concentrated *in vacuo* and then subjected to a usual workup (EtOAc) to give a colourless syrup. Ac_2O (10 cm^3) and I_2 (300 mg) were then added to this syrup, and the mixture was

[1] During the preparation of this manuscript, a report on a retaining 1,3-α-galactosyl transferase by *Gastinel* and coworkers suggested the detection of a glycosyl-enzyme intermediate [63]; without access to the primary crystallographic data (electron density maps), the claim appears somewhat equivocal.

allowed to stand at room temperature (3 h). The mixture was then concentrated *in vacuo*, subjected to a usual workup (EtOAc), and purified using flash chromatography (30% EtOAc/petroleum ether).

First to elute was the D-gluco derivative (748 mg, 28%); m.p.: 75–77°C (Ref. [65]: 78°C), $[\alpha]_D = +147°$ (Ref. [65]: +148°); ^{1}H NMR (300 MHz, $CDCl_3$): $\delta = 2.01, 2.15, 2.22, 2.27$ (4s, 12H, CH_3), 3.96–4.02 (m, H5,6), 4.23 (dd, $J_{5,6} = 5.1$, $J_{6,6} = 11.5$ Hz, H6), 4.61 (ddd, $J_{2,F} = 47.6$, $J_{1,2} = 4.7$, $J_{2,3} = 9.5$ Hz, H2), 5.02 (t, $J_{3,4} \approx J_{4,5} = 9.6$ Hz, H4), 5.48 (dt, $J_{3,F} = 13.1$ Hz, H3), 6.35 (d, H1) ppm.

Next to elute was the D-manno derivative (1.34 g, 50%); m.p.: 68–69°C (Ref. [35]: 66–68°C), $[\alpha]_D = +65°$ (Ref. [35]: +60°); ^{1}H NMR (300 MHz, $CDCl_3$): $\delta = 2.12, 2.15, 2.22, 2.27$ (4s, 12H, CH_3), 4.13 (ddd, $J_{4,5} = 9.6$, $J_{5,6} = 5.1$, 3.1 Hz, H5), 4.15 (dd, $J_{6,6} = 11.9$ Hz, H6), 6.28 (dd, H6), 4.73 (dt, $J_{2,F} = 48.9$, $J_{1,2} \approx J_{2,3} = 2.1$ Hz, H2), 5.23 (ddd, $J_{3,F} = 23.0$, $J_{3,4} = 9.7$ Hz, H3), 5.41 (dd, H4), 6.23 ($J_{1,F} = 8.5$ Hz, H1) ppm.

Di(cyclohexylammonium) 2-deoxy-2-fluoro-α-D-glucopyranosyl phosphate ($C_{18}H_{38}FN_2O_8P$)

Tetra-O-acetyl-2-deoxy-2-fluoro-β-D-glucopyranose [60] (1.15 g) was stirred with molten anhydrous H_3PO_4 (4.0 g) under vacuum (1 mm Hg) at 60°C overnight. The mixture was allowed to cool slightly (but not enough to allow crystallization) and then treated with *THF* (5 cm^3) to give a dark solution. Aqueous 2 *M* LiOH (80 cm^3) was then added to the solution with cooling, and the mixture was left to stir at room temperature overnight. The mixture was filtered and the filtrate passed through a column of Dowex 50W-X8 (H^+) resin into an excess of cyclohexylamine in H_2O. Concentration of the eluent *in vacuo* and subsequent lyophilization gave di(cyclohexylammonium) 2-deoxy-2-fluoro-α-D-glucopyranosyl phosphate as a pale yellow powder (1.32 g, 87%). $[\alpha]_D = +46°$ (Ref. [60]: +48.6°); the ^{1}H NMR spectrum was consistent with that reported [60].

Tri-O-acetyl-2-deoxy-2-fluoro-α-D-mannopyranose ($C_{12}H_{17}FO_8$)

HBr in AcOH (2 cm^3, 30% w/w) was added to tetra-O-acetyl-2-deoxy-2-fluoro-α-D-mannopyranose (1.80 g, 5.4 mmol), and the mixture was left to stir at room temperature (4 h). Usual workup (EtOAc) gave the α-D-mannosyl bromide as a colourless oil (2.1 g) which was used without further purification.

Ag_2CO_3 (1.14 g, 4.13 mmol) was added to a solution of the above bromide (1.4 g, 3.8 mmol) in aqueous acetone (20 cm^3, 1:10), and the mixture left to stir (2 h). The mixture was then filtered through a plug of silica and concentrated *in vacuo*. Usual workup (EtOAc) of the residue followed by flash chromatography (40% EtOAc/petroleum ether) gave the title hemiacetal as a colourless oil (1.15 g, 98%). The spectroscopic data (^{1}H and ^{13}C NMR) were consistent with those reported [39].

Diphenyl (tri-O-acetyl-2-deoxy-2-fluoro-α-D-mannopyranosyl) phosphate ($C_{24}H_{26}FO_{11}P$)

A solution of tri-O-acetyl-2-deoxy-2-fluoro-α-D-mannopyranose (0.76 g, 2.5 mmol) and *DMAP* (0.55 g) in CH_2Cl_2 (20 cm^3) was cooled to −10°C. Diphenyl chlorophosphate (1.0 cm^3, 4.8 mmol) was added dropwise, and the solution was allowed to stir (4 h) under gradual warming to 10°C. The solution was diluted with CH_2Cl_2 (20 cm^3) and washed sequentially with cold H_2O, 0.5 *M* HCl, and saturated $NaHCO_3$ solution, dried ($MgSO_4$), and concentrated *in vacuo*. Flash chromatography (20% EtOAc/petroleum ether) of the residue gave the title phosphate as a colourless oil (1.01 g, 76%).

$[\alpha]_D = +46°$; found: C 53.5, H 4.7; calcd.: C 53.3, H 4.8; ^{1}H NMR (300 MHz, $CDCl_3$): $\delta = 1.95$, 2.00, 2.09 (3s, 9H, CH_3), 3.88 (dd, $J_{5,6} = 1.4$, $J_{6,6} = 12.5$ Hz, H6), 3.98–4.04 (m, H5), 4.13 (dd, $J_{5,6} = 4.2$ Hz, H6), 4.74 (dddd, $J_{2,F} = 48.9$, $J_{1,2} \approx J_{2,3} = 2.2$, $J_{2,P} = 0.5$ Hz, H2), 5.25 (ddd, $J_{3,F} = 27.2$, $J_{3,4} = 10.3$ Hz, H3), 5.37 (ddd, $J_{4,F} = 1.2$, $J_{4,5} = 10.2$ Hz, H4), 5.99 (ddd, $J_{1,F}$, $J_{1,P} = 7.1$, 8.5 Hz, H1), 7.15–7.39 (m, 10H, Ph) ppm; ^{13}C NMR (75.5 MHz, $CDCl_3$): $\delta = 20.47$, 20.55 (3C, CH_3), 61.19 (C6),

64.59, 70.67 (C4,5), 68.75 (d, $J_{3,F}=17$ Hz, C3), 85.98 (dd, $J_{2,F}=182$, $J_{2,P}=12$ Hz, C2), 95.19 (dd, $J_{1,F}$, $J_{1,P}=6.0$, 32 Hz, Cl), 119.90–150.11 (12C, Ph) ppm.

2-Deoxy-2-fluoro-α-D-mannopyranosyl-bis(tributylammonium) phosphate ($C_{30}H_{66}FN_2O_8P$)

PtO_2 (40 mg) was added to a solution of diphenyl 3,4,6-tri-O-acetyl-2-deoxy-2-fluoro-α-D-mannosyl phosphate (400 mg) in MeOH (15 cm^3), and the mixture was stirred vigorously under an atmosphere of H_2 at room temperature (2 d). The mixture was filtered, diluted with H_2O (2 cm^3) and Et_3N (2 cm^3), and the solution was left to stand at 4°C overnight. The mixture was concentrated *in vacuo*, taken up in H_2O (5 cm^3), and passed through a column of Dowex 50 W-X8 (H^+) resin into a vigorously stirred mixture of H_2O and excess tributylamine. Concentration *in vacuo* and subsequent lyophilization gave the title phosphate as a white powder (416 mg).

^{1}H NMR (300 MHz, D_2O): $\delta=0.72$ (t, 18H, CH_3), 1.10–1.21, 1.35–1.51 (2m, 24H, CH_2), 2.85–2.98 (m, 12H, CH_2N), 3.48–3.69 (m, 4H, H4,5,6), 3.76 (ddd, $J_{3,F}=30.8$, $J_{2,3}=2.5$, $J_{3,4}=9.6$ Hz, H3), 4.59 (bddd, $J_{2,F}=48.7$, $J_{1,2}=2.4$ Hz, H2), 5.38 (ddd, $J_{1,P}$, $J_{1,F}=6.0$, 8.1 Hz, H1) ppm; ^{13}C NMR (75.5 MHz, D_2O): $\delta=24.05$ (6C, CH_3), 30.52, 36.66 (12C, CH_2), 63.89 (6C, CH_2N), 71.57 (C6), 77.62, 84.66 (C4,5), 80.40 (d, $J_{3,F}=17.3$ Hz, C3), 100.97 (dd, $J_{2,F}=174.4$, $J_{2,P}=9.2$ Hz, C2), 104.18 (dd, $J_{1,F}=4.5$, $J_{1,P}=30.6$ Hz, C1) ppm.

Uridine 5′-(trihydrogendiphosphate), P′-(2-deoxy-2-fluoro-α-D-glucopyranosyl) ester ($C_{15}H_{23}FN_2O_{16}P_2$)

Di(cyclohexylammonium) 2-deoxy-2-fluoro-α-D-glucopyranosyl phosphate (370 mg, 803 μmol) was dissolved in H_2O (5 cm^3) and passed through a column of Dowex 50 W-X8 (H^+) resin into a vigorously stirred mixture of H_2O and excess tributylamine. The eluent was then concentrated *in vacuo*, coevaporated with MeOH (2 × 10 cm^3), and dried under high vacuum (2 d) to give the *bis*(tributylammonium) salt (410 mg).

^{1}H NMR (300 MHz, D_2O): $\delta=0.72$ (t, 18H, CH_3), 1.07–1.16, 1.25–1.37 (2m, 24H, CH_2), 2.89–2.93 (m, 12H, CH_2N), 3.29 (t, $J_{3,4}\approx J_{4,5}=9.9$ Hz, H4), 3.56 (dd, $J_{5,6}=4.8$, $J_{6,6}=12.3$ Hz, H6), 3.63–3.69 (m, H5,6), 3.80 (dt, $J_{2,3}=9.6$, $J_{3,F}=13.1$ Hz, H3), 4.21 (dddd, $J_{1,2}$, $J_{1,P}=2.3$, 3.7, $J_{2,F}=49.4$ Hz, H2), 6.13 (dd, $J_{1,P}=6.6$ Hz, H1) ppm; ^{13}C NMR (125.8 MHz, D_2O): $\delta=15.14$ (6C, CH_3), 21.61, 27.51 (12C, CH_2), 54.96 (6C, CH_2N), 62.52 (C6), 71.15 (d, $J_{4,F}=8.3$ Hz, C4), 73.37 (d, $J_{3,F}=17.3$ Hz, C3), 74.78 (C5), 91.89 (dd, $J_{2,P}=6.8$, $J_{2,F}=187$ Hz, C2), 94.25 (dd, $J_{1,P}=4.6$, $J_{1,F}=22.3$ Hz, C1) ppm.

The above *bis*(tributylammonium) salt (410 mg) and 4-morpholine-N,N′-dicyclohexylcarboxamidinium uridine 5′-monophosphomorpholidate (662 mg, 964 μmol) were coevaporated with dry pyridine (3 × 2 cm^3). 1*H*-Tetrazole (112 mg, 1.61 mmol) and dry pyridine (4 cm^3) were added, and the solution was then concentrated to half of its original volume and left to stir at room temperature (2 d). The solution was then diluted with H_2O (30 cm^3), washed with Et_2O, and concentrated *in vacuo*. The residue was purified on a Bio-Gel P-2 column (2.5 × 95 cm) eluting with 250 m*M* NH_4HCO_3 solution and lyophilized to give the title compound as a white powder (320 mg, 68%).

^{1}H NMR (500 MHz, D_2O): $\delta=3.34$ (t, $J_{3'',4''}\approx J_{4'',5''}=9.8$ Hz, H4″), 3.59 (dd, $J_{5'',6''}=4.5$, $J_{6'',6''}=12.5$ Hz, H6″), 3.67 (dd, $J_{5'',6''}=2.2$ Hz, H6″), 3.73 (ddd, H5″), 3.85 (dt, $J_{2'',3''}=9.7$, $J_{3'',F}=12.8$ Hz, H3″), 3.97–4.10, 4.16–4.22, 4.27–4.32 (3m, 6H, H2′,3′,4′,5′,2″), 5.59 (dd, $J_{1'',2''}=3.5$, $J_{1'',F}=7.3$ Hz, H1″), 5.77–5.82 (m, H5,1′), 7.78 (d, $J_{5,6}=8.1$ Hz, H6) ppm.

Adenosine 5′-(trihydrogendiphosphate), P′-(2-deoxy-2-fluoro-α-D-glucopyranosyl) ester ($C_{16}H_{24}FN_5O_{14}P_2$)

Following the above procedure, di(cyclohexylammonium) 2-deoxy-2-fluoro-α-D-glucopyranosyl phosphate (285 mg, 618 μmol) was treated with 4-morpholine-N,N′-dicyclohexylcarboxamidinium adenosine 5′-monophosphomorpholidate (527 mg, 742 μmol) and 1*H*-tetrazole (88 mg, 1.2 mmol) in dry pyridine (2 cm^3) to give the title compound as a hygroscopic white solid (257 mg, 71%).

^{1}H NMR (300 MHz, D_2O): $\delta = 3.35$ (dd, $J_{3'',4''} = 11.1$, $J_{4'',5''} = 10.9$ Hz, H4″), 3.56–3.78 (m, 3H, H5″,6″), 3.87 (ddd, $J_{2'',3''} = 11.0$, $J_{3'',F} = 50.9$ Hz, H3″), 4.06–4.41 (m, 6H, H2′,3′,4′,5′,2″), 5.51 (dd, $J_{1'',2''} = 4.1$, $J_{1'',F} = 7.2$ Hz, H1″), 5.80 (d, $J_{1',2'} = 4.5$ Hz, H1′), 7.86, 8.00 (2s, H2,8) ppm.

Guanosine 5′-(trihydrogendiphosphate), P′-(2-deoxy-2-fluoro-α-D-mannopyranosyl) ester ($C_{16}H_{24}FN_5O_{15}P_2$)

Following the previous procedure, *bis*(tributylammonium) 2-deoxy-2-fluoro-*α*-D-mannopyranosyl phosphate (320 mg, 694 μmol) was treated with 4-morpholine-N,N′-dicyclohexylcarboxamidinium guanosine 5′-monophosphomorpholidate (605 mg, 833 μmol) and 1*H*-tetrazole (98 mg, 1.4 mmol) in dry pyridine (2 cm^3) to give the title compound as a white powder (281 mg, 65%). Spectroscopic data (^{1}H and ^{13}C NMR) were consistent with those reported [66].

Acknowledgments

The authors would like to thank the Australian Research Council for financial assistance, as well as Air Products and Chemicals, Inc. for a kind donation of Selectfluor™ and financial assistance.

References

[1] Wong C-H, Halcomb RL, Ichikawa Y, Kajimoto T (1995) Angew Chem Int Ed Engl **34**: 521
[2] Wong C-H, Halcomb RL, Ichikawa Y, Kajimoto T (1995) Angew Chem Int Ed Engl **34**: 412
[3] Kobata A (1993) Acc Chem Res **26**: 319
[4] Dwek RA (1995) Biochem Soc Trans **23**: 1
[5] Dwek RA (1996) Chem Rev **96**: 683
[6] Sears P, Wong C-H (1998) Chem Commun 1161
[7] Varki A (1993) Glycobiology **3**: 97
[8] Koshland DE (1953) Biol Rev Cambridge Philos Soc **28**: 416
[9] Legler G, Bause E (1973) Carbohydr Res **28**: 45
[10] Sinnott ML (1990) Chem Rev **90**: 1171
[11] Sinnott ML (1993) Bioorg Chem **21**: 34
[12] Varrot A, Schülein M, Pipelier M, Vasella A, Davies GJ (1999) J Am Chem Soc **121**: 2621
[13] Davies GJ, Henrissat B (1995) Structure **3**: 853
[14] Davies GJ, Wilson KS, Henrissat B (1997) Biochem J **321**: 557
[15] Withers SG, Street IP, Bird P, Dolphin DH (1987) J Am Chem Soc **109**: 7530
[16] Withers SG, Street IP (1988) J Am Chem Soc **110**: 8551
[17] Withers SG, Rupitz K, Street IP (1988) J Biol Chem **263**: 7929
[18] McCarter JD, Withers SG (1996) J Am Chem Soc **118**: 241
[19] McCarter JD, Withers SG (1994) Curr Opin Struct Biol **4**: 885
[20] McCarter JD (1995) PhD Thesis, The University of British Columbia
[21] Henrissat B (1991) Biochem J **280**: 309
[22] Henrissat B, Bairoch A (1993) Biochem J **293**: 781
[23] Henrissat B, Bairoch A (1996) Biochem J **316**: 695
[24] Heightman TD, Vasella AT (1999) Angew Chem Int Ed **38**: 750
[25] Gebler JC, Aebersold R, Withers SG (1992) J Biol Chem **267**: 11126
[26] Miao S, McCarter JD, Grace ME, Grabowski GA, Aebersold R, Withers SG (1994) J Biol Chem **269**: 10975
[27] Street IP, Kempton JB, Withers SG (1992) Biochemistry **31**: 9970
[28] Braun C, Brayer GD, Withers SG (1995) J Biol Chem **270**: 26778

[29] Withers SG, Aebersold R (1995) Prot Sci **4**: 361
[30] Vocadlo DJ, Davies GJ, Laine R, Withers SG (2001) Nature **412**: 835
[31] Vocadlo DJ, Meyer C, He S, Withers SG (2000) Biochemistry **39**: 117
[32] Numao S, He S, Evjen G, Howard S, Tollersrud OK, Withers SG (2000) FEBS Letters **484**: 175
[33] Mackenzie LF, Wang Q, Warren RAJ, Withers SG (1998) J Am Chem Soc **120**: 5583
[34] Mayer C, Zechel DL, Reid SP, Warren RAJ, Withers SG (2000) FEBS Letters **466**: 40
[35] Adamson J, Foster AB, Hall LD, Johnson RN, Hesse RH (1970) Carbohydr Res **15**: 351
[36] Adamson J, Foster AB, Westwood JH (1971) Carbohydr Res **18**: 345
[37] Diksic M, Jolly D (1986) Carbohydr Res **153**: 17
[38] Burkart MD, Zhang Z, Hung S-C, Wong C-H (1997) J Am Chem Soc **119**: 11743
[39] Ortner J, Albert M, Weber H, Dax K (1999) J Carbohydr Chem **18**: 297
[40] Vincent SP, Burkart MD, Tsai C-Y, Zhang Z, Wong C-H (1999) J Org Chem **64**: 5264
[41] Albert M, Dax K, Ortner J (1998) Tetrahedron **54**: 4839
[42] Stick RV, Watts AG (unpublished results)
[43] Campbell JA, Davies GJ, Bulone V, Henrissat B (1997) Biochem J **326**: 929
[44] Geremia RA, Petroni EA, Ielpi L, Henrissat B (1996) Biochem J **318**: 133
[45] Qiao L, Murray BW, Shimazaki M, Schultz J, Wong C-H (1996) J Am Chem Soc **118**: 7653
[46] Murray BW, Wittmann V, Burkart MD, Hung S-C, Wong C-H (1997) Biochemistry **36**: 823
[47] Murray BW, Takayama S, Schultz J, Wong C-H (1996) Biochemistry **35**: 11183
[48] Cid E, Gomis RR, Geremia RA, Guinovart JJ, Ferrer JC (2000) J Biol Chem **275**: 33614
[49] Davies GJ, Mackenzie L, Varrot A, Dauter M, Brzozowski AM, Schülein M, Withers SG (1998) Biochemistry **37**: 11707
[50] Vrielink A, Ruger W, Driessen HPC, Freemont PS (1994) EMBO J **13**: 3413
[51] Gastinel LN, Cambillau C, Bourne Y (1999) EMBO J **18**: 3546
[52] Charnock SJ, Davies GJ (1999) Biochemistry **38**: 6380
[53] Ünligil UM, Zhou S, Yuwaraj S, Sarkar M, Schachter H, Rini JM (2000) EMBO J **19**: 5269
[54] Ha S, Walker D, Shi Y, Walker S (2000) Protein Sci **9**: 1045
[55] Pedersen LC, Tsuchid K, Kitagawa H, Sugahara K, Darden TA, Negishi M (2000) J Biol Chem **275**: 34580
[56] Persson K, Ly HD, Dieckelmann M, Wakarchuk WW, Withers SG, Strynadka NCJ (2001) Nature Struct Biol **8**: 166
[57] Lin A, Mu J, Yang J, Roach PJ (1999) Arch Biochem Biophys **363**: 163
[58] Alonso MD, Lomako J, Lomako WM, Whelan WJ (1995) FASEB J **9**: 1126
[59] Lomako J, Lomako WM, Whelan WJ (1990) BioFactors **2**: 251
[60] Withers SG, MacLennan DJ, Street IP (1986) Carbohydr Res **154**: 127
[61] Wittmann V, Wong C-H (1997) J Org Chem **62**: 2144
[62] Roach PJ (unpublished results)
[63] Gastinel LN, Bignon C, Misra AK, Hindsgaul O, Shaper JH, Joziasse DH (2001) EMBO J **20**: 638
[64] McAuliffe JC, Stick RV (1997) Aust J Chem **50**: 193
[65] Adam MJ (1982) J Chem Soc Chem Commun 730
[66] Wong C-H (1998) Inhibitors of Glycosyl Transferases, WO 98/25940 (USA), 18th June 1998

Received October 16, 2001. Accepted November 7, 2001

N-Alkylated Derivatives of 1,5-Dideoxy-1,5-iminoxylitol as *β*-Xylosidase and *β*-Glucosidase Inhibitors[a]

Herwig Häusler[1], **Karen Rupitz**[2], **Arnold E. Stütz**[1,*], and **Stephen G. Withers**[2,*]

[1] Glycogroup, Institut für Organische Chemie der Technischen Universität Graz, A-8010 Graz, Austria

[2] Department of Chemistry, University of British Columbia, Vancouver, B.C., V6T 1Z1 Canada

Summary. A range of lipophilic derivatives of the D-xylosidase inhibitor 1,5-dideoxy-1,5-iminoxylitol was synthesized by N-alkylation of the parent compound with different alkyl halides. Inhibitory activities of the products obtained were measured with *β*-glucosidase from *Agrobacterium sp.*, a family 1 enzyme, as well as with *β*-xylosidase from *Thermoanaerobacterium saccharolyticum* belonging to family 39 of the glycohydrolases. Whereas the former enzyme, which has low *β*-xylosidase activity, was only moderately inhibited, with K_i values in the millimolar range, good inhibition was observed with the latter one. Inhibitors with terminally functionalized N-alkyl chains open up opportunities for novel applications as affinity ligands as well as for various types of tagging for diagnostic purposes.

Keywords. 1,5-Dideoxy-1,5-iminoxylitol; *β*-Glucosidase; *β*-Xylosidase; Glycosidase inhibitor.

Introduction

D-Xylans are polysaccharides consisting of *β*-1,4-linked D-xylopyranosyl moieties and can be found in the cell walls of all land plants and most plant parts [1]. Their biodegradation relies on *β*-1,4-xylanases (EC 3.2.1.8) and *β*-xylosidases (EC 3.2.1.37). Xylanases are endo-enzymes attacking internal glycosidic linkages, whereas xylosidases release xylose by attack of the terminal glycosidic bond at the non-reducing end of the oligomer. These enzymes are produced by biodegradative microorganisms such as fungi and bacteria that are important for carbon turnover in nature. For example, 35% of birch wood (dry weight) consists of xylans. In pulp and paper technology, these enzymes are employed in a prebleach treatment to remove residual xylans, thereby reducing the need for expensive and environmentally questionable chemicals in the process. The food

---

[a] Dedicated to Prof. *Günter Legler* on the occasion of his 75th birthday

* Corresponding authors. E-mail: stuetz@orgc.tu-graz.ac.at, withers@chem.ubc.ca

and feed industry has also been identified as a large potential area for application of xylanases.

Low molecular weight inhibitors of xylanases and xylosidases are not only useful tools for enzyme characterization and purification but might also serve other, more 'applied' purposes. For example, powerful and inexpensive inhibitors could be employed in crop protection against xylan degrading plant pathogens or as components in wood preservation coatings and paints. Iminosugars including iminoalditols have been found to be powerful inhibitors of many glycosidases with K_i values in the micromolar and, in quite a few cases, even nanomolar range [2]. Some of these compounds, due to their interference with biochemical pathways, exert interesting biological activities [2]. Many have become valuable diagnostic tools for studies of enzyme active sites and catalytic mechanisms. Despite the abundance of xylosidases and xylanases and the fact that they play important roles in the degradation of hemicelluloses, corresponding inhibitors such as 1,5-dideoxy-1,5-iminoxylitol (**1**) have not yet gained as much attention as the related hexopyranoside mimetics.

1,5-Dideoxy-1,5-iminoxylitol (**1**) is the paradigmatic example of a reversible, xyloside mimicking inhibitor. Its first synthesis has been reported by *Paulsen* in the mid-nineteen sixties (its precursor, 5-amino-5-deoxyxylose, was independently reported by both *Paulsen* and *Hanessian*) [3]. Subsequently, this compound was synthesized by several other groups [4]. Examples of its biochemical applications have remained rare and largely confined to studies of β-glucosidases, despite the fact that this compound could be a useful probe for xylanase/β-xylosidase research.

Based on recent findings with related glycosidase inhibitors, in particular derivatives of the powerful β-glucosidase inhibitor 2,5-dideoxy-2,5-imino-D-mannitol, which indicated that lipophilic chain extensions can improve the inhibitory activity of such derivatives [5], we prepared a small exploratory set of simple N-alkylated derivatives of compound **1** featuring simple hydrocarbon substituents as well as terminally functionalized chains suitable for tagging or immobilization studies.

Results and Discussion

Parent compound **1** was readily accessible from D-xylose as mentioned previously according to Scheme 1. Controlled conversion of D-xylose gave the corresponding furanoid 1,2-O-isopropylidene protected derivative **2** which was regioselectively O-tosylated at the primary position in dichloromethane/pyridine employing the standard procedure. Conventional reaction of 1,2-O-isopropylidene-5-O-tosyl-α-D-xylofuranose (**3**) with sodium azide in N,N-dimethylformamide furnished the corresponding 5-azidodeoxysugar **4** [6]. Deprotection employing acidic ion exchange resin Amberlite IR 120 [H^+] gave the free sugar **5** which was cyclized by conventional catalytic hydrogenation and concomitant intramolecular reductive amination to yield the desired key intermediate **1**.

Its reaction with bromohexane, 3-phenyl-1-bromopropane, methyl 6-bromohexanoate, and 6-bromohexanoic nitrile led smoothly to the corresponding N-alkyl derivatives **6–9**. From **8**, the 6-aminohexyl substituted compound **10** was formed by catalytic hydrogenation over palladium hydroxide on charcoal. Compounds **8–10** can be modified at their terminal functional groups *e.g.* by

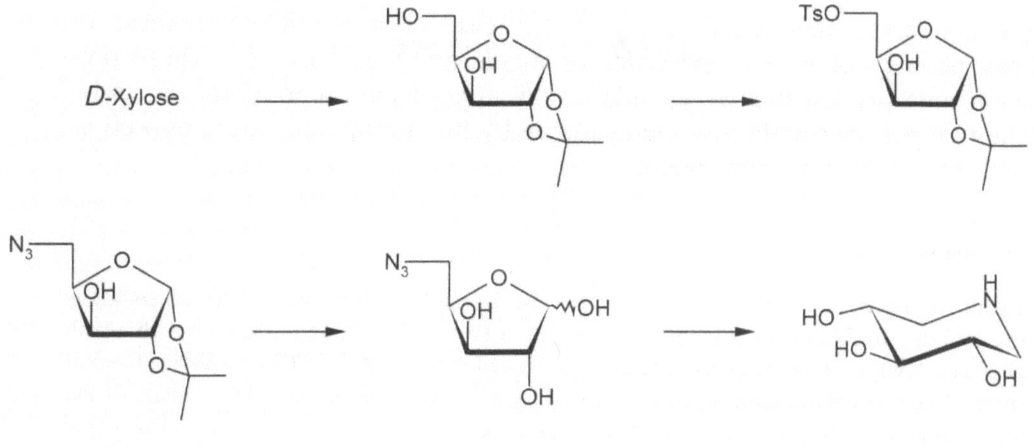

4 5 1

Scheme 1

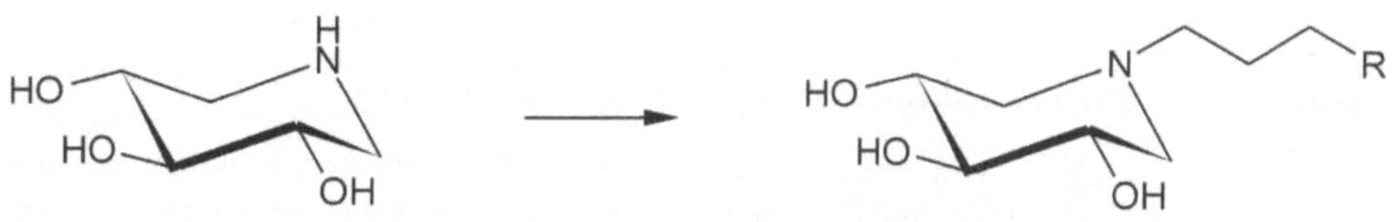

1

6: $R = (CH_2)_2CH_3$
7: R = phenyl
8: $R = (CH_2)_2CN$
9: $R = (CH_2)_2CO_2CH_3$
10: $R = (CH_2)_3NH_2$

Scheme 2

Table 1. Inhibition constants (K_i values, μM) for substituted iminoxylitols with β-glucosidase from *Agrobacterium sp.* at *pH* 7.0 as well as with β-xylosidase from *Thermoanaerobacterium saccharolyticum* at *pH* 5.5

	1	**6**	**7**	**8**	**9**	**10**
β-glucosidase	50	2600	1600	3600	2000	2500
β-xylosidase	206	88	135	25	35	118

attachment onto a polymer support for affinity chromatography or by other types of transformations.

Preliminary screening of inhibitory activities were conducted with β-glucosidase from *Agrobacterium sp.* as well as with β-xylosidase from *Thermoanaerobacterium saccharolyticum* (Table 1). All N-alkylated compounds were found to be moderate inhibitors of the β-glucosidase employed with activities lower than that of parent compound **1**.

Conversely, with the β-xylosidase employed in this study, N-alkylation improved activities by a factor of 2 to 3 (compounds **6**, **7**, and **10**) as compared to the parent compound. Gratifyingly, in the case of nitrile **8**, an increase of inhibitory potency of nearly one order of magnitude was observed. These preliminary results suggest that β-xylosidase inhibitors of the iminoxylitol type with interesting properties should become available by fine-tuning the nature as well as the length of the N-alkyl substituent.

Experimental

Melting points were recorded on a Tottoli apparatus (Büchi) and are uncorrected. Optical rotations were measured on a JASCO Digital Polarimeter or with a Perkin Elmer 341 instrument with a path length of 10 cm. NMR spectra were recorded at 200 (^{1}H) and at 50.29 MHz (^{13}C) on a Varian Gemini spectrometer. Chemical shifts are given in δ relative to the solvent resonances. The signals of the protecting groups were found in the expected regions and are not listed explicitly. Thin-layer chromatography was performed on precoated aluminum sheets (Merck 5554). Plates were stained with a mixture of 10% ammonium molybdate (w/v) in 10% H_2SO_4 containing 0.8% $Ce(SO_4)_2$ (w/v). For column chromatography, Silica Gel 60 (E. Merck) was used employing $CHCl_3$:MeOH: concd. aq. ammonia = 100:100:1 as the solvent system. Elemental analyses agreed favourably with the calculated values.

General method for N-alkylation reactions employing bromoalkyl reagents

To a 5% solution of compound **1** in MeOH, 3 equivalents of the respective bromoalkane and 3 equivalents of pyridine were added, and the mixture was kept at ambient temperature for 48 h. Removal of solvents under reduced pressure followed by chromatography of the residue furnished compounds **6–8**.

N-Hexyl-1,5-dideoxy-1,5-iminoxylitol (**6**; $C_{11}H_{23}NO_3$)

Following the general procedure, 170 mg **1** (1.28 mmol) were reacted with 530 mm^3 1-bromohexane (3.78 mmol). Compound **6** (215 mg) was obtained in 78% yield.

M.p.: 79–82°C; ^{1}H NMR (CD_3OD, δ, 200 MHz): 3.78 (2H, m, H-2, H-4), 3.46 (1H, broad, H-3), 3.33 (2H, m, H-1a, H-5a), 2.98 (2H, m, H-1e, H-5e), 2.80 (2H, broad, H-1′, H-1″), 1.71 (2H, m, H-2′, H-2″), 1.37 (6H, m, H-3′, H-3″, H-4′, H-4″, H-5′, H-5″), 0.92 (3H, t, H-6′, H-6″, H-6‴) ppm; the poor resolution of signals is due to inherent conformational flexibility of the molecule; ^{13}C NMR (CD_3OD, δ, 50 MHz): 68.1 (C-2, C-3, C-4), 57.4 (C-1, C-5), 55.4 (C-1′), 31.3, 26.4, 24.5, 22.4, 13.1 (C-2′, C-3′, C-4′, C-5′, C-6′) ppm.

N-(3-Phenyl)-propyl-1,5-dideoxy-1,5-iminoxylitol (**7**; $C_{14}H_{21}NO_3$)

Employing the general procedure, 142 mg **1** (107 mmol) and 48 mm^3 3-phenyl-1-bromopropane (3.15 mmol) furnished 160 mg of **7** (59%) as a wax.

^{1}H NMR (CD_3OD, δ, 200 MHz): 3.67 (2H, ddd, $J_{1a,2} = J_{4,5a} = 7.5$ Hz, $J_{2,3} = J_{3,4} = 8.4$ Hz, $J_{1e,2} = J_{4,5e} = 4.4$ Hz, H-2, H-4), 3.15 (1H, H-3), 3.01 (2H, dd, $J_{1a,1e} = J_{5a,5e} = 11.4$ Hz, H-1a, H-5a), 2.70–2.12 (6H, m, H-1b, H-5b, H-1′, H-1″, H-3′, H-3″), 1.76 (2H, m, H-2′, H-2″) ppm; ^{13}C NMR (CD_3OD, δ, 50 MHz): 145.1, 132.3, 132.2, 129.9 (phenyl), 79.8 (C-3), 72.9 (C-2, C-4), 60.8 (C-1, C-5), 60.4 (C-1′), 36.7, 31.1 (C-2′, C-3′) ppm.

N-(5-Cyano)-pentyl-1,5-dideoxy-1,5-iminoxylitol (**8**; $C_{11}H_{20}N_2O_3$)

Iminoalditol **1** (170 mg, 1.28 mmol) was reacted with 250 mm^3 6-bromohexanoic nitrile (1.92 mmol) to give 134 mg (46%) **8**.

M.p.: 103–107°C; ^{1}H NMR (CD_3OD, δ, 200 MHz): 3.55 (2H, ddd, $J_{1a,2}=J_{4,5a}=J_{2,3}=J_{3,4}=$ 9.2 Hz, $J_{1e,2}=J_{4,5e}=$ 4.4 Hz, H-2, H-4), 3.18 (1H, H-3), 3.06 (2H, dd, $J_{1a,1e}=J_{5a,5e}=$ 11.0 Hz, H-1a, H-5a), 2.52 (4H, m, H-1′, H-1″, H-5′, H-5″), 2.14 (2H, m, H-1b, H-5b), 1.82–1.38 (6H, m, H-2′, H-2″, H-3′, H-3″, H-4′, H-4″) ppm; ^{13}C NMR (CD_3OD, δ, 50 MHz): 119.7 (CN), 79.0 (C-3), 69.7 (C-2, C-4), 57.6 (C-1, C-5), 57.2 (C-1′), 26.2, 25.3, 25.1, 16.1 (C-2′, C-3′, C-4′, C-5′) ppm.

N-Methoxycarbonylpentyl-1,5-dideoxy-1,5-iminoxylitol (**9**; $C_{12}H_{23}NO_5$)

To a 5% solution of 240 mg of the starting material **1** (1.80 mmol), 294 mm^3 adipic acid methyl ester hemialdehyde (2.7 equivalents) and 120 mg $Pd(OH)_2/C$ (20%) were added, and the mixture was stirred under H_2 for 5 h. After removal of the catalyst, the filtrate was concentrated under reduced pressure. Chromatography of the residue gave 394 mg **9** (89%).

M.p.: 118–119°C; ^{1}H NMR (CD_3OD, δ, 200 MHz): 3.66 (3H, s, OMe), 3.49 (2H, ddd, $J_{1a,2}=J_{4,5a}=$ 10.5 Hz, $J_{1e,2}=J_{4,5e}=$ 4.8 Hz, $J_{2,3}=J_{3,4}=$ 9.2 Hz, H-2, H-4), 3.09 (1H, H-3), 3.00 (2H, dd, $J_{1a,1b}=J_{5a,5b}=$ 11.0 Hz, H-1a, H-5a), 2.37 (4H, m, H-1′, H-1″, H-5′, H-5″), 1.92 (2H, H-1b, H5b), 1.76–1.24 (6H, m, H-2′, H-2″, H-3′, H-3″, H-4′, H-4″) ppm; ^{13}C NMR (CD_3OD, δ, 50 MHz): 174.7 (C-6′), 79.2 (C-3), 70.2 (C-2, C-4), 58.3 (C-1, C-5), 57.6 (C-1′), 50.8 (OCH_3), 33.5 (C-5′), 26.8, 26.2, 24.7 (C-2′, C-3′, C-4′) ppm.

N-(6-Amino)-hexyl-1,5-dideoxy-1,5-iminoxylitol (**10**; $C_{11}H_{24}N_2O_3$)

To a 5% solution of 112 mg **8** (0.49 mmol), 105 mg $Pd(OH)_2/C$ (20%) were added, and the mixture was stirred under H_2 for 5 days. After removal of the catalyst, the filtrate was concentrated under reduced pressure. Chromatography of the residue gave 52 mg of syrupy **10** (46%).

^{1}H NMR (D_2O, δ, 200 MHz): 3.45 (2H, ddd, $J_{1e,2}=J_{4,5e}=$ 4.8 Hz, $J_{2,3}=J_{3,4}=$ 10.1 Hz, H-2, H-4), 3.15 (1H, H-3), 3.05 (2H, dd, $J_{1a,1b}=J_{5a,5b}=$ 11.0 Hz, H-1a, H-5a), 2.81 (2H, t, $J=$ 7.3 Hz, H-1′, H-1″), 2.49 (2H, m, H-6′, H-6″), 2.13 (2H, dd, H-1b, H-5b), 1.60–1.04 (8H, m, H-2′, H-2″, H-3′, H-3″, H-4′, H-4″, H-5′, H-5″) ppm; ^{13}C NMR (D_2O, δ, 50 MHz): 77.4 (C-3), 68.8 (C-2, C-4), 57.0 (C-1, C-5), 56.1 (C-1′), 39.5 (C-6′), 26.7, 26.0, 25.4, 24.8 (C-2′–C-5′) ppm.

Inhibition constants (K_i values, *μM*) for substituted iminoxylitols with *Agrobacterium sp.* β-glucosidase (8.5×10^{-5} mg · cm^{-3}) were determined at 37°C in *pH* 7.0 Na_3PO_4 buffer (45 *mM*) containing 0.1% bovine serum albumin using a fixed concentration of substrate, 4-nitrophenyl-β-*D*-glucopyranoside (0.11 *mM* $= 1.5 \times K_m$). Inhibition constants (K_i values, *μM*) for substituted iminoxylitols with *Thermoanaerobacterium saccharolyticum* β-xylosidase (0.10 mg · cm^{-3}) were determined at 37°C in *pH* 5.5 sodium citrate – phosphate buffer containing 0.01% bovine serum albumin using a fixed concentration of substrate, phenyl-β-*D*-xylopyranoside (1.7 m*M* $= 1.0 \times K_m$).

Acknowledgements

Financial support by the Austrian *Fonds zur Förderung der Wissenschaftlichen Forschung* (FWF), Vienna, Project P 13593 CHE, and by the Protein Engineering Network of Centers of Excellence of Canada are appreciated.

References

[1] For recent reviews, see Saddler JN (1988) Microbiol Rev **52**: 305; Bajpai P (1997) Adv Appl Microbiol **43**: 141; Wong KKY, Tan LUL, Sakka K, Kimura T, Karita S, Ohmiya K (2000) J Biosci Bioeng **90**: 227

[2] Stütz AE (ed) Iminosugars as Glycosidase Inhibitors. Wiley-VCH, Weinheim, 1999

[3] Paulsen H (1965) Liebigs Ann Chem **683**: 187; Paulsen H, Steinert G (1967) Chem Ber **100**: 2467; Hanessian S (1965) Chem Ind (London) 1296

[4] For examples, see Bernotas RC, Papandreou G, Urbach J, Ganem B (1990) Tetrahedron Lett **31**: 3393; Lehmann J, Rob B (1995) Carbohydr Res **272**: C11; Godskesen M, Lundt I, Madsen R, Winchester B (1996) Bioorg Med Chem **4**: 1857; Smith RD, Tomas NR (2000) Synlett 193
[5] Wrodnigg TM, Withers SG, Stütz AE (2001) Bioorg Med Chem Lett **11**: 1063, and references cited therein
[6] de Raadt A, Ebner M, Ekhart CW, Fechter M, Lechner A, Strobl M, Stütz AE (1994) Catal Today **22**: 549

Received October 25, 2001. Accepted November 19, 2001

Synthesis of C-8 Deuterated Glycosides of 3-Deoxy-*D*-manno-oct-2-ulosonic Acid (*Kdo*) Related to Chlamydial Lipopolysaccharides

Paul Kosma[1,*], **Martina Strobl**[1], **Andreas Hofinger**[1], **Jens Ø. Duus**[2], **Bent O. Petersen**[2], **Klaus Bock**[2], and **Helmut Brade**[3]

[1] Institut für Chemie, Universität für Bodenkultur, A-1190 Wien, Austria
[2] Department of Chemistry, Carlsberg Laboratory, DK-2500 Valby, Denmark
[3] Forschungszentrum Borstel, Institut für Medizin und Biowissenschaften, D-23845 Borstel, BRD

Summary. Methyl glycosides of *Kdo* and a $(2 \rightarrow 8)$-linked *Kdo* disaccharide were prepared which contain a deuterium label at C-8 of the reducing unit. The label was introduced in fair diastereoselectivity upon reduction of an aldehyde group using a chiral borane complex derived from N-benzyloxycarbonyl-(*S*)-proline which produced the 8-(*S*)-deuterated derivative as the major isomer. Further coupling with a *Kdo* bromide gave the α-$(2 \rightarrow 8)$-linked disaccharide in good yield. The deprotected disaccharide serves as a model for NMR spectroscopic studies on the side chain conformation of a carbohydrate epitope from the bacterial pathogen *Chlamydia*.

Keywords. Carbohydrate; Reduction; *Kdo*; Deuterium labeling; Chlamydia.

Introduction

Members of the order Chlamydiales are intracellular parasites of increasing biomedical importance [1]. In the outer membrane of the cell, the bacteria contain a lipopolysaccharide (LPS) which constitutes a highly immunogenic epitope and induces inflammatory responses in a host organism [2]. The epitope is composed of a trisaccharide of 3-deoxy-*D*-manno-oct-2-ulopyranosylonic acid (*Kdo*) residues of the sequence α-*Kdop*-$(2 \rightarrow 8)$-α-*Kdop*-$(2 \rightarrow 4)$-α-*Kdop* which may be exploited for the serological diagnosis of chlamydial infections [3]. Previously, the terminal disaccharide unit of the epitope has been synthesized as an allyl glycoside, and the crystal structure has been determined [4, 5]. The structural data revealed the presence of an interglycosidic hydrogen bond extending from the terminal carboxyl group to OH-7 of the reducing *Kdo* unit. Although the side chain linkage between the *Kdo* moieties inherently confers additional conformational freedom, it

* Corresponding author. E-mail: pkosma@edv2.boku.ac.at

nonetheless is part of a specific epitope which is recognized by monoclonal antibodies. The binding interaction of the disaccharide with *Kdo*-specific monoclonal antibodies has been thoroughly studied using NOE experiments [6] and immunochemical methods [7]. Meanwhile, preliminary crystal data have been obtained from a complex of the disaccharide ligand with the *Fab* fragment of a $(2 \rightarrow 8)$-linked *Kdo* disaccharide specific monoclonal antibody [8].

The determination of the solution conformation of the disaccharide ligand, however, has been hampered by the fact that the geminal protons at C-8 of the reducing *Kdo* unit occur at identical chemical shifts in the ^{1}H NMR spectrum [9, 10]. Hence we have set out to synthesize *Kdo* glycosides deuterated at C-8 in order to be able to differentiate between H-8-(*R*) and H-8-(*S*) to be used for NOESY and ROESY experiments and conformational studies of the hydroxymethyl groups of *Kdo*-glycosides [11].

Results and Discussion

Synthesis of glycosides

For the synthesis of the deuterated *Kdo* glycosides, the previously reported 8-O-trityl-derivative **1** [12] was O-benzylated using O-benzyl trichloroacetimidate/triflic acid [13] to give the 7-O-benzyl derivative **2** in 70% yield. Removal of the trityl group with formic acid in MeCN proceeded smoothly in 84% yield. Subsequent *Swern* oxidation [14] of **3** afforded the aldehyde **4** in 87% yield.

Numerous methods have been developed for the stereoselective reduction of carbonyl groups, which in general are far less selective for aldehydes compared to keto groups [15]. Low diastereoselectivities were observed in the reduction of compound **4** with deuterated boranes (Table 1). The outcome of the reaction was based on the integration values of the H-8-(*R*) and H-8-(*S*) signals in the ^{1}H NMR spectra. The use of deuterated borane complexes prepared from $NaBD_4$ and D-mannitol, (-)-menthol, or (*R*)- and (*S*)-mandelic acid did not result in selective formation of the deuterated hydroxymethyl group. The same held true for other chiral borane complexes derived from phenyl 1,3-dioxaborolidine and an (*R*)-proline complex [16, 17]. By contrast, reduction of **4** at $-126°C$ in *THF* with a complex prepared from $NaBD_4$ and N-benzyloxycarbonyl-(*S*)-proline [17] proceeded in excellent yield (99%) and with fair diastereoselectivity ($\sim$56% *de*)

Table 1. Diastereoselectivities observed upon reduction of **4** with deuterated boranes

T/°C	Solvent	Reagent	Ratio **5a**:**5b**
−70	*THF*	$NaBD_4$, (*S*)-mandelic acid	45:55
−126	*THF*	$NaBD_4$, N-*Fmoc*-(*R*)-proline	50:50
−126	*THF*	$NaBD_4$, N-*Fmoc*-(*S*)-proline	37:63
−126	*THF*	$NaBD_4$, N-*Cbz*-(*R*)-proline	50:50
−126	*THF*	$NaBD_4$, N-*Cbz*-(*S*)-proline	22:78 (*de* = 56)
−126	*THF*	^{2}H-Catecholborane/B-*n*-butyl-oxazaborolidine [15]	60:40

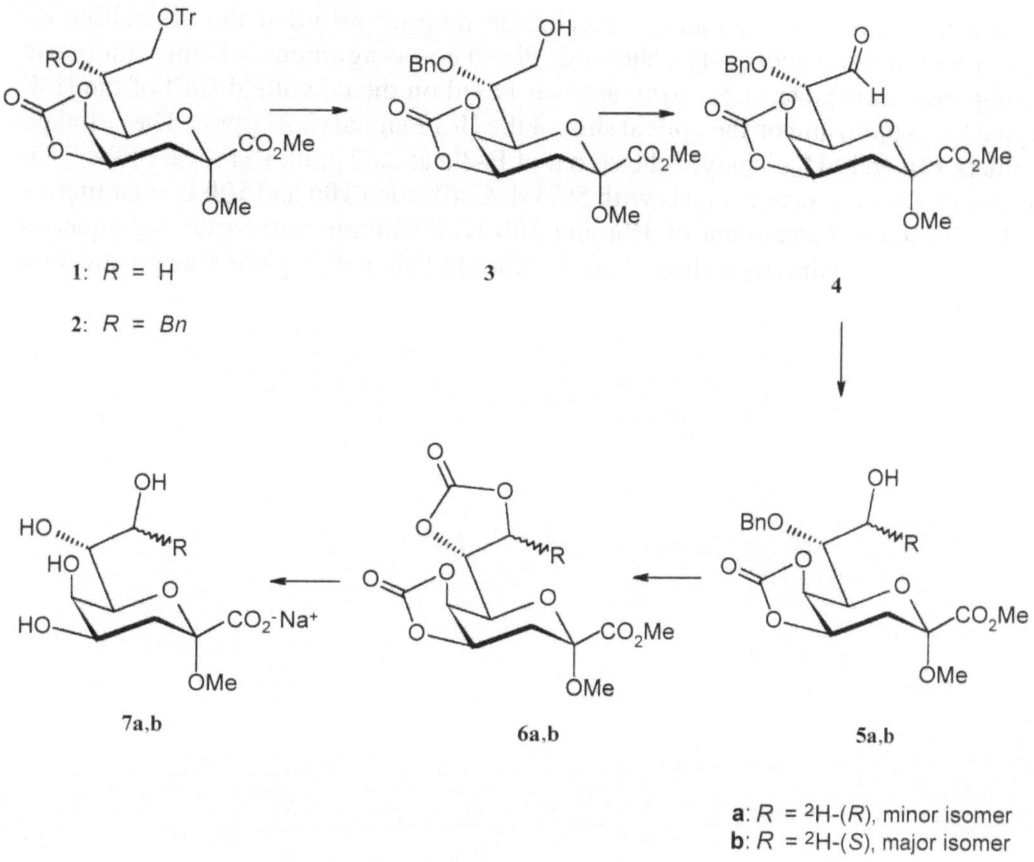

Scheme 1

to give a mixture of **5a** and **5b**. For the assignment of the H-8-(*R*) and H-8-(*S*) protons, the mixture of diastereoisomers **5a** and **5b** was subsequently converted into a cyclic 7,8-O-carbonate ester following quantitative removal of the benzyl group by hydrogenolysis on 5% Pd–C. Reaction of the resulting diol with diphosgene/*sym*-collidine in *THF* at low temperature gave the 4,5;7,8-di-O-carbonyl derivatives **6a** and **6b** in 40% yield (2 steps). Irradiation of H-6 lead to a significant nuclear *Overhauser* enhancement for the H-8 signal observed at 4.62 ppm, whereas the second (minor) H-8 signal at 4.66 ppm remained unaffected. Thus, the major isomer was assigned as the ^{2}H-8-(*S*) substituted isomer **6b**, since only H-8 corresponding to H-8-pro-*R* in the non-labeled compound would be close to H-6. For comparison, compounds **6a** and **6b** were fully deprotected by *Zemplén* de-O-acylation and subsequent hydrolysis of the methyl ester group with aqueous 0.2 *M* NaOH. The ^{1}H NMR data of the methyl ketoside **7b** are in good agreement with reported data for an 8-(*S*) deuterated *Kdo* ammonium salt [18]. In the α-pyranose form of the latter compound, H-8-(*R*) was observed at 3.62 ppm ($J_{7,8(R)} = 5.8$ Hz), which closely matches the corresponding values found for isomer **7b** (3.64 ppm, $J_{7,8(R)} = 6.5$ Hz).

Glycosylation of **5a** and **5b** with 2.5 equivalents of *Kdo* bromide donor **8** using $Hg(CN)_2$ in $MeNO_2$ furnished the glycal ester **12** and the disaccharides **9** and **11** in 85% yield (based on **5**, $\alpha/\beta = 5{:}1$) which were separated by chromatography.

Crystallization of the anomeric disaccharide mixture afforded the crystalline α-$(2 \rightarrow 8)$-linked disaccharides **9a** and **9b**. The assignment of the anomeric configuration of the terminal *Kdo* unit was based on the downfield shift of the H-4′ signal (5.33 ppm) and on the upfield shift of the H-3′*e* signal (2.27 ppm). The β-linked isomers **11a** and **11b** displayed the signal of H-3′*e* at 2.42 ppm. Removal of the 7-O-benzyl group by hydrogenolysis with 5% Pd–C afforded **10a** and **10b** in quantitative yield. Subsequent treatment of **10a** and **10b** with sodium methoxide and aqueous 0.2 *M* NaOH gave the target disaccharides **13a** and **13b** in 95% yield after purification on Bio-Gel P2.

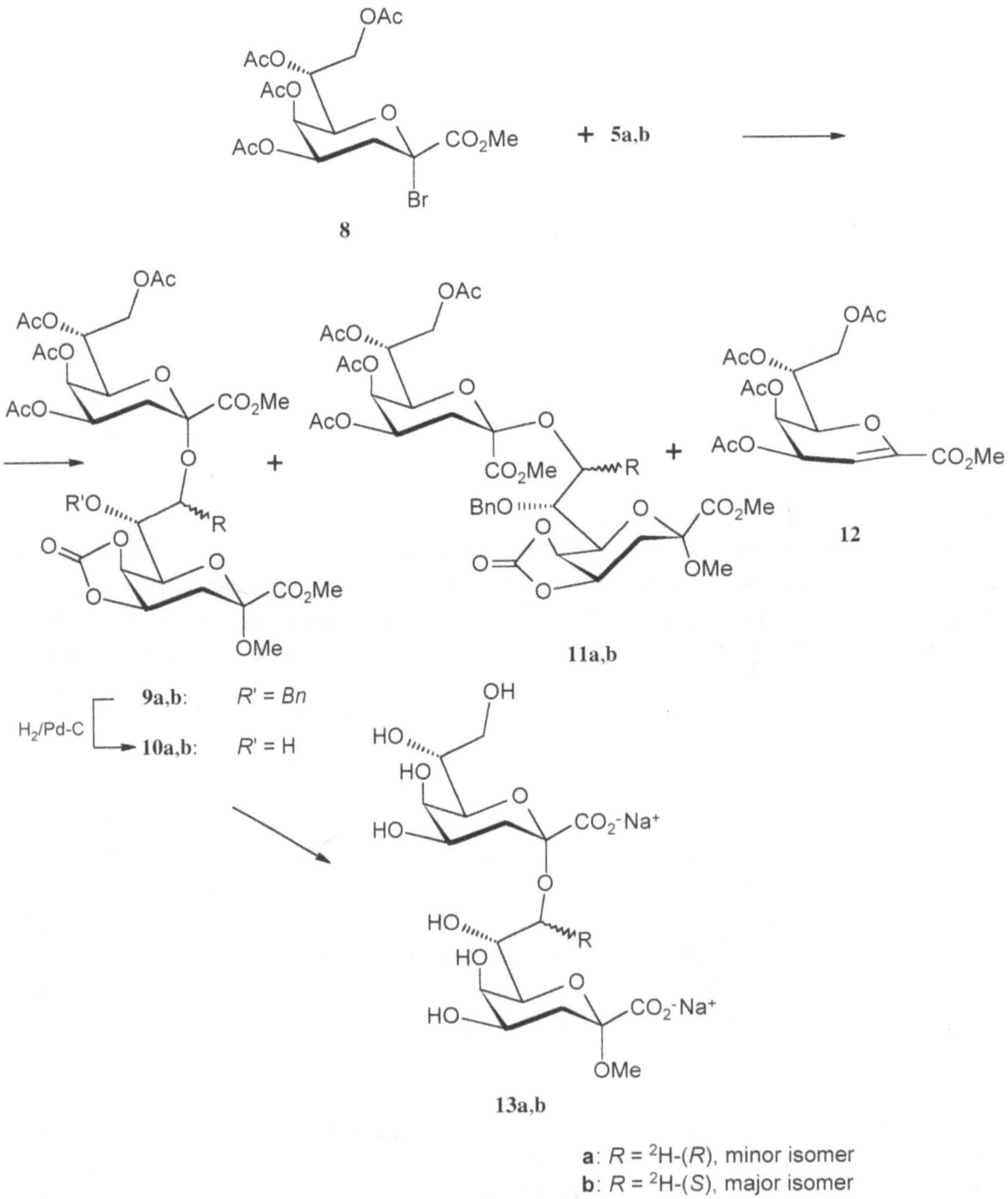

Scheme 2

NMR spectroscopic analysis of the glycosides

The assignment of the mixture of **13a** and **13b** was performed from data obtained at 800 MHz using standard one- and two-dimensional experiments. Even at the high field used severe overlap was observed, and some of the important ^1H–^1H coupling constants could only be obtained from slices of two-dimensional spectra acquired with good digital resolution (*e.g.* $^3J_{7,8}$ of the ^2H-8-(*S*) compound from a slice of an HSQC spectrum, Fig. 1). The identities of the *R* and *S* compounds in the mixtures were assigned based on the relative signal intensity, with the *S* compound as the major component. All chemical shifts except that of H-8 were completely identical for the two compounds; the corresponding data are reported in Table 2.

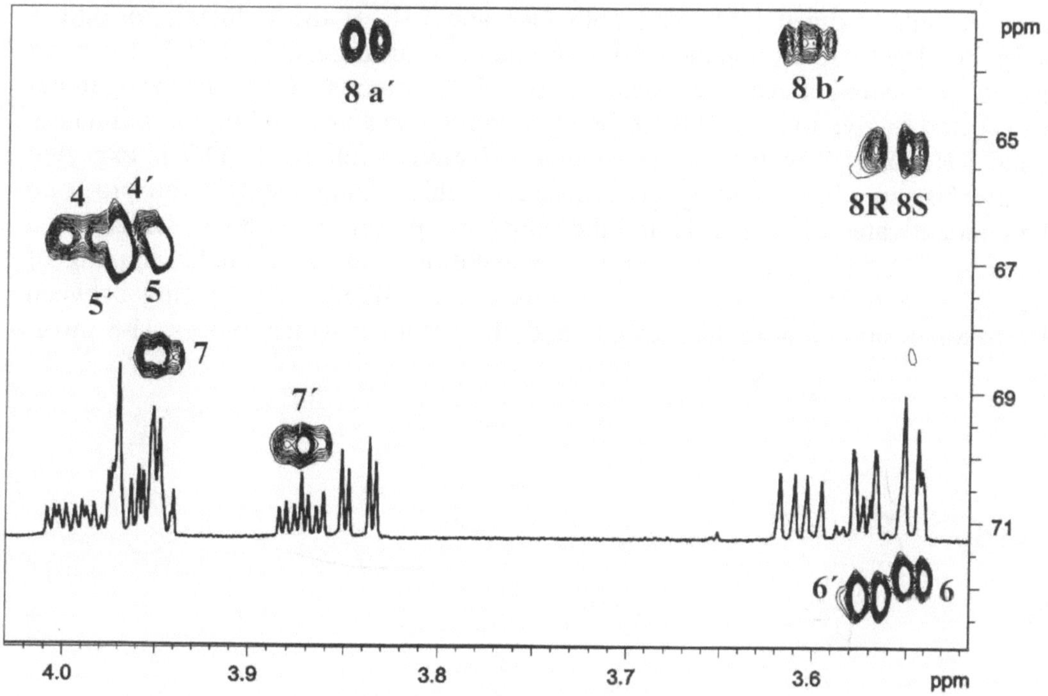

Fig. 1. Two-dimensional plot of the 800 MHz HSQC spectrum of **13a** and **13b** displaying cross-peaks of the side chain protons

Table 2. NMR parameters for **13a** and **13b** assigned at 800 MHz for ^1H and 201 MHz for ^{13}C (the only difference between **13a** and **13b** is observed for H-8)

Position	$\delta(^1H)$/ppm	J/Hz	$\delta(^{13}C)$/ppm	Position	$\delta(^1H)$/ppm	J/Hz	$\delta(^{13}C)$/ppm
MeO	3.107		51.3				
3*a*	1.725	12.2	34.9	3′a	1.754	12.2	34.9
3*e*	1.952	5.1, 13.0		3′e	2.006	5.1, 12.9	
4	3.965		66.7	4′	3.996	3.0	66.6
5	3.970	<1	66.9	5′	3.950	<1	66.9
6	3.547	7.3	71.9	6′	3.571	8.9	72.2
7	3.949		68.5	7′	3.872		69.9
H-8 of ^2H-(*R*)	3.569	<2	65.3	8′a	3.843	2.8	63.6
H-8 of ^2H-(*S*)	3.547	5.5		8′b	3.606	6.5, 11.8	

As expected, a difference in the isotope effect between the two compounds in the mixture is too small to be observed.

The assignment of the H-8 protons allows for a more thorough analysis of the conformational preference around the C-7–C-8 linkage of the reducing residue. Previous studies [6] have been hampered by complete overlap of the H-8-pro-*R* and H-8-pro-*S* protons even at 800 MHz. In the present study, the analysis has solely focussed on this problem associated with the side chain conformation, as previous studies have described the overall conformational properties in detail [6, 9].

The major disaccharide **13b** is (*S*)-configured at C-8, as determined in compounds **6a** and **6b**; that is, the proton observed corresponds to H-8-pro-*R* in a nondeuterated disaccharide and *vice versa*. In the published crystal structure of the corresponding allyl disaccharide [5] the rotamer with synclinal orientation between the glycosidic oxygen (O2′) and both O-7 and C-6 of the reducing residue is observed. The $-sc$ arrangement for C-6 and $+sc$ arrangement of O-7 also gives rise to H-7 being placed between the two H-8 protons. Such an arrangement is expected to give two small three-bond coupling constants [11] of approximately 1 and 3 Hz, as H-8 with an antiperiplanar orientation relative to O-7 is expected to give the smallest coupling. The proton in this arrangement is the pro-*R* in the nondeuterated disaccharide and the remaining proton in ^{2}H-8-(*S*), which has a coupling of 5.5 Hz. This indicates that in solution some rotational averaging of the C-7–C-8 linkage takes place, and this is most likely an averaging between the rotamer observed in the crystal and the rotamer which places H-8-pro-*R*

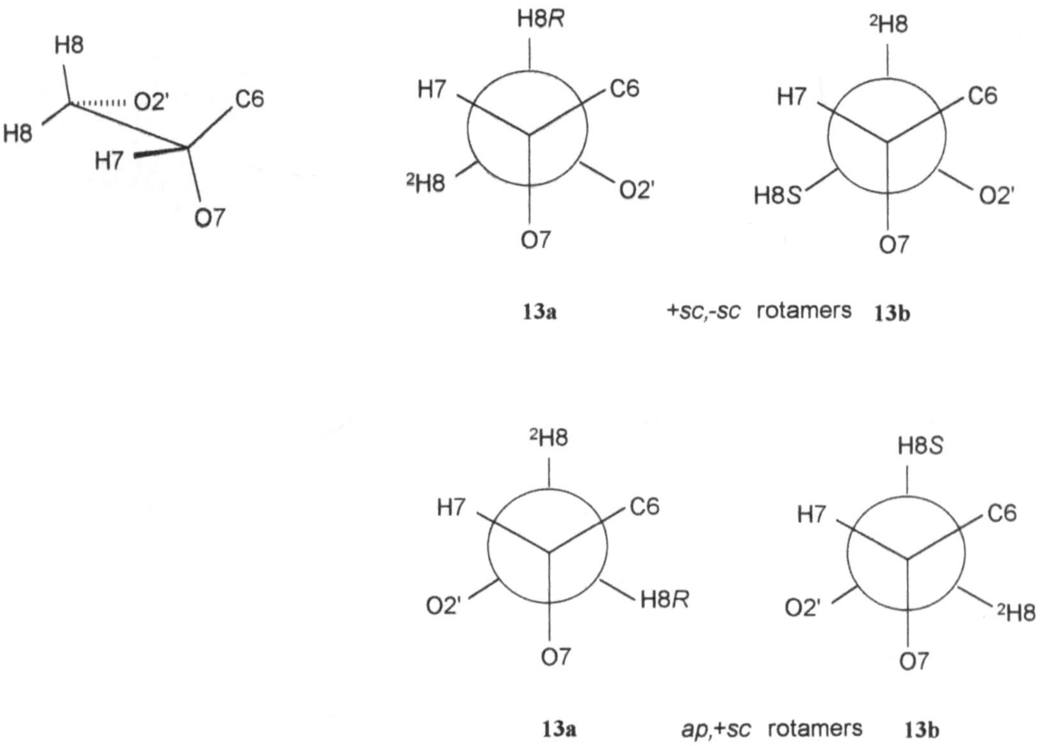

Scheme 3

antiperiplanar to H-7 [11]. Only such an averaging will give a larger value for H-8 in ^{2}H-8-(*S*) and still a small value for H-8 in ^{2}H-8-(*R*). An approximate population of the *sc,sc* and *sc,ap* rotamers of 1:1 would give a coupling constant close to 5.5 Hz for H-8 in ^{2}H-8-(*S*) (Scheme 3).

Unfortunately, no additional information about the conformation could be obtained from the NOESY spectra, partly due to overlap and partly to very small NOEs observed in general. No attempts were made to obtain ROESY spectra.

Experimental

General

Melting points were determined with a hot stage and are uncorrected. Optical rotations were measured with a Perkin-Elmer 243 B polarimeter. ^{1}H NMR spectra were recorded at 297 K with Bruker AC 300F and DPX instruments operating at 300 MHz for ^{1}H using $CDCl_3$ as solvent and *TMS* as internal standard unless stated otherwise. Coupling constants are given in Hz (first order values). Homo- and heteronuclear 2D NMR spectroscopy was performed with Bruker standard software. High-field NMR was performed in 3 mm tubes at 298 K at 799.96 MHz for ^{1}H and 201.12 MHz for ^{13}C on a Varian UNITY INOVA 800 spectrometer, using acetone (2.225 ppm) and 1,4-dioxane (67.4 ppm) as secondary standards for ^{1}H and ^{13}C, respectively.

TLC was performed on E. Merck precoated plates (5×10 cm, layer thickness 0.25 mm, Silica Gel $60F_{254}$); detection was effected by spraying with anisaldehyde-H_2SO_4. For column chromatography, silica gel (40–63 μm) was used. Concentration of solutions was performed at reduced pressure and below 40°C. Elemental analyses were provided by Dr. *J. Theiner*, Mikroanalytisches Laboratorium, Institut für Physikalische Chemie, University of Vienna; the results were in good agreement with the calculated values.

Methyl (methyl 7-O-benzyl-4,5-O-carbonyl-3-deoxy-8-O-triphenylmethyl-α-D-manno-oct-2-ulopyranosid)-onate (**2**; $C_{37}H_{36}O_9$)

A solution of 880 mg (1.64 mmol) **1** and 1.04 g (4.1 mmol) benzyl 2,2,2-trichloroacetimidate in 10 cm^3 dry CH_2Cl_2 was treated with 40 mm^3 trifluoromethane sulfonic acid for 3 h at room temperature. Solid $NaHCO_3$ (2 g) was added, and stirring was continued for 15 min. The suspension was diluted with 50 cm^3 CH_2Cl_2 and washed with satd. aq. $NaHCO_3$. The organic phase was dried ($MgSO_4$) and concentrated. Purification of the residue by flash chromatography on silica gel (toluene: EtOAc = 3:1) gave **2** as a colorless syrup.

Yield: 720 mg (70%); $[\alpha]_{20}^{D} = -14°$ (c = 1.0, $CHCl_3$); ^{1}H NMR ($CDCl_3$, δ, 300 MHz): 7.48–7.20 (m, 20 H, arom. H), 5.06 (dd, 1 H, $J_{5,4}$ = 8.8, $J_{5,6}$ = 1.5 Hz. H-5), 4.98 (t, H-4), 4.86 and 4.62 (AB, J = 10.6 Hz, OCH_2), 4.12 (dd, $J_{6,7}$ = 9.5 Hz, H-6), 4.00 (ddd, $J_{7,8a}$ = 1.6, $J_{7,8b}$ = 4.4 Hz, H-7), 3.77 (s, OMe), 3.64 (dd, $J_{8a,8b}$ = 10.3 Hz, H-8a), 3.29 (dd, H-8b), 2.82 (s, OMe), 2.72 (dd, $J_{3e,4}$ = 3.7, $J_{3e,3a}$ = 16.1 Hz, H-3*e*), 2.02 (dd, $J_{3a,4}$ = 3.2 Hz, H-3*a*) ppm.

Methyl (methyl 7-O-benzyl-4,5-O-carbonyl-3-deoxy-α-D-manno-oct-2-ulopyranosid)-onate (**3**; $C_{18}H_{22}O_9$)

A solution of 52 mg (0.083 mmol) **2** in 5 cm^3 dry CH_3CN was stirred with 2 cm^3 of a mixture of formic acid:Et_2O = 2:1 (v/v), for 15 h at room temperature and for 2 h at 40°C. Solid $NaHCO_3$ (1 g) was added, and the suspension was diluted with 50 cm^3 CH_2Cl_2 and washed with satd. aq. $NaHCO_3$. The organic phase was dried ($MgSO_4$) and concentrated. Purification of the residue by flash chromatography on silica gel (toluene:EtOAc = 1:1) afforded **3** as colorless crystals.

Yield: 27 mg (84%); m.p.: 171–172°C (EtOAc/hexane); $[\alpha]_{20}^{D} = +11°$ ($c = 1.0$, $CHCl_3$); ^{1}H NMR ($CDCl_3$, δ, 300 MHz): 7.40–7.28 (m, 5 arom. H), 5.00 (ddd, $J_{4,5} = 8.5$ Hz, H-4), 4.95 (dd, $J_{5,4} = 8.5$, $J_{5,6} = 1.5$ Hz, H-5), 4.70 and 4.65 (AB, $J = 10.9$ Hz, OCH_2), 4.05 (dd, $J_{6,7} = 9.2$ Hz, H-6), 3.95 (m, H-8a), 3.91 (dt, $J_{7,8a} = 2.5$, $J_{7,8b} = 2.5$ Hz, H-7), 3.83 (s, OMe), 3.77 (m, H-8b), 3.28 (s, OMe), 2.72 (dd, $J_{3e,4} = 3.7$, $J_{3e,3a} = 16.0$ Hz, H-3*e*), 2.09 (dd, $J_{3a,4} = 3.2$ Hz, H-3*a*), 1.82 (dd, $J = 2.9$, $J = 9.6$ Hz, OH) ppm.

Methyl (methyl 2-O-benzyl-4,5-O-carbonyl-6-deoxy-α-D-manno-oct-7-ulo-4,7-pyranosid)-onate (**4**; $C_{18}H_{20}O_9$)

A solution of 112 mm^3 (1.3 mmol) freshly distilled oxalyl chloride in 10 cm^3 dry CH_2Cl_2 was cooled to −40°C. A solution of 222 mm^3 (2.88 mmol) *DMSO* in 1 cm^3 CH_2Cl_2 was added under N_2, followed by addition of 230 mg (0.6 mmol) **3** in 2 cm^3 CH_2Cl_2. After stirring of the solution for 15 min, 0.84 cm^3 (6 mmol) triethylamine in 2 cm^3 CH_2Cl_2 was added, and the temperature was raised to 20°C within 30 min. The suspension was diluted with 50 cm^3 CH_2Cl_2, washed with H_2O, and the organic layer was dried (Na_2SO_4) and concentrated. The residue was chromatographed on silica gel (toluene:EtOAc = 1:1) which furnished **4** as colorless crystals.

Yield: 200 mg (87%); m.p.: 169–172°C (dec., CH_2Cl_2/pentane); $[\alpha]_{20}^{D} = +15°$ ($c = 0.6$, $CHCl_3$); ^{1}H NMR ($CDCl_3$, δ, 300 MHz): 9.80 (d, $J_{1,2} = 1.5$ Hz, H-1), 7.38–7.35 (m, 5 arom. H), 5.00 (ddd, $J_{6e,5} = 3.6$, $J_{6a,5} = 3.1$ Hz, H-5), 4.93 (dd, $J_{5,4} = 8.7$, $J_{4,3} = 1.5$ Hz, H-4), 4.74 and 4.62 (AB, $J =$ 10.9 Hz, OCH_2), 4.30 (d, $J_{1,2} = 1.5$ Hz, H-2), 4.06 (dd, $J_{3,2} = 9.4$ Hz, H-3), 3.83 (s, OMe), 3.19 (s, OMe), 2.81 (dd, $J_{6e,5} = 3.6$, $J_{6e,6a} = 16.2$ Hz, H-6*e*), 2.02 (dd, $J_{6a,5} = 3.1$ Hz, H-6*a*) ppm.

Methyl (methyl 7-O-benzyl-4,5-O-carbonyl-3-deoxy-8-(R)-[8-^{2}H]-α-D-manno-oct-2-ulopyranosid)-onate (**5a**; $C_{18}H_{21}DO_9$) and *methyl (methyl 7-O-benzyl-4,5-O-carbonyl-3-deoxy-8-(S)-[8-^{2}H]-α-D-manno-oct-2-ulopyranosid)-onate* (**5b**; $C_{18}H_{21}DO_9$)

N-Benzyloxycarbonyl-(*S*)-proline (1.3 g, 5.23 mmol) was lyophilized twice from 5 cm^3 D_2O and dissolved in 10 cm^3 dry *THF*. $NaBD_4$ (73 mg, 1.74 mmol) was added, and the solution was stirred under Ar for 3 h at room temperature. The reaction vessel was cooled to −126°C (methylcyclohexane and liquid N_2 as coolant), and a solution of 60 mg **4** (0.158 mmol) in 5 cm^3 *THF* was added over 30 min. After additional 30 min at −126°C the reaction was allowed to attain room temperature. The solution was concentrated and purified by column chromatography (toluene: EtOAc = 1:1) which afforded 60 mg of a mixture of **5a** and **5b** (99%) as a syrup.

Selected ^{1}H NMR data ($CDCl_3$, δ, 300 MHz): **5a**: 3.95 (d, $J_{7,8\text{-}(S)\text{-H}} = 3.0$ Hz, H-8-(*S*)); **5b**: 3.76 (d, $J_{7,8\text{-}(R)\text{-H}} = 2.5$ Hz, H-8-(*R*)).

Methyl (methyl 4,5;7,8-di-O-carbonyl-3-deoxy-8-(R)-[8-^{2}H]-α-D-manno-oct-2-ulopyranosid)-onate (**6a**; $C_{12}H_{13}DO_{10}$) and *methyl (methyl 4,5;7,8-di-O-carbonyl-3-deoxy-8-(S)-[8-^{2}H]-α-D-manno-oct-2-ulopyranosid)-onate* (**6b**; $C_{12}H_{13}DO_{10}$)

A suspension of 15 mg (0.04 mmol) **5a** and **5b** and 20 mg 5% Pd–C in 5 cm^3 MeOH was hydrogenated at atmospheric pressure for 2 h at room temperature. The suspension was filtered over a pad of Celite, and the filtrate was evaporated to dryness. A solution of the residue in 5 cm^3 pyridine was cooled to −20°C. Diphosgene (15 mm^3, 0.12 mmol) was added, and the mixture was stirred for 30 min at −20°C. MeOH (2 cm^3) was added, and the solution was coevaporated twice with toluene. The residue was chromatographed on silica gel (toluene:EtOAc = 1:3) which afforded a mixture of **6a** and **6b** as colorless crystals.

Yield: 5 mg (40%); m.p.: 154–156°C (EtOAc/pentane); $[\alpha]_{20}^{D} = +57°$ ($c = 0.2$, $CHCl_3$); ^{1}H NMR ($CDCl_3$, δ, 300 MHz): 5.06 (dt, $J_{5,4} = 8.6$, Hz, H-4), 5.00 (m, H-7), 4.85 (dd, $J_{5,6} = 1.7$ Hz, H-5), 4.66 (br d, 0.3 H, H-8-(*S*)), 4.62 (br d, 0.7 H, $J_{7,8} = 6.5$ Hz, H-8-(*R*)), 4.12 (dd, $J_{6,7} = 5.1$ Hz, H-6), 3.85 (s, OMe), 3.29 (s, OMe), 2.84 (dd, $J_{3e,3a} = 16.2$ Hz, H-3*e*), 2.10 (dd, $J_{3a,4} = 3.4$ Hz, H-3*a*) ppm.

Sodium (methyl 3-deoxy-8-(R)-[8-^{2}H]-α-D-manno-oct-2-ulopyranosid)-onate(**7a**; $C_9H_{14}DO_8Na$) and *sodium (methyl 3-deoxy-8-(S)-[8-^{2}H]-α-D-manno-oct-2-ulopyranosid)-onate* (**7b**; $C_9H_{14}DO_8Na$)

A solution of 2.5 mg of **6a** and **6b** in 4 cm^3 dry MeOH was stirred with 0.3 cm^3 of an 0.1 *M* methanolic sodium methoxide solution for 4 h at room temperature. The *pH* of the solution was adjusted to 7.5 by adding Dowex-50 H^+ resin. The suspension was filtered, and the filtrate was concentrated. The residue was dissolved in 2 cm^3 H_2O and stirred with 2 cm^3 0.2 *M* NaOH for 12 h at 0°C. The *pH* of the solution was adjusted to 7.5 by adding Dowex-50 H^+ resin, the resin was filtered off, and the filtrate was lyophilized giving 2 mg of **7a** and **7b** (100%) as an amorphous solid.

$[\alpha]_{20}^{D} = +66°$ ($c = 0.2$, H_2O); ^{1}H NMR (D_2O, δ, 300 MHz): 4.02 (ddd, $J_{5,4} = 3.0$, $J_{4,3e} = 5.0$, $J_{4,3a} = 11.5$ Hz, H-4), 4.00 (dd, $J_{4,5} = 3.0$ Hz, H-5), 3.94 (m, H-7), 3.91 (m, 0.3 H, H-8-(*S*)), 3.64 (dd, 0.7 H, $J_{8,7} = 6.5$ Hz, H-8-(*R*)), 3.55 (dd, $J_{7,6} = 9.0$ Hz, H-6), 3.14 (s, OMe), 2.01 (m, H-3*e*), 1.77 (dd, $J_{3e,3a} = 13.2$ Hz, H-3*a*) ppm.

Methyl ((methyl 4,5,7,8-tetra-O-acetyl-α-D-manno-oct-2-ulopyranosyl)-onate)-(2→8)-(methyl 7-O-benzyl-4,5-O-carbonyl-3-deoxy-8-(R)-[8-^{2}H]-α-D-manno-oct-2-ulopyranosid)-onate (**9a**; $C_{35}H_{43}DO_{20}$) and *methyl ((methyl 4,5,7,8-tetra-O-acetyl-α-D-manno-oct-2-ulopyranosyl)-onate)-(2→8)-(methyl 7-O-benzyl-4,5-O-carbonyl-3-deoxy-8-(S)-[8-^{2}H]-α-D-manno-oct-2-ulopyranosid)-onate* (**9b**; $C_{35}H_{43}DO_{20}$)

A suspension of 65 mg (0.17 mmol) **5a** and **5b**, 60 mg (0.24 mmol) $Hg(CN)_2$, and 0.3 g molecular sieve 4 Å in 4 cm^3 dry $MeNO_2$ was stirred for 15 min under N_2. A solution of 320 mg (0.66 mmol) **8** in 2 cm^3 dry $MeNO_2$ was added dropwise during 30 min at room temperature, and stirring was continued for 12 h. The suspension was diluted with 50 cm^3 CH_2Cl_2 and filtered over a pad of Celite. The filtrate was washed sequentially with aq. 5% KI and satd. aq. $NaHCO_3$, dried ($MgSO_4$), and concentrated. The residue was chromatographed on silica gel (toluene:EtOAc = 3:2) which furnished first glycal ester **12**, followed by 112 mg of the disaccharides **11a,b** and **9a,b** as a syrup (85%). Crystallization from EtOAc/ *n*-pentane furnished 72 mg of **9a,b** as colorless crystals.

M.p.: 119–126°C; $[\alpha]_{20}^{D} = +51°$ ($c = 0.5$, $CHCl_3$); ^{1}H NMR ($CDCl_3$, δ, 300 MHz): 7.38–7.28 (m, 5 arom. H), 5.33 (ddd, $J_{4',5'} = 3.1$, $J_{4',3'e} = 5.0$, $J_{4',3'a} = 11.8$ Hz, H-4′), 5.30 (br s, H-5′), 5.23 (ddd, $J_{7',6'} = 10.0$, $J_{7',8'a} = 2.6$, $J_{7',8'b} = 4.4$ Hz, H-7′), 5.03–4.93 (m, H-4, H-5), 4.80 and 4.63 (AB, $J = 10.7$ Hz, OCH_2), 4.55 (dd, $J_{8'a,8'b} = 12.3$ Hz, H-8′a), 4.15 (dd, $J_{6',5'} = 1.3$ Hz, H-6′), 4.14 (dd, H-8′b), 4.00 (m, H-7), 3.93 (dd, ~0.3 H, $J_{8(S),7} = 1.4$ Hz, H-8-(*S*)), 3.84 (dd, $J_{6,7} = 9.5$, $J_{6,5} = 1.4$ Hz, H-6), 3.83 (s, CO_2Me), 3.74 (s, CO_2Me), 3.65 (dd, ~0.7 H, $J_{8(R),7} = 6.3$ Hz, H-8-(*R*)), 3.27 (s, OMe), 2.77 (dd, $J_{3e,4} = 3.8$, $J_{3e,3a} = 16.0$ Hz, H-3*e*), 2.27 (dd, $J_{3'e,3'a} = 12.5$ Hz, H-3′*e*), 2.09 (s, Ac), 2.09 (t, $J_{4',3'a} = 11.8$ Hz, H-3′*a*), 2.06 (dd, $J_{3a,4} = 3.3$ Hz, H-3*a*), 1.99 (s, 3×Ac) ppm.

Methyl ((methyl 4,5,7,8-tetra-O-acetyl-α-D-manno-oct-2-ulopyranosyl)-onate)-(2→8)-(methyl 4,5-O-carbonyl-3-deoxy-8-(R)-[8-^{2}H]-α-D-manno-oct-2-ulopyranosid)-onate (**10a**; $C_{28}H_{37}DO_{20}$) and *methyl ((methyl 4,5,7,8-tetra-O-acetyl-α-D-manno-oct-2-ulopyranosyl)-onate)-(2→8)-(methyl 4,5-O-carbonyl-3-deoxy-8-(S)-[8-^{2}H]-α-D-manno-oct-2-ulopyranosid)-onate* (**10b**; $C_{28}H_{37}DO_{20}$)

A solution of 68 mg (0.086 mmol) **9a** and **9b** in 20 cm^3 dry MeOH was stirred with 75 mg 5% Pd–C under H_2 at atmospheric pressure for 2 h at room temperature. The catalyst was filtered off, and the filtrate was taken to dryness to afford **10a** and **10b** as a syrup.

Yield: 58 mg (quant.); $[\alpha]_{20}^{D} = +65°$ ($c = 1.2$, $CHCl_3$); ^{1}H NMR ($CDCl_3$, δ, 300 MHz): 5.35–5.27 (m, 2 H, H-5′, H-4′), 5.23 (ddd, $J_{7',6'} = 10.0$, $J_{7',8'a} = 2.4$, $J_{7',8'b} = 3.7$ Hz, H-7′), 5.03 (m, H-4, H-5), 4.54 (dd, $J_{8'a,8'b} = 12.4$ Hz, H-8′a), 4.17 (dd, $J_{7',8'b} = 3.7$ Hz, H-8′b), 4.16–4.06 (m, H-7, H-6′), 3.90 (dd, $J_{6,7} = 9.4$ Hz, H-6), 3.87–3.72 (m, H-8), 3.87 (s, CO_2Me), 3.81 (s, CO_2Me), 3.49 (br s, OH), 3.33

(s, OMe), 2.73 (dd, $J_{3e,4} = 3.3$, $J_{3e,3a} = 15.7$ Hz, H-3*e*), 2.20–2.02 (m, H-3′*e*, 3′*a*, 3*a*), 2.11, 2.08, 2.05, 2.00, 1.99 (5s, 5×Ac) ppm.

Methyl (sodium (3-deoxy-α-D-manno-oct-2-ulopyranosyl)-onate)-(2 → 8)-(sodium 3-deoxy-8-(R)-[8-^{2}H]-α-D-manno-oct-2-ulopyranosid)-onate (**13a**) and *methyl (sodium (3-deoxy-α-D-manno-oct-2-ulopyranosyl)-onate)-(2 → 8)-(sodium 3-deoxy-8-(S)-[8-^{2}H]-α-D-manno-oct-2-ulopyranosid)-onate* (**13b**; $C_{17}H_{24}DO_{15}Na_2$)

A solution of 10 mg of (0.013 mmol) **10a** and **10b** in 5 cm^3 dry MeOH was stirred with 0.5 cm^3 0.1 *M* methanolic sodium methoxide solution for 3 h at room temperature. The *pH* of the solution was adjusted to 7.5 by adding Dowex-50 H^+ resin. The suspension was filtered, and the filtrate was concentrated. The residue was dissolved in 2 cm^3 H_2O and stirred with 0.5 cm^3 0.2 *M* NaOH for 15 h at 0°C. The *pH* of the solution was adjusted to 8.5 by adding Dowex-50 H^+ resin, and the resin was filtered off. The filtrate was concentrated and purified on a column of Bio-Gel P2 (100×2.6 cm, EtOH:H_2O = 5:95) which afforded **13a** and **13b** as amorphous solids.

Yield: 5.5 mg (95%); NMR data: see Table 2.

Acknowledgments

Financial support by a grant from FWF (P-13843 CHE) is gratefully acknowledged. The 800 MHz spectra were obtained using the Varian Unity Inova spectrometer of the Danish Instrument Center for NMR Spectroscopy of Biological Macromolecules.

References

[1] Moulder JW (1991) Microbiol Rev **55**: 143
[2] Nurminen M, Leinonen M, Saikku P, Mäkelä PH (1983) Science **220**: 1279
[3] Brade H, Brade L, Nano FE (1987) Proc Natl Acad Sci USA **84**: 2508
[4] Kosma P, Schulz G, Brade H (1988) Carbohydr Res **183**: 183
[5] Mikol V, Kosma P, Brade H (1994) Carbohydr Res **263**: 35
[6] Sokolowski T, Haselhorst T, Scheffler K, Weisemann R, Kosma P, Brade H, Brade L, Peters T (1998) J Biomol NMR **12**: 123
[7] Fu Y, Baumann M, Kosma P, Brade L, Brade H (1992) Infect Immun **60**: 1314
[8] Nguyen H, Evans SL (personal communication)
[9] Bock K, Thomsen U, Kosma P, Christian R, Holst O, Brade H (1992) Carbohydr Res **229**: 213
[10] Maaheimo H, Kosma P, Brade L, Brade H, Peters T (2000) Biochemistry **39**: 12778
[11] Bock K, Duus J (1994) J Carbohydr Chem **13**: 513
[12] Kosma P, Schulz G, Unger FM (1984) Carbohydr Res **132**: 261
[13] Wessel HP, Iversen T, Bundle DR (1985) J Chem Soc Perkin Trans I, 2247
[14] Mancuso AJ, Huang SL, Swern D (1978) J Org Chem **43**: 2480
[15] Nishizawa M, Noyori R (1991) In: Trost BM, Fleming I (eds) Comprehensive Organic Synthesis, vol 8. Pergamon Press, Oxford, New York Seoul Tokyo, p 159
[16] Corey EJ, Link JO (1989) Tetrahedron Lett **30**: 6275
[17] Hayashi M, Terashima S, Koga K (1981) Tetrahedron **37**: 2797
[18] Hori H, Morita M, Ito H, Nishida Y, Ohrui H, Meguro H (1990) XVth Int Carbohydr Symposium, August 12–17, Yokohama, Abstr C030

Received September 17, 2001. Accepted October 17, 2001

Carbohydrate Auxiliaries in Stereoselective Syntheses of Decahydroquinoline Alkaloids[a]

Markus Weymann, Martin Schulz-Kukula, Stephan Knauer, and Horst Kunz*

Institut für Organische Chemie, Universität Mainz, D-55099 Mainz, Germany

Summary. Using tetra-O-pivaloyl-β-*D*-galactopyranosylamine as the chiral auxiliary, both *trans*- and *cis*-annelated decahydroquinoline alkaloids can be synthesized stereoselectively. This methodology of asymmetric synthesis is based on the effect that both enantiomers of 2,6-disubstituted piperidin-4-ones are selectively and alternatively accessible using the auxiliary as the identical stereodifferentiating tool. In addition, the carbohydrate auxiliary controls the stereoselective protonation of enolates formed by conjugate addition of cuprates to N-galactosyl octahydroquinolin-4-ones. The syntheses of *trans*-4a-*epi*-pumiliotoxin C and *cis*-4a-*epi*-perhydro-219A illustrate this concept of asymmetric synthesis of decahydroquinoline alkaloids.

Keywords. Asymmetric synthesis; Decahydroquinoline alkaloids; Glycosylamine auxiliary; Domino *Mannich-Michael* reactions; Conjugate curprate addition.

Introduction

In contrast to their extensive application as enantiopure starting materials in ex-chiral-pool syntheses [1], the utility of carbohydrates as chiral auxiliaries in asymmetric synthesis was recognized only during the past 10 years [2]. Owing to their high content of stereogenic centers and functional groups, carbohydrates facilitate arrangements of a reacting group relative to shielding or coordinating groups which give rise to a distinct stereodifferentiation during the attack of a reagent.

In this sense, glycosylamines proved to be efficient chiral auxiliaries in reactions of imines with nucleophiles, *e.g.* in *Strecker* syntheses of α-amino nitriles [3], *Ugi* reactions [4, 5], syntheses of α-amino phosphonic acids [6], *Mannich* reactions [7], and domino *Mannich-Michael* reaction sequences [8] by which glycosylimines **1** are converted into chiral piperidinone derivatives **2** (Scheme 1).

Taking advantage of the excellent stereocontrol in these reactions and subsequent conjugate additions of cuprates to compounds **2**, we have stereoselectively synthesized alkaloids of the decahydroquinoline series. Such alkaloids have been isolated from glandular secretions of South American frogs of the family *Dendrobates*. Some of these compounds show marked biological effects, *e.g.* on

[a] Dedicated to Professor Dr. *Joachim Thiem* on the occasion of his 60th birthday

* Corresponding author. E-mail: hokunz@mail.uni-mainz.de

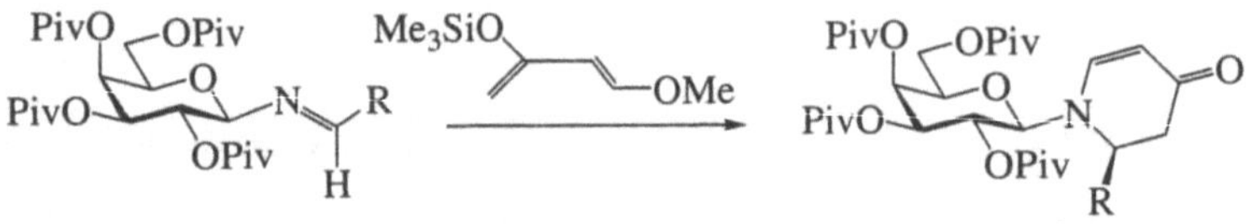

1 2

Scheme 1

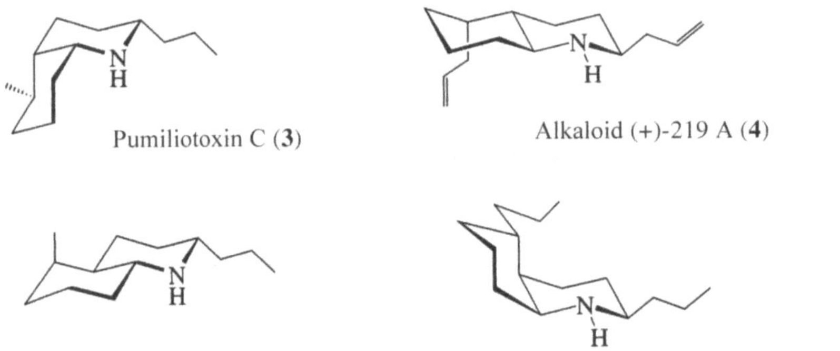

Pumiliotoxin C (**3**) Alkaloid (+)-219 A (**4**)

trans-epi-Pumiliotoxin C (**5**) *cis-epi*-(+)-perhydro-219 A (**6**)

Scheme 2

the transport of ions through membranes and on the transduction of stimuli in the nervous system [9]. Interestingly, decahydroquinoline alkaloids from *Dendrobates pumilio* such as pumiliotoxin C (**3**) [10] have a *cis*-annelated decahydroquinoline structure, whereas those from *Dendrobates histrionicus* exclusively have *trans*-configuration, such as the alkaloid (+)-219A (**4**) [11].

Stereoselective syntheses of natural (−)-(**3**) and (+)-pumiliotoxin C have already been reported [12]. An auxiliary-controlled asymmetric synthesis of **3** was achieved by *Comins et al.* [13] who applied nucleophilic addition reactions to N-alkoxycarbonyl-3-trimethylsilyl-4-trimethylsilyloxy pyridinium salts derived from 8-phenylmenthol in the key step [13].

A stereoselective synthesis of alkaloid 219A (**4**) has also been described [14]. It is based on a reductive conversion of an enol triflate and subsequent hydrogenation of the formed chiral alkene (diastereomeric ratio 87:13). Perhydro 219A and its *cis*-annelated epimer **6** have been obtained from chiral trisubstituted cyclohexanes [15].

We here describe an alternative and uniform concept for the stereoselective syntheses of both *cis*- and *trans*-annelated decahydroquinoline alkaloids. In this strategy, 2,3,4,6-tetra-O-pivaloyl-galactosylamine **7** [3, 7] as the auxiliary not only controls the stereoselective formation of the stereogenic centers in 2-, 5-, and 8a-position, but also the diastereoselective protonation of an enolate at position C-4a which can be steered to either form the *trans*- or the *cis*-annelated bicyclic compounds. The principle is demonstrated for the syntheses of *epi-trans*-pumiliotoxin C [16] (**5**) and *epi-cis*-perhydro-219A (**6**), having opposite configuration to each other at C-2 and representing the *cis-trans*-annelated epimers of the natural alkaloids **3** and **4**, respectively.

Results and Discussion

Stereoselective syntheses of both enantiomers of 2-propyl-octahydroquinolin-4-one using the identical galactosylamine auxiliary **7**

Using galactosylamine **7**, both enantiomers of 2,6-*cis*-disubstituted piperidones can selectively be obtained depending upon the choice of the aldehyde in the formation of the imine **1** and the cuprate in the conjugate addition to product **2** [8].

Condensation of **7** with hex-5-enal gave the imine **8a**, whereas the analogous reaction with butyraldehyde formed the imine **8b** [8]. Domino *Mannich-Michael* reactions of imines **8** furnished the dehydropiperidiones **9a** and **9b** (Scheme 3).

Conjugate addition of the propylcuprate BF_3 complex [17] to **9a** resulted in the formation of N-galactosyl-(2*S*,6*R*)-2-propyl-6-pentenyl-piperidone **10a** with excellent stereoselectivity. On the other hand, 1,4 addition of *bis*-(4-(1-ethoxy-ethyloxy)-butyl)-cuprate to **9b** in presence of trimethylchlorosilane and subsequent treatment of the intermediate silyl enolether with fluoride gave (2*R*,6*S*)-2-propyl-6-(ethoxyethyloxy)-butyl-piperidinone **10b** in good diastereoselectivity. It should be noticed that reactions of the enaminone **9b** with pentenyl or functionalized butylcuprate/BF_3 complexes produced the 1,4 adducts in distinctly lower diastereoselectivity.

Oxidative cleavage of the olefinic double bound of **10a** using potassium osmate/sodium periodate and subsequent cyclizing aldol condensation of the formed aldehyde **11a** furnished the octahydroquinoline **12a** with (2*S*,8a*R*)-configuration. In contrast, removal of the (1-ethoxy)-ethyl (*EE*) group from **10b** using pyridinium *p*-toluenesulfonate (*PPTS*) followed by oxidation of the alcohol with tetrapropylammonium perruthenate (*TPAP*) [18] gave aldehyde **11b**, which under aldol condensation yielded the N-galactosyl octahydroquinolinone **12b** with opposite enantiomeric configuration of the N-heterocyclic system compared to that in **12a**. These syntheses illustrate that both enantiomers of such hydroquinoline derivatives can stereoselectively be obtained by this strategy using the identical glycosylamine auxiliary.

It should be noted that, apart from usual chromatographic purification, no separation of a minor diastereomer was required during the synthesis to obtain the pure diastereomers **11** and **12**.

Stereoselective synthesis of trans-annelated decahydroquinolines: trans-epi-4a-pumiliotoxin C (**5**)

For the conversion of the (2*S*)-propyl-octahydroquinolinone diastereomer **12a** into pumiliotoxin C (**3**), 1,4 addition of lithium dimethylcuprate to the enone in presence of trimethylsilyl chloride was carried out (Scheme 4). Cleavage of the intermediate silyl enolether resulted in the formation of the adduct **13** with excellent diastereoselectivity and the expected configuration [13] at C-5 (5*R*).

However, **13** proved to be a *trans*-decahydroquinolinone derivative as was confirmed by the ^{1}H NMR signal of H-8a ($\delta = 3.26$ ppm (dt, $J_{8a,4a} = J_{8a,8'} = 11.2$ Hz, $J_{8a,8} = 3.5$ Hz)) and by X-ray analysis [16]. This result was unexpected, since *Comins et al.* [13] had carried out 1,4-methylcuprate addition to a octahydroquinolinone analogous to **12a** which carries a N-phenoxycarbonyl group instead of the

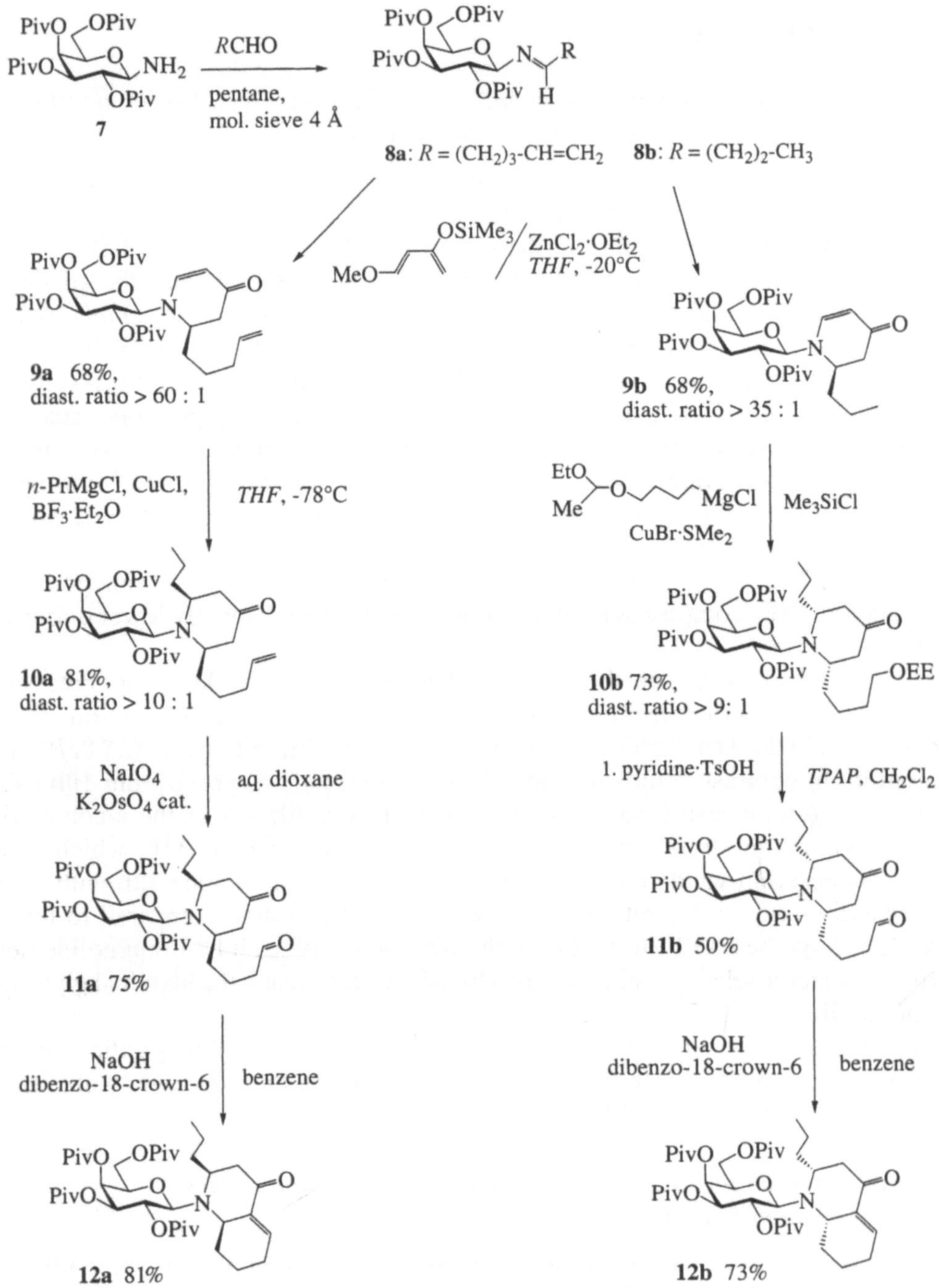

Scheme 3

N-galactosyl moiety and noted the formation of the *cis*-annelated decahydroquinolinone with high preference (97:3). In order to clarify the contradictory situation, the galactosyl auxiliary of **12a** was cleaved off by treatment with dilute HCl in methanol and substituted by the phenoxycarbonyl group (**14**). Compound **14** was identical with the precursor of the optically active pumiliotoxin C described by *Comins et al.* [13]. Conjugate addition of lithium dimethylcuprate to **14** in presence of

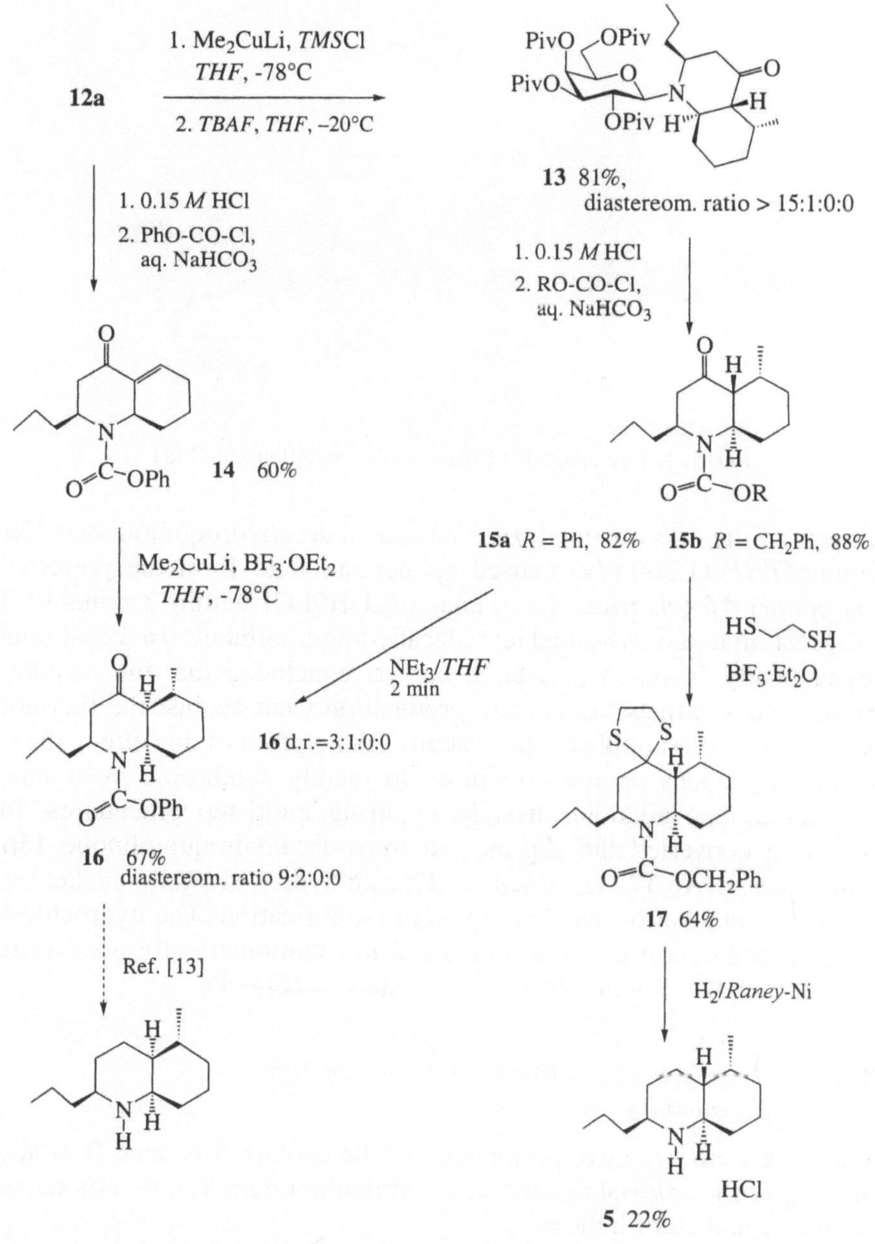

Scheme 4

trimethylchlorosilane and subsequent fluoridolysis of the intermediate silyl enolether preferentially yielded, in contrast to **13**, the *cis*-annelated decahydro-quinolinone **16**.

To prove the hypothesis that the carbohydrate auxiliary actually steers the protonation of the enolate during the cleavage of the intermediate silyl enolether obtained by 1,4-addition of the cuprate to **12a**, the galactosyl moiety was also removed from product **13** and substituted by the phenoxycarbonyl (**15a**) or benzyloxycarbonyl group (**15b**).

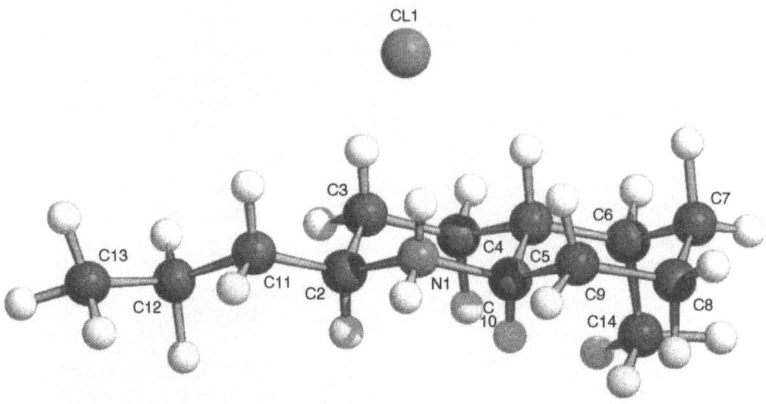

Fig. 1. X-Ray analysis of *trans*-4a-*epi*-pumiliotoxin C (**5**)

Treatment of the thus obtained *trans*-annelated decahydroquinolinone **15a** with triethylamine/*THF* (1:200 v/v) caused epimerization to form the preferred *cis*-annelated epimer **16** (*cis*:*trans* = 3:1, analytical HPLC) within 2 minutes. These results suggest that the *cis*-annelated decahydroquinolinone **16** constitutes the thermodynamically favoured epimer. It can be concluded that the carbohydrate auxiliary obviously directs the enolate protonation even against the thermodynamical preference in this simple cyclic system. Taking profit of this effect, the carbohydrate auxiliary opens up the possibility to readily synthesize *trans*-annelated decahydroquinolinone alkaloids usually requiring multistep procedures. In this sense, we have converted the Z-protected *trans*-decahydroquinolinone **15b** into its dithiolane derivative **17**. Treatment of **17** with hydrogen/*Raney* nickel resulted in simultaneous removal of the Z group and desulfurization. The hydrochloride of *trans*-4a-*epi*-pumiliotoxin C (**5**) was obtained in enantiomerically pure form. The configuration of **5** was confirmed by X-ray analysis (Fig. 1).

Interpretation of the stereoselective enolate protonation steered by the N-galactosyl auxiliary

The astonishing stereoselective protonation of the enolate **A** formed from the silyl enolether to give the *trans*-annelated decahydroquinolinone can be rationalized by stereoelectronic and steric effects.

The X-ray analysis of decahydroquinolinone **13** [16] gives evidence that the *exo*-anomeric effect, *i.e.* the delocalization of the nitrogen lone pair into the σ^*-orbital of the ring C–O bond, stabilizes the N-galactosyl heterocycle in conformation **A** with carbohydrate and heterocyclic ring systems arranged almost perpendicular to each other. Approach of the proton from the front side reduces the steric strain during ketone formation, since the carboxylic ring moves backwards, *i.e.* away from the bulky 2-pivaloyloxy group. If this steric hindrance is not present as in the case of the N-phenoxycarbonyl substituted enolate **B**, protonation preferentially leads to the *cis*-annelated decahydroquinoline **16**.

The *cis*-annelated compound **16** is the weaker CH-acid since its H-4a is located almost in the σ-plane of the carboxyl group. In contrast, the C_{4a}–H_{4a} bond of the *trans*-compound is located coaxial with the π^*-orbital of the carbonyl group and,

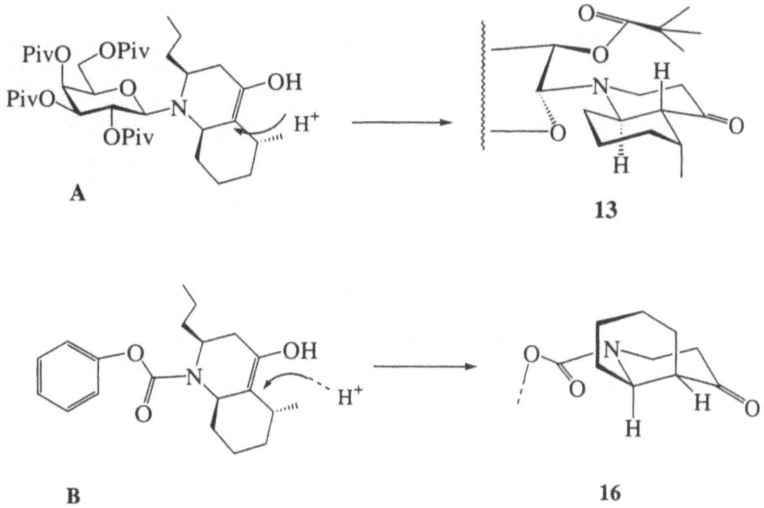

Scheme 5

therefore, the compound is considerably more acidic. If the outlined interpretation is correct, 1,4-cuprate additions to the enone in the N-galactosyl octahydroquinolinone **12b** should logically give *cis*-annelated decahydroquinolinones.

Stereoselective synthesis of cis-perhydro 219A

Reaction of N-galactosyl octahydroquinolinone **12b** with propylcuprate in the presence of trimethylsilyl chloride and subsequent cleavage of the intermediate silyl enolether gave the *cis*-annelated decahydroquinolinone **18** as predicted (Scheme 6).

The diastereoselectivity of the reaction was high, but the yield of **18** was only moderate. Obviously, the addition reaction is sterically more hindered compared to that with **12a** (Scheme 4). In the ^{1}H NMR spectrum of **18**, the double triplet with diaxial couplings $J_{h8a,H4a} = J_{H8a,8} \sim 11$ Hz characteristic for H-8a in *trans*-decahydroquinolinones is not observed. Instead, H-8a causes a multiplet at $\delta =$ 3.54 ppm which is not well structurized.

The galactosyl decahydroquinolinone **18** (as well as **13**) resists conversion into a dithiolane. Since reduction with $NaBH_4$ also failed, transformation to the corresponding enol triflate was applied. The only reagent to achieve this reaction was N-trifluoromethanesulfonyl-5-chloro-pyrid-2-yl-trifluoromethanesulfonamide (Cl*Pyr*N*Tf*$_2$) [19]. Unfortunately, the desired enol triflate **19** was formed in a mixture with its undesired regioisomer **20**. Separation of **19** was achieved by chromatography. Hydrogenation of **19** followed by acidolytic cleavage of the N-glycosidic bond gave the hydrochloride of *cis*-4a-*epi*-perhydro-219A (**6**). Comparison of the NMR data of this hydrochloride and those of the free amine **6** with those reported in the literature for the racemic [20] and optically active *cis*-perhydro-219A [15] confirm the structure of **6** and the correct stereochemistry at C-5.

The stereoselective syntheses of *trans*-*epi*-pumiliotoxin C (**5**) and *cis*-*epi*-perhydro-219A (**6**) illustrate the versatility of the galactosylamine **7** as the chiral auxiliary in stereoselective syntheses of *trans*- and of *cis*-annelated

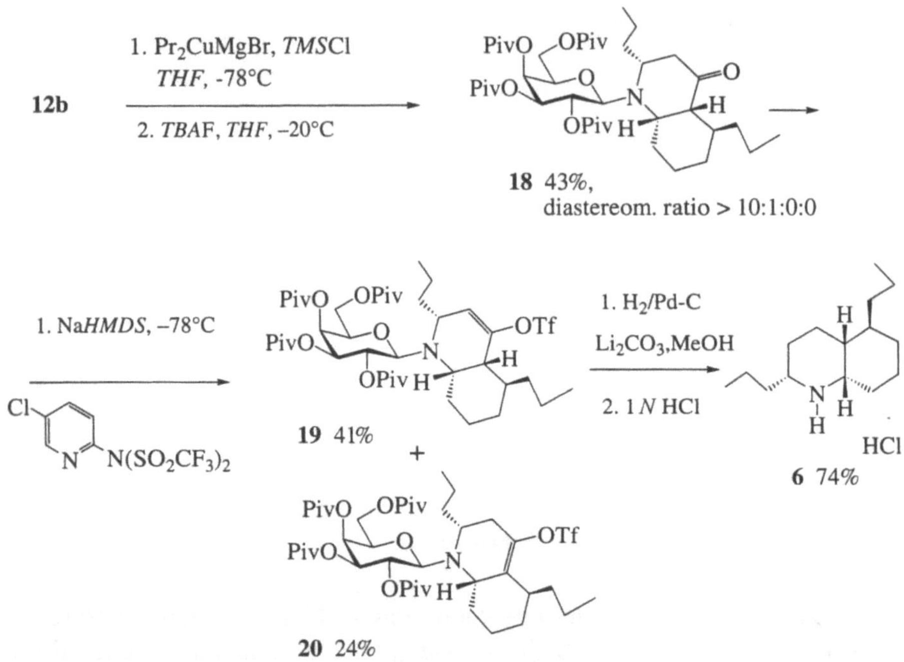

Scheme 6

decahydroquinoline alkaloids. Moreover, *cis*-annulated decahydroquinolines of enantiomeric configuration opposite to **6** are also selectively accessible *via* the detachment and epimerization pathway (Scheme 4). The value of this methodology can be expanded even more if tri-O-pivaloyl-β-*D*-arabinopyranosylamine [21] is applied as the virtual enantiomer of the *D*-galactopyranosyl amine **7**. By this way, enantiomers of *trans*-annelated decahydroquinolines, like **5**, can selectively be constructed.

Experimental

NMR spectra were recorded on Bruker AM 400 (400 MHz ^{1}H, 100.6 MHz ^{13}C) and Bruker WT 200 (200 MHz ^{1}H, 50.3 MHz ^{13}C) NMR spectrometers. Chemical shifts are given in ppm relative to *TMS*. For N-galactosylated heterocyclic compounds, protons and carbons of the carbohydrate moiety are marked with a superscribed dash, *e.g.* H-2′. Analytical HPLC was carried out using a Pharmacia LBK 2150 unit with diode array detection. Columns: A: Eurospher 100, RP 18, 5 μ, 250 × 4 mm, Knauer Eurochrom, B: Sperisorb ODS II, RP 18, 5 μ, 250 × 4 mm, Bischoff; flow rate: 1 cm^3/min. Thin layer chromatography (TLC) was performed on Merck silica gel 60_{F254}, column chromatography on silica gel 60 (0.06–0.2 mm, Baker). FAB and FD mass spectra were measured on a Finnigan MAT 95 spectrometer, optical rotations on a Perkin Elmer 241 polarimeter. Melting points were taken on a Büchi Dr. Tottoli apparatus and are uncorrected. Elemental analysis data of new compounds were in agreement with the calculated values.

Schiff bases **8** of commercially available galactosylamine **7** (Merck-Schuchardt) were prepared according to a described procedure [22]. The synthesis of compound **9b** is described in Ref. [8].

(2R)-N-(2,3,4,6-Tetra-O-pivaloyl-β-D-galactopyranosyl)-2-(4-pentenyl)-5,6-didehydropiperidin-4-one (**9a**; $C_{36}H_{57}NO_{10}$)

9a was synthesized according to the general procedure described in Ref. [8] and purified by chromatography (petroleum ether:ethyl acetate = 4:1).

Yield: 68%; m.p.: 155–158°C; $[\alpha]_D^{22} = -61.5$ ($c = 1$, $CHCl_3$); $R_f = 0.30$ (petroleum ether: EtOAc = 3:1); *d.r.* = 65:1 (HPLC, column A; MeOH/H_2O 4:1); ^{1}H NMR (400 MHz, δ, $CDCl_3$): 1.09, 1.15 and 1.26 (4s, 4 × 9H, piv-CH_3), 1.41 (m, 1H, CH_2), 1.60 (m, 2H, CH_2), 1.88 (m, 1H, CH_2), 2.00 (m, 2H, CH_2), 2.34 (d, 1H, $J_{gem} = 16.6$ Hz, H-3*ax*), 2.60 (dd, 1H, $J_{vic} = 6.3$ Hz, $J_{gem} = 16.6$ Hz, H-3*eq*), 3.70 (m, 1H, H-2), 3.95 (dd, 1H, $J_{6a',5'} = 6.9$ Hz, $J_{6a',6b'} = 10.5$ Hz, H-6a′), 4.02 (dd, 1H, $J_{5',6a'} =$ 6.8 Hz, $J_{5',6b'} = 6.4$ Hz, H-5′), 4.14 (dd, 1H, $J_{6b',5'} = 6.3$ Hz, $J_{6b',6a'} = 10.5$ Hz, H-6b′), 4.54 (d, 1H, $J_{1',2'} = 9.1$ Hz, H-1′), 4.96 (m, 3H, H-5 and CH=CH_2), 5.16 (dd, 1H, $J_{3',4'} = 3.1$ Hz, $J_{3',2'} =$ 10.1 Hz, H-3′), 5.41 (d, 1H, $J_{4',3'} = 3.0$ Hz, H-4′), 5.52 (dd, 1H, $J_{2',3'} = 9.8$ Hz, $J_{2',1'} = 9.5$ Hz, H-2′), 5.71 (dddd, 1H, $J_1 = 6.7$ Hz, $J_2 = 10.2$ Hz, –CH=CH_2), 6.88 (dd, 1H, $J_1 = 0.9$ Hz, $J_2 = 7.7$ Hz, H-6) ppm; ^{13}C NMR (100.6 MHz, δ, $CDCl_3$): 25.08 (CH_2), 26.98, 27.08, 27.14 (piv-CH_3), 30.22, 33.34 (CH_2), 38.65, 38.71, 38.85, 39.03 (piv-C_{quart}), 53.60 (Alkyl-CHN), 60.81 (C-6′), 65.77, 66.52, 71.28, 72.88 (C-2′, C-3′, C-4′, C-5′), 91.56 (C-1′), 99.96 (C-5), 115.01, 137.85 (–CH=CH_2), 149.75 (C-6), 176.42, 176.99, 177.03, 177.63 (pivC=O), 191.91 (C=O) ppm.

(2R,6S)-N-(2,3,4,6-Tetra-O-pivaloyl-β-D-galactopyranosyl)-2-(5-pentenyl)-6-propyl-piperidin-4-one (**10a**; $C_{39}H_{65}NO_{10}$)

10a was synthesized from **9a** (0.66 g, 1 mmol) according to the general procedure for conjugate addition of RCu · BF_3 described in Ref. [8] and purified by column chromatography (petroleum ether: EtOAc = 4:1).

Yield: 0.57 g (81%); m.p.: 122°C; $[\alpha]_D^{22} = -15.9$ ($c = 1$, $CHCl_3$); $R_f = 0.56$ (petroleum ether:EtOAc = 3:1); *d.r.*≫10:1; (^{1}H NMR); ^{1}H NMR (200 MHz, δ, $CDCl_3$): 0.86 (t, 3H, $J = 6.8$ Hz, CH_3), 1.10, 1.14, 1.15 and 1.24 (4s, 4 × 9H, piv-CH_3), 1.38 (m, 7H, CH_2), 1.61 (m, 1H, CH_2), 1.99 (m, 2H, CH_2), 2.26 (dd, 1H, $J_{vic} = 7.0$ Hz, $J_{gem} = 15.5$ Hz, CH_2C=O), 2.35 (d, 1H, $J_{gem} = 15.6$ Hz, CH_2C=O′), 2.61 (dd, 1H, $J_{vic} = 5.6$ Hz, $J_{gem} = 15.0$ Hz, CH_2C=O), 2.69 (dd, 1H, $J_{vic} = 6.6$ Hz, $J_{gem} = 15.4$ Hz, CH_2C=O), 3.32 (m, 1H, H-2), 3.46 (m, 1H, H-6), 3.85–3.96 (m, 2H, H-5′, H-6a′), 4.09 (dd, 1H, $J_{6b',5'} = 9.2$ Hz, $J_{6b',6a'} = 13.1$ Hz, H-6b′), 3.88 (d, 1H, $J_{1',2'} = 9.2$ Hz, H-1′), 4.91–5.01 (m, 2H, olefin-CH_2), 5.08 (dd, 1H, $J_{3',4'} = 2.9$ Hz, $J_{3',2'} = 9.9$ Hz, H-3′), 5.35 (d, 1H, $J_{4',3'} = 2.9$ Hz, H-4′), 5.45 (t, 1H, $J = 9.6$ Hz, H-2′), 5.71 (ddd, 1H, $J_1 = 6.5$ Hz, $J_2 = 10.2$ Hz, $J_3 = 16.9$ Hz, *CH*=CH_2) ppm; ^{13}C NMR (100.6 MHz, δ, $CDCl_3$): 13.91 (CH_3), 20.65, 25.75 (CH_2), 27.05, 27.16, 27.25 (piv-CH_3), 33.59, 37.17 (CH_2), 38.71, 39.02 (piv-C_{quart}), 40.77 (CH_2), 44.12, 44.43 (C-3, C-5), 61.66 (C-6′), 65.91, 67.22, 71.90, 72.55 (C-2′, C-3′, C-4′, C-5′), 94.07 (C-1′), 114.99, 138.11 (–CH–CH_2), 176.66, 176.77, 177.37, 177.84 (piv-C=O) ppm.

(2S,6R)-N-(2,3,4,6-Tetra-O-pivaloyl-β-D-galactopyranosyl)-2-(4-(1-ethoxyethyloxy)-butyl)-6-propyl-piperidin-4-one (**10b**; $C_{42}H_{73}NO_{12}$)

10b was prepared from 3.56 cm^3 **9b** [8] (5.6 mmol) according to the general procedure for conjugate addition of R_2CuMgX in presence of Me_3SiCl as described in Ref. [8] and purified by column chromatography (petroleum ether:EtOAc = 5:1).

Yield: 3.19 g (73%); colorless oil; $[\alpha]_D^{22} = -11.8$ ($c = 1$, $CHCl_3$); $R_f = 0.61$ (petroleum ether: EtOAc = 3:1); *d.r.* > 9:1 (^{1}H NMR); ^{1}H NMR (400 MHz, δ, $CDCl_3$): 0.82 (t, 3H, $J = 7.0$ Hz, propyl-CH_3), 1.01–1.43 (m, 48H, piv-CH_3, CH_2, *EE*O-CH_3), 1.45–1.57 (m, 4H, CH_2), 2.19 (dd, 1H, $J_{vic} = 7.0$ Hz, $J_{gem} = 15.5$ Hz, CH_2C=O), 2.31 (d, 1H, $J_{gem} = 15.4$ Hz, CH_2C=O), 2.55 (dd, 1H, $J_{vic} = 5.9$ Hz, $J_{gem} = 15.3$ Hz, CH_2C=O′), 2.63 (dd, 1H, $J_{vic} = 6.4$ Hz, $J_{gem} = 15.2$ Hz, CH_2C=O′), 3.27–3.60 (m, 6H, H-2, H-6, *R*CH_2O, CH_3*CH*$_2$O), 3.81–3.89 (m, 2H, H-5′, H-6a′), 4.03 (dd, 1H, $J_{6b',5'} = 9.3$ Hz, $J_{6b',6a'} = 13.2$ Hz, H-6b′), 4.33 (d, 1H, $J_{1',2'} = 9.3$ Hz, H-1′), 4.59 (m, 1H, OCHO), 5.04 (dd, 1H, $J_{3',2'} = 9.9$ Hz, $J_{3',4'} = 3.0$ Hz, H-3′), 5.30 (d, 1H, $J_{4',3'} = 2.8$ Hz, H-4′), 5.38 (t, 1H, $J = 9.6$ Hz, H-2′) ppm; ^{13}C NMR (100.6 MHz, δ, $CDCl_3$): 13.87 (propyl-CH_3), 15.27 (OCH_2*CH*$_3$), 19.66 (CH_2), 19.83 (OCHO*CH*$_3$), 24.13, 24.17 (CH_2), 27.03, 27.16, 27.21 (piv-CH_3), 29.73, 29.84 (CH_2), 38.35, 38.65, 38.69, 39.00 (piv-C_{quart}), 44.00, 44.46 (C-3, C-5), 52.02 (NCH*R*), 60.60 (C-6′),

61.65 (RCH_2O), 65.01 (CH_3CH_2O), 66.03, 67.26, 71.96, 72.56 (C-2′, C-3′, C-4′, C-5′), 94.20 (C-1′), 99.51 (OCHO), 176.57, 176.62, 177.25, 177.72 (piv-C=O), 210.00 (C=O) ppm.

(2R,6S)-N-(2,3,4,6-Tetra-O-pivaloyl-β-D-galactopyranosyl)-2-(4-oxy-butyl)-6-propyl-piperidin-4-one (**11a**; $C_{38}H_{63}NO_{11}$)

To a solution of 0.58 g (0.81 mmol) **10a** in 20 cm^3 oxygen-free 1,4-dioxane under Ar, a solution of 10 cm^3 $K_2OsO_4 \cdot 2H_2O$ (3.5 mol%) in 5 cm^3 oxygen-free H_2O was added. The yellowish-brown solution was stirred for 15 min, and 0.87 g (4.1 mmol) $NaIO_4$ were added. After stirring for 15 h, 20 cm^3 diethyl ether and 40 cm^3 H_2O were added, and the aqueous layer was separated and extracted with 3×30 cm^3 diethyl ether. The combined organic solutions were washed with 2×20 cm^3 10% Na_2SO_3 solution and dried with Na_2SO_4. After evaporation of the solvent *in vacuo*, the crude product was purified by column chromatography (petroleum ether:EtOAc = 3:1).

Yield: 0.43 g (73%); amorphous solid; $[\alpha]_D^{22} = -11.9$ ($c = 1$, $CHCl_3$); $R_f = 0.57$ (petroleum ether: EtOAc = 2:1); ^1H NMR (400 MHz, δ, $CDCl_3$): 0.84 (t, 3H, $J = 7.2$ Hz, CH_3), 1.07, 1.12 and 1.22 (4s, $4 \times 9H$, piv-CH_3), 1.27 and 1.38 (m, 5H, CH_2), 1.51–1.74 (m, 3H, CH_2), 2.25 (dd, 1H, $J_{vic} = 6.7$ Hz, $J_{gem} = 15.4$ Hz, CH_2C=O), 2.33 (d, 1H, $J_{gem} = 15.0$ Hz, CH_2C=O′), 2.40 (m, 2H, CH_2CHO), 2.60–2.70 (m, 2H, H-3, H-5), 3.34 (m, 1H, H-2), 3.46 (m, 1H, H-6), 3.84–3.92 (m, 2H, H-5′, H-6a′), 4.08 (dd, 1H, $J_{6b',5'} = 9.2$ Hz, $J_{6b',6a'} = 13.0$ Hz, H-6b′), 4.36 (d, 1H, $J_{1',2'} = 9.2$ Hz, H-1′), 5.07 (dd, 1H, $J_{3',4'} = 2.9$ Hz, $J_{3',2'} = 9.9$ Hz, H-3), 5.34 (d, 1H, $J_{4',3'} = 2.9$ Hz, H-4′), 5.44 (t, 1H, $J = 9.6$ Hz, H-2′), 9.71 (s, 1H, CHO) ppm; ^{13}C NMR (100.6 MHz, δ, $CDCl_3$): 13.85 (CH_3), 19.16, 20.62 (CH_2), 27.06, 27.17, 27.24 (piv-CH_3), 37.09 (CH_2), 38.69, 38.72, 39.03 (piv-C_{quart}), 40.67 (CH_2), 43.59, 44.14, 44.43 (C-3, C-5, CH_2CHO), 61.70 (C-6′), 66.01, 67.30, 72.09, 72.57 (C-2′, C-3′, C-4′, C-5′), 94.34 (C-1′), 176.60, 176.69, 177.30, 177.78 (piv-C=O), 201.22 (CHO), 209.70 (C=O) ppm.

(2S,6R)-N-(2,3,4,6-Tetra-O-pivaloyl-β-D-galactopyranosyl)-2-(4-hydroxy-butyl)-6-propyl-piperidin-4-one ($C_{38}H_{65}NO_{11}$)

To a solution of 2.93 g (3.73 mmol) **10b** in dry 70 cm^3 CH_2Cl_2, 15.6 g (62 mmol) pyridinium *p*-toluenesulfonate were added in 5 portions after every 8 h. Monitoring by TLC is necessary in order to terminate the reaction before cleavage of the N-glycosidic bond occurs (3 d, conversion about 85%). 200 cm^3 CH_2Cl_2 were added, and the solution was extracted with 50 cm^3 H_2O, 50 cm^3 sat. $NaHCO_3$, and 80 cm^3 brine. After drying with Na_2SO_4 and evaporation of the solvent *in vacuo*, the remaining residue was purified by chromatography (petroleum ether:EtOAc = 2:1).

Yield: 1.59 g (60%); amorphous solid; $[\alpha]_D^{22} = -18.3$ ($c = 1$, $CHCl_3$); $R_f = 0.15$ (petroleum ether: EtOAc = 3:1); ^1H NMR (400 MHz, δ, $CDCl_3$): 0.85 (t, 3H, $J = 7.2$ Hz, CH_3), 1.07–1.46 (m, 42H, piv-CH_3, CH_2), 1.52–1.62 (m, 2H, CH_2), 2.23 (dd, 1H, $J_{vic} = 7.1$ Hz, $J_{gem} = 15.4$ Hz, CH_2C=O), 2.36 (d, 1H, $J_{gem} = 15.6$ Hz, CH_2C=O′), 2.62 (dd, 1H, $J_{vic} = 5.4$ Hz, $J_{gem} = 15.3$ Hz, CH_2C=O), 2.70 (dd, 1H, $J_{vic} = 6.7$ Hz, $J_{gem} = 15.4$ Hz, CH_2C=O′), 3.11 (s, 1H, OH), 3.30 (m, 1H, NCHR), 3.44 (m, 1H, NCH*R*′), 3.57 (dd, 2H, $J_1 = 6.5$ Hz, $J_2 = 5.3$ Hz, CH_2OH), 3.87–3.92 (m, 2H, H-5′, H-6a′), 4.08 (dd, 1H, $J_{6b',5'} = 9.3$ Hz, $J_{6b',6a'} = 13.1$ Hz, H-6b′), 4.37 (d, 1H, $J_{1',2'} = 9.2$ Hz, H-1′), 5.08 (dd, 1H, $J_{3',4'} = 3.0$ Hz, $J_{3',2'} = 9.9$ Hz, H-3′), 5.34 (d, 1H, $J_{4',3'} = 2.9$ Hz, H-4′), 5.42 (t, 1H, $J = 9.6$ Hz, H-2′) ppm; ^{13}C NMR (100.6 MHz, δ, $CDCl_3$): 13.86 (CH_3), 19.60, 23.38 (CH_2), 27.04, 27.06, 27.18 (piv-CH_3), 32.02, 37.93 (CH_2), 38.68, 38.73, 39.04 (piv-C_{quart}), 39.89 (CH_2), 44.19, 44.47 (C-3, C-5), 47.50, 54.65 (C-2, C-6), 61.62 (CH_2OH), 61.67 (C-6′), 66.38, 67.22, 71.97, 72.49 (C-2′, C-3′, C-4′, C-5′), 94.15 (C-1′), 176.61, 177.30, 177.36, 177.80 (piv-C=O) ppm.

(2S,6R)-N-(2,3,4,6-Tetra-O-pivaloyl-β-D-galactopyranosyl)-2-(4-oxo-butyl)-6-propyl-piperidin-4-one (**11b**; $C_{38}H_{63}NO_{11}$)

To a solution of 15 g (2.1 mmol) of the afore-described hydroxybutyl piperidone in 40 cm^3 CH_2Cl_2, molecular sieve (4 Å, 5 g) was added. After stirring for 10 min, 0.5 g dry N-methylmorpholine

(4.1 mmol) and 74 cm^3 tetrapropylammonium perruthenate (*TPAP*) [18] (10 mol%) were added, and the solution was stirred in the dark at room temperature. After 3 h, the catalyst was filtered off through celite/silica gel and washed with 200 cm^3 CH_2Cl_2. The solvent was evaporated from the combined organic solutions, and the remainder was purified by chromatography (petroleum ether: EtOAc = 5:2) to give pure **11b**.

Yield: 1.23 g (83%); colorless amorphous solid; $[\alpha]_D^{22} = -10.2$ ($c = 1.1$, $CHCl_3$); $R_f = 0.14$ (petroleum ether:EtOAc = 3:1); ^{1}H NMR (400 MHz, δ, $CDCl_3$): 0.85 (t, 3H, $J = 7.1$ Hz, CH_3), 1.07–1.40 (m, 42H, piv-CH_3, CH_2), 1.46–1.66 (m, 2H, CH_2), 2.22 (dd, 1H, $J_{vic} = 6.9$ Hz, $J_{gem} = 15.4$ Hz, $CH_2C{=}O$), 2.33–2.39 (m, 3H, *CH₂*CHO, $CH_2C{=}O'$), 2.61 (dd, 1H, $J_{vic} = 5.7$ Hz, $J_{gem} = 15.5$ Hz, $CH_2C{=}O$), 2.61 (dd, 1H, $J_{vic} = 6.7$ Hz, $J_{gem} = 15.5$ Hz, $CH_2C{=}O'$), 3.31 (m, 1H, NCH*R*), 3.47 (m, 1H, NCH*R'*), 3.90 (m, 2H, H-5′, H-6a′), 4.08 (dd, 1H, $J_{6b',5'} = 9.3$ Hz, $J_{6b',6a'} = 13.1$ Hz, H-6b′), 4.36 (d, 1H, $J_{1',2'} = 9.3$ Hz, H-1′), 5.08 (dd, 1H, $J_{3',4'} = 3.0$ Hz, $J_{3',2'} = 9.9$ Hz, H-3′), 5.34 (d, 1H, $J_{4',3'} = 2.9$ Hz, H-4′), 5.41 (t, 1H, $J = 9.6$ Hz, H-2′), 9.71 (t, 1H, $J = 1.5$ Hz, CHO) ppm; ^{13}C NMR (100.6 MHz, δ, $CDCl_3$): 13.82 (CH_3), 19.61, 20.00 (CH_2), 27.01, 27.04, 27.16 (piv-CH_3), 37.95 (CH_2), 38.66, 39.01 (piv-C_{quart}), 39.89 (CH_2), 43.46, 44.01, 44.38 (C-3, C-5, *CH₂*CHO), 61.62 (C-6′), 66.02, 67.21, 72.02, 72.46 (C-2′, C-3′, C-4′, C-5′), 94.26 (C-1′), 176.54, 176.85, 177.25, 177.72 (piv-C=O), 201.86 (CHO), 209.65 (C=O) ppm.

(2S,8aR)-N-(2,3,4,6-Tetra-O-pivaloyl-β-D-galactopyranosyl)-2-propyl-4a,5-didehydro-decahydroquinolin-4-one (**12a**; $C_{38}H_{61}NO_{10}$)

In 40 cm^3 benzene, 0.17 g (4.25 mmol) powdered NaOH and dibenzo [18] crown [6] were heated under reflux. A solution of 1.4 g (1.97 mmol) **11a** in 20 cm^3 dry benzene was added, and the reflux was continued until completion of the conversion (1.5 h). After cooling, the reaction mixture was poured into 50 cm^3 sat. NH_4Cl solution and extracted with 200 cm^3 diethyl ether. The organic layer was washed with 40 cm^3 sat. $NaHCO_3$ solution and 60 cm^3 brine, dried over $MgSO_4$, and concentrated *in vacuo* to 50% of its volume. After filtration and complete evaporation of the solvent, the remaining residue was purified by chromatography (petroleum ether:EtOAc = 4:1).

Yield: 1.10 g (81%); colorless amorphous solid; $[\alpha]_D^{22} = -28.0$ ($c = 0.5$, $CHCl_3$); $R_f = 0.57$ (petroleum ether:EtOAc = 3:1); ^{1}H NMR (400 MHz, δ, $CDCl_3$): 0.81 (t, 3H, $J = 6.8$ Hz, CH_3), 1.09, 1.12, 1.14, 1.24 (4s, 4 × 9H, piv-CH_3), 1.26–1.32 (m, 4H, H-7, H-8), 1.40, 1.52, 1.71, 1.97 (4m, 4 × 1H, CH_2), 2.24 (m, 2H, H-6), 2.48 (dd, 1H, $J_{vic} = 2.3$ Hz, $J_{gem} = 16.4$ Hz, H-3*ax*), 2.87 (dd, 1H, $J_{vic} = 4.8$ Hz, $J_{gem} = 16.4$ Hz, H-3*eq*), 3.36 (m, 1H, H-2), 3.82–3.90 (m, 3H, H-8a, H-5′, H-6a′), 4.12 (dd, 1H, $J_{6b',5'} = 9.5$ Hz, $J_{6b',6a'} = 13.6$ Hz, H-6b′), 4.37 (d, 1H, $J_{1',2'} = 9.4$ Hz, H-1′), 5.06 (dd, 1H, $J_{3',4'} = 3.0$ Hz, $J_{3',2'} = 9.9$ Hz, H-3′), 5.34 (d, 1H, $J_{4',3'} = 2.9$ Hz, H-4′), 5.43 (t, 1H, $J = 9.6$ Hz, H-2′), 6.91 (d, 1H, $J = 2.7$ Hz, H-5) ppm; ^{13}C NMR (100.6 MHz, δ, $CDCl_3$): 13.72 (CH_3), 19.62, 20.14, 25.87 (CH_2), 27.03, 27.07, 27.17, 27.25 (piv-CH_3), 37.02 (CH_2), 38.67, 38.69, 38.72, 39.04 (piv-C_{quart}), 39.34 (CH_2), 43.27 (C-3), 54.18 (C-2), 61.35 (C-6′), 65.65, 67.25, 71.91, 72.80 (C-2′, C-3′, C-4′, C-5′), 92.39 (C-1′), 135.65, 137.76 (C-4a, C-5), 176.60, 176.67, 177.37, 177.77 (piv-C=O), 198.13 (C=O) ppm.

(2R,8aS)-N-(2,3,4,6-Tetra-O-pivaloyl-β-D-galactopyranosyl)-2-propyl-4a,5-didehydro-decahydroquinolin-4-one (**12b**; $C_{38}H_{61}NO_{10}$)

12b was obtained from 1.10 g (1.54 mmol) **11b** according to the afore-described procedure for the synthesis of its diastereomer **12a**.

Yield: 0.78 g (73%); colorless amorphous solid; $[\alpha]_D^{22} = +37.6$ ($c = 1$, $CHCl_3$); $R_f = 0.50$ (petroleum ether:EtOAc = 3:1); ^{1}H NMR (400 MHz, δ, $CDCl_3$): 0.82 (t, 3H, $J = 7.0$ Hz, CH_3), 1.07, 1.11, 1.13, 1.21 (4s, 9H, piv-CH_3), 1.25–1.38 (m, 4H, CH_2), 1.51–1.59 (m, 2H, CH_2), 1.80, 1.93 (2m, 1H, CH_2), 2.22 (m, 2H, CH_2), 2.29 (d, 1H, $J_{gem} = 18.0$ Hz, H-3*ax*), 2.85 (d, 1H, $J_{gem} = 13.0$ Hz, H-3*eq*), 3.13 (m, 1H, H-2), 3.87–3.98 (m, 4H, H-8a, H-5′, H-6a′, H-6b′), 4.39 (d, 1H, $J_{1',2'} = 9.2$ Hz,

H-1′), 5.01 (dd, 1H, $J_{3',4'}=3.0$ Hz, $J_{3',2'}=9.8$ Hz, H-3′), 5.28 (t, 1H, $J=9.5$ Hz, H-2′), 5.31 (d, 1H, $J_{4',3'}=2.7$ Hz, H-4)′, 6.66 (d, 1H, $J=2.9$ Hz, H-5) ppm; ^{13}C NMR (100.6 MHz, δ, $CDCl_3$): 13.70 (CH_3), 20.12, 21.39, 25.65 (CH_2), 26.92, 26.98, 27.03 (piv-CH_3), 32.52, 38.32 (CH_2), 38.50, 38.56, 38.61, 38.87 (piv-C_{quart}), 43.81 (C-3), 61.92 (C-6′), 66.70, 67.21, 71.50, 72.64 (C-2′, C-3′, C-4′, C-5′), 94.45 (C-1′), 137.18 (C-5.), 137.67 (C-4a), 176.49, 176.73, 177.16, 177.70 (piv-C=O), 198.73 (C=O) ppm.

(2S,4aR,5R,8aR)-N-(2,3,4,6-Tetra-O-pivaloyl-β-D-galactopyranosyl)-2-propyl-5-methyl-decahydroquinolin-4-one (**13**; $C_{39}H_{65}NO_{10}$)

To a suspension of 0.28 g (1.47 mmol) CuI in 10 cm^3 dry *THF* under argon cooled to −28°C, 1.8 cm^3 of a 1.6 *M* solution of methyllithium in diethyl ether (2.88 mmol) were added. After 30 min of stirring, the now clear colorless solution was cooled to −78°C. A solution of 0.5 g (0.72 mmol) of the enone **12a** and 0.73 cm^3 (5.76 mmol) trimethyl chlorosilane in 25 cm^3 dry *THF* was added. The color changed to yellow. After 2–3 h (TLC monitoring) the reaction was complete. A mixture of 15 cm^3 conc. NH_4OH/NH_4Cl (1:1 v/v) was added. The solution was warmed to room temperature, 100 cm^3 diethyl ether were added, and the separated organic layer was washed with conc. NH_4OH/NH_4Cl (1:1 v/v) until the blue color in the aqueous layer vanished. The collected aqueous solutions were extracted with 2 × 100 cm^3 diethyl ether, and the combined organic solutions were washed with 50 cm^3 brine and dried over $MgSO_4$. After removal of the solvent *in vacuo*, the crude silyl enolether ($R_f=0.67$, petroleum ether:EtOAc = 3:1, *d.r.* > 10:1, ^{1}H NMR) was obtained; yield: 0.56 g (quantitative), yellowish amorphous solid.

In 3 cm^3 *THF*, 0.10 g (0.128 mmol) of this silyl enolether were dissolved and cooled to −20°C. 0.3 cm^3 of a 1 *M* solution of tetrabutylammonium fluoride (0.3 mmol) in *THF* was added, and the mixture was stirred for 36 h (TLC monitoring). After addition of 2 × 20 cm^3 diethyl ether, the organic solutions were washed with 40 cm^3 brine and dried over $MgSO_4$. After evaporation of the solvent, the remaining compound **13** was purified by chromatography (petroleum ether:EtOAc = 5:1).

Yield: 73 mg (81%); m.p. 168°C (MeOH); $[\alpha]_D^{22} = -60.0$ ($c=1$, $CHCl_3$); $R_f=0.67$ (petroleum ether:EtOAc = 3:1); *d.r.* > 15:1:0:0 (^{1}H NMR); ^{1}H NMR (400 MHz, δ, $CDCl_3$): 0.77 (d, 3H, $J=7.1$ Hz, CH_3), 0.82 (t, 3H, $J=6.8$ Hz, propyl-CH_3), 1.07–1.53 (m, 45 H, piv-CH_3, CH_2), 2.04 (m, 1H, CH_2), 2.19 (dd, 1H, $J_{vic}=3.7$ Hz, $J_{vic'}=12.0$ Hz, H-4a), 2.47 (d, 1H, $J_{gem}=16.2$ Hz, H-3*ax*), 2.63 (m, 2H, H-3*eq*, H-5), 3.26 (dt, 1H, $J_{8a,8}=3.5$ Hz, $J_{8a,4a}=J_{8a,8'}=11.2$ Hz, H-8a), 3.36 (m, 1H, H-2), 3.64 (m, 2H, H-5′, H-6a′), 4.10 (dd, 1H, $J_{6b',5'}=9.5$ Hz, $J_{6b',6a'}=13.2$ Hz, H-6b′), 4.43 (d, 1H, $J_{1',2'}=9.3$ Hz, H-1′), 5.04 (dd, 1H, $J_{3',4'}=2.9$ Hz, $J_{3',2'}=9.9$ Hz, H-3′), 5.34 (d, 1H, $J_{4',3'}=3.0$ Hz, H-4′), 5.35 (t, 1H, $J=9.5$ Hz, H-2′) ppm; ^{13}C NMR (100.6 MHz, δ, $CDCl_3$): 13.62, 13.83 (propyl-CH_3, CH_3), 19.59, 20.16 (CH_2), 27.08, 27.19, 27.25 (piv-CH_3), 28.34 (C-5), 32.12, 36.17 (CH_2), 38.69, 39.04 (piv-C_{quart}), 41.59 (CH_2), 44.22 (C-3), 49.67 (C-4a), 52.11, 54.23 (H-8a, C-2), 61.81(C-6′), 65.84, 67.47, 71.94, 72.86 (C-2′, C-3′, C-4′, C-5′), 90.42 (C-1′), 176.64, 176.69, 177.41, 177.84 (piv-C=O), 211.04 (C=O) ppm; X-ray: see Ref. [16].

(2S,8aR)-N-Phenoxycarbonyl-2-propyl-4a,5-didehydro-quinolin-4-one (**14**; $C_{19}H_{23}NO_3$)

To a solution of 0.4 g (0.58 mmol) **12a** in 5 cm^3 MeOH, 1 cm^3 of 1 *M* HCl was added, the mixture was stirred at room temperature for 48 h, the MeOH was evaporated *in vacuo*, and 20 cm^3 diethyl ether were added to the remaining residue. This mixture was extracted five times with 10 cm^3 H_2O. After evaporation of the H_2O, the hydrochloride of the quinolinone remained (133 mg, >95%), which was dissolved in 5 cm^3 H_2O. 5 cm^3 of sat. $NaHCO_3$ solution were added, the mixture was stirred for 1 h, and then 0.1 cm^3 (0.68 mmol) phenyl chloroformate were added. After stirring for 2 h, the mixture was extracted with 3 × 30 cm^3 diethyl ether. The organic layer was dried over $MgSO_4$ and the solvent removed *in vacuo*. The remaining crude **14** was purified by chromatography (petroleum ether: EtOAc = 10:1).

Yield: 110 mg (60%); colorless oil; $[\alpha]_D^{22} = -145.6$ ($c = 1$, $CHCl_3$) (Ref. [13]: $[\alpha]_D^{22} = -164$ ($c = 3$, $CHCl_3$)); $R_f = 0.49$ (petroleum ether:EtOAc = 3:1); ^1H NMR (400 MHz, δ, $CDCl_3$): 0.92 (t, 3H, CH_3), 1.16–1.52 (m, 5H, CH_2), 1.53–1.95 (3m, 3H, CH_2), 2.32 (dd, 2H, $J_1 = 3.8$ Hz, H-6), 2.49 (dd, 1H, $J_{gem} = 17.6$ Hz, $J_{vic} = 1.6$ Hz, H-3a), 2.60 (m, 1H, CH_2), 2.74 (dd, 1H, $J_{gem} = 16.7$ Hz, H-3*eq*), 4.68 (m, 2H, H-2, H-8a), 6.91 (d, 1H, H-5), 7.09 (d, 2H, $J = 7.9$ Hz, phenyl-H), 7.17 (t, 1H, $J = 7.3$ Hz, phenyl-H), 7.34 (t, 2H, $J = 7.0$ Hz, phenyl-H) ppm; ^{13}C NMR (100.6 MHz, δ, $CDCl_3$): 13.7 (CH_3), 19.64, 20.72, 20.92, 25.79, 36.49 (CH_2), 44.28 (C-3), 50.41, 52.61 (C-2, C-8a), 121.61, 125.22, 129.27, 129.54 (phenyl), 134.33, 139.54 (C-5, C-4a), 151.37 (N–COO–Ph), 195.88 (C=O) ppm.

(2S,4aR,5R,8aR)-N-Phenoxycarbonyl-2-propyl-5-methyl-decahydroquinolin-4-one (**15a**; $C_{20}H_{27}NO_3$)

Removal of the N-galactosyl group from 400 mg (0.56 mmol) quinolinone **13** was carried out as described for **12a**, as well as the subsequent acylation with phenyl chloroformate. Purification of the crude product by chromatography (petroleum ether:EtOAc = 12:1) gave **15a** (*d.r.* 7(*trans*): 1(*cis*):0:0).

Yield: 150 mg (82%); colorless oil; $[\alpha]_D^{22} = -160.3$ ($c = 1$, $CHCl_3$); $R_f = 0.48$ (petroleum ether:EtOAc = 5:1); ^1H NMR (400 MHz, δ, $CDCl_3$): 0.96 (d, 3H, $J = 7.0$ Hz, CH_3), 0.97 (t, 3H, $J = 7.1$ Hz, propyl-CH_3), 1.28 (m, 1H, H-6), 1.49 (m, 3H, H-8*ax*, CH_2), 154–1.66 (m, 4H, H-7, H-7′, H-6, CH_2), 1.84 (m, 1H, CH_2), 2.48 (dd, 1H, $J_{vic} = 1.5$ Hz, $J_{gem} = 18.8$ Hz, H-3*ax*), 2.64 (dd, 1H, $J_{vic} = 7.0$ Hz, $J_{gem} = 19.0$ Hz, H-3*eq*), 2.65–2.72 (m, 2H, H-5, H-8*eq*), 2.73 (dd, 1H, $J_{4a,8a} = 12.0$ Hz, $J_{4a,5} = 3.2$ Hz, H-4a), 4.02 (dt, 1H, $J_{8a,4a} = J_{8a,8} = 11.5$ Hz, $J_{8a,8} = 3.1$ Hz, H-8a), 4.68 (ddt, 1H, $J_{2,3} = 1.5$ Hz, $J_{vic} = 7.6$ Hz, H-2), 7.10 (m, 2H, phenyl-H), 7.19 (m, 1H, phenyl-H), 7.36 (m, 2H, phenyl-H) ppm; ^{13}C NMR (100.6 MHz, δ, $CDCl_3$): 13.25, 13.85 (propyl-CH_3, CH_3), 20.22 (CH_2), 27.73 (C-5), 32.19, 35.98, 39.39, 44.33 (CH_2), 51.34, 52.03, 53.93 (C-2, C-4a, C-8a), 121.65, 125.36, 129.29 (phenyl) 151.37 (NCOOPh), 154.82 (phenyl-O), 207.25 (C=O) ppm.

(2S,4aR,5R,8aR)-N-Benzylycarbonyl-2-propyl-5-methyl-decahydroquinolin-4-one (**15b**; $C_{21}H_{29}NO_3$)

15b was synthesized in a completely analogous manner to **15a** from 0.4 g (0.58 mmol) **13** and 0.1 cm^3 benzyl chloroformate (0.68 mmol) and purified by chromatography (petroleum ether: EtOAc = 10:1).

Yield: 0.17 g (88%); colorless oil; $[\alpha]_D^{22} = -132.1$ ($c = 1$, $CHCl_3$); $R_f = 0.46$ (petroleum ether:EtOAc = 5:1); FD-MS: $m/z = 342.9$ [M^+]; ^1H NMR (400 MHz, δ, $CDCl_3$): 0.9 (t, 3H, $J = 7.3$ Hz, propyl-CH_3), 0.91 (d, 3H, $J = 7.0$ Hz, CH_3), 1.19 (m, 1H, CH_2), 1.3–1.65 (m, 7H, CH_2, CH), 1.75 (m, 1H, CH_2), 2.39 (dd, 1H, $J_{cis} = 1.5$ Hz, $J_{gem} = 19.5$ Hz, H-3*ax*), 2.51 (dd, 1H, $J_{trans} = 6.8$ Hz, $J_{gem} = 19.5$ Hz, H-3*eq*), 2.59 (m, 2H, H-5, H-8), 2.66 (dd, 1H, $J_{4a,5} = 3.6$ Hz, $J_{4a,8a} = 11.3$ Hz, H-4a), 3.91 (dt, 1H, $J_{8a,4a} = J_{8a,8} = 11.3$ Hz, $J_{8a,8'} = 2.9$ Hz, H-8a), 4.55 (m, 1H, H-2), 5.12 (m, 2H, benzyl-CH_2), 7.33 (m, 5H, phenyl-H) ppm; ^{13}C NMR (100.6 MHz, δ, $CDCl_3$): 13.16, 13.75 (propyl-CH_3, CH_3), 27.59 (*C*H-CH_3), 32.12, 35.94, 39.13, 43.53, 44.21 (CH_2), 50.71, 51.51, 53.81, (C-4a, C-2, C-8a), 67.18 (benzyl-CH_2), 127.65, 127.9, 128.42 (phenyl), 136.58 (phenyl-*ipso*), 156.08 (N–CO–O), 207.51 (C=O) ppm.

(2S,4aS,5R,8aR)-N-Phenyloxycarbonyl-2-propyl-5-methyl-decahydroquinolin-4-one (**16**; $C_{20}H_{27}NO_3$)

For the preparation of **16**, of also Ref. [13].

To a stirred suspension of 0.28 g (1.47 mmol) CuI in 10 cm^3 dry *THF* under argon at −30°C, 1.8 cm^3 of a 1.6 *M* solution of methyllithium in diethyl ether (2.88 mmol) were added. The clear solution was cooled to −78°C, and 1.6 cm^3 $BF_3 \cdot OEt_2$ (12 mmol) were added dropwise. After 20 min, a solution of 11 mg (0.035 mmol) **15a** in 2 cm^3 *THF* was added, and the mixture was stirred for 30 min at −78°C and for 2 h at −18°C. At this temperature, 8 cm^3 MeOH and, at room temp., 50 cm^3 diethyl ether were added. The organic solution was washed with conc. NH_4OH/NH_4Cl (1:1, v/v) solution until blue color vanished. The organic layer was washed with brine and dried over $MgSO_4$. After evaporation of the solvent, the crude product **16** was purified by chromatography (petroleum ether:EtOAc = 15:1).

Yield: 7 mg (67%); colorless oil; $[\alpha]_D^{22} = +30.9$ ($c = 1$, $CHCl_3$); $R_f = 0.64$ (petroleum ether: EtOAc = 3:1); *d.r.* (**16**:**15a**) = 9:2:0:0; ^{1}H NMR (400 MHz, δ, $CDCl_3$): 0.91 (t, 3H, $J = 7.3$ Hz, propyl-CH_3), 1.03 (d, 3H, $J = 7.3$ Hz, CH_3), 1.24–1.70 (m, 9H, CH_2), 1.93 (d, 1H, $J = 9.4$ Hz, H-8*endo*), 2.34 (dd, 1H, $J_{vic} = 2.1$ Hz, $J_{gem} = 14.1$ Hz, H-3*eq*), 2.60 (m, 1H, H-5), 2.65 (d, 1H, $J = 6.5$ Hz, H-4a), 2.78 (dd, 1H, $J_{vic} = 8.2$ Hz, $J_{gem} = 14.1$ Hz, H-3*eq*), 4.71 (m, 1H, H-2), 4.86 (m, 1H, H-8a), 7.11 (m, 2H, phenyl-H), 7.20 (m, 1H, phenyl-H), 7.36 (m, 2H, phenyl-H) ppm; ^{13}C NMR (100.6 MHz, δ, $CDCl_3$): 13.75, 18.36 (propyl-CH_3, CH_3), 18.08, 26.62 (C-4a, C-5), 20.08, 20.28, 27.27, 36.60, 39.76, 43.66 (CH_2), 53.57, 54.22 (C-2, C-8a), 121.68, 125.40, 129.30 (phenyl), 151.44 (NCOOPh), 154.17 (phenyl-*ipso*), 208.84 (C=O) ppm.

Epimerization of **15a**

To a solution of N-phenoxycarbonyl decahydroquinolin-4-one, 5 mg **15a** (*trans*:*cis* = 7:1) in 3 cm^3 *THF* and two drops of triethylamine (~6 mg) were added. The mixture was stirred at room temperature. After 2 min, an analytical sample was poured into a mixture of diethyl ether/1 *M* HCl. The solvent was evaporated, and the remaining residue was investigated by analytical HPLC, column B; ratio of diastereomers: **15a**(*trans*):**16**(*cis*) = 1:3; eluent: acetonitrile:water, gradient: 60/40→75/25 (0–20 min); R_t(**15a**) = 13.5 min, R_t(**16**) = 14 min.

(2S,4aR,5R,8aR)-N-Benzyloxycarbonyl-2-propyl-5-methyl-decahydroquinolin-4-one-ethane-1,2-diyl dithioacetal (**17**; $C_{23}H_{33}NO_2S_2$)

To a solution of 160 mg (0.38 mmol) **15b** and 0.1 cm^3 (0.9 mmol) ethane dithiol in 5 cm^3 dry CH_2Cl_2 at 0°C, 0.3 cm^3 borontrifluoride etherate (2.5 mmol) were added dropwise. After 2 h at 0°C the mixture was stirred at room temperature for 15 h, diluted with 30 cm^3 CH_2Cl_2, and washed with 2 × 20 cm^3 sat. $NaHCO_3$ solution and H_2O. After drying over $MgSO_4$, the solvent was evaporated *in vacuo*.

Yield: 120 mg (64%); colorless oil; $[\alpha]_D^{22} = -24.8$ ($c = 1$, $CHCl_3$); $R_f = 0.35$ (petroleum ether: EtOAc = 7:1); FD-MS: $m/z = 419.9$ [M^+]; ^{1}H NMR (400 MHz, δ, $CDCl_3$): 0.85 (t, 3H, $J = 7.3$ Hz, propyl-CH_3), 1.08 (d, 3H, $J = 7.0$ Hz, CH_3), 1.26 (m, 3H, CH_2), 1.39–1.66 (m, 6H, CH_2), 1.91 (dd, 1H, $J_{gem} = 13.5$ Hz, $J_{cis} = 3.8$ Hz, H-3*eq*), 2.09 (dd, 1H, $J_{4a,8a} = 12$ Hz, $J_{4a,5} = 3.5$ Hz, H-4a), 2.18 (m, 1H, CH_2), 2.62 (m, 1H, H-5), 2.76 (dd, 1H, $J_{gem} = 13.3$ Hz, $J_{trans} = 11.4$ Hz, H-3*ax*), 3.22 (m, 4H, S-CH_2), 4.10 (dt, $J_{8a,4a} = J_{8a,8} = 11.7$ Hz, $J_{8a,8'} = 3.5$ Hz, H-8a), 4.39 (m, 1H, H-2), 5.16 (m, 2H, benzyl-CH_2), 7.32 (m, 5H, phenyl-H) ppm; ^{13}C NMR (100.6 MHz, δ, $CDCl_3$): 13.42, 13.81 (propyl-CH_3, CH_3), 19.3, 20.57 (CH_2), 31.35 (CH), 34.64, 34.97 (CH_2), 38.33, 38.67 (S-CH_2), 42.79 (CH_2), 49.33, 50.07, 54.29 (CH, NCH*R*), 52.42 (CH_2), 66.89 (benzyl-CH_2), 127.38, 127.59, 128.32 (phenyl-C), 137.29 (quart. phenyl-C), 156.33 (N–CO–O) ppm.

(2S,4aR,5R,8aR)-2-Propyl-5-methyl-decahydroquinoline hydrochloride (trans-4a-epi-pumiliotoxin C) (**5**; $C_{13}H_{26}NCl$)

To a solution of 90 mg (0.22 mmol) of the dithioacetal **17** in 20 cm^3 dry isopropanol, 1.5 g freshly prepared and neutral washed *Raney* nickel were added. A gentle stream of hydrogen was passed through the solution at room temperature for 2 h. After stirring for additional 14 h, the catalyst was filtered off. The catalyst was dissolved in aqueous HCl, neutralized with conc. aq. ammonia, and the solution was extracted with $2 \times 30\,cm^3$ $CHCl_3$. The organic layer was extracted with aq. HCl, and the acidic aq. solution was evaporated to dryness *in vacuo*. Impurities were removed from the crude hydrochloride **5** by silica layer chromatography with EtOAc as the eluent.

Yield: 11 mg (22%); m.p. 256–258°C; $[\alpha]_D^{22} = -11.3$ ($c = 0.5$, $CHCl_3$); FD-MS: $m/z = 195.8$ $[M-Cl]^+$; ^1H NMR (400 MHz, δ, $CDCl_3$): 0.82 (d, 3H, $J = 7.3$ Hz, CH_3), 0.88 (t, 3H, $J = 7.3$ Hz, propyl-CH_3), 1.2–2.0 (m, 14H, m CH_2), 2.05, 2.36 (2m, 2H, CH), 2.87 (m, 2H, N-CH*R*) ppm; ^{13}C NMR (100.6 MHz, δ, $CDCl_3$): 12.68, 13.73 (propyl-CH_3, CH_3), 18.79, 19.0, 30.31 (CH_2), 31.96 (CH), 32.16, 35.1 (CH_2), 41.45 (C-4a), 56.85, 58.25 (C-2, C-8a) ppm; X-ray analysis: $P2_12_12_1$ (orthorhombic), $a = 7.2852(2)$, $b = 8.7387(3)$, $c = 22.0953(5)$ Å, $V = 1406.67(7)$ Å^3, $z = 4$, $F(000) = 512$, Cu-K_α, SIR-92, SHELXL-97 [23].

(2R,4aR,5S,8aS)-N-(2,3,4,6-Tetra-O-pivaloyl-β-D-galactopyranosyl)-2,5-di-propyl-decahydroquinolin-4-one (**18**; $C_{41}H_{69}NO_{10}$)

To 0.34 g (1.6 mmol) $CuBr \cdot SMe_2$ in $20\,cm^3$ dry *THF* at −78°C under argon, $1.2\,cm^3$ of a 2 *M* solution of *n*-propyl magnesium chloride in diethyl ether (3.2 mmol) were added. The pale yellow suspension was warmed to −55°C within 60 min and became clear. This solution was cooled again to −78°C, and a solution of 0.39 g (0.56 mmol) enone **12b** and $0.85\,cm^3$ trimethylchlorosilane (0.67 mmol) in $20\,cm^3$ dry *THF* was added through a syringe. The color of the reaction mixture changed first to deep red and then to orange. After 45 min stirring, only traces of the starting material could be detected (monitoring by *TLC*). After 1.5 h, $20\,cm^3$ of conc. NH_4OH/NH_4Cl (1:1, v/v) were added. The mixture was warmed up to room temperature, $140\,cm^3$ diethyl ether were added, and the organic layer was extracted with conc. NH_4OH/NH_4Cl (1:1, v/v) as long as a blue color occurred in the aqueous layer. The collected aq. solutions were extracted with $2 \times 130\,cm^3$ diethyl ether, the combined organic solutions were washed with $70\,cm^3$ brine, dried over $MgSO_4$, and the solvent was evaporated *in vacuo*. The remaining crude silyl enolether was dissolved in $7\,cm^3$ *THF*, cooled to −20°C, and $0.8\,cm^3$ (0.8 mmol) of a 1 *M* solution of tetrabutylammonium fluoride in *THF* were added. After 24 h the solvent was evaporated, and the remaining residue was dissolved in $55\,cm^3$ diethyl ether and washed with $60\,cm^3$ H_2O. The aq. layer was extracted with $2 \times 25\,cm^3$ diethyl ether. The combined organic solutions were washed with $50\,cm^3$ brine and dried over $MgSO_4$. After evaporation of the solvent *in vacuo*, the crude **18** was purified by chromatography (petroleum ether:EtOAc = 7:1).

Yield: 0.18 g (43%); colorless amorphous solid; $[\alpha]_D^{22} = -46.7$ ($c = 1.1$, $CHCl_3$); $R_f = 0.51$ (petroleum ether:EtOAc = 5:1); *d.r.* > 10:1:0:0 (^1H NMR); ^1H NMR (400 MHz, δ, $CDCl_3$): 0.87 (t, 6H, $J = 6.9$ Hz, CH_3, CH_3'), 1.08–1.42 (m, 48 H, piv-CH_3, CH_2), 1.54 (m, 1H, CH_2), 1.65 (m, 1H, CH_2), 2.25 (dd, 1H, $J_{vic} = 9.7$ Hz, $J_{vic'} = 16.6$ Hz, H-3*ax*), 2.44 (m, 1H, H-5), 2.43 (m, 2H, $CH_2C{=}O$, $CHC{=}O$), 3.28, 3.54 (2m, 1H, H-2, H-8a), 3.93 (m, 2H, H-5′, H-6a′), 4.10 (dd, 1H, $J_{6b',5'} = 9.5$ Hz, $J_{6b',6a'} = 13.6$ Hz, H-6b′), 4.38 (d, 1H, $J_{1',2'} = 9.4$ Hz, H-1′), 5.09 (dd, 1H, $J_{3',4'} = 3.2$ Hz, $J_{3',2'} = 10.0$ Hz, H-3′), 5.38 (d, 1H, $J_{4',3'} = 3.4$ Hz, H-4′), 5.38 (dd, 1H, $J_{2',1'} = 9.3$ Hz, $J_{2',3'} = 9.9$ Hz, H-2′) ppm; ^{13}C NMR (100.6 MHz, δ, $CDCl_3$): 14.00, 14.07 (CH_3, CH_3'), 20.83, 20.92, 25.07 (CH_2), 27.04, 27.10, 27.23, 27.29 (piv-CH_3), 32.13 (C-5), 33.40, 35.00 (CH_2), 38.70, 38.74, 39.03 (piv-C_{quart}), 40.27 (CH_2), 44.29 (C-3), 54.44, 56.23 (C-2, H-8a), 61.46 (C-6′), 66.04, 67.28, 72.12, 72.69 (C-2′, C-3′, C-4′, C-5′), 176.73, 176.82, 177.32, 177.79 (piv-C=O), 211.34 (C=O) ppm.

(2R,4aR,5S,8aS)-N-(2,3,4,6-Tetra-O-pivaloyl-β-D-galactopyranosyl)-2,5-di-propyl-4-trifluoromethane-sulfonyloxy-3,4-didehydro-decahydroquinoline (**19**; $C_{42}H_{68}NO_{12}SF_3$)

To a solution of 0.15 g (0.20 mmol) **18** in $6\,cm^3$ dry *THF* at 78°C, $0.3\,cm^3$ (3.0 mmol) of a 1 *M* solution of sodium hexamethyldisilazane (Na*HMDS*) in *THF* were added. After stirring for 1 h, $0.86\,cm^3$ (0.43 mmol) of a 0.5 *M* solution of 5-chloro-pyridine-2-N,N-*bis*-trifluoromethanesulfonimide (dried with molecular sieve 4 Å) were added, and stirring was continued for 2 h at −78°C. After additional stirring at $-20^\circ C$ for 12 h, 5 cm^3 sat. NH_4Cl solution was added, followed by addition of 50 cm^3 diethyl ether at room temperature. The organic layer was washed with 30 cm^3 brine and dried over $MgSO_4$. The solvent was evaporated *in vacuo*. Column chromatography with petroleum ether:EtOAc = 8:1 gave 25 mg of the starting material, 41 mg of the regioisomer **20**, and the product **19**.

Yield: 71 mg (41%); oil containing traces of impurities; $[\alpha]_D^{22} = +4.82$ ($c = 0.5$, $CHCl_3$); $R_f = 0.59$ (petroleum ether:EtOAc = 5:1); ^1H NMR (400 MHz, δ, $CDCl_3$): 0.81–0.91 (m, 6H, CH_3, CH_3'), 1.07–1.61 (m, 50 H, piv-CH_3, CH_2), 2.14 (m, 1H, CH_2), 2.82 (br s, 1H, H-4a), 3.43 (m, 1H, H-8a), 3.80 (dd, 1H, $J_{5',6a'} = 6.9$ Hz, $J_{5',6b'} = 6.4$ Hz, H-5′), 3.89 (m, 2H, H-6a′, H-1′), 4.03 (dd, 1H, $J_{6b',5'} = 6.6$ Hz, $J_{6b',6a'} = 17.6$ Hz, H-6b′), 4.23 (m, 1H, H-2), 5.08 (dd, 1H, $J_{3',4'} = 3.3$ Hz, $J_{3',2'} = 10.0$ Hz, H-3′), 5.34 (d, 1H, $J_{4',3'} = 3.4$ Hz, H-4′), 5.48 (t, 1H, $J = 9.6$ Hz, H-2′), 5.69 (t, 1H, $J = 3.0$ Hz, H-3) ppm; ^{13}C NMR (100.6 MHz, δ, $CDCl_3$): 13.88, 14.14 (CH_3, CH_3'), 20.17, 20.87 (CH_2), 27.07, 27.11, 27.24, 27.28 (piv-CH_3), 30.14 (CH_2), 33.20 (C-5), 34.09 (CH_2), 38.33, 38.65, 38.73, 39.01 (piv-C_{quart}), 43.94 (C-4a), 61.20 (C-6′), 65.48, 67.22, 72.06, 72.51 (C-2′, C-3′, C-4′, C-5′), 96.10 (C-1′), 119.21 (C-3), 148.66 (C-4), 176.64, 176.84, 176.94, 177.20 (piv-C=O) ppm.

(2R,4aR,5S,8aS)-2,5-Di-propyl-decahydroquinoline (cis-4a-epi-(+)-perhydro 219A) (**6**; $C_{15}H_{29}N$)

To a solution of 50 mg (0.57 mmol) of enol triflate **19** in 10 cm^3 dry MeOH, 100 mg lithium carbonate and 10 mg palladium on charcoal (5%) were added. The mixture was vigorously stirred under hydrogen at room temperature for 24 h. After filtration through celite and washing of the solid with 50 cm^3 MeOH, the solvent was evaporated from the combined MeOH solutions. The remaining residue was dissolved in 30 cm^3 diethyl ether, washed with 25 cm^3 H_2O and 25 cm^3 brine, and dried over $MgSO_4$. After evaporation of the solvent, the residue was dissolved in 3 cm^3 MeOH; subsequently, 0.1 cm^3 1 *N* HCl were added, and the mixture was stirred for 24 h at room temperature. After evaporation of the solvents, 15 cm^3 diethyl ether were added to the residue, and the solution was extracted with 5×10 cm^3 H_2O. Evaporation of H_2O *in vacuo* gave 14 mg of the crude hydrochloride **6**. The residue was stirred with 0.5 cm^3 sat. Na_2CO_3 solution and extracted with 3×10 cm^3 CH_2Cl_2 to give the free amine **6** (yield: 9 mg, 74%). NMR spectroscopic data of both the free amine **6** and its hydrochloride are in agreement with those reported in Refs. [15] and [20].

Hydrochloride of **6**: $[\alpha]_D^{22} = +12.2$ ($c = 0.7$, MeOH); ^{13}C NMR (100.6 MHz, δ, $CDCl_3$): 13.81, 14.42 (CH_3, CH_3'), 18.75, 19.21, 20.66, 23.39, 25.02, 29.13, 31.11, 31.43, 34.71, 35.10, 38.89 (CH_2), 58.64, 60.27 (C-2, C-8a) ppm.

Free amine **22**: ^{13}C NMR (100.6 MHz, δ, $CDCl_3$): 14.09, 14.46 (CH_3, CH_3'), 18.99, 19.11, 21.03, 26.64, 27.09, 31.55, 32.01, 32.99, 35.46, 39.29 ($9CH_2$, CH), 40.27 (C-4a), 56.24, 57.78 (C-2, C-8a) ppm.

Acknowledgements

This work was supported by the *Deutsche Forschungsgemeinschaft* and by the *Fonds der Chemischen Industrie*.

References

[1] Reviews: a) Berecibar A, Grandjean C, Sirivardena A (1999) Chem Rev **99**: 779; b) Fraser-Reid B (1996) Acc Chem Res **29**: 57; c) Giese B, Koppins B, Göbel T, Dickhaut J, Thomas G, Kulicke K, Trach F (1996) Org React **48**: 301; d) Hanessian S (1983) Total Syntheses of Natural Products. In: The Chiron Approach. Pergamon Press, New York

[2] Reviews: a) Hultin PG, Earle MA, Sudharshan M (1997) Tetrahedron **53**: 14833; b) Kunz H, Rück-Braun K (1993) Angew Chem Int Ed Engl **32**: 336

[3] Kunz H, Sager W (1987) Angew Chem Int Ed Engl **26**: 557
[4] Kunz H, Pfrengle W (1988) J Am Chem Soc **110**: 651
[5] Lindermann RJ, Binet J, Petrich SR (1999) J Org Chem **64**: 336
[6] Laschat S, Kunz H (1992) Synthesis 90
[7] Kunz H, Burgard A, Schanzenbach D (1997) Angew Chem Int Ed Engl **36**: 386
[8] Weymann M, Pfrengle W, Schollmeyer D, Kunz H (1997) Synthesis 1151
[9] a) Daly JW (1998) J Nat Prod **61**: 162; b) Tokuyama T, Tsujita T, Shimata A, Garraffo HM, Spande TF, Daly JW (1991) Tetrahedron **47**: 5401 and references cited therein
[10] Daly JW, Tokuyama T, Habermehlg, Karle IL, Witkop B (1969) Liebigs Ann Chem **729**: 198
[11] Tokuyama T, Nishimori N, Karle IL, Edwards NW, Daly JW (1987) Tetrahedron **43**: 643
[12] a) Naruse M, Aoyagi S, Kibayashi C (1994) Tetrahedron Lett **35**: 9213; b) Murahashi S, Sasao S, Saito S, Naota T (1992) J Org Chem **57**: 2521; c) Murahashi S, Sasao S, Saito E, Naota T (1993) Tetrahedron **49**: 8805; d) Comins DL, Dehghani A (1993) J Chem Soc Chem Commun 1839; e) Schultz AG, McCloskey PJ, Court JJ (1987) J Am Chem Soc **109**: 6493; f) Bonin M, Royer J, Grierson DS, Husson HP (1986) Tetrahedron Lett **27**: 1569; g) Review: Daly JW, Garraffo HM, Spande TF (1993) In: Cordell GA (ed) The Alkaloids. Academic Press, San Diego, pp 185–288
[13] Comins DL, Dehghani A (1993) J Chem Soc Chem Commun 1839
[14] Comins DL, Dehghani A (1995) J Org Chem **60**: 794
[15] McCloskey PJ, Schultz AG (1988) J Org Chem **53**: 1380
[16] Weymann M, Schultz-Kukula M, Kunz H (1998) Tetrahedron Lett **39**: 7835
[17] Yamamoto Y, Maruyama K (1978) J Am Chem Soc **100**: 3240
[18] Ley SV, Norman J, Griffith WP, Marsden SP (1994) Synthesis 639
[19] Comins DL, Dehghani A (1992) Tetrahedron Lett **33**: 6299
[20] Overman LE, Jessup PJ (1978) J Am Chem Soc **100**: 5179
[21] Kunz H, Pfrengle W, Rück K, Sager W (1991) Synthesis 1039
[22] Kunz H, Sager W, Schanzenbach D, Decker M (1991) Liebigs Ann Chem 649
[23] Further details are available on request from the Cambridge Crystallographic Data Center quoting the deposit number CCDC 170250

Received September 7, 2001. Accepted October 16, 2001

Analysis of Carbohydrate Mixtures by Diffusion Difference NMR Spectroscopy

Hanspeter Kählig[1,*], **Katharina Dietrich**[1], and **Silke Dorner**[2]

[1] Institut für Organische Chemie, Universität Wien, A-1090 Wien, Austria
[2] Institut für Medizinische Biochemie, Universität Wien, A-1030 Wien, Austria

Summary. Diffusion difference NMR spectroscopy can be used as a fast and powerful tool to separate carbohydrates in a mixture by their translational diffusion properties. This method is of general interest for analyzing natural and synthetic mixtures and for monitoring and optimizing synthetic reactions. The proposed subtraction procedure applied to two varying diffusion encoded spectra can be combined with homo- and heteronuclear as well as with one- and multi-dimensional NMR experiments.

Keywords. NMR spectroscopy; Carbohydrates; Diffusion difference; BPP-LED.

Introduction

The structural investigation of carbohydrates by nuclear magnetic resonance (NMR) spectroscopy is still a very challenging field even at the highest available magnetic field strengths (for a recent review, see Ref. [1]). Additionally, nowadays not only purified samples containing one isolated compound are analyzed; there is an increasing demand on the determination of the composition of complex mixtures by spectroscopic techniques, its main reason being the effort to increase efficiency and throughput in drug discovery processes. Combinatorial chemistry techniques have been developed to create complex libraries of molecules with the necessity of testing and identifying the individual compounds as mixtures; structure activity relationships (SAR) by NMR is an example for a strategy developed for screening mixtures for biological activity [2]. In a similar way, natural products, which are applied as a rich source of useful compounds, involve the identification of interesting molecules in a biological matrix. Analytical tools for controlling the success of a synthetic process at a very early stage are also desirable in 'classical' organic chemistry. Thus, the main issue is avoiding expensive and time consuming physical separation of the mixtures prior to analysis.

In contrast to the majority of spectroscopic techniques, NMR spectroscopy in combination with pulsed field gradients (PFG) can lead to a non-invasive structural investigation of unknown compounds in complex mixtures. These experiments rely on the attenuation behaviour of individual resonances under the influence of linear field gradients due to translational diffusion [3]. Since diffusion, in addition to

* Corresponding author. E-mail: hanspeter.kaehlig@univie.ac.at

effects exerted by bulk parameters, is a property of a molecule as a whole, individual mixture components can be resolved based on the variance of their diffusion rates using diffusion ordered (DOSY; for a recent review, see Ref. [4]) or diffusion decoded (DECODES, [5–8]) NMR spectroscopy. After calibration of the PFG amplitudes and fitting of the resulting lineshapes, the individual diffusion coefficients can be derived for different sized molecules or, because of differences in hydrodynamic radii, even for molecules of the same molecular weight [9].

A few applications of DOSY experiments have already been reported in the field of carbohydrate NMR spectroscopy. These include complexation studies of monosaccharides with lanthanide cations [10], the investigation of silylated mono- and disaccharides [11], and measurement of the diffusion constants of nucleic acids for assessing whether a monomer or a duplex is present [12]. A qualitative approach has been published for identifying low molecular weight impurities in synthetic pentasaccharide samples [13]. In this paper we present diffusion difference NMR spectroscopy as our contribution to carbohydrate mixture analysis resulting from the demand for a fast, non-invasive, and easily workable spectroscopic tool of very broad applicability.

Results and Discussion

The concept of diffusion difference NMR spectroscopy

In fact, for the practical use in the routine analysis of carbohydrate mixtures the above mentioned DOSY or DECODES related experiments exhibit several drawbacks. First of all, they are rather time consuming as a series of one- or two-dimensional NMR spectra have to be recorded to resolve the molecules by their diffusion properties. New approaches recently published are very promising, accomplishing *e.g.* diffusion measurements within a single scan (DIFFTRAIN) [14] or a one-dimensional DOSY by using a nonuniform magnetic field gradient to encode the diffusion information [15]. Nevertheless, the analysis of the resulting spectra, which consists of applying appropriate line fitting algorithms to individual peaks belonging to different compounds for decoding the diffusion constants, still remains. Secondly, the gradient amplitude has to be calibrated very accurately; otherwise, only relative diffusion coefficients can be measured. Convection currents resulting from small temperature gradients within the sample tube can cause additional signal decay [16, 17], thus hampering the quantitative interpretation of the spectra.

As for the NMR analysis of mixtures the accurate determination of the diffusion coefficients of the individual components is not of main interest, we propose a general and faster qualitative diffusion difference approach by combining an old concept [18] with modern stimulated echo experiments. In this case, instead of a series only two NMR spectra will be acquired, using as a diffusion encoding element the bipolar-gradient pulses longitudinal eddy-current delayed sequence (BPP-LED) [19]. The first spectrum will be recorded with a very weak gradient amplitude, the second one with a somewhat stronger, regardless of its absolute value. The resulting spectra, showing differently attenuated peaks due to the distinct diffusion properties of individual compounds, are then subtracted. Depending on the proper scaling of their relative intensities, the difference trace gives the low or the high molecular weight compounds, respectively. As main advantages over DOSY type experiments, diffusion difference NMR spectroscopy is fast, no calibration of the gradient amplitude is necessary, and the postprocessing consists only of a rather simple subtraction of two spectra. Additionally, the receiver gain can be optimized for every single experiment depending on the chosen gradient amplitude, thus giving better signal-to-noise ratios. This fact is of major importance, since the overall signal intensity of the whole spectrum will be attenuated in any case by increasing the gradient amplitude. The following selected examples for the analysis of natural or synthetically derived carbohydrate mixtures using modern one- (1D) as well as two-dimensional (2D) NMR techniques will demonstrate the general and fast applicability of diffusion difference NMR spectroscopy.

Analysis of a mixture of homoglycans

A mixture of the α-(1→4)-linked glucose oligomers maltose and maltotriose of about equal amounts on the milligram scale should illustrate the concept of the diffusion difference NMR procedure. The resonance signals for the disaccharide in a stimulated echo ^{1}H NMR spectrum should be significantly reduced compared to the signals of the trisaccharide. As the structural similarity of these two homoglycans, which differ only in chain length, leads to almost identical NMR spectra, high demands have to be made on the experimental conditions in order to resolve the components by a subtraction method. For spectrometer stability reasons, the bipolar-gradient pulses LED sequence [19] was chosen for diffusion encoding. Figure 1 shows expansions of the resulting spectra recorded with 32 scans using 5% and 80% gradient amplitudes, respectively. Scaling the signal intensities for the higher molecular weight compound to an equal level and subtracting the spectrum derived with higher gradient strength from that obtained with the lower one results in a positive difference trace for all faster moving compounds, in this case maltose and the solvent. Figure 1 shows also a comparison between the diffusion difference NMR spectrum (trace c) and a normal ^{1}H NMR spectrum of pure maltose (trace d). Signals resonating very closely together like the non-reducing anomeric protons and especially H-4 (assignments taken from Ref. [20]) from the non-reducing end glucopyranoses with a chemical shift difference between maltose and maltotriose of only 1.8 Hz can be resolved in the mixture without any distortions.

Peptidoglycan analysis

An illustrative example for the application of this fast one-dimensional NMR carbohydrate mixture analysis procedure derives from the characterization of the linkage region between a bacterial cell wall polysaccharide and the surface-layer (S-layer) protein of *Thermoanaerobacterium thermosaccharolyticum* D120-70 [21]. After proteolytic degradation of the S-layer glycoprotein, tyrosine was found as the linking amino acid; however, the signals of the repeating units consisting of six residues obscured the information of the linkage sugar to the protein. In order to reduce the polysaccharide background, a mild acidic degradation of the glycopeptide was performed, yielding – after a crude separation – an NMR sample suitable for a fast diffusion edited experiment. After scaling the intensities of the

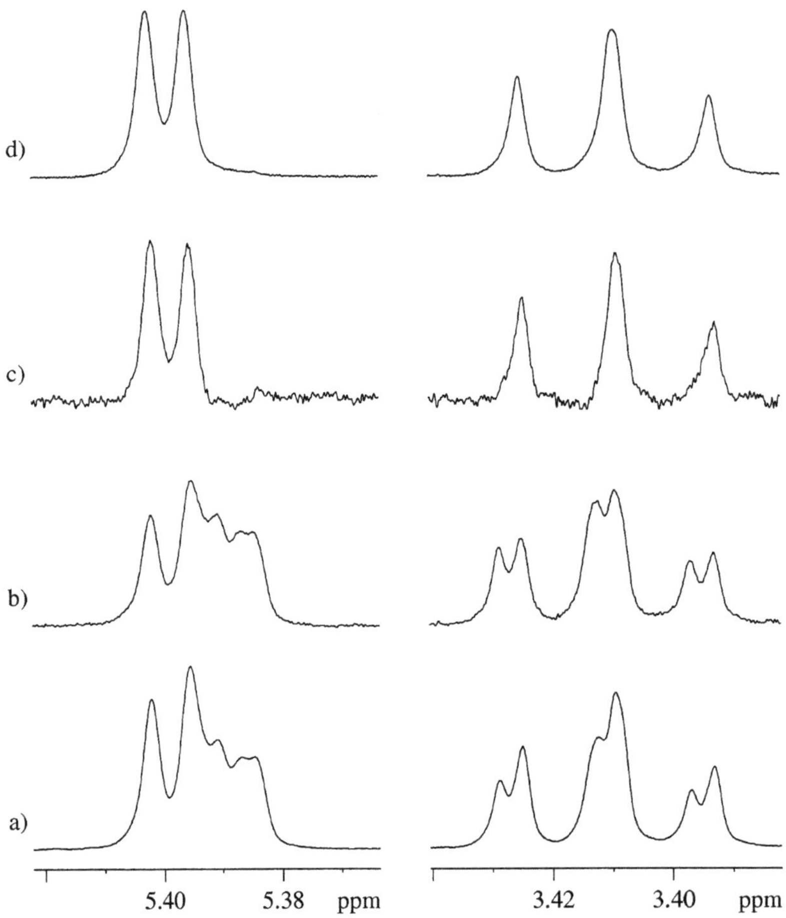

Fig. 1. BPP-LED spectra of a maltose/maltotriose mixture showing the non-reducing anomeric protons in the left and H-4 from the non-reducing end glucopyranose in the right column; a) recorded with 5%, b) with 80% gradient amplitudes, c) diffusion difference spectrum, d) normal ^{1}H NMR spectrum of pure maltose

signals for the central repeats to equal height, the diffusion difference experiment results in a clean and straightforwardly assignable ^{1}H NMR spectrum of the O-glycosidic linkage region *β*-*D*-glucopyranose-O→tyrosine (Fig. 2).

The BPP-LED pulse sequence together with the subtraction procedure is also suitable for addressing another problem connected to natural polysaccharide samples. The chain length of the aforementioned intact peptidoglycan can be determined by integration of the ^{1}H NMR spectrum. By comparing the intensities of the well-separated anomeric signals with those of the aromatic protons of tyrosine it could be deduced that the peptidoglycan consists of seven repeating units. However, the diffusion difference spectrum obtained after scaling the intensities of signals belonging to terminal sugar residues and to the amino acid to an equal level still shows the resonance lines of the central repeats (Fig. 3). Thus, the chain length of the polysaccharide is only on the average seven repeating units, and the BPP-LED NMR spectra supply evidence for a natural microheterogeneity of the isolated peptidoglycan. A matrix-assisted laser desorption ionization time-of-flight

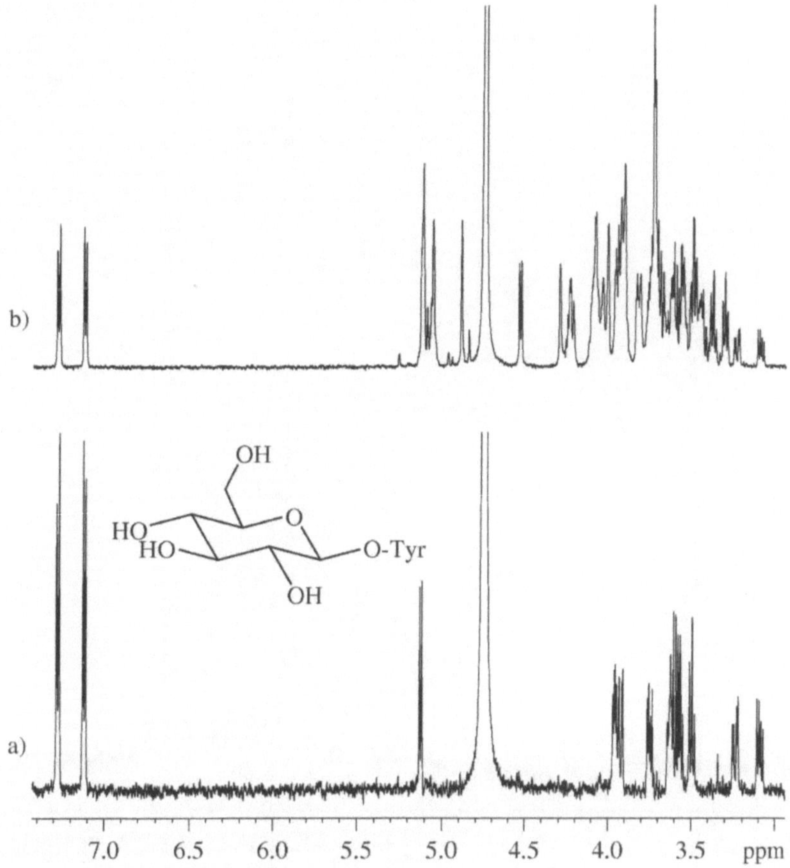

Fig. 2. Partly degraded bacterial cell wall peptidoglycan: a) diffusion difference NMR spectrum (5% and 50% gradient amplitudes), b) ^{1}H NMR spectrum

(MALDI-TOF) mass spectrum confirmed this chain length variability, showing five to nine hexasaccharide repeats linked to the tyrosine [21].

Derivatization of adenosine monophosphate (AMP)

Diffusion difference NMR spectroscopy proved to be a powerful tool for the quick monitoring of the success of the chemical synthesis of 2′(3′)-O-(N-acylaminoacyl)-nucleotide derivatives of adenosine-5′-monophosphate as well as of the corresponding *AMP*-3′-thio analogues. These modified mononucleotides were used to investigate the peptidyl transferase activity in ribosomes. Simple aminoacylated analogues of the 3′-terminal adenosine of transfer ribonucleic acid (t*RNA*) possess donor activity in the peptidyl transferase reaction, provided that they are used at very high concentrations (about 1 m*M*) [22]. To study the substrate specificity of the peptidyl transferase center as well as to investigate functional groups which might be involved in the chemical step, several differently modified mononucleotide substrates were synthesized. The success of the preparation of aminoacylated *AMP* derivatives and the optimization of the synthetic process can be investigated only by mixture analysis, as even after reversed phase high performance

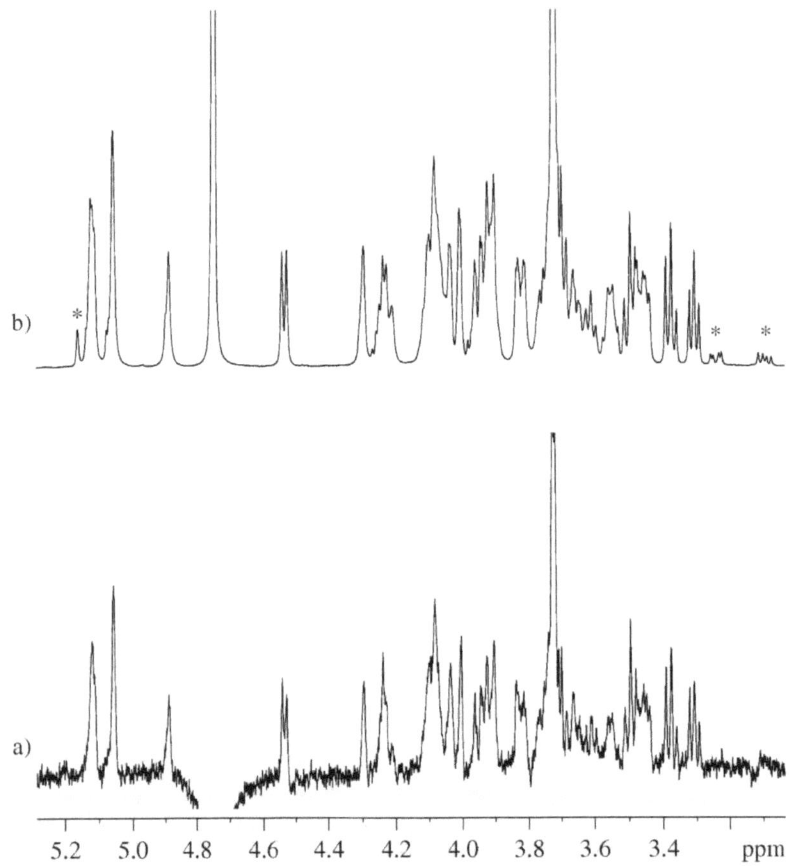

Fig. 3. Intact peptidoglycan: a) diffusion difference NMR spectrum (5% and 55% gradient amplitudes, terminal signal intensities scaled to equal level), b) ^{1}H NMR spectrum (signals from terminal residues marked with $*$)

liquid chromatography (HPLC) purification no fraction contained only one single product. This problem does not arise mainly from comigration on the column but is rather due to the low stability of the ester bond between amino acid and nucleotide. Furthermore, the interconversion between the 2′- and 3′-hydroxyl groups of adenosine is rapid in aqueous buffer solution. An even more complex pattern was detected by the use of a sulfur containing mononucleotide resulting from dimerization products due to disulfide formation.

Nevertheless, many of the possible components in the reaction mixture are of different molecular weight and should be distinguishable by their translational diffusion properties. Therefore, diffusion difference NMR spectroscopy can be used as a fast and easy tool to prove whether *e.g.* the ester bond between *AMP* and the acylamino acid is formed by the chosen reaction conditions or not. Figure 4 gives an example of the mixture obtained after a successful acylation of *AMP* with N-acetylphenylalanin cyanomethylester. In the chemical shift region for the anomeric protons five resonance signals can be seen, indicating five different *AMP* species. In the diffusion difference NMR spectrum (trace b), only a positive trace for the excessive free N-acetylphenylalanin and for the fast moving solvents (HDO and

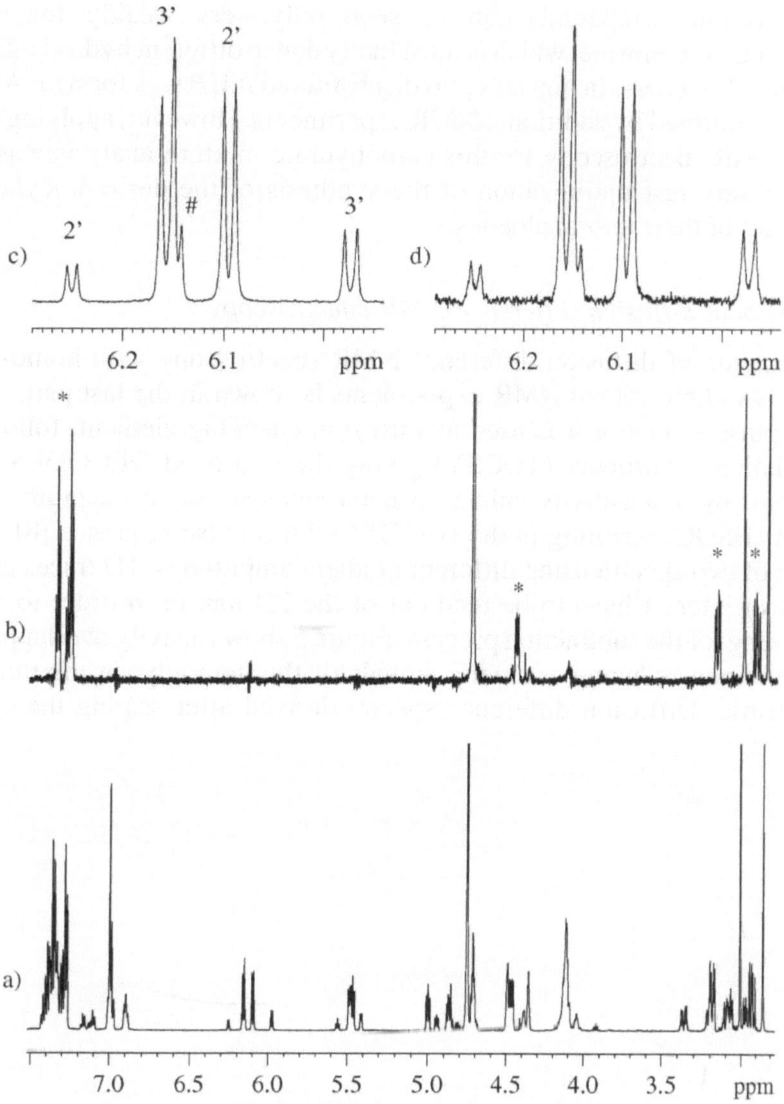

Fig. 4. Reaction mixture from the derivatization of *AMP* with N-acetyl-phenylalanine (Ac-*Phe*); a) ^{1}H NMR spectrum, b) diffusion difference spectrum showing free Ac-*Phe* (signals marked with *), c) BPP-LED spectrum with 5% and d) with 40% gradient amplitudes (2′- and 3′-substituted *AMP* with racemization of Ac-*Phe*; free *AMP* marked with #)

DMF) can be observed after scaling the anomeric signals to equal height. Looking closer to the H-1′ shift region, the weak doublet at 6.15 ppm decays significantly faster by applying stronger gradients than the other four signals (traces c and d). Although this signal as well as the rest of the related ribose protons cannot easily be seen in the difference trace due to their very low intensity and the poor signal-to-noise ratio after the subtraction, this anomeric proton belongs to the faster moving free *AMP*. The other four H-1′ signals with almost identical diffusion behaviour are part of the monoacylated ribose 2′- and 3′-derivatives, including racemization products of the amino acid. Negative signals in the difference trace b indicating higher molecular weight compounds can be seen only very weakly for the main 3′-aminoacylated adenosine, which is most likely due to different hydrodynamic radii of the 2′- and 3′-isomers. In any case, no disubstituted *AMP* was formed. All assignments were confirmed by additional NMR experiments. However, applying diffusion difference NMR spectroscopy for this carbohydrate mixture analysis was the best choice for a very fast optimization of the synthesis of the desired acylated *AMP* substrates and of their thio-analogues.

Two-dimensional diffusion difference NMR spectroscopy

The combination of diffusion difference NMR spectroscopy with homo- and heteronuclear two-dimensional NMR experiments is shown in the last part. Again the BPP-LED pulse sequence was used as diffusion encoding element, followed by a total correlation experiment (TOCSY) giving the improved DECODES pulse sequence [8] or by a sensitivity enhanced heteronuclear single quantum coherence experiment (HSQC) resulting in the HETDECODES pulse sequence [8]. After the acquisition of two spectra using different gradient amplitudes, 1D traces containing the signals of interest have to be read out of the 2D matrix in order to adjust the vertical scaling for the subtraction process. Figure 5 shows heavily overlapping cross-peaks for the non-reducing anomeric signals of the homoglycan mixture maltose and maltotriose. Diffusion difference spectra derived after scaling the cross-peak

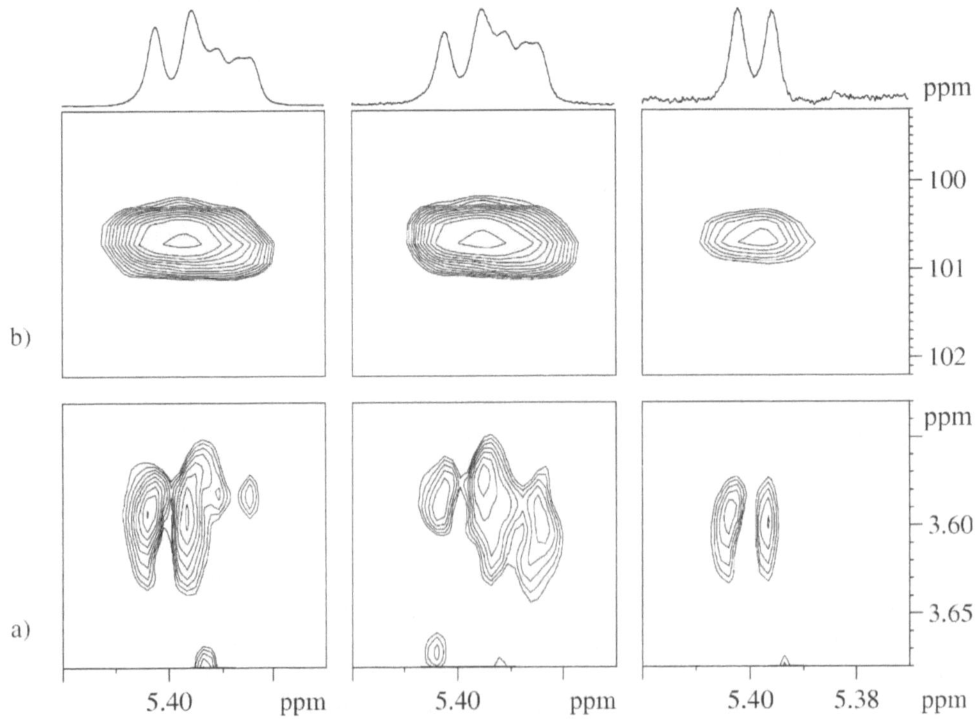

Fig. 5. Expansions from 2D diffusion edited NMR spectra of the non-reducing anomeric signals of a maltose/maltotriose mixture; a) improved DECODES experiment, b) HETDECODES experiments with 3% (left) and 50% (middle) gradient amplitudes and diffusion difference spectra (right)

intensities for the higher molecular weight compound to equal height and subtraction of two DECODES or of two HETDECODES spectra results in clean cross-peaks for the maltose, thus demonstrating the power of this technique even for the mixture analysis of very similar structures with almost identical ^{1}H and ^{13}C chemical shifts.

Conclusions

Diffusion difference NMR spectroscopy was shown to be a powerful tool in carbohydrate mixture analysis. It is suitable for addressing a wide variety of problems, like characterization of known and unknown compounds, identification of microheterogeneity in natural polysaccharide samples, and controlling and optimization of synthetic reactions. It is a very fast non-invasive tool covering 1D as well as 2D homo- and heteronuclear NMR experiments. Moreover, its application is not limited to carbohydrate mixture analysis.

Experimental

All NMR spectra were recorded on a Bruker Avance DRX 600 NMR spectrometer using a 5 mm inverse triple probe (^{1}H, ^{13}C, broad-band) with triple axis gradient coils. The samples were dissolved in 0.6 cm^3 D_2O and were measured at a temperature of 300 K with a resonance frequency of 600.13 MHz for ^{1}H and 150.90 MHz for ^{13}C. The amounts of the samples were on the milligram scale (homoglycan mixture: 3.3 mg maltose, 3.5 mg maltotriose; intact peptidoglycan sample: 4.3 mg, partial degraded peptidoglycan mixture: 1 mg; mixtures of the *AMP* derivatization reactions: approximately 10 mg). The spectra were referenced for ^{1}H to the internal HDO solvent signal ($\delta =$ 4.75 ppm), for ^{13}C to external dioxane ($\delta = 67.40$ ppm).

In all experiments, the diffusion encoding was performed with the BPP-LED sequence [19] with 1 ms sinusoidal bipolar-gradient pulse pairs; the diffusion delay was set to 150 ms. The gradient amplitudes are given directly in the figure captions (Acustar II gradient amplifier, 3×10 A, *z*-gradient: $g_{max} = 60$ G/cm). The number of scans for the 1D spectra was in the range from 32 to 512 depending on the relative intensities of the signals from the individual mixture components, using more scans if weak signals were of interest. The TOCSY experiment was performed using the improved DECODES pulse sequence [8] with a spin lock time of 100 ms collecting 2 k data points in f2 and 256 increments with 8 scans each in f1. For the HSQC experiment, the HETDECODES pulse sequence [8] was used collecting 2 k data points in f2 and 128 increments with 32 scans each in f1.

All processing was done off-line on a Silicon Graphics O2 workstation using the Bruker software Xwinnmr 2.6. The 1D spectra were weighted with an exponential function using a 0.2 Hz line broadening factor. The 2D spectra were multiplied with a squared sine function shifted by $\pi/2$. In the indirect dimension, the number of data points was increased by a factor of two using forward linear prediction with 64 coefficients. The subtraction of 1D as well as of 2D spectra was performed using the supplied features within the Bruker software.

Acknowledgements

The peptidoglycan samples derive from a collaboration with Dr. *C. Schäffer* and Prof. Dr. *P. Messner* from the *Zentrum für Ultrastrukturforschung, Universität für Bodenkultur*, A-1180 Wien which is gratefully acknowledged. *S. Dorner* was funded by grant P13651-GEN from the Austrian Science Foundation (FWF) to Dr. *Andrea Barta*.

References

[1] Duus JØ, Gotfredsen CH, Bock K (2000) Chem Rev **100**: 4589
[2] Shuker SB, Hajduk PJ, Meadows RP, Fesik SW (1996) Science **274**: 1531
[3] Stejskal EO, Tanner JE (1965) J Chem Phys **42**: 288
[4] Johnson Jr CS (1999) Proc Nucl Magn Reson **34**: 203
[5] Lin M, Shapiro MJ (1996) J Org Chem **61**: 7617
[6] Lin M, Shapiro MJ, Wareing JR (1997) J Am Chem Soc **119**: 5249
[7] Bleicher K, Lin M, Shapiro MJ, Wareing JR (1998) J Org Chem **63**: 8486
[8] Williamson RT, Chapin EL, Carr AW, Gilbert JR, Graupner PR, Lewer P, McKamey P, Carney JR, Gerwick WH (2000) Org Lett **2**: 289
[9] Derrick TS, Larive CK (1999) Appl Spectrosc **53**: 1595
[10] Diaz DM, Berger S (2000) Carbohydr Res **329**: 1
[11] Schraml J, Blechta V, Soukupova L, Petrakova E (2001) J Carbohydr Chem **20**: 87
[12] Lapham J, Rife JP, Moore PB, Crothers DM (1997) J Biomol NMR **10**: 255
[13] Kellenbach E, Burgering M, Kaspersen F (1999) Org Process Res Dev **3**: 141
[14] Stamps JP, Ottink B, Visser JM, Van Duynhoven JPM, Hulst R (2001) J Magn Reson **151**: 28
[15] Loening NM, Keeler J, Morris GA (2001) J Magn Reson **153**: 103
[16] Jerschow A, Müller N (1997) J Magn Reson **125**: 372
[17] Jerschow A, Müller N (1998) J Magn Reson **132**: 13
[18] Stilbs P (1981) Anal Chem **53**: 2135
[19] Wu D, Chen A, Johnson Jr CS (1995) J Magn Reson Ser A **115**: 260
[20] Sugiyama H, Nitta T, Horii M, Motohashi K, Sakai J, Usui T, Hisamichi K, Ishiyama J (2000) Carbohydr Res **325**: 177
[21] Schäffer C, Dietrich K, Unger B, Scheberl A, Rainey FA, Kählig H, Messner P (2000) Eur J Biochem **267**: 5482
[22] Cerna J, Rychlik I, Krayevsky AA, Gottikh BP (1973) FEBS Lett **37**: 188

Received November 26, 2001. Accepted November 30, 2001

MIX
Papier aus verantwortungsvollen Quellen
Paper from responsible sources
FSC® C105338

If you have any concerns about our products,
you can contact us on
ProductSafety@springernature.com

In case Publisher is established outside the EU,
the EU authorized representative is:
Springer Nature Customer Service Center GmbH
Europaplatz 3, 69115 Heidelberg, Germany

Printed by Libri Plureos GmbH
in Hamburg, Germany